Walker's

Bats of the World

Walker's

Bats of the World

Ronald M. Nowak

Introduction by
Thomas H. Kunz and Elizabeth D. Pierson

The Johns Hopkins University Press
Baltimore and London

Printed in the United States of America on acid-free paper
03 02 01 00 99 98 97 96 95 5 4 3 2

Portions of this book have been adapted from
Walker's Mammals of the World, 5th edition, by
Ronald M. Nowak, © 1991 by the Johns Hopkins University Press

The Johns Hopkins University Press
2715 North Charles Street
Baltimore, Maryland 21218-4319
The Johns Hopkins Press Ltd., London

Library of Congress Cataloging-in-Publication Data will be found
at the end of this book.

A catalog record for this book is available from the British Library.

ISBN 0-8018-4986-1 (pbk.)

Contents

Walker's

Bats of the World

Bats of the World: An Introduction

Thomas H. Kunz and Elizabeth D. Pierson

Bats (order Chiroptera) are among the most diverse and geographically dispersed kinds of living mammals. They form the largest mammalian aggregations known and may be the most abundant as measured by number of individuals. Only the rodents (order Rodentia) have more species. Flight, which sets bats apart from other mammals, has undoubtedly been an important factor in their wide distribution. Flight also accounts, in part, for their diverse feeding and roosting habits, reproductive strategies, and social behavior. Bats are known to occur on all continents except Antarctica and are found from the southern tip of South America to northern Scandinavia. They are absent only from the polar regions and a few isolated oceanic islands. Their roosting habitats include foliage, caves, rock crevices, hollow trees, crevices beneath exfoliating bark, and an assortment of human-made structures. The diversity of bats' diet, which includes insects, fruits, leaves, flowers, nectar, pollen, fish, other vertebrates, and blood, is unparalleled among living mammals. Bats that live in highly seasonal environments may hibernate during the cold seasons or migrate to warmer climates. Their reproductive patterns range from seasonal monestry to polyestry, with some species characterized by sperm storage and delayed fertilization and others by delayed implantation or delayed development. The social organization of bats is based on promiscuous, monogamous, or polygynous mating systems. Despite their adaptiveness and abundance, bats are highly susceptible to environmental disruption, and many species have declined drastically in response to human activity.

Taxonomy and Distribution

The order Chiroptera is divided into two suborders, the Megachiroptera and the Microchiroptera (Koopman 1993). The suborder Megachiroptera comprises just one family, the Pteropodidae, with 42 genera and 166 species.* This exclusively Old World family of fruit- and nectar-feeding bats ranges from subtropical and tropical regions of Africa and the eastern end of the Mediterranean, eastward across the southern Arabian Peninsula and the Indian Ocean islands to India, southeast Asia, Australia, Indonesia, Malaysia, the Philippines, southern Japan, Melanesia, and all but the easternmost islands of the central and South Pacific (Koopman 1970; Rainey and Pierson 1992). The more ecologically diverse Microchiroptera, with 16 families, 135 genera, and 759 species, has a virtually worldwide distribution and overlaps geographically with the Mega-

*Number of families, genera, and species after Koopman 1993.

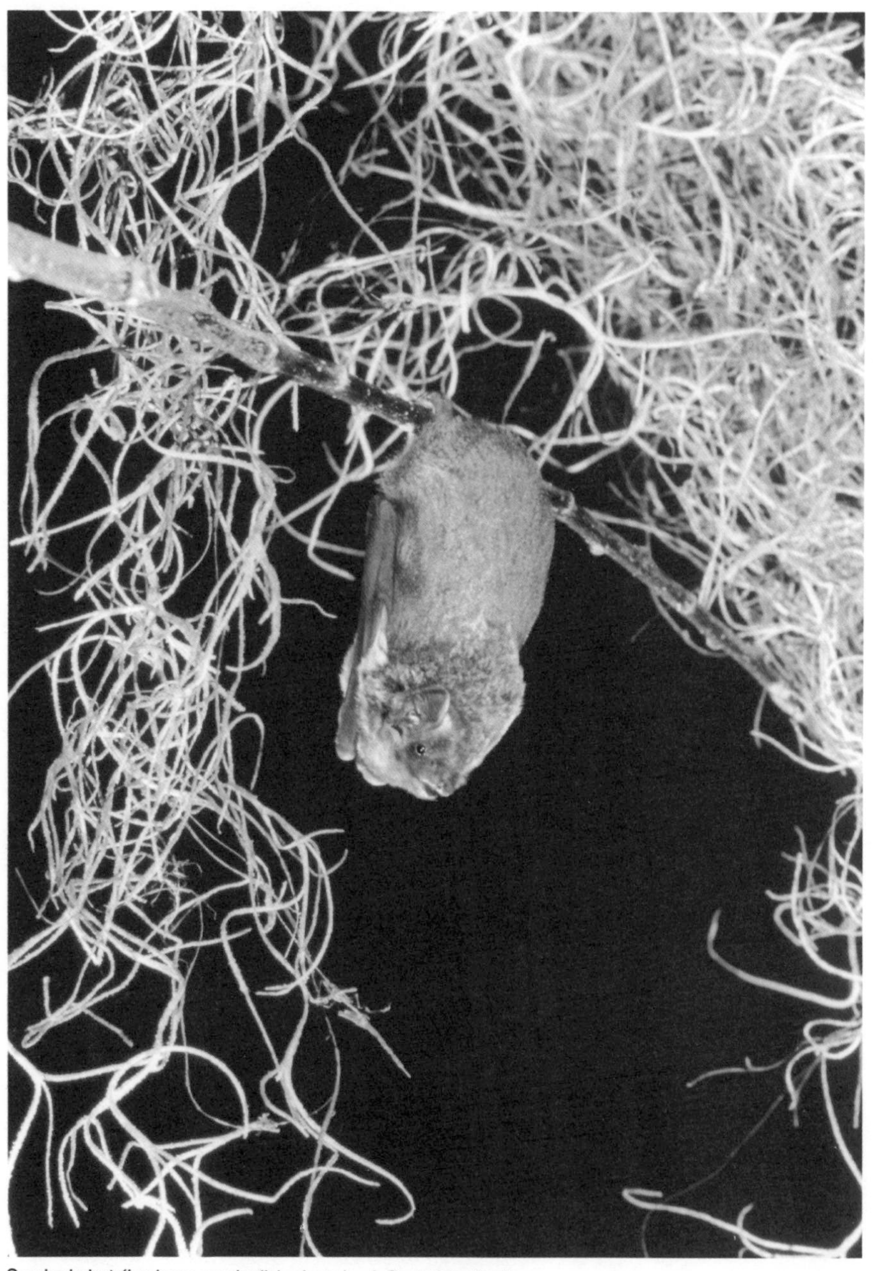

Seminole bat *(Lasiurus seminolis),* photo by J. Scott Altenbach.

chiroptera except on a few islands in the Indian Ocean and the central Pacific, where only Megachiroptera occur.

The greatest diversity of Microchiroptera occurs in tropical regions, with the richness of species decreasing with increasing latitude (Findley and Wilson 1983; Willig and Selcer 1989). Nine families of the Microchiroptera (about 130 species) may coexist in parts of the Central and South American tropics, but only a few members of the family Vespertilionidae extend into the subpolar regions (e.g., 6 species in Alaska, 1 in southwestern Labrador, and 16 in Sweden) (Hill and Smith 1984).

The sequence of families and genera presented herein largely follows the arrangement of Koopman (1984), who recognized 4 superfamilies within the Microchiroptera: Emballonuroidea, Rhinolophoidea, Phyllostomoidea, and Vespertilionoidea. Of the 16 microchiropteran families, 8 are restricted to the Old World, 6 to the New World, and 3 have representatives in both hemispheres. Most families have a tropical distribution, but 4 (Molossidae, Mystacinidae, Rhinolophidae, and Vespertilionidae) have penetrated into the cool temperate regions. Only the Vespertilionidae (the largest family, with about 300 species) has a nearly global distribution. The Emballonuridae and Molossidae occur in both hemispheres but are more restricted latitudinally. Of the widely distributed but exclusively Old World families, only the Rhinolophidae extends into the temperate zone; the Hipposideridae, Rhinopomatidae, Nycteridae, and Megadermatidae, while broadly distributed, are confined to tropical and subtropical regions. Several monotypic Old World families have very limited distributions—the Mystacinidae are found only in New Zealand, the Myzopodidae only in Madagascar, and the Craseonycteridae only in Thailand. The 6 exclusively New World families all have a relatively broad tropical distribution but vary in size, from the diverse Phyllostomidae (about 140 species) to the 5 smaller families (2 to 8 species each), the Furipteridae, Natalidae, Thyropteridae, Noctilionidae, and Mormoopidae.

As the names imply, the Megachiroptera are generally larger than the Microchiroptera. There is, however, considerable overlap in size. The Megachiroptera weigh from about 10 grams to more than 1,500 grams and have a forearm length of 36–228 mm. The Microchiroptera weigh from about 2 grams to 196 grams and have a forearm length of 22–115 mm. Among the smallest megachiropterans are three little-known Indo-Malaysian species (the spotted-winged fruit bat, *Balionycteris maculata,* the black-capped fruit bat, *Chironax melanocephalus,* and the pygmy fruit bat, *Aethalops alecto*), and the Southeast Asian and Australian species of blossom bats *(Syconycteris)* and long-tongued fruit bats *(Macroglossus).* The largest is the Indonesian large fruit bat, *Pteropus vampyrus,* with a wingspan of about 1,700 mm. The majority of megachiropterans are relatively small. Twenty-four (57 percent) of the 42 genera contain species with a maximum forearm length of less than 70 mm. Only 7 genera (16.7 percent) *(Eidolon, Pteropus, Acerodon, Pteralopex, Dobsonia, Aproteles,* and *Hypsignathus)* contain species with a maximum forearm length in excess of 110 mm. Even in the genus *Pteropus,* 11 of the 58 species (19 percent) have a maximum forearm length of less than 110 mm (Pierson and Rainey 1992). The smallest microchiropteran, Kitti's hog-nosed bat, of Thailand, *Craseonycteris thonglongyai*, is one of the smallest mammals in the world (only a few species of shrews are smaller) having a body weight of about 2 grams. The largest microchiropteran as measured by forearm length is the false vampire bat, *Vampyrum spectrum*, of Central and South America; and the largest by weight is the naked bat, *Cheiromeles torquatus*, of Indo-Malaysia and the Philippines (Corbet and Hill 1992).

Most of what is known about the Microchiroptera is based on studies of temperate species that form large maternity and hibernating colonies (Hill and Smith 1984; Kunz 1982; Ransome 1990). Increased interest in subtropical and tropical environments has greatly improved our understanding of several common species, including *Miniopterus schreibersi, Carollia perspicillata, Artibeus jamaicensis, Phyllostomus hastatus,* and *Desmodus rotundus* (see, e.g., Dwyer 1966; Fleming 1988; Greenhall and Schmidt 1988; Handley, Wilson, and Gardner 1991; McCracken and Bradbury 1977, 1981; Morrison 1979; Morrison and Morrison 1981; and Wilkinson 1984, 1985*a*, 1985*b*). Little is known, however, about most of the tropical microchiropterans.

Research on the Megachiroptera has focused mostly on the large, colonial "flying foxes," which belong to the genus *Pteropus*. There has been a growing interest, however, in the study of some of the smaller, lesser-known taxa, including *Haplonycteris, Cynopterus, Macroglossus, Nyctimene,* and *Rousettus* (Bhat 1995; Bhat and Kunz 1995; Gould 1978; Heideman and Heaney 1989; Spencer and Fleming 1989).

Faunal and autecological studies continue to improve our understanding of individual species and the composition of bat communities, e.g., in Africa, Australia, Britain and Europe, India, North and Central America, South America, the Philippines, and Southeast Asia and the Indo-Malaysian region (on Africa, see Aldridge and Rautenbach 1987; Bradbury 1977*a*; Fenton, Brigham, et al. 1985; Fenton, Cumming, et al. 1987; Fenton and Rautenbach 1986; Kingdon 1974; LaVal and LaVal 1977; McWilliam 1987; O'Shea 1980; Rosevear 1965; Thomas 1983; Vaughan 1976; Vaughan and Vaughan 1986; Wickler and Seibt 1976; and Wickler and Uhrig 1969. On Australia, see Crome and Richards 1988; Dwyer 1966, 1970; Hall and Richards 1979; Hamilton-Smith 1974; Kulzer et al. 1984; McKenzie and Rolfe 1986; Nelson 1964*b*, 1965; Richards 1986, 1987, 1990*b*, 1990*c*, 1995; Richardson 1977; Strahan 1983; and Tidemann et al. 1985. On Britain and Europe, see Gaisler 1963*a*, 1963*b*, 1979; Gaisler, Hanak, and Dunsel 1979; Jones and Rayner 1988, 1991; Kalko and Schnitzler 1989; Racey and Swift 1985; Rydell 1986, 1989; Schoeber and Grimmberger 1989; Speakman 1990, 1991, 1992; Stebbings 1988*a*, 1988*b*; and Swift and Racey 1985. On India, see Advani 1981; Balasingh, Koilraj, and Kunz 1995; Bhat 1995; Bhat and Kunz 1995; Brosset 1962; Fiedler 1979; Krishna and Dominic 1983; and Marimuthu 1984. On North and Central America, see Barbour and Davis 1969; Bonaccorso 1979; Fleming 1986; Gardner, LaVal, and Wilson 1970; Handley, Wilson, and Gardner 1991; Humphrey 1975; Humphrey and Bonaccorso 1979; Jones 1966; LaVal and LaVal 1980; Schmidly 1991; Silva Taboada 1979; Van Zyll de Jong 1985; and Villa-R. 1966. On South

America, see Albuja 1982; Barquez, Giannini, and Mares 1993; Goodwin and Greenhall 1962; Graham 1988; Husson 1962; Linares 1986; Myers 1977; Sazima and Sazima 1978; Willig 1983, 1989; and Willig and Moulton 1989. On the Philippines, see Heaney 1991; Heaney and Heideman 1987; Heideman and Heaney 1989; Ingle and Heaney 1992; Utzurrum 1992; and Utzurrum and Heideman 1991. And on Southeast Asia and the Indo-Malaysian region, see Bergmans and Rozendaal 1988; Corbet and Hill 1992; Davidson and Akbar 1987; Francis 1990, 1994; Francis et al. 1994; Goodwin 1979; Gould 1978; Lekagul and McNeely 1977; Medway 1972; Phillips 1968; Van Peenen 1968; Kitchener, Charlton, and Maharadatunkamsi 1989; Kitchener, Gunnell, and Maharadatunkamsi 1990; Tidemann, Kitchener, et al. 1990; Tidemann, Priddel, et al. 1985; and Zubaid 1993).

Distinctive Characteristics

Bats are distinguished among mammals by their capacity for true flight. All other so-called flying mammals (e.g., flying squirrels or flying lemurs) are gliders and lack the morphological and physiological adaptations that enable bats to fly. As the name Chiroptera, derived from the Greek roots *cheir* (hand) and *pteron* (wing), implies, the bat wing is a highly modified hand. Elongated fingers and forelimb bones provide the structural supports for the thin, elastic membrane that originates on the side or back of the body, extends from the hindlimbs to the forearm (plagiopatagium), and forms the triangular membranes between the fingers (chiropatagium). An additional membrane (propatagium) extends from the shoulders to the thumb along the leading edge of the wing and varies considerably in size among different taxa. In many species, particularly of the Microchiroptera, there is also a membrane (uropatagium) between the hind limbs, encompassing all or part of a tail. These membranes comprise two layers of skin between which are sandwiched a thin layer of vascularized and enervated connective tissue and muscle fibers.

Flight involves both lift and thrust, created by the movements of the airfoil section of wings, by powerful pectoral and wing muscles (Altenbach and Hermanson 1987; Vaughan 1970). Lift allows the bat to remain aloft by resisting the force of gravity, whereas thrust propels the bat forward by countering the effect of drag, or friction, as the bat moves through the air. Most of the lift is generated by negative pressure created when air flows more rapidly over the upper surface of the wing than it does across the lower surface. Thrust is generated by the distal parts of the wing membrane, located between the fingers of the hand wing. The rate of the air flow across the wings, and hence lift, is influenced by the change in the camber of the wings, caused by movements of the wrist, fingers, and hind feet. Lift and thrust during flight are created as the wings move through a series of wing-beat cycles. Wings are raised and extended backward at the beginning of the wing-beat cycle and then moved downward and forward in the downstroke. At the end of the downstroke, the bat rapidly rotates its arms and wrist, so that the wing partially folds or collapses during the upstroke, thus returning the wing to the same position as at the beginning of the wing-beat cycle (Fenton 1992; Norberg 1990).

The power for flight in bats is generated by the contraction of flight muscles located in the upper arms and chest. Two opposing muscles provide most of the power for flight in birds, but several muscles, some of them unique, power the flight of bats. Three pairs of pectoral (chest) muscles power the downstroke, and several smaller muscles located on the back are responsible for the upstroke (Altenbach and Hermanson 1987; Vaughan 1970). The shoulder girdle, with its mobile scapula and modified clavicle and humerus, is also highly adapted for flight (Fenton 1992; Hill and Smith 1984).

Wing shape and flight speed vary among the Chiroptera and are highly correlated with diet and foraging behavior (Fenton 1992; Norberg 1990; Norberg and Rayner 1987). The wing shape of a bat (or bird) is best described by its aspect ratio, determined as the wing area divided by the wingspan squared. Generally, a high aspect ratio indicates a long, narrow wing, whereas a low aspect ratio indicates a short, broad wing. Generally, bats with high aspect ratios are fast flyers that feed in open habitats, whereas bats with low aspect ratios have slow, maneuverable flight and more often feed in cluttered environments (Baagøe 1987; Fenton 1992; Norberg 1990; Norberg and Rayner 1987). Some species hover to glean insects from surfaces, feed on nectar, or navigate in close quarters.

Wing loading, expressed as Newtons per square meter, is an aspect of wing design that is determined by dividing the wing area by the body weight (Mg/S). Typically, bats with broad wings have low wing loading and those with high wing loading are characteristic of species with high aspect ratios. Horizontal flight in bats requires a great deal of energy (Speakman, Anderson, and Racey 1989; Winter et al. 1993), and hovering flight requires even more (Norberg et al. 1993; Winter et al. 1993).

The hind limbs of bats are unique among mammals in being rotated 180°, causing the knees to point backward. This arrangement facilitates steering during flight and a head-down roosting posture. Bats generally hang by one or both feet, using the sharp, recurved claws of five toes to support their weight. Except in the sucker-footed bats (Thyropteridae and Furipteridae), hanging is aided by specialized, locking tendons in the toes, which allow bats to hang without expending energy (Bennett 1993; Quinn and Baumel 1993; Schutt 1993. See also Howell and Pylka 1977). The bat's thumb, usually with a sharp, recurved claw, is generally used for food manipulation and also aids in crawling and clinging to surfaces. Modified cervical vertebrae (absent in the Megachiroptera) contribute to the flexibility of the neck in Microchiroptera, making it possible for bats to arch their heads backward from a typical roosting posture (Crerar and Fenton 1984).

The faces and ears of bats are among their most striking features. Many taxa have fleshy ornaments associated with the nose and mouth (Fenton 1992). These vary greatly in shape and size and can resemble such structures as leaves (Phyllostomidae, Megadermatidae, Hipposideridae) and horseshoes (Rhinolophidae). Some of the most elaborate modifications are found in the Hipposideridae *(Rhinonycteris, Triaenops)*, Phyllostomidae *(Centurio)*, and Mormoopidae *(Mormoops)*. Facial ornaments are greatly reduced or absent in other Microchiroptera. Some of the Megachiroptera *(Nyctimene* and *Paranyctimene)* and Microchiroptera *(Murina* and *Harpiocephalus)* have conspicuous tubelike nostrils. The most unusual megachiropteran is *Hypsignathus monstrosus,* in which the male has enlarged vocal organs and ornamented lips (see Fenton 1992).

The pinnae (external ears) are particularly important to bats, especially in those species that rely on hearing for orientation and to detect prey (Orbst et al. 1993). Many have special folds and crenulations that are thought to play a role in sound reception. Microchiropterans also have a tragus and/or an anti-tragus (fleshy projections on the anterior edge of the ear opening), which are lacking in the Megachiroptera (Vaughan 1986). The tragus is thought to aid echolocating bats in the horizontal localization of targets (Lawrence and Simmons 1982). Several genera of microchiropterans, including some vespertilionids *(Euderma, Histiotus, Idionycteris, Laephotis, Nyctophilus,* and *Plecotus),* phyllostomids *(Macrotus, Micronycteris),* nycterids *(Nycteris),* and megadermatids *(Lavia, Cardioderma, Macroderma,* and *Megaderma),* have unusually large pinnae. The shapes and sizes of pinnae are often highly correlated with feeding habits, especially in species that take prey from the ground or from surfaces (Orbst et al. 1993). Bats with large pinnae are able to detect sounds produced by their prey (Bell 1982, 1985; Fenton 1990; Guppy and Coles 1988; Vaughan 1976; Vaughan and Vaughan 1986).

The Megachiroptera are distinguished from the Microchiroptera in other ways as well, most of which relate to differences in flight and sensory capabilities (Pettigrew et al. 1989; Suthers 1970). These differences have led some investigators to suggest that the Megachiroptera evolved independently of the Microchiroptera (diphyletic origins) and may be more closely allied to Primates than to the Microchiroptera (Pettigrew 1986, 1995; Pettigrew et al. 1989). This hypothesis remains highly controversial, however (see, e.g., Bailey, Slightom, and Goodman 1992; Baker, Novacek, and Simmons 1991; Bennett et al. 1988; and Mindell, Dick, and Baker 1991; Simmons 1995), since other molecular and morphological data sets support the hypothesis that bat origins are monophyletic. The Microchiroptera usually have a prominent tail and tail membrane (uropatagium), which are generally lacking or are highly reduced in the Megachiroptera. The Megachiroptera usually also retain a claw on the second digit (lacking in living Microchiroptera), which is presumably a primitive characteristic of the order.

Most authorities believe that bats evolved from arboreal insectivorous ancestors (Norberg 1990). This view is consistent with the fossil record, which shows that bats were plentiful and well developed at least as long ago as the Eocene (about 50 million years before the present). This suggests that bats probably originated several million years before they were first reported as fossils.

Early fossils of bats indicate that they were quite similar to present-day bats (Habersetzer and Storch 1989; Jepson 1970; Novacek 1985). The oldest known bats from Eocene deposits of Wyoming and the middle Eocene of Germany show certain basiocranial features that suggest refinement of ultrasonic echolocation comparable to that of modern microchiropterans.

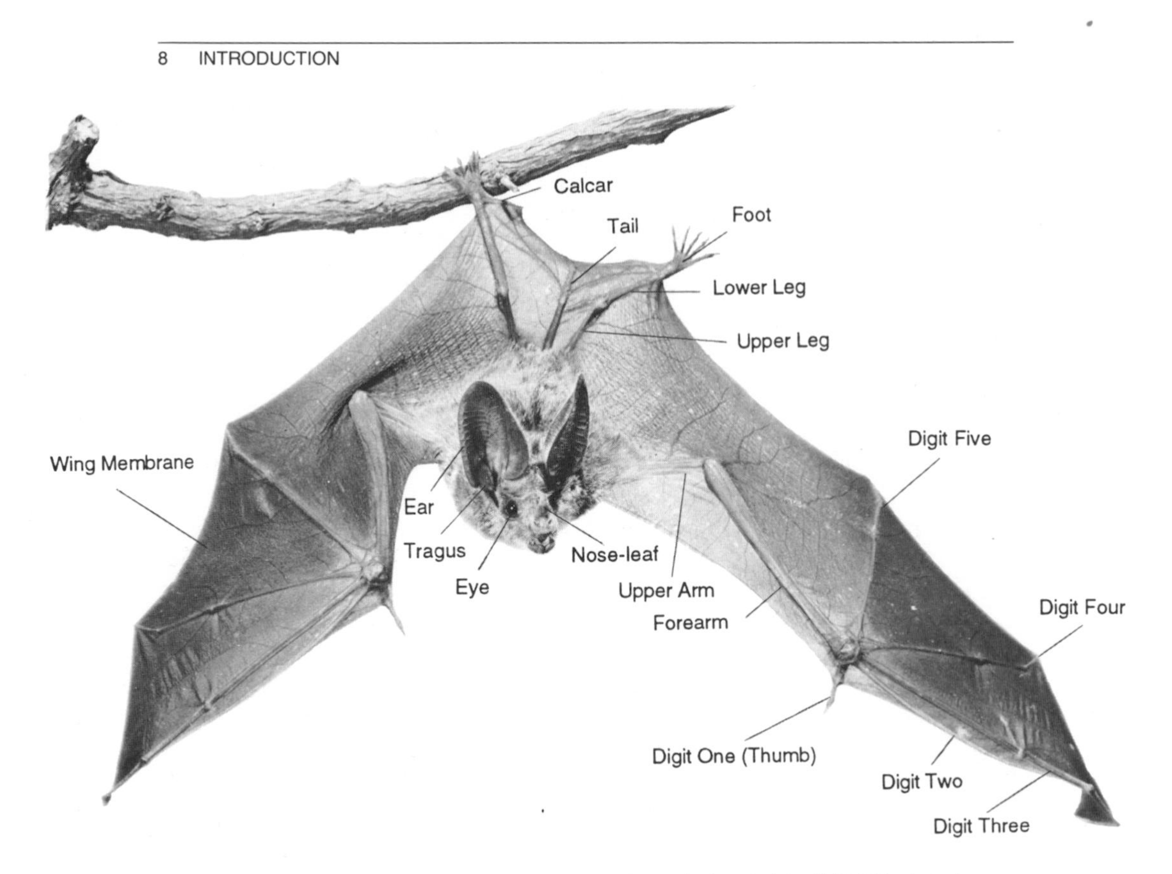

Top: A Mexican big-eared bat *(Macrotus waterhousii mexicanus),* about to take flight. This photo, by Ernest P. Walker, illustrates many of the parts of the bat's external form. Bottom: Himalayan leaf-nosed bat *(Hipposideros armiger),* photo by Klaus-Gerhard Heller.

Echolocation, Hearing, Vision, and Olfaction

Among the approximately 900 species of bats, more than half orient and capture their prey by means of echolocation. Echolocation involves an active sonar (the transmission and reception of brief, generally ultrasonic calls) with a number of unique morphological and physiological adaptations that allow bats essentially to "see" with sound. The ability to echolocate—present in all microchiropterans but apparently absent in all but one genus of the Megachiroptera—is one of the primary characteristics distinguishing the two suborders. Although echolocation is used by bats to detect obstacles and find roost sites, it is used primarily in foraging. All species of Microchiroptera appear to use echolocation for orientation and prey capture, but its use and importance is highly variable (Fenton 1992; Neuweiler 1984, 1989, 1990). Although most microchiropteran species probably locate their prey by echolocation, others rely on vision and passive listening (Bell 1982, 1985; Fenton 1990; Suthers 1970). Among the Megachiroptera, only members of the genus *Rousettus* are known to produce audible echolocation calls (using tongue clicks). They use these calls only for orientation, and like other megachiropterans, *Rousettus* relies on vision and olfaction for locating food.

Microchiropteran echolocation sounds are produced by the contractions of muscles in the larynx (Novick and Griffin 1961) and can be characterized by their frequency, or pitch (measured in kilohertz, kHz), duration, and intensity, or loudness (measured in decibels, dB) (Fenton 1992; Griffin 1958). The strength of a sound is expressed as loudness or intensity, which is how sound is perceived by the listener. Sounds typically are emitted through the mouth, although the Old and New World leaf-nosed bats emit sounds through their nostrils, with the fleshy ornaments functioning, at least in some species, as "acoustic lenses," or megaphones (see Hartley and Suthers 1987). Most echolocation calls are ultrasonic (> 20 kHz), so described because they are produced at frequencies beyond the range of "normal" human hearing.

Time also is an important source of information for echolocating bats. Many bats use echoes from the time-frequency patterns of emitted vocalizations to determine their relative distance to a target (Griffin 1958); others can detect the altered time domain of echoes produced by the wing-beat frequencies of their prey (Bell and Fenton 1984; Henson, Wilson, and Hansen 1987; Neuweiler 1990). The ears and neurons of the brain of echolocating bats are tuned to the frequencies of the emitted sounds and their returning echoes (Novick 1977; Pollak and Casseday 1989), and bats collect detailed information about targets by comparing the spectral characteristics of the original signal with its echo.

Call patterns vary considerably both within and among species, often making it difficult to establish species-specific vocal signatures. Nevertheless, the Rhinolophidae and Hipposideridae can be easily distinguished by their relatively invariable constant-frequency (CF) components, and many vespertilionids, emballonurids, and molossids have unique calls. Echolocation calls often include both CF and frequency-modulated (FM) components, the latter usually containing several harmonics (Simmons and Dear 1992). Frequency-modulated sounds can vary from more than 200 kHz to 10 kHz, with intensities ranging upward from 50 dB to 120 dB (Fenton 1992). Species that produce low-intensity sounds (members of the Phyllostomidae, Megadermatidae, Nycteridae, and some Vespertilionidae) are known as "whispering bats" (Griffin 1958) and are nearly impossible to detect in the field using currently available microphones and bat detectors (Ahlen 1990).

So-called FM bats emit volleys of calls through their open mouths, with call durations ranging from 0.5 ms to 10 ms. The duration of each call and the interval between calls will vary depending on whether the bat is in a search phase, an approach phase, a track phase, or a terminal phase. For example, the common North American vespertilionid bat *Eptesicus fuscus* produces echolocation calls that simultaneously sweep through a range of frequencies (from about 50 kHz to 22 kHz in the first harmonic and from about 100 kHz to 44 kHz in the second). This and other microchiropterans that produce pure FM calls use their sounds as broadband signals and thus derive acoustic images by integrating information from echoes across a range of frequencies (Neuweiler 1990; Schmidt 1988; Simmons and Dear 1992). Using echolocation, an FM bat can compare the spectral characteristics of the emitted calls with the spectral characteristics of the returning echo, which makes it possible for the bat to distinguish the characteristics of its target. FM bats are able to measure the distance to a target by assessing the temporal delay of echoes relative to the outgoing calls. Insectivorous bats, as they close in on their prey, typically increase the rate of echolocation calls, with the calls terminating in what biologists describe as a "feeding buzz," the high pulse repetition rate associated with an attack on its prey.

Horseshoe and Old World leaf-nosed bats (Rhinolophidae and Hipposideridae) emit signals through their nostrils, with the basal portion of the leaflike structures functioning as a megaphone. Their calls typically include a relatively long CF component followed or preceded by a short FM component. CF calls have narrow band widths and are ideal for detecting moving targets. For example, the European horseshoe bat, *Rhinolophus ferrumequinum,* emits CF components as narrow-band signals, the individual frequency serving as a carrier for modulations in frequency and amplitude that the flutter from an insect may impose on the echo (Schnitzler 1987; Schuller 1985). Old World rhinolophids and hipposiderids and one New World mormoopid *(Pternotus parnellii)* detect the flutter of moth wings by registering the frequency of Doppler-shifted echoes (Bell and Fenton 1984; Henson, Wilson, and Hansen 1987; Neuweiler 1989, 1990).

As in other mammals that depend on acoustic signals, the ears of bats are very important. The prominence of external ears (pinnae) on echolocating bats in particular attests to their important role in hearing (Orbst et al. 1993). The size and shape of the pinnae of echolocating bats appear to be modified for the amplification of sounds from returning echoes or the actual sounds produced by prey. The auditory systems of bats are mechanically and neurologically tuned to the frequencies of their dominant echolocation calls. As in most other mammals, sound vibrations are transmitted mechanically to the oval window of the cochlea by the auditory ossicles from the tympanic membrane.

Two muscles of mammals, the tensor tympani and the stapedius of the inner ear, are known to lessen the ability of the auditory ossicles to transmit vibrations when the ear is subjected to loud sounds. These muscles are especially well developed in echolocating bats. The contraction of the tensor tympani changes the tension of the tympanic membrane, and the stapedius changes the angle at which the stapes contacts the oval window, thus reducing the sensitivity of echolocating bats to outgoing pulses. The contraction of the middle ear muscles only a few milliseconds after the contraction of the crycothyroid muscles of the larynx attenuates the intensity of auditory self-stimulation by at least 25 percent (Jen and Suga 1976). Further neural attenuation occurs in the brain (Pollak and Casseday 1989; Suga and Schlegel 1972). Additionally, the otic capsule housing the middle and inner ears of echolocating bats is insulated from the rest of the skull by blood-filled sinuses or fatty tissues (Vaughan 1986).

Some flying insects can detect the ultrasonic cries of bats and take evasive action. The ears of these insects are tuned to the same frequencies emitted by echolocating bats, and thus these insects can detect approaching bats before bats can detect them (see, e.g., Fullard 1987; May 1991; Miller and Olesen 1979; and Roeder and Treat 1961). Depending upon the type of insect and its distance from an approaching bat, different kinds of evasive actions have been observed. For example, if a noctuid moth hears an approaching bat before it is detected by the bat, the moth may fly away from the oncoming signal. By contrast, if the moth or other insect (e.g., lacewing) is within the target range of the bat (and the echolocation call is perceived by the insect as being more intense or louder), the insect may take evasive action by making a passive or power dive to the ground or else may engage in erratic looping flights to avoid being captured. Other insects (tiger moths; Arctiidae) avoid capture by producing ultrasonic clicks that appear to confuse echolocating bats.

In addition to echolocation, which involves a highly specialized sense of hearing, bats use their ears for maintaining equilibrium and detecting audible sounds. Detection of audible sounds allows bats to locate and capture potential prey, to respond to other bats in a social context, or to increase their awareness to approaching predators (Henson 1970). Among the most remarkable features of the chiropteran ear are the variability in size, shape, and elaboration of the pinnae and associated accessory appendages. The simplest pinnae are found in the Megachiroptera, and the most complex are found among the echolocating Microchiroptera. The funnel-shaped pinnae of *Mormoops,* each with a complex array of folds, crenulations, and leaflike structures, are among the most elaborate belonging to any mammal. A few bats with enormous pinnae are known to hunt insects and small vertebrates resting or moving on surfaces (e.g., *Plecotus, Antrozous, Cardioderma, Macrotus, Megaderma, Nycteris, Trachops*). The ears of some bat species are highly tuned to their echolocation pulses, whereas the ears of other species are more closely tuned to audible sounds produced by their prey. Some insectivorous bats may exploit other audible sounds present in their environment, such as frog calls, to orient themselves to feeding sites where insects may be abundant (Buchler and Childs 1981). Audible sounds probably can be detected by most species and may be used, for example, in communica-

Little big-eared bat *(Micronycteris megalotis)*, photo by B. Ronacher.

tion during courtship (Bradbury 1977*b*; Fenton 1985) and in mother-pup recognition (Balcombe and McCracken 1992; Gelfand and McCracken 1986).

Most bats have well-developed eyes, comparable in sensitivity to those of other mammals. Notwithstanding, compared with other senses, such as hearing and olfaction, vision appears to play a supplementary role in the daily lives of microchiropterans. Most of the megachiropterans have large eyes, which suggests that their vision is highly developed, whereas the fact that most of the Microchiroptera have small eyes seems to indicate that vision is relatively less important for them. The megachiropteran eye has unique choroidal papillae with contours that are matched by the retina (Suthers 1970), as well as highly specialized visual pathways from the retina to the midbrain. These neural pathways, also found in primates and some carnivores, have been used by some authors to distinguish the megachiropterans from microchiropterans (Pettigrew 1986; Pettigrew et al. 1989).

Generally, the eyes of bats are specialized for functioning under conditions of low light, as their retinae consist almost entirely of rod cells. While both megachiropterans and microchiropterans can see in dim light, color vision is known only among the former—a characteristic that may be associated with their largely diurnal, tree-roosting habits. No species of bats are blind, but visual acuity tends to vary widely among species. Some of the microchiropterans are capable of pattern discrimination, which suggests that this aspect of visual discrimination is used by some species (e.g., plant-visiting bats) for food selection. Bats that roost in caves, mines, and other darkened habitats tend to rely mostly on hearing and echolocation for prey detection and orientation, and most have smaller eyes than do bats that roost in foliage and other open-roosting situations. Vision appears to play an important role in orientation during nocturnal flights and perhaps in predator surveillance, especially in the megachiropterans (Suthers 1970). Vision in the Microchiroptera appears to function mostly for the regulation of daily activity rhythms, seasonal reproductive cycles, and predator surveillance, especially among tree-roosting species (Suthers 1970). Vision in echolocating bats may also be important for detecting objects beyond the relatively short detection range of echolocation and thus may be used for orientation during flight (Suthers and Wallis 1970).

As in most other mammals, olfaction, along with taste, is important to bats (Suthers 1970). In addition to well-developed taste receptors, olfactory epithelium, and olfactory bulbs, many species have large vomeronasal organs, each with specialized ducts connecting to the mouth or buccal cavity (Bhatnagar 1980; Cooper and Bhatnagar 1976). These paired organs, located in cartilages of the anterior nasal septum and above the hard palate and innervated by neurons leading to the olfactory bulb, pump saliva containing dissolved chemicals from the mouth cavity into the organ via paired ducts located posterior to the incisors. Highly developed senses of smell and taste allow bats to distinguish food sources (Suthers 1970) and recognize conspecifics during courtship and mating (Bradbury 1977*a*, 1977*b*), promote mother-pup interactions (Suthers 1970; McCracken and Gustin 1991), and perhaps help identify roosting sites. Olfaction is especially well developed in Old World and New World plant-visiting bats, including fruit- and nectar-feeding species, although other species, including vampire bats and insectivorous and carnivorous species, may have to rely on odor and taste in the detection and ultimate selection of their prey. The highly developed facial and skin glands and secretions from such structures (Quay 1970) appear to have important social functions, including scent marking of objects and conspecifics (Bradbury 1977*b*).

Roosting Habits

Bats spend more than half their lives in their roost environment and exhibit a variety of adaptations that may reflect compromises associated with mode of flight, social behavior, diet, group size, and reproduction (Bradbury 1977*b*; Gaisler 1979; Kunz 1982; Racey 1982). Roost sites include caves and mines, crevices in rock and tree bark, buttresses, cavities in tree trunks and branches, unmodified foliage, foliage modified as tents, and various human-made structures (Kunz 1982; Kunz, Fujita, et al. 1994; Stebbings 1988*a*, 1988*b*; Verschuren 1957). Some colonially roosting megachiropterans (primarily *Pteropus*, *Acerodon*, and *Eidolon*) aggregate on exposed tree branches; many megachiropterans and most microchiropterans seek concealed places during roosting periods. Some megachiropterans, such as *Dobsonia*, *Aproteles*, *Eonycteris*, *Notopteris*, and *Rousettus*, prefer to roost in caves.

Most of what we know about the roosting habits of the Microchiroptera comes from intensive studies on a few species many of which form maternity and hibernating colonies in caves, large hollow trees, and human-made structures. Maternity roosts of some gregarious taxa (e.g., *Myotis*, *Pipistrellus*,

Miniopterus, Nyctalus, Rhinolophus, Hipposideros, Tadarida, Emballonura, Taphozous, and *Macroderma*) provide warm, dark, sheltered environments (Barbour and Davis 1969; Dwyer 1966; Gaisler 1979; Gaisler, Hanak, and Dunsel 1979; Kunz 1973, 1974; Licht and Leitner 1967; Tuttle 1975, 1976*a*; Wilson 1971), where natural temperatures, or temperatures generated by metabolic heat from the animals, often increase the temperature of the roost environment.

Some species appear to have very specific roosting requirements; for example, *Lasiurus cinereus, L. borealis,* and some species of *Nyctophilus* have been found roosting only in foliage, and *Euderma maculatum* only in cliff faces. Others, such as *Plecotus townsendii,* historically found in caves in the western United States, frequently use old mine tunnels or spacious attics of buildings. Still others, such as *Eptesicus fuscus, Artibeus jamaicensis, Noctilio leporinus,* and *Nyctalus noctula,* which often roost in tree hollows, may also be found in caves and human-made structures, depending on habitat and the availability of alternate roost sites (Brigham and Fenton 1986; Gaisler 1979; Kunz, August, and Burnett 1983; Morrison 1979). In Europe, Asia, and Australia many cave-roosting species, especially rhinolophids and hipposiderids, also utilize human-made mines.

Many tropical Microchiroptera have other unusual roosting and social habits. For example, the New World disk-winged bat *(Thyroptera tricolor)* roosts in unfurled *Heliconia* leaves; Old World *Pipistrellus nanus, Myotis bocagei,* and *Myzopoda aurita* roost in similar types of leaves (Kunz 1982); *Tylonycteris* roosts in the insect-modified internodes of bamboo culm in Asia (Medway 1972); and the New and Old World tent-making bats roost in a variety of plants species, in which they modify the leaves and stems into tents (Balasingh, Koilraj, and Kunz 1995; Bhat and Kunz 1995; Brooke 1987; Kunz 1982; Kunz, Fujita, et al. 1994; Timm 1987).

In highly seasonal environments, where insect abundance may be greatly reduced, bats deposit fat in the late summer and early autumn in preparation for migration or hibernation (Racey, Speakman, and Swift 1987). Hibernating bats use many kinds of roosts, including caves, mines and bunkers, rock crevices, tree cavities, and various other human-made structures. These sites typically offer bats a stable microclimate, with ambient temperatures above freezing, allowing their body temperatures to drop within a few degrees of the ambient temperature (Davis 1970; Racey, Speakman, and Swift 1987; Ransome 1990). This allows bats to conserve important fat reserves, which provide the sole source of energy during long periods without food (Brigham 1987; Speakman and Racey 1989). While in hibernation, bats may arouse periodically to drink and urinate, but usually their metabolic rate and activities are greatly depressed (Davis 1970).

Bats that hibernate in caves and mines in cold months are highly susceptible to disturbance; thus, human entry into these sites should be minimized during this energetically critical period. Many important hibernating sites (caves and mines) in Europe and North America have been placed under protection by private and government agencies. These organizations often distribute interpretive materials, post warning signs, and sometimes erect physical barriers such as fences and steel gates to discourage would-be intruders (Stebbings 1988*b*).

New Zealand short-tailed bat *(Mysticina tuberculata)*, photo by E. D. Pierson.

Migratory species typically move to warmer climates (Fenton and Thomas 1985; Griffin 1970), where they continue to feed; some may migrate relatively short distances but also hibernate (Davis and Hitchcock 1965; Griffin 1970; Tuttle 1976*b*). Species known or suspected of making long migrations to warmer climates include the American lasiurine bats *(Lasiurus borealis* and *L. cinereus)* and *Lasionycteris noctivagans* (Barbour and Davis 1969). Some highly gregarious species, including the Mexican free-tailed bat, *Tadarida brasiliensis,* in North America (Cockrum 1969) and *Miniopterus schreibersi* in Australia (Dwyer 1966), undertake long migrations. Several species in Europe and Asia also make long migrations, including *Pipistrellus nathusii, Nyctalus noctula, N. leisleri,* and *Pipistrellus pipistrellus* (Massing 1988; Roer 1980; Schober and Grimmberger 1989; Stebbings 1988*b*; Strelkov 1969).

Some Old and New World nectarivorous and frugivorous species, including several Australian *Pteropus* species (Eby 1991; McWilliam 1985–86) and the phyllostomid *Leptonycteris curasoae* (Fleming 1991; Fleming, Nuñez, and Sternberg 1993), follow the flowering and fruiting seasons of their food resources.

Group Size and Social Organization

A mother and her dependent young are the basic social unit of the Chiroptera (Bradbury 1977*b*). In species that are considered "solitary" the mother appears to raise her young alone. More typically, however, females aggregate during the maternity season, forming colonies, which, depending on the species, may range in size from fewer than a dozen individuals to many million.

Among the Megachiroptera, strongly colonial behavior is documented for only nine genera—*Eidolon, Rousettus, Boneia, Pteropus, Acerodon, Dobsonia, Aproteles, Eonycteris,* and *Notopteris* (Marshall 1983)—and has been reported for about one-third of the *Pteropus* species (Pierson and Rainey 1992). Large aggregations in trees, known as "camps," are formed by *Eidolon, Acerodon,* and several species of *Pteropus*. Some species can be found year-round at the same roosting sites, whereas others disperse from their camps on a seasonal basis (Baker and Baker 1936; Eby 1991; Nelson 1965; Richards 1990*b;* Spencer, Palmer, and Parry-Jones 1991; Thomas 1983; Tidemann 1985).

The taxa *Epomophorus, Cynopterus, Thoopterus, Penthetor,* and *Ptenochirus* and some *Pteropus* species form small, less conspicuous colonies (Balasingh, Koilraj, and Kunz 1995; Bhat and Kunz 1995; Pierson and Rainey 1992). Most little-known species are assumed to be noncolonial; they form inconspicuous groups or aggregate only during maternity periods (Nelson 1964*a*). Information on roosting and foraging habits and behavior is lacking for more than half of the 58 *Pteropus* species (Pierson and Rainey 1992). A number of these are relatively large, with a forearm measurement greater than 100 mm, and probably are noncolonial, canopy-dwelling species (e.g., *Pteropus leucopterus* and *P. pohlei*).

Among the Microchiroptera, the largest aggregations are formed by a few molossid and vespertilionid species (e.g., *Tadarida brasiliensis* and *Miniopterus schreibersii*), although colony size can vary greatly among populations. For example, the North American *T. brasiliensis* typically forms colonies of from only a few hundred to a few thousand in California but up to 20 million in Texas (Davis, Herreid, and Short 1962; McCracken and Gustin 1991). A new species of *Chaerephon,* from Cambodia, is known from one colony in a building that numbers from 1.5 million to 2 million individuals (G. C. Richards, pers. comm.). Some of the most intensively studied species are also the most gregarious (e.g., *Miniopterus schreibersi* in Africa and Australia, *Pipistrellus pipistrellus* in Europe, *Myotis lucifugus* and *Tadarida brasiliensis* in North America). Many species, particularly the families Emballonuridae, Hipposideridae, Rhinolophidae, Phyllostomidae, Megadermatidae, Molossidae, and Vespertilionidae, form colonies of up to several hundred bats, while others form very small groups, of only a few individuals. The aggregation pattern can vary widely among species within the same genus. For example, *Rhinopoma microphyllum* and *Rhinolophus megaphyllus* form colonies of many thousands, while *R. hardwickei* roosts alone or in small groups (Kingdon 1974; Lekagul and McNeely 1977). Similarly, among North American *Myotis* species, *M. lucifugus* forms colonies of up to several thousand, *M. grisescens* forms colonies numbering in the hundreds of thousands, *M. evotis* colonies number a few dozen, and *M. californicus* lives in small family groups consisting of mother-pup pairs (Barbour and Davis 1969). Males are usually solitary or occur in small numbers at maternity sites, although a few species appear to roost in mixed-gender associations during the maternity period. In some species, numbers of bats can vary considerably from maternity to hibernating colonies. For example, maternity colonies of *M. sodalis* are in the range of a few dozen (Humphrey, Richter, and Cope 1977; Kurta et al. 1993), whereas hibernating colonies may number upwards to several hundred thousand (Hall 1962).

Social organization and mating systems have been examined in relatively few species, but they are highly variable, ranging from monogamy to various forms of polygyny (Bradbury 1977*b*). The types of interactions observed within a given roosting group depends on the mating system as well as the size, composition, and stability of the group. Much of the variation in mating systems can in turn be attributed to variation in female dispersion patterns, particularly to the number of females at a defensible resource such as a roost or a feeding site (Bradbury 1977*b*; Wilkinson 1987).

Monogamy, a relatively rare phenomenon in mammals (Kleiman 1977), is found in a number of bat species (Bradbury 1977*b*); it is best documented in *Vampyrum spectrum* (Vehrencamp, Stiles, and Bradbury 1977), *Lavia frons* (Vaughan and Vaughan 1986), *Hipposideros beatus* (Brosset 1982), *Pteropus samoensis* (Cox 1983), and *Rhinolophus sedulus* (Heller, Achmann, and Witt 1993). A lek mating system, in which males set up breeding stations at which they perform courtship displays to attract females, is found in two African megachiropterans, *Hypsignathus monstrosus* (Bradbury 1977*a*) and *Epomophorus wahlbergi* (Wickler and Seibt 1976), and in the microchiropteran *Mystacina tuberculata* (Daniel and Pierson 1987) and likely in *Miniopterus minor* (McWilliam 1990). A promiscuous mating system, with multiple mating and multiple mates, is assumed for many of the highly colonial species, particularly the temperate-zone vespertilionids, which appear to mate a number of times over a period of several months during the fall and winter—for example, *Myotis lucifugus* (Thomas, Fenton, and Barclay 1979).

Resource defense polygyny, in which an individual harem male gains exclusive reproductive access to a group of females, is perhaps the most prevalent mating system in bats. In some cases these groups appear to be transient (*Pipistrellus pipistrellus* in Sweden [Gerell and Lundberg 1985]) or compositionally labile (*Pipistrellus nanus* in Kenya [O'Shea 1980]). In other cases the composition of the female group is more stable over time (*Myotis adversus* in Australia [Dwyer 1970]). Females of the neotropical sac-winged bat, *Saccopteryx bilineata*, form small groups that commonly roost on tree trunks and in buttress cavities, and single dominant males recruit and maintain reproductive access to small groups of females both at the roost site and in foraging areas (Bradbury and Emmons 1974; Bradbury and Vehrencamp 1976).

A similar type of social organization has been observed in the greater spear-nosed bat, *Phyllostomus hastatus* (McCracken and Bradbury 1977, 1981), which typically roosts in groups where females outnumber males. Single harem males defend these stable groups for up to three years and usually father most of the synchronously produced young. Each year young females disperse from roosting clusters to form new groups. Roosting sites of *P. hastatus* (typically in caves) also house groups of nonharem males, some of whom may attempt to usurp the position of a dominant male during the mating season. Individual females maintain exclusive foraging areas during most of the year (McCracken and Bradbury 1981).

The social organization of the common vampire bat, *Desmodus rotundus* (Wilkinson 1984, 1985*a*, 1985*b*), appears to be based mostly on shared use of roosts by females and their young and the use of nonoverlapping foraging ranges with neighboring groups (Wilkinson 1987). The mating system of this species is based on resource defense polygyny, in which adult males fight to gain access to preferred female roost sites and engage in up to 80 percent of all copulations. Females suckle their own young and also regurgitate blood to close relatives. This altruistic behavior observed among roost mates in *D. rotundus* appears to be the result of both kin selection and reciprocity. Moreover, kin or roost mates appear to share wound sites on their prey, and reciprocal grooming is most frequent among these individuals (Wilkinson 1987).

Similar types of social organizations and mating systems have been observed in *Carollia perspicillata* (Fleming 1988) and *Artibeus jamaicensis* (Kunz, August, and Burnett 1983; Kunz and McCracken 1995; Morrison 1979; Morrison and Morrison 1981). Adult female *C. perspicillata* form small roosting groups in caves and hollow trees, where roost areas are defended by harem males for up to four years (Fleming 1988). Females show little faithfulness to roost sites, and thus the paternity of harem males is reduced considerably (Fleming 1988). There is no evidence that members of roosting groups share information or cooperate while foraging. Neither *C. perspicillata* nor *A. jamaicensis* exhibit altruistic or cooperative behaviors.

Polygynous mating systems have also been observed in *Myotis adversus* (Dwyer 1970), *Myotis bocagei* (Brosset 1976), *Coleura afra* (McWilliam 1987), and several members of the tent-making Phyllostomidae, including *Ectophylla alba* and *Uroderma bilobatum* (Brooke 1987; Kunz and McCracken 1995; Lewis 1992), as well as several megachiropterans, including *Cynopterus sphinx* (Balasingh, Koilraj, and Kunz 1995; Bhat and Kunz 1995) and a number of "camp"-forming *Pteropus* species (Pierson and Rainey 1992).

Many species of bats are highly vocal. Their vocalizations are usually associated with social interactions, including courtship, mating, and harem defense (Bradbury 1977*b*; McCracken and Bradbury 1981; Morrison and Morrison 1981; Wickler and Seibt 1976), spacing (Nelson 1964*b*; Richards 1990*c*), and mother-infant communication (Balcombe and McCracken 1992; Brown 1976; Fenton 1985; Gelfand and McCracken 1985). Species that form leks *(Hypsignathus monstrosus)* or defend feeding territories (e.g., *Pteropus poliocephalus* and *Cardioderma cor*) often engage in physical displays at night (e.g., wing flapping or erected hair patches) accompanied by loud vocalizations (Bradbury 1977*a*; Richards 1990*c*; Wickler and Seibt 1976).

Reproduction and Life History Characteristics

The reproductive patterns of bats range from seasonal monestry to polyestry (O'Brien 1993; Racey 1982). Most temperate species and many tropical species are monestrus, producing one litter per year, although some tropical species are polyestrus, producing two and sometimes three litters per year (see, e.g., Fleming 1971; Krishna and Dominic 1983; Myers 1977; O'Brien 1993; Wilson 1979; and Wilson and Findley 1970).

Although the length of gestation is fixed and species-specific in most mammals, in bats it can vary in response to environmental conditions, both among and within species (Racey 1981, 1982). To achieve synchrony between birth peaks and food availability, bats employ a variety of mechanisms for adjusting the timing or length of pregnancy. Most notable is the strategy of delayed fertilization (or delayed ovulation), known only among temperate-zone vespertilionids and rhinolophids (Racey 1982). Typically, these bats mate in the fall and winter; sperm is stored for several months in the female reproductive tract; and ovulation and fertilization occur in spring. Gestation proceeds and pups are born in early summer, when insects are abundant, although low temperatures and shortages of food are known to support fetal development (Racey 1973; Racey and Swift 1981).

Delayed implantation occurs in some species. Fertilization is followed by blastocyst formation, but the blastocyst remains free in the female reproductive tract until environmental conditions become favorable for implantation at a later time (Racey 1982). This pattern of delayed development occurs in the African megachiropteran *Eidolon helvum*, with blastocyst implantation coinciding with peak rainfall and the period of maximum fruit availability (Mutere 1980). Delayed implantation also occurs in the widely distributed Old World species *Miniopterus schreibersi* and *M. fraterculus* (Kimura and Uchida 1983. See also Racey 1982). Gestation length varies geographically in *M. schreibersi*, indicating that differences in the timing of implantation may reflect local environmental conditions (Racey 1982). Observed differences in the timing of implantation may also reflect differences in the endocrine control of implantation in different populations (see Bernard, Bojarski, and Millar 1991; Crichton, Seamark, and Krutzsch 1989; Kimura, Takeda, and Uchida 1987; Kimura and Uchida 1983; Richardson 1977; Van Aarde, Van der Merwe, and Skinner 1994; and Van der Merwe and Van Aarde 1989).

In some species mating is followed immediately by fertilization and implantation, but development of the fetus is arrested until the following spring, when gestation proceeds to term (Racey 1982). Delayed development has been reported in the New World phyllostomid *Macrotus californicus* (Bradshaw 1961, 1962) and in one of the two annual cycles in *Artibeus jamaicensis* (Fleming 1971). Embryonic diapause occurs during the dry season, so that pups are born immediately before the peak of fruit production. Delayed development has also been reported in the Old World insectivorous bat *Miniopterus australis* (Medway 1972; Richardson 1977). The prolonged gestation in this species has been attributed to the variability and the unpredictable availability of insect populations. Other species that reportedly undergo delayed development include *Rhinolophus rouxi, Cynopterus sphinx, Eptesicus furinalis,* and *Myotis albescens* (see Racey 1982). Some of these species are polyestrus, and delayed development appears to be associated with only one of two annual pregnancies.

In temperate regions the general pattern for females is to produce one young per year, in the late spring or early summer. Births tend to be fairly synchronous within populations for a given year (Racey 1982). In tropical regions some species have two birth peaks, which tend to be correlated with patterns of rainfall (Krishna and Dominic 1983; O'Brien 1993; Racey 1982). Some megachiropterans appear to have continuous breeding patterns, but the majority are seasonal breeders (O'Brien 1993). Some tropical species, including the New World phyllostomid *Anoura geoffroyi* and the Old World pteropodid *Pteropus scapulatus,* show no sensitivity to photoperiod (Heideman, Deorj, and Bronson 1992; O'Brien, Curlewis, and Martin 1993), although one population of *P. poliocephalus* from a high-latitude tropical environment is responsive to photoperiod (McGuckin and Blackshaw 1992). Thus, it appears that seasonal reproduction in tropical bats is influenced primarily by endogenous factors (O'Brien 1993).

In tropical regions the patterns of parturition are highly variable.. While some microchiropteran species are monestrus, bimodal polyestry (two birth peaks per year) is quite common, especially in the tropics (Wilson 1979). Births in both the Megachiroptera and the Microchiroptera tend to coincide with times of greatest food availability (Fleming 1988; O'Brien 1993; Racey 1982; Richards 1989; Wilson 1979). For some species, such as the common vampire bat *Desmodus rotundus* and some pteropodids for which food supply is abundant year-round, births are aseasonal (O'Brien 1993; Pierson and Rainey 1992; Wimsatt and Trapido 1952).

Bats are unusual among small mammals in having a slow rate of fetal growth (pregnancies typically last three to six months), a low reproductive rate (generally one to two young per year), and a long life span (15 years is not unusual, and the record is 31 years, for a banded male *Myotis lucifugus*) (see Racey 1982; and Tuttle and Stevenson 1982). Most bats have only one young per litter, although some species (most typically the vespertilionids) have twins, and there are records of some species producing litters of up to five (e.g., *Lasiurus borealis*). Young bats are exceptionally large at birth, with microchiropteran pups averaging about 25 percent of their mother's weight (Kurta and Kunz 1987) and megachiropterans in the range of 12–15 percent. Litter size in bats may be constrained by the pregnant mother's need to fly (Barclay 1995).

Female bats suckle their young for an extended period relative to most other mammals. Although most mammals wean their young when they reach 40 percent of adult size (or less), bats nurse their young until they are almost adult size (Barclay 1995; Kunz 1987). This difference reflects the fact that bats are not capable of feeding on their own until their wings have approached adult dimensions (Kunz 1987; Kunz and Allgaier 1995).

Many microchiropterans are born in the breech position (feet first), whereas most megachiropterans are born in the head-first position typical of other mammals. The developmental state at birth varies among taxa, with Megachiroptera and some Microchiroptera (Phyllostomidae) born with well-developed pelage and open eyes (Orr 1970; Tuttle and Stevenson 1982). Most other microchiropterans are born naked and with their eyes closed. Pup growth is relatively rapid in bats, with temperate species generally growing faster than tropical species (Kunz and Allgaier 1995). Small species tend to grow rapidly and are weaned within three to six weeks, whereas larger species (mostly megachiropterans) tend to grow more slowly. Microchiropteran pups are usually weaned when they reach about 90 percent of adult linear size and about 80 percent of adult body weight, whereas megachiropteran pups are weaned at a proportionately smaller size and body weight (Barclay 1995; Kunz and Stern 1995). Rates of postnatal growth in bats may be limited by both quality and quantity of food (Barclay 1995; Kunz 1987; Kunz and Stern 1995). Differences in growth rates between insectivorous and frugivorous species may reflect differences in the composition of milk supplied to pups during the postnatal period (Kunz, Oftedal, et al. 1994). Reduced food supply during inclement weather may depress milk production in mothers and reduce growth rates in pups. Calcium may be the limiting nutrient for successful reproduction in bats (Barclay 1995; Studier and Kunz 1995).

Feeding Habits

Bats exhibit a dietary diversity unparalleled among living mammals. Approximately 260 species are primarily frugivorous or nectarivorous; they are represented by the Pteropodidae in the Old World and several subfamilies of the Phyllostomidae in the New World. The majority of species (about 625) found in all the microchiropteran families are insectivorous. The few remaining species are either carnivorous or sanguinivorous. Frugivorous, nectarivorous, carnivorous, and sanguinivorous species are confined to tropical and subtropical regions, whereas insectivorous species are found at all inhabited latitudes. All bats living above about 38°N and below 40°S are insectivorous.

Frosted bat *(Vespertilio murinus)*, photo by Klaus-Gerhard Heller.

The megachiropteran diet consists predominantly of fruit, nectar, flower parts, and occasionally leaves, insects, and insect exudates (Bhat 1995; Francis 1990; Gould 1977, 1978; Heideman and Heaney 1989; Kunz and Diaz 1995; McWilliam 1985–86; Marshall 1985; Richards 1995; Spencer and Fleming 1989; Wiles and Fujita 1992). Many species feed on both fruit and nectar but are usually referred to as either "fruit" or "blossom" bats, depending on their dietary preference (Gould 1977; Marshall 1985; Wiles and Fujita 1992). Large megachiropterans (e.g., *Acerodon* and the larger *Pteropus* species) tend to feed on the fruits and flowers of large canopy trees, whereas small megachiropterans are more likely to feed in the understory (Marshall 1985; Rainey et al. 1995; Richards 1995; Wiles 1987*b*). Some are specialized for feeding on nectar and pollen (Marshall 1985), while others feed almost exclusively on small subcanopy and cauliflorous fruits (Marshall and McWilliam 1982; Spencer and Fleming 1989).

Folivory, by leaf fractionation (chewing and extracting the soluble juices from leaves), has been documented for several megachiropteran and a few microchiropteran species (Kunz and Diaz 1995; Lowry 1989). Although no bats feed exclusively on leaves, occasional folivory may allow bats to extract valuable nutrients without being constrained during flight by a highly specialized gut and the added weight of bulky leaf fiber (Kunz and Ingalls 1994). Leaf-eating may be highly adaptive during times of food shortage following typhoons (Pierson et al. 1995). Although some pteropodids ingest insects, whether deliberately or accidentally (Parry-Jones and Augee 1992), insectivory appears to be relatively uncommon among the Megachiroptera. Both insects and soluble extracts from leaves potentially provide a rich source of protein not available in an exclusively frugivorous or nectarivorous diet (Richards 1995).

The New World fruit- and nectar-feeding phyllostomids differ in several ways from their Old World counterparts (Fleming 1993; Rainey et al. 1995). Although a number of species, particularly in the subfamily Glossophaginae, are specialized for nectar consumption (having, for example, elongate rostra, long papillated tongues, and reduced dentition), phyllostomids, including some glossophagines, tend to be more eclectic in their diets (many also eat insects) (Gardner 1977; Helversen 1993). Also, because the generally smaller phyllostomids focus more on small-seeded fruits, particularly *Ficus*, and understory plants, they may play a less significant role than pteropodids in the dispersal of large canopy fruits (Rainey et al. 1995). Except for the primarily insectivorous *Mystacina tuberculata* from New Zealand, which is phylogenetically associated with the New World phyllostomids, there are no fruit- or nectar-feeding Microchiroptera in the Old World (Daniel 1976; Pierson et al. 1986).

The microchiropteran insectivorous assemblage is very diverse and can be divided into a number of guilds, including sallying and gleaning insectivores, forest and clearing aerial insectivores, water-surface foragers, and open-air aerial insectivores (see Findley 1993). Throughout their cosmopolitan range, "insectivorous" bats feed on a wide array of insects and other arthropods, including spiders, scorpions, and small crustaceans (Black 1974; Kunz, Wadanoli, and Whitaker 1995; Poulton 1929; Ross 1967; Rydell 1986; Swift and

Racey 1985; Warner 1985; Whitaker and Black 1976; Whitaker, Maser, and Keller 1977). During periods of peak energy demand, lactating females may consume more than their body mass nightly (Kurta, Bell, et al. 1989, 1990).

Lepidopterans are highly favored in the diets of many species, and coleopterans in the diets of others (Black 1974), but a wide range of soft-bodied to hard-bodied insects or other arthropods may be eaten. Prey size can vary from very small dipterans (e.g., midges and mosquitoes) to large coleopterans (beetles), orthopterans (crickets and grasshoppers), and scorpionids. In many cases general preferences can be predicted from jaw morphology—animals with more robust jaws, for example, tend to feed on hard-bodied insects, such as Coleoptera (Freeman 1981, 1988). Although some species appear to be quite specialized (e.g., the Australian *Kerivoula papuensis,* which has a preference for orb spiders, which it plucks from webs [Richards 1990*a*]; or the North American *Plecotus townsendii,* which feeds almost exclusively on moths [Dalton, Brack, and McTeer 1986; Robinson 1990]), other species appear to be relatively opportunistic and will feed on a variety of insects available within their selected foraging habitat. A bat's body weight and echolocation frequency, however, set limits on the prey species a bat can detect and handle (Barclay and Brigham 1991). Some species appear to be quite restricted to particular habitats (for example, *Myotis keenii* is known only from old-growth forest in the North American Pacific Northwest); others may feed in a range of habitats, from open desert to dense forest (e.g., *Plecotus townsendii* or *Antrozous pallidus* in western North America [Hermanson and O'Shea 1983; Kunz and Martin 1982]).

Carnivorous species, which feed primarily on small vertebrates (birds, rodents, lizards, frogs, and smaller bats), are relatively few; they are represented primarily by megadermatids *(Macroderma gigas, Cardioderma cor,* or *Megaderma lyra)* and nycterids *(Nycteris thebaica)* in the Old World (Fenton, Cumming, et al. 1987; Fenton, Rautenbach, et al. 1993; Marimuthu 1984; Vaughan 1976) and by phyllostomids *(Vampyrum spectrum, Chrotopterus auritus, Trachops cirrhosus, Mimon cozumelae)* in the New World (Fenton 1990, 1992; Tuttle and Ryan 1981; Vehrencamp, Stiles, and Bradbury 1977). Several species, most notably the noctilionid *Noctilio leporinus* (Bloedel 1955; Brooke 1994) and the vespertilionids *Myotis vivesi* and *Myotis adversus* (Jones and Rayner 1991; Robson 1984) sometimes feed on small fish and other aquatic arthropods. The three sanguinivorous species, all phyllostomids, are confined to the New World tropics: *Desmodus rotundus* specializes in mammals, and *Diaemus youngi* and *Diphylla ecaudata* feed primarily on birds. *Desmodus rotundus* populations have increased in number with the proliferation of horse and cattle ranches, which provide an abundance of food (Greenhall and Schmidt 1988).

Activity Patterns

Most bats rest during the day and feed at night, dispersing from their day roosts around dusk (Erkert 1982; Kunz 1982). The distance traveled varies greatly with species, habitat, colony size, and food availability. Many microchiropterans that have been followed by radiotelemetry appear to feed within 10–15 km of their day roost (see, e.g., Audet 1990; Barclay 1985, 1989; and

Long-tongued fruit bat *(Macroglossus sobrinus)*, photo by Klaus-Gerhard Heller.

Fenton and Rautenbach 1986), although some range upwards to 80 km or more nightly (Davis, Herreid, and Short 1962). Large megachiropterans are also known to travel up to 50 km, sometimes crossing coastal waters from offshore island roosts to mainland feeding areas (Eby 1991; Nelson 1965; Richards 1987; Spencer, Palmer, and Parry-Jones 1991). Although both megachiropterans and microchiropterans generally return to their day roosts around dawn, a number of microchiropteran species aggregate in night roosts, which are often located some distance from the day roost and closer to the foraging area (Kunz 1982). Although lactating pteropodids often carry nonvolant young with them when they forage, most microchiropterans leave their young behind in the day roost, returning to nurse them during the night (Anthony, Stack, and Kunz 1981; Kunz 1974; Anthony et al. 1981; Kunz, Wadanoli, and Whitaker 1995).

Although most bat species are strictly nocturnal, a few are partially or predominantly diurnal. Diurnality has been observed most frequently in island-dwelling pteropodids (e.g., *Pteropus melanotus* on Christmas Island [Tidemann 1987], *P. samoensis* in Samoa [Cox 1983], *P. tonganus* on Niue [Wodzicki and Felten 1975], and *P. insularis* and *P. molossinus* in Micronesia [Bruner and Pratt 1979]), but it also occurs in a few island microchiropterans (e.g., *Emballonura sulcata* on Pohnpei [Bruner and Pratt 1979] and *Nyctalus azoreum* on the Azores [Stoddart 1975]). Diurnality has also been observed during summer months at high latitudes in several microchiropterans (e.g., *Myotis mystacinus*

Samoan flying fox, or Samoan fruit bat *(Pteropus samoensis)*, photo by W. E. Rainey.

and *M. daubentoni* [Nyholm 1965] and *Pipistrellus pipistrellus* [see Speakman 1990, 1991]). Even among the nocturnal foragers, differences have been observed in emergence times relative to sunset (Erkert 1982). While many species wait until almost total darkness before emerging, others (the small Central American forest-dwelling emballonurid, *Saccopteryx bilineata* [Bradbury and Vehrencamp 1976], and the diminutive desert-dwelling vespertilionid, *Pipistrellus hesperus* [Barbour and Davis 1969]), begin their nightly activity long before sunset.

The Ecological and Economic Importance of Bats

Old World pteropodids and New World phyllostomids are important pollinators and seed dispersers for a number of ecologically and economically important plants (see, e.g., Cox et al. 1991, 1992; Dobat 1985; Fujita and Tuttle 1991; McWilliam 1985–86; Marshall 1985; Rainey et al. 1995; and Richards 1990*a*, 1990*c*, 1991). The New World plant-visiting bat *Leptonycteris curasoae* appears to be the major pollinator of two primary cactus species of the Sonoran Desert, the cardon *(Pachycereus pringlei)* and the organ pipe *(Stenocereus thurberi)* (Fleming 1989). The Old World bats *Rousettus aegyptiacus, Epomophorus wahlbergi,* and *Eidolon helvum* pollinate baobab *(Adansonia digitata)*, an economically important tree in the African savannah (Duxoux 1983; Start 1972). On faunally depauperate Old World oceanic islands, pteropodids may be the sole pollinators of plants that are known to have multiple pollinators on mainland areas (Elmqvist et al. 1992), and they are often the only vertebrates large enough to carry large-seeded fruits. Thus, in these communities bats may play a "keystone role" in structuring the forest community (Cox et al. 1991; Rainey et al. 1995). As frequent dispersers of "pioneer" species, like *Solanum* and *Piper*, bats also play an important role in the revegetation of cleared areas (Fleming 1988; Gorchov et al. 1993).

Fujita and Tuttle (1991) identified 186 paleotropical plant species utilized by flying foxes that were of economic importance to people for a variety of products, including food, medicines, dyes, fibers, ornamental plants, and timber. For example, pteropodids are the primary pollinators of two plants that are extremely important to the local economies of Southeast Asia, durian *(Durio zibethinus)* and petai *(Parkia speciosa* and *P. javanica)*, and play a role in both pollination and seed dispersal of a number of valuable timber species.

Insectivorous species are the primary consumers of nocturnal insects. Given the relatively large volumes consumed (up to 100 percent of body weight per night) and the long distances traveled (several km per night), these bats are thought to play a major role in regulating nocturnal insect populations and intransporting nutrients across the landscape, particularly from stream corridors to tree roosts (Kunz 1982; Rainey et al. 1992). Although mosquitoes are often considered to be an important dietary item of some insectivorous bats, the overwhelming numbers and diversity of insects eaten by bats are represented by other groups, including lepidopterans, coleopterans, homopterans, hemipterans, and tricopterans (see, e.g., Anthony and Kunz 1977; Black 1974; Dalton, Brack, and McTeer 1986; Kunz 1974; Kunz, Wadanoli, and Whitaker 1995; Ross 1967; Rydell 1986; Swift and Racey 1985; Warner 1985; Whitaker and Black 1976; and Whitaker, Maser, and Keller 1977). Bats are predators on a number of economically important insects, including cucumber beetles *(Diabrotica)*, June bugs *(Phyllophaga)*, and corn borers and Jerusalem crickets, which are important agricultural pests on such crops as corn, cotton, and potatoes (Whitaker 1993).

Conservation Issues

Bat populations appear to be declining almost everywhere in the world. Several species have apparently become extinct, including the megachiropterans *Pteropus brunneus* from Australia, *P. pilosus* from Pelau, *P. subniger* from the Mascarene Islands, *P. tokudae* from Guam, and *Dobsonia chapmani* from the Philippines and the microchiropterans *Mystacina robusta* from New Zealand and *Nyctophilus howensis* from Lord Howe Island (Heaney and Heideman 1987; Hill and Daniel 1985; McKean 1975; Moutou 1982; Richards 1995; Wiles 1987*b*, 1992). For many other species, ranges are contracting, numbers are declining, and only remnant populations remain—for example, the megachiropterans *Aproteles bulmerae* in Papua New Guinea, *Pteropus livingstonii* in the Comoro Islands, and *P. mariannus* on Guam and the microchiropterans *Craseonycteris thonglongyai* in Thailand, *Emballonura semicaudata* on several Pacific islands, *Eumops glaucinus* in the United States, *Rhinolophus hipposideros* in much of Europe, and *Rhinolophus ferrumequinum* in Great Britain (Belwood 1992; Flannery 1994; Hill and Smith 1991; Mickleburgh, Hutson, and Racey 1992; Stebbings 1988*b*; Wiles 1987*a*). Other species usually considered abundant have also experienced declines. For example, the numbers of

Pipistrellus pipistrellus, the most common species in Britain, decreased by 62 percent between 1978 and 1986 (Stebbings 1988*b*).

Declines in the Megachiroptera, particularly the colonial, tree-roosting species, are largely the consequence of extensive deforestation and exploitation of the animals for food (Fujita and Tuttle 1991; Mickleburgh, Hutson, and Racey 1992; Pierson and Rainey 1992). Large megachiropterans have been important dietary items in a number of Pacific island cultures. The most severe pressure on flying foxes in recent years has come from their commercial exploitation for a luxury food market in Guam and the North Marianna Islands (Wiles 1992; Wiles and Payne 1986). The scale of this trade and its impact led in 1989 to the protective listing of seven Pacific island *Pteropus* species on Appendix I of the Convention on International Trade of Endangered Species (CITES) and all remaining *Pteropus* and *Acerodon* species on Appendix II (Brautigam and Elmqvist 1990). Even exploitation for local use, which is important to many subsistence economies, has become unsustainable in some areas with expanding human populations and increasing availability of firearms. Loss of forest habitat, due largely to traditional timber harvesting practices and agricultural conversion, has also had serious consequences for many species, for example, *Pteropus rodricensis* in the Mascarenes, *P. voeltzkowi* on Pemba Island, *P. livingstonii* in the Comoros, *P. tonganus* on Niue and the Cook Islands, and several species in the Philippines (Pierson and Rainey 1992; Utzurrum 1992). Most of the smaller megachiropterans appear to be less vulnerable to direct human interference, with the exception of some colonial taxa, such as the now extinct hollow-tree-roosting *P. subniger* (Moutou 1982). Megachiropterans have also come into conflict with people in agricultural areas. These problems are most acute when forest clearing or adverse weather (e.g., drought) reduces the availability of native flowers and fruits (Loebel and Sanewski 1987).

Cave-dwelling colonial species are at very high risk wherever they occur. Although few megachiropterans are known to occupy caves, those that do are generally experiencing population declines, most likely resulting from hunting or vandalism (e.g., *Aproteles bulmerae* in Papua New Guinea and *Notopteris macdonaldi* in Vanuatu and New Caledonia) (Flannery 1994; Rainey 1992). In the majority of cases, declines of microchiropteran cave-dwelling bats can be attributed to human disturbance of roosts, particularly maternity and hibernating sites (McCracken 1988; Tuttle 1979). Disturbances may be inadvertent (recreational caving), or they may involve deliberate vandalism or even government-sponsored eradication programs (vampire control measures used in Mexico; fruit bat controls in Israel) (Greenhall and Schmidt 1988; Makin and Mendelssohn 1985). Inadequate cave management policies in the United States help explain why all five species on the U.S. federal list of endangered species are cave-dwelling taxa. Only with the installation of protective gates and exclusion of human visitors have colonies begun to recover in some areas in the United States (Pierson, Rainey, and Koontz 1991).

In many parts of the world cave-dwelling bats have also occupied cave analogs, such as abandoned mines and military bunkers. In North America, north of Mexico, 29 species of cave-dwelling bats (65 percent of the recognized species) are known to use mines as roosts (Pierson, Rainey, and Koontz 1991). For the endangered *Leptonycteris curasoae*, the only known roosts in the United States are in mines. In Europe and the United States many bat colonies are destroyed when mine openings collapse or are sealed for hazard abatement (Belwood and Waugh 1991; Stebbings 1988*b*). Although some steps have been taken to ensure protection of bat populations in mines on public lands in the United States (Pierson and Brown 1992), in many areas there are government-sponsored programs that sanction mine closings without biological assessment, thus placing colonially roosting bats at high risk. Recently in Australia a limestone quarry company, with authorization from the Queensland government, destroyed one of the few known roosts of the rare endemic *Macroderma gigas*. Conversely, measures taken by a few mining companies in the western United States (Brown, Berry, and Brown 1993; Pierson, Rainey, and Koontz 1991) have led to protection of bat populations resident in old mines and are serving as positive models for management of these important habitats.

Timber harvesting and agricultural practices have adversely affected bat populations in many parts of the world. Clearing of rain forests or temperate-zone old-growth forests (with selective harvesting of snags) has resulted in the loss of crevice, cavity, and foliage roosts, as well as important foraging habitats. Consequences are likely to be most serious for species such as the neotropical *Vampyrum spectrum* or the North American *Lasionycteris noctivagans*, which appear to require tree hollows for roosting (Parsons, Smith, and Whittam 1986; Vehrencamp, Stiles, and Bradbury 1977). Several species that in the past most likely occupied hollow trees now regularly use human-made structures (Kunz 1982), where often they are not welcome and may face exclusion or eradication. In Europe, treatment of wood timbers in buildings with insecticides, often lethal to bats, has been responsible for population declines in some building-dwelling species (Stebbings 1988*b*). Efforts to attract some specialized tree-roosting species and those excluded from buildings to artificial bat houses have had limited to moderate success in Europe (Stebbings and Walsh 1988*a*) and North America (Tuttle and Hensley 1993). Currently available bat houses attract mostly crevice-dwelling species, and not the cavity-roosting taxa, whose populations are at highest risk. None of the species considered threatened or endangered in the United States have been known to occupy artificial bat houses.

Although it appears that loss of roosting habitats accounts for the most serious threats to bat populations (McCracken 1988), contamination and alteration of feeding habitats—siltation of rivers and lakes from agricultural runoff, channeling of rivers for flood control, clearing of riparian and hedgerow habitat, extensive use of insecticides, and overgrazing of grasslands—may be increasingly important. Environmental contamination from a wide range of pesticides and pollutants has been implicated in the decline of some North American species (Clark 1981). There are others, however, whose populations appear to have increased in disturbed and/or managed areas (e.g., *Desmodus rotundus* [Turner 1975]). The most vulnerable species are likely to be those with highly specialized roosting or feeding requirements and dependence on the successional stages of natural communities.

It is possible to reverse current population trends and ensure long-term viability for many species. For those that are limited by the availability of roosts and for which declines can be attributed to disturbance or destruction of roosting habitat, the identification and protection of key roosting sites can go a long way toward protecting populations. This has been demonstrated in Great Britain, where sanctuary status has been granted to a number of roost sites (Stebbings 1988*b*), and in North America, with the protection (gating) of hibernating sites for *Myotis grisescens* and *Myotis sodalis* (U.S. Fish and Wildlife Service 1982, 1983) and maternity sites for *Plecotus townsendii* (Pierson, Rainey, and Koontz 1991; Stihler and Hall 1994). In parts of the world where hunting pressure is the primary cause of decline, limits on hunting (through bans and firearms controls) are needed. In other areas, limiting the clearing of native forest will be required to protect both foraging and roosting habitat. The primary issue, however, is that whatever type of protection is required, it too often depends on the collective efforts of many individuals and organizations rather than on government policies or legislative mandates. Bat populations will not be secure until their ecological importance is acknowledged, their habitat requirements are understood, and the protection of roosting sites and foraging areas is incorporated into land management policies.

Acknowledgments

We wish to thank M. Brock Fenton, Karl F. Koopman, Ronald M. Nowak, Paul A. Racey, Gregory C. Richards, and an anonymous reviewer for making helpful suggestions and comments on an earlier version of this text.

Literature Cited

Advani, R. 1981. Seasonal fluctuations in the feeding ecology of the Indian false vampire, *Megaderma lyra lyra* (Chiroptera: Megadermatidae) in Rajasthan. Z. Saugetierk. 46:90–93.

Ahlen, I. 1990. Identification of bats in flight. Swed. Soc. Conserv. Nat., Stockholm, 50 pp.

Albuja, L. 1982. Murcielagos del Ecuador. Escuela Politecnica Nacional, Quito, 285 pp.

Aldridge, H. D. N., and I. L. Rautenbach. 1987. Morphology, echolocation, and resource partitioning in insectivorous bats. J. Anim. Ecol. 56:763–78.

Altenbach, J. S., and J. W. Hermanson. 1987. Bat flight muscle function and the scapulohumeral lock. *In* Fenton, M. B., P. A. Racey, and J. M. V. Rayner, eds., Recent advances in the study of bats, Cambridge Univ. Press, Cambridge, pp. 100–121.

Anthony, E. L. P., and T. H. Kunz. 1977. Feeding strategies of the little brown bat, *Myotis lucifugus*, in southern New Hampshire. Ecology 58:755–86.

Anthony, E. L. P., M. H. Stack, and T. H. Kunz. 1981. Night roosting and the nocturnal time budget of the little brown bat, *Myotis lucifugus:* effects of reproductive status, prey density, and environmental conditions. Oecologia 51:151–56.

Audet, D. 1990. Foraging behavior and habitat use by a gleaning bat, *Myotis myotis*. J. Mamm. 71:420–24.

Baagøe, H. J. 1987. The Scandinavian bat fauna: adaptive wing morphology and free flight in the field. *In* Fenton, M. B., P. A. Racey, and J. M. V. Rayner, eds., Recent advances in the study of bats, Cambridge Univ. Press, Cambridge, pp. 47–74.

Bailey, W. J., J. L. Slightom, and M. Goodman. 1992. Rejection of the "flying primate" hypothesis by phylogenetic evidence from the ϵ-globin gene. Science 256:86–89.

Baker, J. R., and Z. Baker. 1936. The seasons in a tropical rainforest (New Hebrides). Part 3. Fruit bats (Pteropidae). Zool. J. Linnean Soc. 40:123–41.

Baker, R. J., M. J. Novacek, and N. B. Simmons. 1991. On the monophyly of bats. Syst. Zool. 40:216–31.

Balasingh, J., J. Koilraj, and T. H. Kunz. 1995. Tent construction by the short-nosed fruit bat, *Cynopterus sphinx* (Chiroptera: Pteropodidae) in southern India. Ethology 100:210–29.

Balcombe, J. P., and G. F. McCracken. 1992. Vocal recognition in Mexican free-tailed bats: do pups recognize mothers? Anim. Behav. 43:89–94.

Barbour, R. W., and W. H. Davis. 1969. Bats of America. Univ. Press of Kentucky, Lexington, 286 pp.

Barclay, R. M. R. 1985. Foraging strategies of the tropical bat *Scotophilus leucogaster*. Biotropica 17:65–70.

———. 1989. The effect of reproductive condition on the foraging behavior of female hoary bats, *Lasiurus cinereus*. Behav. Ecol. Sociobiol. 24:31–37.

———. 1995. Constraint on reproduction by bats: energy or calcium? *In* Racey, P. A., and S. M. Swift, eds., Ecology, behaviour, and evolution of bats, Symp. Zool. Soc. Lond., Oxford Univ. Press, Oxford, in press.

Barclay, R. M. R., and R. M. Brigham. 1991. Prey detection, dietary niche breadth, and body size in bats: why are aerial insectivorous bats so small? Am. Nat. 137:693–703.

———. 1994. Constraints on optimal foraging: a field test of prey discrimination by echolocating insectivorous bats. Anim. Behav. 144:1021–31.

Barclay, R. M. R., M. B. Fenton, and D. W. Thomas. 1979. Social behavior of the little brown bat, *Myotis lucifugus*. II. Vocal communication. Behav. Ecol. Sociobiol. 6:137–46.

Barquez, R. M., N. P. Giannini, and M. A. Mares. 1993. Guide to bats of Argentina. Okla. Mus. Nat. Hist., Univ. Okla., Norman, 119 pp.

Bell, G. P. 1982. Behavioral and ecological aspects of gleaning by a desert insectivorous bat, *Antrozous pallidus* (Chiroptera: Vespertilionidae). Behav. Ecol. Sociobiol. 10: 217–23.

———. 1985. The sensory basis of prey location by the California leaf-nosed bat *Macrotus californicus* (Chiroptera: Hipposideridae). Behav. Ecol. Sociobiol. 16:343–47.

Bell, G. P., and M. B. Fenton. 1984. The use of Doppler-shifted echoes as a flutter detection and clutter rejection system: the echolocation and feeding behavior of *Hipposideros ruber* (Chiroptera: Hipposideridae). Behav. Ecol. Sociobiol. 15: 109–14.

Belwood, J. J. 1992. Florida mastiff bat, *Eumops glaucinus floridanus*. Pp. 216–23. *In* Humphrey, S. R., ed., Rare and endangered biota of Florida, vol. 1, Mammals, Univ. Press of Florida, Gainesville, pp. 216–23.

Belwood, J. J., and R. J. Waugh. 1991. Bats and mines: abandoned does not always mean empty. Bats 9:13–16.

Bennett, B. 1993. Structural modifications involved in the fore and hind limb grip of some flying foxes (Chiroptera: Pteropodidae). J. Zool., Lond. 229:237–48.

Bennett, S., L. J. Alexander, R. H. Crozier, and A. G. Mackinlay. 1988. Are megabats flying primates? Contrary evidence from a mitochondrial DNA sequence. Austral. J. Biol. Sci. 41:327–32.

Bergmans, W., and F. G. Rozendaal. 1988. Notes on collections of fruit bats from Sulawesi and some off-lying islands (Mammalia, Megachiroptera). Zool. Verhand., no. 248, 74 pp.

Bernard, R. T. F., C. Bojarski, and R. P. Millar. 1991. Plasma progesterone and lutenizing hormone concentrations and the role of the corpus luteum and LH gonadotrophs in the control of delayed implantation in Schreibers' long-fingered bat *(Miniopterus schreibersii)*. J. Reprod. Fertil. 93:31–42.

Bhat, H. R. 1995. Observations on the food and feeding behavior of *Cynopterus sphinx* (Chiroptera, Pteropodidae). Mammalia 58:368–70.

Bhat, H. R., and T. H. Kunz. 1995. Fruit and flower clusters used as roosts by the short-nosed fruit bat, *Cynopterus sphinx*, near Pune, India. J. Zool. 235:597–604.

Bhatnagar, K. P. 1980. The chiropteran vomeronasal organ: its relevance to the phylogeny of bats. *In* Wilson, D. E., and A. L. Gardner, eds., Proceedings of the Fifth International Bat Research Conference, Texas Tech Univ. Press, Lubbock, pp. 289–316.

Black, H. L. 1974. A north temperate bat community: structure and prey populations. J. Mamm. 55:138–57.

Bloedel, P. 1955. Hunting methods of fish-eating bats, particularly *Noctilio leporinus*. J. Mamm. 36:390–99.

Bonaccorso, F. J. 1979. Foraging and reproductive ecology in a Panamanian bat community. Bull. Florida State Mus. Biol. Sci. 24:359–408.

Bradbury, J. W. 1977*a*. Lek mating behavior in the hammer-headed bat. Z. Tierpsychol. 45:225–55.

———. 1977*b*. Social organization and communication. *In* Wimsatt, W. A., ed., Biology of bats, vol. 3, Academic Press, New York, pp. 1–72.

Bradbury, J. W., and L. H. Emmons. 1974. Social organization of some Trinidad bats. I. Emballonuridae. Z. Tierpsychol. 36:137–83.

Bradbury, J. W., and S. L. Vehrencamp. 1976. Social organization and foraging in emballonurid bats. I. Field Studies. Behav. Ecol. Sociobiol. 1:337–81.

Bradshaw, G. V. R. 1961. Le cycle des reproduction des *Macrotus californicus* (Chiroptera, Phyllostomatidae). Mammalia 25:117–19.

———. 1962. Reproductive cycle of the California leaf-nosed bat, *Macrotus californicus*. Science 136:645–46.

Brautigam, A., and T. Elmqvist. 1990. Conserving Pacific island flying foxes. Oryx 24:81–89.

Brigham, R. M. 1987. The significance of winter activity by the big brown bat *(Eptesicus fuscus)*: the influence of energy reserves. Can. J. Zool. 65:1240–42.

Brigham, R. M., and M. B. Fenton. 1986. The influence of roost closure on the roosting and foraging behaviour of *Eptesicus fuscus* (Chiroptera: Vespertilionidae). Can. J. Zool. 64:1128–33.

Brooke, A. P. 1987. Tent selection, roosting ecology, and social organization of the tent-making bat, *Ectophylla alba*, in Costa Rica. J. Zool., Lond. 221:11–19.

———. 1994. Diet of the fishing bat, *Noctilio leporinus* (Chiroptera: Noctilionidae). J. Mamm. 75:212–18.

Brosset, A. 1962. The bats of central and western India. J. Bombay Nat. Hist. Soc. 59:1–57, 583–624, 707–46.

———. 1976. Social organization of the African bat, *Myotis boccagei*. Z. Tierpsychol. 42:50–56.

———. 1982. Structure sociale de chiroptere *Hipposideros beatus*. Mammalia 46:3–9.

Brown, P. 1976. Vocal communication in the pallid bat, *Antrozous pallidus*. Z. Tierpsychol. 41:34–54.

Brown, P. E., R. D. Berry, and C. Brown. 1993. Bats and mines: finding solutions. Bats 11:12–13.

Bruner, P. L., and H. D. Pratt. 1979. Notes on the status and natural history of Micronesian bats. Elepaio 40:1–4.

Buchler, E. R., and S. B. Childs. 1981. Orientation to distant sounds by foraging big brown bats *(Eptesicus fuscus)*. Anim. Behav. 29:428–32.

Clark, D. R., Jr. 1981. Bats and environmental contaminants: a review. USFWS, Spec. Sci. Rept., no. 235, pp. 1–27.

Cockrum, E. L. 1969. Migration of the guano bat *Tadarida brasiliensis*. *In* Jones, J. K., Jr., ed., Contributions in mammalogy, Misc. Publ., Univ. Kans. Mus. Nat. Hist., no. 51, pp. 303–36.

Cooper, J. G., and K. P. Bhatnagar. 1976. Comparative anatomy of the vomeronasal organ complex in bats. J. Anat. 122:571–601.

Corbet, G. B., and J. E. Hill. 1992. The mammals of the Indomalayan region: a systematic review. Brit. Mus. Publ., Oxford Univ. Press, Oxford, 488 pp.

Cox, P. A. 1983. Observations of the natural history of Samoan bats. Mammalia 47:519–23.

Cox, P. A., T. Elmqvist, E. D. Pierson, and W. E. Rainey. 1991. Flying foxes as strong interactors in South Pacific island ecosystems: a conservation hypothesis. Conserv. Biol. 5:448–54.

———. 1992. Flying foxes as pollinators and seed dispersers in Pacific island ecosystems. *In* Wilson, D. E., and G. Graham, eds., Proceedings of the Pacific Island Flying Fox Conservation Conference, USFWS, Biol. Rept. no. 90, Washington, D.C., pp. 19–23.

Crerar, L. M., and M. B. Fenton. 1984. Cervical vertebrae in relation to roosting posture in bats. J. Mamm. 65:395–403.

Crichton, E. G., R. F. Seamark, and P. H. Krutzsch. 1989. The status of the corpus luteum during pregnancy in *Miniopterus schreibersii* (Chiroptera: Vespertilionidae) with emphasis on its role in developmental delay. Cell Tiss. Res. 258:183–201.

Crome, F. H. J., and G. C. Richards. 1988. Bats and gaps: microchiropteran community structure in a Queensland rain forest. Ecology 69:1960–69.

Dalton, V. M., V. Brack, Jr., and P. M. McTeer. 1986. Food habits of the big-eared bat, *Plecotus townsendii virginianus*, in Virginia. Virginia J. Sci. 37:248–54.

Daniel, M. J. 1976. Feeding by the short-tailed bat *(Mystacina tuberculata)* on fruit and possibly nectar. New Zealand J. Zool. 3:391–98.

Daniel, M. J., and E. D. Pierson. 1987. A lek mating system in New Zealand's short-tailed bat, *Mystacina tuberculata*. Bat Res. News 28:33.

Davidson, G. W. H., and Z. Akbar. 1987. The abundance of bats in selected forest types at Ula Endua, Johre, Malaysia. Malay. Nat. J. 41:441–46.

Davis, R. B., C. F. Herreid II, and H. L. Short. 1962. Mexican free-tailed bats in Texas. Ecol. Monogr. 32:311–46.

Davis, W. H. 1970. Hibernation: ecology and physiological ecology. *In* Wimsatt, W. A., ed., Biology of Bats, vol. 1, Academic Press, New York, pp. 265–300.

Davis, W. H., and H. B. Hitchcock. 1965. Biology and migration of the bat, *Myotis lucifugus*, in New England. J. Mamm. 46:296–313.

DeBlase, A. F. 1980. The bats of Iran: systematics, distribution, ecology. Fieldiana Zool., n.s., no. 4, 424 pp.

Dobat, K. 1985. Blüten und Fledermäuse. Verlag von Waldemar Kramer, Frankfurt, 370 pp.

Duxoux, E. 1983. La pollisation des fleurs de Baobab est-elle seulement la fait des mâles de la Rousette Paillée *Eidolon helvum*? Rev. Ecol. 38:229–31.

Dwyer, P. D. 1966. The population pattern of *Miniopterus schreibersii* (Chiroptera) in north-eastern New South Wales. Aust. J. Zool. 14:1073–1137.

———. 1970. Social organization of the bat *Myotis adversus*. Science 168:1006–8.

Eby, P. 1991. Seasonal movements of grey-headed flying foxes, *Pteropus poliocephalus* (Chiroptera: Pteropodidae), from two maternity camps in northern New South Wales. Wildl. Res. 18:547–59.

Elmqvist, T., P. A. Cox, W. E. Rainey, and E. D. Pierson. 1992. Restricted pollination on oceanic islands: pollination of *Ceiba pentandra* by flying foxes in Samoa. Biotropica 24:15–23.

Erkert, H. G. 1982. Ecological aspects of bat activity rhythms. *In* Kunz, T. H., ed., Ecology of bats, Plenum Press, New York, pp. 201–42.

Fenton, M. B. 1982. Echolocation, insect hearing, and feeding ecology of insectivorous bats. *In* Kunz, T. H., ed., Ecology of bats, Plenum Press, New York, pp. 261–85.

———. 1985. Communication in the Chiroptera. Univ. of Indiana Press, Bloomington, 161 pp.

———. 1990. Foraging behaviour and ecology of animal eating bats. Can. J. Zool. 68:411–22.

———. 1992. Bats. Facts on File, New York, 207 pp.

Fenton, M. B., N. G. Boyle, T. M. Harrison, and D. J. Oxley. 1977. Activity patterns, habitat use, and prey selection by some African insectivorous bats. Biotropica 9: 73–85.

Fenton, M. B., R. M. Brigham, A. M. Mills, and I. L. Rautenbach. 1985. The roosting and foraging areas of *Epomophorus wahlbergi* (Pteropodidae) and *Scotophilus viridis* (Vespertilionidae) in Kruger National Park, South Africa. J. Mamm. 66:461–68.

Fenton, M. B., D. H. M. Cumming, J. M. Hutton, and C. M. Swanepoel. 1987. Foraging and habitat use by *Nycteris grandis* (Nycteridae) in Zimbabwe. J. Zool., Lond. 211:709–16.

Fenton, M. B., and I. L. Rautenbach. 1986. A comparison of the roosting and foraging behaviour of three species of African insectivorous bats (Rhinolophidae, Vespertilionidae, and Molossidae). Can. J. Zool. 64:2860–67.

Fenton, M. B., I. L. Rautenbach, D. Chipese, D. H. M. Cumming, M. K. Musgrave, J. S. Taylor, and T. Volpers. 1993. Variation in foraging behaviour, habitat use, and diet of large slit-faced bats *(Nycteris grandis)*. Z. Saugetierk. 58:65–74.

Fenton, M. B., and D. W. Thomas. 1985. Migrations and dispersal of bats (Chiroptera). *In* Rankin, M. A., ed., Migration: mechanisms and adaptive significance, Univ. Texas Contr. Mar. Sci., Austin, pp. 490–524.

Fiedler, J. 1979. Prey catching with and without echolocation in the Indian false vampire *(Megaderma lyra)*. Behav. Ecol. Sociobiol. 6:155–60.

Findley, J. S. 1993. Bats: a community perspective. Cambridge Univ. Press, Cambridge, 167 pp.

Findley, J. S., and D. E. Wilson. 1983. Are bats rare in tropical Africa? Biotropica 15:299–303.

Flannery, T. F. 1994. The rediscovery of Bulmer's fruit bat. Bats 12:3–5.

Fleming, T. H. 1971. *Artibeus jamaicensis:* delayed embryonic development in a neotropical bat. Science 171:402–4.

———. 1986. The structure of neotropical bat communities: a preliminary analysis. Rev. Chilena Hist. Nat. 59:135–50.

———. 1988. The short-tailed fruit bat. Univ. of Chicago Press, 365 pp.

———. 1989. Climb every cactus. Bats 7:3–6.

———. 1991. Following the nectar trail. Bats 9:4–7.

———. 1993. Plant-visiting bats. Amer. Sci. 81:460–67.

Fleming, T. H., R. A. Nuñez, and L. Sternberg. 1993. Seasonal changes in the diets of migrant and non-migrant nectarivorous bats as revealed by carbon stable isotope analysis. Oecologia 94:72–75.

Francis, C. M. 1990. Trophic structure of bat communities in the understorey of lowland dipterocarp rainforest in Malaysia. J. Trop. Ecol. 6:421–31.

———. 1994. Vertical stratification of fruit bats (Pteropodidae) in lowland dipterocarp rain forest in Malaysia. J. Trop. Ecol. 10:523–30.

Francis, C. M., E. L. P. Anthony, J. Brunton, and T. H. Kunz. 1994. Lactation in male fruit bats. Nature 367:691–92.

Freeman, P. W. 1981. Correspondence of food habits and morphology in insectivorous bats. J. Mamm. 62:166–73.

———. 1988. Frugivorous and animalivorous bats (Microchiroptera): dental and cranial adaptations. Biol. J. Linnean Soc. 33:249–72.

Fujita, M. S., and M. D. Tuttle. 1991. Flying foxes (Chiroptera: Pteropodidae): threatened animals of key economic importance. Conserv. Biol. 5:455–63.

Fullard, J. H. 1987. Sensory ecology and neuroethology of moths and bats: interaction in a global perspective. *In* Fenton, M. B., P. A. Racey, and J. M. V. Rayner, eds., Recent advances in the study of bats, Cambridge Univ. Press, Cambridge, pp. 244–77.

Gaisler, J. 1963*a*. The ecology of lesser horseshoe bat (*Rhinolophus hipposideros hipposideros* Bechstein, 1980) in Czechoslovakia. Part I. Acta Soc. Zool. Bohem. 27: 211–33.

———. 1963*b*. The ecology of lesser horseshoe bat (*Rhinolophus hipposideros hipposideros* Bechstein, 1980) in Czechoslovakia. Part II: ecological demands, problems of synanthropy. Acta Soc. Zool. Bohem. 28:322–27.

———. 1979. Ecology of bats. *In* Stoddart, D. M., ed., Ecology of small mammals, Chapman and Hall, London, pp. 281–342.

Gaisler, J., V. Hanak, and J. Dunsel. 1979. A contribution to the population ecology of *Nyctalus noctula* (Mammalia: Chiroptera). Acta Sci. Nat. Brno 13:1–38.

Gardner, A. L. 1977. Feeding habits. *In* Baker, R. J., J. K. Jones, Jr., and D. C. Carter, eds., Biology of bats of the New World family Phyllostomatidae, Spec. Publ., Mus. Texas Tech Univ., Lubbock, pp. 293–350.

Gardner, A. L., R. K. LaVal, and D. E. Wilson. 1970. The distributional status of some Costa Rican bats. J. Mamm. 51:712–29.

Gelfand, D. L., and G. F. McCracken. 1986. Individual variation in isolation calls of Mexican free-tailed bat pups *(Tadarida brasiliensis)*. Anim. Behav. 34:1078–86.

Gerell, R., and K. Lundberg. 1985. Social organization in the bat *pipistrellus pipistrellus*. Behav. Ecol. Sociobiol. 16:177–84.

Goodwin, G. G., and A. M. Greenhall. 1961. A review of the bats of Trinidad and Tobago. Bull. Amer. Mus. Nat. Hist. 122:187–302.

Goodwin, R. E. 1979. The bats of Timor: systematics and ecology. Bull. Amer. Mus. Nat. Hist. 163:73–122.

Gorchov, D. L., F. Cornejo, C. Ascorra, and M. Jaramillo. 1993. The role of seed dispersal in the natural revegetation of rain forest after strip-cutting in the Peruvian Amazon. *In* Fleming, T. H., and A. Estrada, eds., Frugivory and seed dispersal: ecological and evolutionary aspects. Kluwer Academic Publ., Dordrecht, pp. 339–49.

Gould, E. 1977. Foraging behavior of *Pteropus vampyrus* on the flowers of *Durio zibethinus*. Malay. Nat. J. 30:53–57.

———. 1978. Foraging behavior of some Malaysian nectar feeding bats. Biotropica 10:184–93.

Graham, G. L. 1988. Interspecific associations among Peruvian bats at diurnal roosts and roost sites. J. Mamm. 69:711–20.

Greenhall, A. M., and U. Schmidt, eds. 1988. Natural history of vampire bats. CRC Press, Boca Raton, 246 pp.

Griffin, D. R. 1958. Listening in the dark. Yale Univ. Press, New Haven, 413 pp.

———. 1970. Migrations and homing of bats. *In* Wimsatt, W. A., ed., Biology of bats, vol. 1, Academic Press, New York, pp. 233–65.

Guppy, A., and R. B. Coles. 1988. Acoustical and neural aspects of hearing in the Australian gleaning bats, *Macroderma gigas* and *Nyctophilus gouldi*. J. Comp. Physiol. 162:653–68.

Habersetzer, J., and G. Storch. 1989. Ecology and echolocation of the Eocene Messel bats. *In* Hanák, V., I. Horáček, and J. Gaisler, eds., European Bat Research 1987, Charles Univ. Press, Prague, pp. 213–33.

Hall, J. S. 1962. A life history and taxonomic study of the Indiana bat, *Myotis sodalis*. Reading Publ. Mus. Art. Galley Publ. 12:1–68.

Hall, L. S., and G. C. Richards. 1979. Bats of eastern Australia. Queensland Mus. Booklet 12:1–66.

———. 1991. Flying fox camps. Wildl. Austral. 28:19–22.

Hamilton-Smith, E. 1974. The present knowledge of Australian Chiroptera. Austral. Mammal. 2:95–108.

Handley, C. O., Jr., D. E. Wilson, and A. L. Gardner, eds. 1991. Demography and natural history of the common fruit bat, *Artibeus jamaicensis*, on Barro Colorado Island, Panama. Smithsonian Contr. Zool., no. 511, Washington, D.C., 173 pp.

Hartley, D. J., and R. A. Suthers. 1987. The sound emission pattern and the acoustical role of the noseleaf in the echolocating bat, *Carollia perspicillata*. J. Acoust. Soc. Amer. 82:1892–1900.

Heaney, L. R. 1991. An analysis of patterns of distribution and species richness among Philippine fruit bat (Pteropoididae). Bull. Amer. Mus. Nat. Hist. 206:145–67.

Heaney, L. R., and P. D. Heideman. 1987. Philippine fruit bats: endangered and extinct. Bats 5:3–5.

Heaney, L. R., P. D. Heideman, E. R. Rickart, R. C. Utzurrum, and J. H. S. Klompen. 1989. Elevational zonation of mammals in the central Philippines. J. Trop. Ecol. 5:259–80.

Heideman, P. D., P. Deorj, and F. H. Bronson. 1992. Seasonal reproduction of a tropical bat, *Anoura geoffroyi*, in relation to photoperiod. J. Reprod. Fert. 96:765–73.

Heideman, P. D., and L. R. Heaney. 1989. Population biology and estimates of abundance of fruit bats (Pteropodidae) in Philippine submontane rainforest. J. Zool., Lond. 218:565–86.

Heller, K.-G., R. Achmann, and K. Witt. 1993. Monogamy in the bat *Rhinolophus sedulus?* Z. Säugetierk. 58:376–77.

Helversen, O. von. 1993. Adaptations of flowers to the pollination by glossophagine bats. *In* Barthlott, W., et al., eds., Plant-animal interactions in tropical environments, Mus. Alexander Koenig, Bonn, pp. 41–59.

Henson, M., B. Wilson, and R. Hansen. 1987. Biosonar imaging of insects by *Pteronotus parnellii*, the moustached bat. Nat. Geogr. Res. 3:82–101.

Henson, O. W., Jr. 1970. The ear and audition. *In* Wimsatt, W. A., ed., Biology of bats, vol. 2, Academic Press, New York, pp. 181–263.

Hermanson, J. W., and T. J. O'Shea. 1983. Antrozous pallidus. Mammalian Species, no. 213, 8 pp.

Hill, J. E., and M. J. Daniel. 1985. Systematics of the New Zealand short-tailed *Mystacina* Gray, 1843 (Chiroptera: Mysticinidae). Bull. Brit. Mus. Nat. Hist., Zool. 48:279–300.

Hill, J. E., and J. D. Smith. 1981. Craseonycteris thonglongyai. Mammalian Species, no. 160, 4 pp.

———. 1984. Bats, a natural history. Univ. of Texas Press, Austin, 243 pp.

Howell, D. J., and J. Pylka. 1977. Why bats hang upside down: a biomechanical hypothesis. J. Theoret. Biol. 69:625–31.

Humphrey, S. R. 1975. Nursery roosts and community diversity of Nearctic bats. J. Mamm. 56:321–46.

Humphrey, S. R., and F. J. Bonaccorso. 1979. Population and community ecology. *In* Baker, R. J., J. K. Jones, Jr., and D. C. Carter, eds., Biology of bats of the New World family Phyllostomidae, III, Spec. Publ., Mus. Texas Tech Univ., Lubbock, pp. 409–41.

Humphrey, S. R., A. R. Richter, and J. B. Cope. 1977. Summer habitat and ecology of the endangered Indiana bat, *Myotis sodalis*. J. Mamm. 58:334–46.

Husson, M. 1962. The bats of Surinam. Zool. Verhand., no. 58, 282 pp.

Ingle, R. N., and L. R. Heaney. 1992. A key to the bats of the Philippine Islands. Fieldiana Zool., n.s., 69:1–44.

Jen, P. H.-S., and N. Suga. 1976. Coordinated activities of middle-ear and laryngeal muscles in echolocating bats. Science 191:950.

Jepson, G. L. 1970. Bat origins and evolution. *In* Wimsatt, W. A., ed., Biology of bats, vol. 1, Academic Press, New York, pp. 1–64.

Jones, G., and J. M. V. Rayner. 1988. Flight performance, foraging tactics, and echolocation in free-living Daubenton's bat, *Myotis daubentoni*. J. Zool., Lond. 215:113–32.

———. 1991. Flight performance, foraging tactics, and echolocation in the trawling bat, *Myotis adversus*. J. Zool., Lond. 225:393–412.

Jones, J. K., Jr. 1966. Bats from Guatemala. Univ. Kans. Publ. Mus. Nat. Hist. 16: 439–72.

Kalko, E., and H.-U. Schnitzler. 1989. The echolocation and hunting behaviour of Daubenton's bat, *Myotis daubentoni*. Behav. Ecol. Sociobiol. 24:225–38.

Kimura, K. A., A. Takeda, and T. A. Uchida. 1987. Changes in progesterone concentrations in the Japanese long-fingered bat, *Miniopterus schreibersii fulginosis*. J. Reprod. Fertil. 80:59–63.

Kimura, K. A., and T. A. Uchida. 1983. Ultrastructural observations of delayed implantation in the Japanese long-fingered bat, *Miniopterus schreibersii fulginosis*. J. Reprod. Fertil. 69:187–93.

Kingdon, J. 1974. East African mammals: an atlas of evolution in Africa, vol. 2, pt. A, Insectivores and bats, Academic Press, London, 341 pp.

Kitchener, D. H., B. L. Charlton, and Maharadatunkamsi. 1989. The wild mammals of Lombok Island, Nusa Tenggara, Indonesia: systematics and natural history. Rec. West. Austral. Mus., Suppl., no. 33, 129 pp.

Kitchener, D. H., A. Gunnell, and Maharadatunkamsi. 1990. Aspects of the feeding biology of fruit bats (Pteropodidae) on Lombok Island, Nusa Tenggara, Indonesia. Mammalia 54:561–78.

Kleiman, D. G. 1977. Monogamy in mammals. Q. Rev. Biol. 42:39–69.

Koopman, K. F. 1970. Zoogeography of bats. *In* Slaughter, B. H., and D. W. Walton,

eds., About bats, a chiropteran symposium, Southern Methodist Univ. Press, Dallas, pp. 29–50.
———. 1979. Zoogeography of mammals from islands off the northeastern coast of New Guinea. Amer. Mus. Novit., no. 2690, 17 pp.
———. 1984. A synopsis of the families of bats. Part VII. Bat Res. News 25:25–27.
———. 1993. Chiroptera. *In* Wilson, D. E., and D. M. Reeder, eds., Mammalian species of the world, 2d ed., Smiths. Inst. Press, Washington, D.C., pp. 137–241.
Krishna, A., and C. J. Dominic. 1983. Growth of young and sexual maturity in three species of Indian bats. J. Anim. Morph. Physiol. 30:162–68.
Kulzer, E., J. E. Nelson, J. L. McKean, and F. P. Mohres. 1984. Prey-catching behaviour and echolocation in the Australian ghost bat, *Macroderma gigas* (Microchiroptera: Megadermatidae). Austral. Mammal. 7:37–50.
Kunz, T. H. 1973. Population studies of the cave bat *(Myotis velifer)*: reproduction, growth, and development. Occas. Pap. Mus. Nat. Hist. Univ. Kans. 15:1–43.
———. 1974. Feeding ecology of a temperate insectivorous bat, *Myotis velifer*. Ecology 55:693–711.
———. 1982. Roosting ecology of bats. *In* Kunz, ed., Ecology of bats, Plenum Press, New York, pp. 1–55.
———. 1987. Post-natal growth and energetics of suckling bats. *In* Fenton, M. B., P. Racey, and J. M. V. Rayner, eds., Recent advances in the study of bats, Cambridge Univ. Press, Cambridge, pp. 394–420.
Kunz, T. H., and A. A. Stern. 1995. Maternal investment and post-natal growth in the Chiroptera. *In* Racey, P. A., and S. M. Swift, eds., Ecology, behaviour, and evolution of bats, Symp. Zool. Soc. Lond., Oxford Univ. Press, Oxford, in press.
Kunz, T. H., P. V. August, and C. D. Burnett. 1983. Harem social organization in cave roosting *Artibeus jamaicensis* (Chiroptera: Phyllostomidae). Biotropica 15:133–38.
Kunz, T. H., and C. A. Diaz. 1995. Folivory in fruit-eating bats, with new evidence from *Artibeus jamaicensis* (Chiroptera: Phyllostomidae). Biotropica 27:106–20.
Kunz, T. H., M. S. Fujita, A. P. Brooke, and G. F. McCracken. 1994. Convergence in tent architecture and tent-making behavior among neotropical and paleotropical bats. J. Mamm. Evol. 2:57–78.
Kunz, T. H., and K. A. Ingalls. 1994. Folivory in bats: an adaptation derived from frugivory. Funct. Ecol. 8:665–68.
Kunz, T. H., and G. F. McCracken. 1995. Tents and harems: apparent defense of foliage roosts by tent-making bats. J. Trop. Ecol., in press.
Kunz, T. H., and R. A. Martin. 1982. Plecotus townsendii. Mammalian Species, no. 175, 6 pp.
Kunz, T. H., O. T. Oftedal, S. R. Robson, M. B. Kretzman, and C. Kirk. 1995. Changes in milk composition during lactation in three species of insectivorous bats. J. Comp. Physiol. B 164:543–51.
Kunz, T. H., M. D. Wadanoli, and J. O. Whitaker, Jr. 1995. Dietary energetics of the Mexican free-tailed bat *(Tadarida brasiliensis)* during pregnancy and lactation. Oecologia 101:407–415.
Kurta, A., G. P. Bell, K. A. Nagy, and T. H. Kunz. 1989. Energetics of pregnancy and lactation in free-ranging little brown bats *(Myotis lucifugus)*. Physiol. Zool. 62: 804–18.
———. 1990. Energetics and water flux of free-ranging big brown bats *(Eptesicus fuscus)* during pregnancy and lactation. J. Mamm. 71:59–65.
Kurta, A., D. King, J. A. Teramino, J. M. Stribley, and K. J. Williams. 1993. Summer

roosts of the endangered Indiana bat *(Myotis sodalis)* in the northern edge of its range. Amer. Midl. Nat. 129:132–38.

Kurta, A., and T. H. Kunz. 1987. Size of bats at birth and maternal investment during pregnancy. Symp. Zool. Soc. Lond. 57:79–106.

LaVal, R. K., and M. L. LaVal. 1977. Reproduction and behavior of the African banana bat, *Pipistrellus nanus*. J. Mamm. 48:403–10.

———. 1980. Ecological studies and management of Missouri bats, with emphasis on cave-dwelling species. Terrestrial ser. 8, Missouri Dept. Conserv., Jefferson City, 52 pp.

Lawrence, B. D., and J. A. Simmons. 1982. Echolocation in bats: the external ear and perception of the vertical positions of targets. Science 218:481–83.

Lekagul, B., and J. A. McNeely. 1977. Mammals of Thailand. Assoc. Conserv. Wildl., Bangkok, 758 pp.

Lewis, S. E. 1992. Behavior of Peter's tent making bat, *Uroderma bilobatum*, at maternity roosts in Costa Rica. J. Mamm. 73:541–46.

Licht, P., and P. Leitner. 1967. Behavioral responses to high temperature in three species of California bats. J. Mamm. 48:52–61.

Linares, O. J. 1986. Murcielagos de Venezuela. Cuadernos Lagoven, Filial de Petroleos de Venezuela, 119 pp.

Loebel, M. R., and G. Sanewski. 1987. Flying-foxes (Chiroptera: Pteropodidae) as orchard pests. Austral. Mammal. 10:147–50.

Lowry, F. B. 1989. Green-leaf fractionation by fruit bats: is this feeding behaviour a unique nutritional strategy for herbivores? Austral. Wildl. Res. 16:203–6.

McCracken, G. F. 1988. Who's endangered and what can we do? Bats 6:5–9.

McCracken, G. F., and J. W. Bradbury. 1977. Paternity and genetic heterogeneity in the polygynous bat *Phyllostomus hastatus*. Science 198:303–6.

———. 1981. Social organization and kinship in the polygynous bat *Phyllostomus hastatus*. Behav. Ecol. Sociobiol. 8:11–34.

McCracken, G. F., and M. K. Gustin. 1991. Nursing behavior in Mexican free-tailed bat maternity colonies. Ethology 89:305–21.

McGuckin, M. A., and A. W. Blackshaw. 1992. Effects of photoperiod on the reproductive physiology of male flying foxes. Reprod. Fert. Dev. 92:339–46.

McKean, J. L. 1975. The bats of Lord Howe Island with the description of a new nyctophiline bat. Austral. Mammal. 1:329–32.

McKenzie, N. L., and J. K. Rolfe. 1986. Structure of bat guilds in the Kimberley mangroves, Australia. J. Anim. Ecol. 55:401–20.

McLean, J. L. 1975. The bats of Lord Howe Island with the description of a new nyctophiline bat. Austral. Mammal. 1:329–32.

McWilliam, A. N. 1985–86. The feeding ecology of *Pteropus* in northeastern New South Wales, Australia. Myotis 23–24:201–8.

———. 1987. The reproductive and social biology of *Coleura afra* in a seasonal environment. *In* Fenton, M. B., P. Racey, and J. M. V. Rayner, eds., Recent advances in the study of bats, Cambridge Univ. Press, Cambridge, pp. 324–50.

———. 1990. Mating system of the bat *Miniopterus minor* (Chiroptera: Vespertilionidae) in Kenya, East Africa: A lek? Ethology 85:302–12.

Makin, D., and H. Mendelssohn. 1985. Insectivorous bats victims of Israeli campaign. Bats 2:1–2.

Marimuthu, G. 1984. Seasonal changes in the precision of the circadian clock of a tropical bat under natural photoperiod. Oecologia 61:352–58.

Marshall, A. G. 1983. Bats, flowers, and fruit: evolutionary relationships in the Old World. Biol. J. Linnean Soc. 20:115–35.

———. 1985. Old World phytophagous bats (Megachiroptera) and their food plants: a survey. Zool. J. Linnean Soc. 83:351–69.

Marshall, A. G., and A. N. McWilliam. 1982. Ecological observations on epomophorine fruit-bats (Megachiroptera) in West Africa savanna woodland. J. Zool., Lond. 198: 53–67.

Massing, M. 1988. Long-distance flights of *Pipistrellus nathusii* banded or recaptured in Estonia. Myotis 26:159–64.

May, M. 1991. Aerial defense tactics of flying insects. Amer. Sci. 79:316–28.

Medway, Lord. 1972. Reproductive cycles of the flat-headed bats *Tylonycteris pachypus* and *T. robustula* (Chiroptera: Vespertilionidae) in a humid equatorial environment. Zool. J. Linnean Soc. 51:33–62.

Mickleburgh, S. P., A. M. Hutson, and P. A. Racey. 1992. Old world fruit bats: an action plan for their conservation. IUCN, Gland, Switzerland, 252 pp.

Miller, L. A., and J. Olesen. 1979. Avoidance behavior in green lacewings. I. Behavior of free flying lacewings to hunting bats and ultrasound. J. Comp. Physiol. 131:113–20.

Mindell, D. P., C. W. Dick, and R. J. Baker. 1991. Phylogenetic relationships among megabats, microbats, and primates. Proc. Natl. Acad. Sci. 88:10322–26.

Morrison, D. W. 1979. Apparent male defense of tree hollows in the fruit bat, *Artibeus jamaicensis*. J. Mamm. 60:11–15.

Morrison, D. W., and S. H. Morrison. 1981. Economics of harem maintenance by a neotropical bat. Ecology 62:864–66.

Moutou, F. 1982. Note sur les chiroptères de l'île de la Réunion (Ocean Indian). Mammalia 46:35–51.

Mutere, F. A. 1980. *Eidolon helvum* revisited. *In* Wilson, D. E., and A. L. Gardner, eds., Proceedings of the Fifth International Bat Research Conference, Texas Tech Univ. Press, Lubbock, pp. 145–50.

Myers, P. 1977. Patterns of reproduction of four species of vespertilionid bats in Paraguay. Univ. Calif. Publ. Zool. 107:701–11.

Nelson, J. E. 1964*a*. Notes on *Syconycteris australis*, Peters, 1867 (Megachiroptera). Mammalia 28:429–32.

———. 1964*b*. Vocal communication in Australian flying foxes (Pteropidae: Megachiroptera). Z. Tierpsychol. 21:857–70.

———. 1965. Movements of Australian Pteropidae (Megachiroptera). Austral. J. Zool. 13:53–73.

Neuweiler, G. 1984. Foraging, echolocation, and audition in bats. Naturwissen. 71: 446–55.

———. 1989. Foraging ecology and audition in echolocating bats. Trends Ecol. Evol. 4:160–66.

———. 1990. Auditory adaptations for prey capture in echolocating bats. Physiol. Rev. 70:615–41.

Norberg, U. M. 1990. Vertebrate flight. Springer-Verlag, Berlin, 291 pp.

Norberg, U. M., T. H. Kunz, J. F. Steffensen, Y. Winter, and O. von Helversen. 1993. The cost of hovering and forward flight in a nectar-feeding bat, *Glossophaga soricina*, estimated from aerodynamic theory. J. Exp. Biol. 182:207–27.

Norberg, U. M., and J. M. V. Rayner. 1987. Ecological morphology and flight in bats (Mammalia; Chiroptera): wing adaptations, flight performance, foraging strategy and echolocation. Phil. Trans. R. Soc. Lond. B, Biol. Sci. 316:335–427.

Novacek, M. J. 1985. Evidence for echolocation in the oldest known bat. Nature 315:140–41.

Novick, A. 1977. Acoustic orientation. *In* Wimsatt, W. A., ed., Biology of bats, vol. 3, Academic Press, New York, pp. 74–289.

Novick, A., and D. R. Griffin. 1961. Laryngeal mechanisms in bats for the production of sounds. J. Exp. Zool. 148:125–45.

Nyholm, E. S. 1965. Zur Ökologie von *Myotis mystacinus* (Leisl.) und *M. daubentoni* (Leisl.) (Chiroptera). Ann. Zool. Fenn. 2:77–123.

O'Brien, G. M. 1993. Seasonal reproduction in flying foxes, review in the context of other tropical mammals. Reprod. Fertil. Develop. 5:499–521.

O'Brien, G. M., J. D. Curlewis, and L. Martin. 1993. Effect of photoperiod on the annual cycle of testis growth in a tropical mammal, the little red flying fox, *Pteropus scapulatus*. J. Reprod. Fert. 98:121–27.

Orbst, M. K., M. B. Fenton, J. L. Eger, and P. A. Schlegel. 1993. What ears do for bats: a comparative study of pinna sound pressure transformation in Chiroptera. J. Exp. Biol. 180:119–52.

Orr, R. T. 1970. Development: prenatal and postnatal. *In* Wimsatt, W. A., ed., Biology of bats, vol. 1, Academic Press, New York, pp. 217–31.

O'Shea, J. T. 1980. Roosting, social organization, and the annual cycle in a Kenya population of the bat *Pipistrellus nanus*. Z. Tierpsychol. 53:171–95.

Parry-Jones, K., and M. Augee. 1992. Insects in flying fox diets. Bat Res. News 33:9–11.

Parry-Jones, K., and L. Martin. 1987. Open forum on movements and feeding patterns of flying foxes (Chiroptera: Pteropodidae). Austral. Mammal. 10:129–32.

Parsons, H. J., D. A. Smith, and R. F. Whittam. 1986. Maternity colonies of silver-haired bats, *Lasionycteris noctivagans*, in Ontario and Saskatchewan. J. Mamm. 67:578–600.

Pettigrew, J. D. 1986. Flying primates? Megabats have the advanced pathway from eye to midbrain. Science 231:1304–6.

———. 1995. Flying primates: crashed or crashed through? The case for diphyly. *In* Racey, P. A., and S. M. Swift, eds., Ecology, behaviour, and evolution of bats, Symp. Zool. Soc. Lond., Oxford Univ. Press, Oxford, in press.

Pettigrew, J. D., B. G. M. Jamieson, S. K. Robson, L. S. Hall, I. I. McAnally, and J. M. Cooper. 1989. Phylogenetic relations between microbats (Mammalia: Chiroptera and Primates). Phil. Trans. R. Soc. Lond. B, Biol. Sci. 325:489–559.

Phillips, C. J. 1968. Systematics of the megachiropteran bats in the Solomon Islands. Spec. Publ., Univ. Kans. Mus. Nat. Hist. 16:777–837.

Pierson, E. D., and P. E. Brown. 1992. Saving old mines for bats. Bats 10:11–13.

Pierson, E. D., T. Elmqvist, W. E. Rainey, and P. A. Cox. 1995. Effects of tropical cyclonic storms on flying fox populations on the South Pacific islands of Samoa. Conserv. Biol., in press.

Pierson, E. D., and W. E. Rainey. 1992. The biology of flying foxes of the genus *Pteropus:* a review. *In* Wilson, D. E., and G. Graham, eds., Proceedings of the Pacific Island Flying Fox Conservation Conference, USFWS, Biol. Rept. No. 90, Washington, D.C., pp. 1–17.

Pierson, E. D., W. E. Rainey, and D. M. Koontz. 1991. Bats and mines: experimental mitigation for Townsend's big-eared bat at the McLaughlin Mine in California. *In* Corner, R. D., P. R. Davis, S. Q. Foster, C. V. Grat, S. Rush, O. Thorne III, and J. Todd, eds., Issues and technology in management of impacted wildlife, Proc. 5th Thorne Ecol. Inst. Symp., April 8–10, 1991, Snowmass, Colo., pp. 31–42.

Pierson, E. D., V. M. Sarich, J. M. Lowenstein, M. J. Daniel, and W. E. Rainey. 1986. A molecular link between the bats of New Zealand and South America. Nature 323: 60–63.

Pollak, G. D., and J. H. Casseday. 1989. Neural basis of echolocation in bats. Springer-Verlag, Berlin, 143 pp.

Poulton, E. B. 1929. British insectivorous bats and their prey. Proc. Zool. Soc. Lond. 129:277–302.

Quay, W. B. 1970. Integument and derivatives. *In* Wimsatt, W. A., ed., Biology of bats, vol. 2, Academic Press, New York, pp. 1–56.

Quinn, T. H., and J. J. Baumel. 1993. Chiropteran tendon locking mechanism. J. Morph. 216:197–208.

Racey, P. A. 1973. Environmental factors affecting the length of gestation in heterothermic bats. J. Reprod. Fertil., Suppl., 19:175–89.

———. 1981. Environmental factors affecting the length of gestation in mammals. *In* Gilmore, D. P., and B. Cook, eds., Environmental factors in Mammalian Reproduction, Macmillan, London, pp. 199–213.

———. 1982. Ecology of bat reproduction. *In* Kunz, T. H., ed., Ecology of bats, Plenum Press, New York, pp. 57–104.

Racey, P. A., J. R. Speakman, and S. M. Swift. 1987. Reproductive adaptations of heterothermic bats on the northern borders of their distribution. Suid-Afrik. Tydsk. Wetensk. 83:635–38.

Racey, P. A., and S. M. Swift. 1981. Variation in gestation length in a colony of pipistrelle bats *(Pipistrellus pipistrellus)* from year to year. J. Reprod. Fert. 61: 123–29.

———. 1985. Feeding ecology of *Pipistrellus pipistrellus* (Chiroptera: Vespertilionidae) during pregnancy and lactation. I. Foraging behaviour. J. Anim. Ecol. 54:205–15.

Rainey, W. E. 1992. *Notopteris macdonaldii* Species Account. *In* Mickleburgh, Hutson, and Racey (1992), 81–83.

Rainey, W. E., and E. D. Pierson. 1992. Distribution of Pacific island flying foxes. *In* Wilson, D. E., and G. Graham, eds., Proceedings of the Pacific Island Flying Fox Conservation Conference, USFWS, Biol. Rept. no. 90, Washington, D.C., pp. 111–21.

Rainey, W. E., E. D. Pierson, N. Colbene, and J. H. Barclay. 1992. Bats in hollow redwoods: seasonal uses and role in nutrient transfer into old growth communities. Bat Res. News 33:71.

Rainey, W. E., E. D. Pierson, T. Elmqvist, and P. A. Cox. 1995. The role of pteropodids in oceanic island ecosystems of the Pacific. *In* Racey, P. A., and S. M. Swift, eds., Ecology, behaviour, and evolution of bats, Symp. Zool. Soc. Lond., Oxford Univ. Press, Oxford, in press.

Ransome, R. 1990. The natural history of hibernating bats. Christopher Helm, London, 235 pp.

Richards, G. C. 1986. Notes on the natural history of the Queensland tube-nosed bat, *Nyctimene robinsoni*. Macroderma 2:64–67.

———. 1987. Aspects of the ecology of spectacled flying foxes, *Pteropus conspicillatus* (Chiroptera: Pteropodidae) in tropical Queensland. Austral. Mammal. 10:87–88.

———. 1989. Nocturnal activity pattern of insectivorous bats relative to temperature and prey availability in tropical Queensland. Austral. Wildl. Res. 16:151–58.

———. 1990*a*. Rainforest bat conservation: unique problems in a unique environment. Austral. Zool. 26:44–46.

———. 1990*b*. The spectacled flying-fox, *Pteropus conspicillatus* (Chiroptera: Pteropodidae) in north Queensland. 1. Roost sites and distribution patterns. Austral. Mammal. 13:17–24.

———. 1990*c*. The spectacled flying-fox, *Pteropus conspicillatus* (Chiroptera: Pteropodidae) in north Queensland. 2. Diet, seed dispersal, and feeding ecology. Austral. Mammal. 13:25–31.

———. 1991. The conservation of forest bats in Australia: do we really know the problems and solutions? *In* Lunney, D. L., ed., Conservation of forest fauna in Australia, R. Zool. Soc., New South Wales, Mosman, Australia, pp. 81–90.

———. 1995. A review of ecological interactions of fruit bats in Australian ecosystems. *In* Racey, P. A., and S. M. Swift, eds., Ecology, behaviour, and evolution of bats, Symp. Zool. Soc. Lond., Oxford Univ. Press, Oxford, in press.

Richardson, E. G. 1977. The biology and evolution of the reproductive cycle of *Miniopterus schreibersii* and *M. australis* (Chiroptera: Vespertilionidae). J. Zool., Lond. 183:353–75.

Robinson, M. 1990. Prey selectivity by the brown big-eared bat *(Plecotus auritis)*. Myotis 28:5–18.

Robson, S. K. 1984. *Myotis adversus* (Chiroptera: Vespertilionidae): Australia's fish-eating bat. Austral. Mammal. 7:51–52.

Roeder, K. D., and A. E. Treat. 1961. The detection and evasion of bats by moths. Amer. Sci. 49:135–48.

Roer, H. 1980. Zur Bestandsentwicklung einiger Fledermäuse in Mittleeuropa. Myotis 18:60–68.

Rosevear, D. R. 1965. The bats of West Africa. Brit. Mus. Nat. Hist., London, 418 pp.

Ross, A. 1967. Ecological aspects of food habits of insectivorous bats. Proc. West. Found. Vert. Zool. 1:205–63.

Rydell, J. 1986. Foraging and diet of the northern bat *Eptesicus nilssoni* in Sweden. Holarc. Ecol. 9:272–76.

———. 1989. Feeding activity of the northern bat *Eptesicus nilssoni* during pregnancy and lactation. Oecologia 80:562–65.

Sazima, I., and M. Sazima. 1978. Bat pollination in the passion flower *Passiflora mucronata*, in southern Brazil. Biotropica 10:100–109.

Schmidly, D. J. 1991. The bats of Texas. Texas A&M Univ. Press, College Station, 185 pp.

Schmidt, S. 1988. Evidence for a spectral basis of texture perception in bat sonar. Nature 331:617–19.

Schnitzler, H.-U. 1987. Echoes of fluttering insects: information for echolocating bats. *In* Fenton, M. B., P. Racey, and J. M. V. Rayner, eds., Recent advances in the study of bats, Cambridge Univ. Press, Cambridge, pp. 226–43.

Schober, W., and E. Grimmberger. 1989. A guide to the bats of Britain and Europe. Hamlyn Publ. Group, London, 224 pp.

Schuller, G. 1985. Natural ultrasonic echoes from wing beating insect are encoded by collicular neurones in the CF-FM bat, *Rhinolophus ferrumequinum*. J. Comp. Physiol. 155:121–28.

Schutt, W. A., Jr. 1993. Digital morphology in the Chiroptera: the passive digital lock. Acta Anat. 148:219–27.

Silva Taboada, G. 1979. Los murciélagos de Cuba. Editorial Academia, Havana, 423 pp.

Simmons, J. A., and S. P. Dear. 1992. Through a bat's ear. IEEE Spectrum, March, 46–48.

Simmons, N. B. 1993. Morphology, function, and phylogenetic significance of pubic nipples in bats (Mammalia: Chiroptera). Amer. Mus. Novit., no. 3077, 37 pp.

Simmons, N. B. 1995. Bat relationships and the origin of flight. *In* Racey, P. A., and

S. M. Swift, eds., Ecology, behaviour, and evolution of bats, Symp. Zool. Soc. Lond., Oxford Univ. Press, Oxford, in press.

Speakman, J. R. 1990. The function of daylight flying in British bats. J. Zool., Lond. 220:101–13.

———. 1991. Why do insectivorous bats in Britain not fly in daylight more frequently? Funct. Ecol. 5:518–24.

———. 1992. The impact of predation by birds on bat populations in the British Isles. Mammal. Rev. 21:123–42.

Speakman, J. R., M. E. Anderson, and P. A. Racey. 1989. The energy cost of echolocation in pipistrelle bats *(Pipistrellus pipistrellus)*. J. Comp. Physiol. A 165:679–85.

Speakman, J. R., and P. A. Racey. 1989. Hibernal ecology of the pipistrelle bat: energy expenditure, water requirements and mass loss, implications for survival and the function of winter emergence fights. J. Anim. Ecol. 58:797–813.

Spencer, H. J., and T. H. Fleming. 1989. Roosting and foraging behaviour of the Queensland tube-nosed bat, *Nyctimene robinsoni* (Pteropodidae): preliminary radiotracking observations. Wildl. Res. 16:413–20.

Spencer, H. J., C. Palmer, and K. Parry-Jones. 1991. Movements of fruit bats in eastern Australia, determined by using radiotracking. Wildl. Res. 18:463–68.

Start, A. N. 1972. Pollination of the baobab (*Adansonia digitata* L.) by the fruit bat *Rousettus aegyptiacus* Geoffroy. E. Afr. Wildl. J. 10:71–72.

Stebbings, R. E., and S. T. Walsh. 1988*a*. Bat boxes; a guide to their history, function, construction, and use in the conservation of bats. 2d ed. Flora and Fauna Preserv. Soc., London, 24 pp.

———. 1988*b*. Conservation of European bats. Christopher Helm, London, 246 pp.

Stihler, C. W., and J. S. Hall. 1994. Endangered bat populations in West Virginia cave: gated or fenced to reduce human disturbance. Bat Res. News 24, in press. 34:130.

Stoddart, D. M. 1975. The diurnal flight of the Azorean bat *(Nyctalus azoreum)* and the avifauna of the Azores. J. Zool., Lond. 177:483–506.

Strahan, R. 1983. The Australian Museum complete book of Australian mammals. Angus and Robertson, Sydney, 529 pp.

Strelkov, P. P. 1969. Migratory and stationary bats (Chiroptera) of the eastern part of the Soviet Union. Acta Zool. 16:393–435.

Studier, E. H., and T. H. Kunz. 1995. Mineral budgets in suckling insectivorous bats *(Myotis velifer* and *Tadarida brasiliensis)*. J. Mamm. 76:32–42.

Suga, N., and P. Schlegel. 1972. Neural attenuation of responses to emitted sounds in echolocating bats. Science 177:82–84.

Suthers, R. A. 1970. Vision, olfacation, and taste. *In* Wimsatt, W. A., ed., Biology of bats, vol. 2, Academic Press, New York, pp. 265–309.

Suthers, R. A., and N. E. Wallis. 1970. Optics of the eyes of echolocating bats. Vision Res. 10:1165–73.

Swift, S. M., P. A. Racey, and M. I. Avery. 1985. Feeding ecology of *Pipistrellus pipistrellus* (Chiroptera: Vespertilionidae) during pregnancy and lactation. II. Diet. J. Anim. Ecol. 54:217–25.

Thomas, D. W. 1983. The annual migrations of three species of West African fruit bats (Chiroptera: Pteropodidae). Can. J. Zool. 61:2266–72.

Thomas, D. W., M. B. Fenton, and R. M. R. Barclay. 1979. Social behaviour of the little brown bat, *Myotis lucifugus*. I. Mating behaviour. Behav. Ecol. Sociobiol. 6:129–36.

Tidemann, C. R. 1985. A study of the status, habitat requirements, and management of the two species of bats on Christmas Island (Indian Ocean). Austral. Natl. Parks Wildl. Ser., 78 pp.

———. 1987. Notes on the flying-fox, *Pteropus melanotus* (Chiroptera: Pteropodidae), on Christmas Island, Indian Ocean. Austral. Mammal. 10:89–91.

Tidemann, C. R., D. J. Kitchener, R. A. Zann, and I. W. B. Thornton. 1990. Recolonization of the Krakatau Islands and adjacent areas of West Java, Indonesia, by bats (Chiroptera) 1883–1986. Phil. Trans. R. Soc. Lond. B, Biol. Sci. 38:123–30.

Tidemann, C. R., D. M. Priddel, J. E. Nelson, and J. D. Pettigrew. 1985. Foraging behavior of the Australian ghost bat, *Macroderma gigas* (Microchiroptera: Megadermatidae). Austral. J. Zool. 33:705–13.

Timm, R. M. 1987. Tent construction by bats of the genera *Artibeus* and *Uroderma*. Fieldiana Zool. 39:187–212.

Turner, D. C. 1975. The vampire bat. Johns Hopkins Univ. Press, Baltimore, 145 pp.

Tuttle, M. D. 1975. Population ecology of the gray bat *(Myotis grisescens):* factors influencing early growth and development. Occas. Pap. Mus. Nat. Hist. Univ. Kans. 36:1–24.

———. 1976*a*. Population ecology of the gray bat *(Myotis grisescens):* factors influencing growth and survival of newly volant young. Ecology 57:587–95.

———. 1976*b*. Population ecology of the gray bat *(Myotis grisescens):* philopatry, timing, and patterns of movement, weight loss during migration, and seasonal adaptive strategies. Occas. Pap. Mus. Nat. Hist. Univ. Kans. 54:1–38.

———. 1979. Status, causes of decline, and management of endangered gray bats. J. Wildl. Mgmt. 43:1–17.

Tuttle, M. D., and D. Hensley. 1993. Bat houses: the secrets of success. Bats 11:3–14.

Tuttle, M. D., and M. J. Ryan. 1981. Bat predation and the evolution of frog vocalizations in the neotropics. Science 214:677–78.

Tuttle, M. D., and D. Stevenson. 1982. Growth and survival of bats. *In* Kunz, T. H., ed., Ecology of bats, Plenum Press, New York, pp. 105–50.

U.S. Fish and Wildlife Service. 1982. Gray bat recovery plan. Washington, D.C., 61 pp.

———. 1983. Recovery plan for the Indiana bat. Washington, D.C., 94 pp.

Utzurrum, R. C. B. 1992. Conservation status of Philippine fruit bats (Pteropodidae). Silliman J. 36:27–45.

Utzurrum, R. C. B., and P. D. Heideman. 1991. Differential ingestion of viable vs. nonviable *Ficus* seeds by fruit bats. Biotropica 23:311–12.

Van Aarde, R. J., M. Van der Merwe, and D. C. Skinner. 1994. Progesterone concentrations and contents in the plasma, ovary, adrenal gland, and placenta of the pregnant Natal clinging bat *Miniopterus schreibersii natalensis*. J. Zool., Lond. 232:457–64.

Van der Merwe, M., and R. J. van Aarde. 1989. Plasma progesterone concentrations in the female Natal clinging bat *(Miniopterus schreibersii natalensis)*. J. Reprod. Fertil. 87:665–69.

Van Peenen, P. F. D. 1968. A guide to the fruit bats of South Vietnam. Formosan Sci. 22:95–107.

Van Zyll de Jong, C. G. 1985. Handbook of Canadian mammals: 2, Bats. Nat. Mus. Nat. Sci., Ottawa, 212 pp.

Vaughan, T. A. 1970. The muscular system. *In* Wimsatt, W. A., ed., Biology of bats, vol. 1, Academic Press, New York, pp. 98–139.

———. 1976. Nocturnal behaviour of the African false vampire bat *(Cardioderma cor)*. J. Mamm. 57:227–48.

———. 1986. Mammalogy. 3d ed. Saunders, New York, 576 pp.

Vaughan, T. A., and R. P Vaughan. 1986. Seasonality and behavior of the African yellow-winged bat. J. Mamm. 67:91–102.

Vehrencamp, S. L., F. G. Stiles, and J. W. Bradbury. 1977. Observations on the foraging

behavior and avian prey of the neotropical carnivorous bat, *Vampyrum spectrum*. J. Mamm. 58:469–78.
Verschuren, J. 1957. Ecologie, biologie, et systématique des chiroptères. Exploration du Parc National de la Garamba No. 7 (Mission H de Saeger). Inst. Parcs nat. Congo Belge, Brussels, 473 pp.
Villa-R., B. 1966. Los murcielagos de Mexico. Univ. Auton. Mexico, Mexico City, 491 pp.
Warner, R. M. 1984. Interspecific and temporal dietary variation in an Arizona bat community. J. Mamm. 66:45–51.
Whitaker, J. O., Jr. 1993. Bats, beetles, and bugs. Bats 11:23.
Whitaker, J. O., Jr., and H. L. Black. 1976. Food habits of cave bats from Sabia, Africa. J. Mamm. 57:56–65.
Whitaker, J. O., Jr., C. Maser, and L. E. Keller. 1977. Food habits of bats of western Oregon. Northwest. Sci. 51:46–55.
Wickler, W., and U. Seibt. 1976. Field studies of the African fruit bat, *Epomophorus wahlbergi*, with special reference to male calling. Z. Tierpsychol. 40:345–76.
Wickler, W., and D. Uhrig. 1969. Verhalten und ökologische Nische der Gelbflugelfledermus, *Lavia frons* (Geoffroy) (Chiroptera, Megadermatidae). Z. Tierpsychol. 26:726–36.
Wiles, G. S. 1987*a*. Current research and future management of Mariannas fruit bats (Chiroptera: Pteropodidae) on Guam. Austral. Mammal. 10:93–95.
———. 1987*b*. The status of fruit bats on Guam. Pacific Sci. 41:148–57.
———. 1992. *Pteropus pilosus* species account. *In* Mickleburgh, Hutson, and Racey, pp. 121–22.
Wiles, G. J., and M. S. Fujita. 1992. The food plants and economic importance of flying foxes on Pacific Islands. *In* Wilson, D. E., and G. Graham, eds., Proceedings of the Pacific Island Flying Fox Conservation Conference, USFWS, Biol. Rept. No. 90, Washington, D.C., pp. 24–35.
Wiles, G. J., and N. H. Payne. 1986. The trade in fruit bats, *Pteropus* spp. on Guam and other Pacific islands. Biol. Conserv. 38:143–61.
Wilkinson, G. S. 1984. Reciprocal food sharing in the vampire bat. Nature 308:181–84.
———. 1985*a*. The social organization of the common vampire bat. I. Pattern and cause association. Behav. Ecol. Sociobiol. 17:111–21.
———. 1985*b*. The social organization of the common vampire bat. II. Mating system, genetic structure, and relatedness. Behav. Ecol. Sociobiol. 17:123–34.
———. 1987. Altruism and co-operation in bats. *In* Fenton, M. B., P. A. Racey, and J. M. V. Rayner, eds., Recent advances in the study of bats, Cambridge Univ. Press, Cambridge, pp. 299–323.
Williams, T. C., L. C. Ireland, and J. M. Williams. 1973. High altitude flights of the free-tailed bat, *Tadarida brasiliensis*, observed with radar. J. Mamm. 14:807–21.
Willig, M. R. 1983. Composition, microgeographic variation, and sexual dimorphism in Caatingas and Cerrado bat communities from Northeast Brazil. Bull. Car. Mus. 23:1–123.
———. 1986. Bat community structure in south America: a tenacious chimera. Rev. Chilena Hist. Nat. 59:151–68.
———. 1989. A comparison of bat assemblages from phytogeographic zones of Venezuela. Spec. Publ., Mus. Texas Tech Univ. 38:59–67.
Willig, M. R., and M. P. Moulton. 1989. The role of stochastic and deterministic processes in structuring neotropical bat communities. J. Mamm. 70:323–29.
Willig, M. R., and K. W. Selcer. 1989. Bat species density gradients in the New World: a statistical assessment. J. Biogeogr. 16:189–95.

Wilson, D. E. 1971. Ecology of *Myotis nigricans* (Mammalia: Chiroptera) on Barro Colorado Island, Panama Canal Zone. J. Zool., Lond. 163:1–13.

———. 1979. Reproductive patterns. *In* Baker, R. J., J. K. Jones, Jr., and D. C. Carter, eds., Biology of bats of the New World family Phyllostomidae, III, Spec. Publ., Mus. Texas Tech Univ., Lubbock, pp. 317–78.

Wilson, D. E., and J. S. Findley. 1970. Reproductive cycle of a neotropical insectivorous bat *Myotis nigricans*. Nature 225:1155.

Wimsatt, W. A., and H. Trapido. 1952. Reproduction and the female reproductive cycle in the tropical American vampire bat, *Desmodus rotundus rotundus*. Amer. J. Anat. 91:415–45.

Winter, Y., O. von Helversen, U. M. Norberg, T. H. Kunz, and J. F. Steffensen. 1993. Flight costs and economy of nectar-feeding in the bat *Glossophaga soricina* (Phyllostomidae: Glossophaginae). *In* Barthlott, W., et al., eds., Plant-animal interactions in tropical environments. Mus. Alexander Koenig, Bonn, pp. 167–74.

Wodzicki, K., and H. Felten. 1975. The peka, or fruit bat *(Pteropus tonganus tonganus)* (Mammalia, Chiroptera) of Niue Island, South Pacific. Pacific Sci. 29:131–38.

Zubaid, A. 1993. A comparison of the bat fauna between a primary and fragmented secondary forest in Peninsular Malaysia. Mammalia 57:201–6.

Bats
Chiroptera

A. Four female Mexican fruit bats *(Artibeus jamaicensis)* with their young, suspended from the underside of a leaf frond, photo by Cory T. de Carvalho. B. Rough-legged bats *(Myotis dasycneme)* clustered in a church attic—the way bats frequently assemble in caves and buildings for daily sleeping or for hibernation, photo by P. F. Van Heerdt.

CHIROPTERA; **Family PTEROPODIDAE**

Old World Fruit Bats

This family of 42 living genera and 173 species is found in the tropical and subtropical regions of the Old World, east to Australia and the Caroline and Cook islands. There are two living subfamilies: Pteropodinae, for most of the genera; and Macroglossinae, for the genera *Eonycteris, Megaloglossus, Macroglossus, Syconycteris, Melonycteris,* and *Notopteris* (Koopman and Jones 1970).

Head and body length varies from 50 to 400 mm according to the species. The tail is short, rudimentary, or absent, except in *Notopteris,* in which it has 10 vertebrae. The tail membrane is only a narrow border. Adults range in weight from approximately 15 grams for the smallest nectar- and pollen-feeding members of the family to over 1,500 grams for the largest fruit eaters. This family contains the largest living bats, some species of *Pteropus* and *Acerodon* having a wingspread of 1.7 meters. The fur is as much as 3 cm in length, dense, and variable in color but with prevailing tinges of dark brown. The genera *Epomophorus, Micropteropus, Epomops, Scotonycteris, Nanonycteris, Hypsignathus, Casinycteris,* and *Plerotes,* among others, exhibit secondary sexual characters. These include much greater size, glandular pouches with tufts of hair on the shoulders, and large pharyngeal sacs in the chest region, all in males. The males of *Hypsignathus* have greatly folded lips, and the males of some genera have erectile hairs on the nape. All the members of the family

A "bare-backed" fruit bat or flying fox (*Pteropus neohibernicus* [?]) from New Guinea walking along a branch, photo by Sten Bergman.

Pteropodidae except the genera *Dobsonia, Eonycteris, Notopteris,* and *Neopteryx* have a claw on the second finger in addition to the claw on the thumb; all the members of the suborder Microchiroptera lack a claw on the second finger. The external ear of Old World fruit bats is elongate, oval, and rather simple; the margin of the ear forms a complete ring, and a tragus is absent. In the Microchiroptera the external ear is often complex; the margin does not form a closed tube, and a tragus is usually present. Two genera of Pteropodidae, *Nyctimene* and *Paranyctimene,* have tubular nostrils that open laterally. The members of this family have large, well-developed eyes. The surface area for the rods, the receptors in the eye for black-and-white vision, is greatly increased by small fingerlike projections, enclosing blood vessels, of the inner coat of the eyeball that penetrate the outer layer of the retina; these projections apparently are lacking in the Microchiroptera. Only the genus *Rousettus* is known to emit ultrasonic sounds during flight, in addition to utilizing its eyes for vision. The remaining genera of Pteropodidae are believed to guide themselves visually, unless perhaps the tube-nosed fruit bats *Nyctimene* and *Paranyctimene* orient by means of echolocation.

The penis resembles that of some Primates. Females have one pair of mammae in the chest region.

The dental formula varies from (i 2/2, c 1/1, pm 3/3, m 2/3) × 2 = 34 in *Pteropus* and *Rousettus* to (i 1/0, c 1/1, pm 3/3, m 1/2) × 2 = 24 in *Nyctimene* and *Paranyctimene.* The incisors are small, and the canines are always present, even in those forms in which the dentition tends to be reduced. The back teeth are low, elongate, and widely separated. The crowns of the molars are smooth and marked with a longitudinal groove except in *Pteralopex* and *Harpyionycteris,* which have cuspidate molars. The dentition is adapted for a soft diet. The palate of the bats of this family generally has eight transverse ridges against which the tongue crushes the food. The bony palate gradually narrows behind the last molars, whereas in the Microchiroptera the bony palate is not continued behind the last molar. The tongue is covered with well-developed papillae in some forms, and it can be protruded far beyond the end of the mouth in the nectar and pollen feeders of the subfamily Macroglossinae.

These bats are active mostly in the evening and at night but have been observed flying in the daytime. All fruit bats are somewhat irregular in their presence in a region, because they often leave areas where fruit is not available. They often make long flights between their roosting and feeding areas. The large forms are slow but powerful fliers; some of the large fruit bats may fly as far as 15 km to their feeding areas. *Eidolon,* which has long narrow wings, is adapted to flying long distances. The bats of this family climb actively in trees for fruit and in their roosts in trees and caves and under the eaves of buildings. When they are at rest, hanging by their feet, their head is at a right angle to the axis of the body. The larger forms are often gregarious, roosting in large groups, whereas the smaller forms are generally solitary.

Most species of this family locate food by smell. Many if not all of these bats crush ripe fruit pulp in the mouth, swallow the juice, and spit out most of the pulp and seeds, often pressed into almost uniformly shaped pieces; some of the softer pulp is swallowed. Material that is swallowed goes through the very simple digestive tract in a very short time, perhaps half an hour on the average. In *Epomophorus* and related genera, the structure of the lips, windpipe, and gullet forms a type of suction apparatus that probably aids in feeding on the softer parts of fruits. Bats can bite into fruit while hovering, or they may hold onto a branch with one foot and press the fruit to the chest with the other foot in order to bite into it, or sometimes they carry small fruit to a branch and hang head downward while eating. The members of this family also chew flowers to obtain nectar and juices. The members of the subfamily Macroglossinae seem to feed mainly on pollen and nectar. *Nyctimene* sometimes includes insects in its diet.

Some forms appear to breed throughout the year, but most members of this family probably have well-defined breeding seasons. Pregnant females of several species shelter apart from the males. There is usually only one young in a birth. Individuals of this family have lived in captivity for over 20 years.

The geological range of this family is early Oligocene in Europe, early Miocene to Recent in Africa, Pleistocene to Recent in Madagascar and the East Indies, and Recent in other parts of its range (Koopman 1984*c*).

CHIROPTERA; PTEROPODIDAE; **Genus EIDOLON**
Rafinesque, 1815

Straw-colored Fruit Bat

The single species, *E. helvum,* occurs in the southwestern Arabian Peninsula, most of the forest and savannah zones of Africa south of the Sahara, and Madagascar (Corbet 1978; Hayman and Hill, *in* Meester and Setzer 1977). It is the most widely distributed of the African fruit bats.

Head and body length is about 143–215 mm, tail length is 4–20 mm, forearm length is 109–32 mm, and wingspan can reach 762 mm. Fayenuwo and Halstead (1974) reported that during the breeding season in Nigeria, adult weight increased from about 230 to 330 grams in males and from 240 to 350 grams in females. The hairs on the neck are longer and more woolly than those on the body, and the interfemoral membrane is hairy in the middle of its upper surface. Coloration is yellowish brown or brownish above and tawny olive or brownish below. The cinnamon hairs on the glandular skin on the foreneck and sides of the neck are most conspicuous in adult males; with this exception, the sexes are much the same in color and size.

A. Fruit bats *(Pteropus tonganus)* roosting near the village of Tonga, Fiji Islands. This picture illustrates the characteristic manner in which the fruit bats sleep during the day. Photo from Public Relations Office, Fiji, through R. A. Derrick. B. A group of African fruit bats *(Rousettus aegyptiacus)* hanging in a cave in the same manner as many of the insectivorous bats, which sometimes congregate by the thousands. Photo by Don Davis.

Straw-colored fruit bat *(Eidolon helvum)*, photo by John Visser.

In contrast to *Epomophorus* and similar genera, *Eidolon* has a more pointed head and lacks the white patches at the base of the ear. The wings are long and narrow, adapted to flying long distances. The wings are also used in climbing about branches in the roosts. When hanging by the hind feet, this genus turns the second phalanges of the third and fourth digits inward, so that they fold against the lower surface of the wing, as in the vespertilionid genus *Miniopterus*.

This bat inhabits forest and savannah country and is found at elevations of up to 2,000 meters in the Ruwenzori Mountains (Kingdon 1974*a*). It is gregarious and prefers to roost in tall trees by day but has also been found in lofts and in caves in rocks. During the daytime it is often noisy and restless and even flies about from place to place. At night groups fly out of the roosts in search of ripe fruit. The large roosts are seldom closer than 60 km to one another, suggesting a foraging range of at least 30 km for a colony. Thomas (1983) determined that *Eidolon* also makes extensive seasonal migrations: a colony of about 500,000 individuals left its roost in the southern forest of the Ivory Coast following the birth of young in February, moved northward into the savannah zone, and migrated at least as far as the Niger River Basin by the middle of the wet season in July. It is probable that even greater distances (a round trip of over 2,500 km) are covered by some colonies in East Africa.

The roosts of *Eidolon* are within easy reach of forests or fruit plantations. Juices of various fruits are the preferred food, though this bat also feeds on the blossoms and perhaps young shoots of the silk-cotton tree *(Ceiba)*. *Eidolon* will eat directly into the fruit of the palm *Borassus* and has the unusual habit of chewing into soft wood, apparently to obtain moisture.

The straw-colored fruit bat sometimes occurs in enormous colonies of 100,000 to 1,000,000 individuals of both sexes. Studies in Nigeria and Uganda (Fayenuwo and Halstead 1974; Kingdon 1974*a;* Mutere 1967) indicate that mating occurs in the colonies from April to June and that there is delayed implantation. Most of the major roosts are abandoned from June to September, apparently because social bonds then break down. There is a reconvergence of the colonies in September–October, which is followed in the females by synchronous implantation of the ova in the uteri. Births occur in February and March, in the midst of clusters formed by hundreds of females and some males. There is a single young per female, and one newborn weighed 50 grams. It appears that the reproductive cycle is geared to the rainfall pattern so as to ensure that the young are weaned when conditions are optimal. Implantation coincides exactly with the start of the dry season, and births with the beginning of the wet season. According to DeFrees and Wilson (1988), the total minimum gestation period is 9 months, though the actual period of embryonic development is 4 months. The known longevity record for *Eidolon* is 21 years and 10 months (Jones 1982).

In some areas *Eidolon* is hunted and eaten by humans, but in others it is traditionally protected (Happold and Happold 1978*b*). In the Ivory Coast *Eidolon* is thought of as a threat to introduced pine plantations, because the bats gnaw the bark, wood, and leaves, causing death to the trees (Malagnoux and Gautun 1976). *Eidolon* may also eat and destroy dates to such a degree that protective measures are required.

CHIROPTERA; PTEROPODIDAE; **Genus ROUSETTUS**
Gray, 1821

Rousette Fruit Bats

There are 10 species (Aggundey and Schlitter 1984; Baeten, Van Cakenberghe, and De Vree 1984; Bergmans 1977; Bergmans and Hill 1980; Chasen 1940; Corbet 1978; Ellerman and Morrison-Scott 1966; Ghose and Ghosal 1984; Hayman and Hill, *in* Meester and Setzer 1977; Kock 1978*a;* Koopman 1979, 1986; Laurie and Hill 1954; Lekagul and McNeely 1977; McKean 1972; Nader 1975; Rookmaaker and Bergmans 1981; Sung 1976):

R. aegyptiacus, Turkey and Cyprus to Pakistan, Arabian Peninsula, Egypt, most of Africa south of the Sahara;
R. obliviosus, Grand Comoro and Anjouan islands between Africa and Madagascar;
R. madagascariensis, lowland forests throughout Madagascar;
R. amplexicaudatus, southern Burma, Thailand, Cambodia, Viet Nam, Malaysia, Indonesia, Philippines, New Guinea, Bismarck Archipelago, Solomon Islands;
R. spinalatus, northern Sumatra, northern Borneo;
R. leschenaulti, Pakistan, India, Nepal, Sikkim, Bhutan, Burma, Thailand, Indochina, extreme southeastern China, Sumatra, Java, Bali;
R. seminudus, Sri Lanka;
R. lanosus, Sudan, northeastern Zaire, Kenya, Uganda, Rwanda;
R. celebensis, Sulawesi, Sanghir Islands;

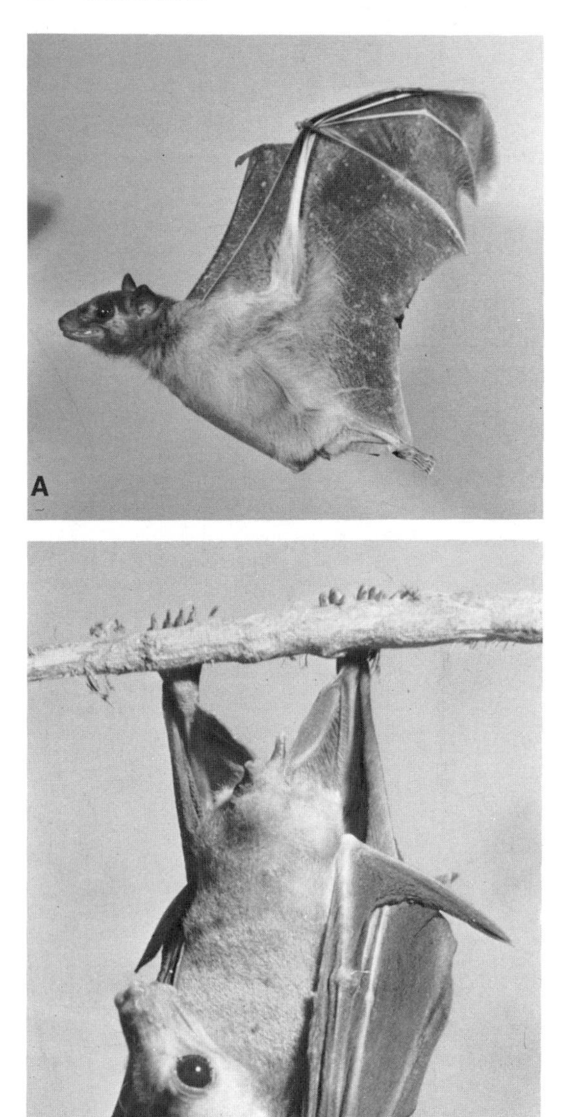

Rousette fruit bat (*Rousettus* sp.): A. Photo from New York Zoological Society; B. Photo by Lim Boo Liat.

R. angolensis, Guinea to Kenya, and south to Angola and Zimbabwe.

The species *R. angolensis* often is placed in a separate genus or subgenus, *Lissonycteris* Andersen, 1912 (see Bergmans 1979*a*). The species *R. lanosus* often is placed in the subgenus *Stenonycteris* Andersen, 1912. Hill (1983*b*) suggested that *R. celebensis* also might belong in the latter subgenus.

Head and body length is about 95–177 mm, tail length is 10–22 mm, and forearm length is 65–103 mm. Reported weights are about 81–171 grams for *R. aegyptiacus* (Jones 1971*a*), 54–75 grams for *R. amplexicaudatus*, and 45–106 grams for *R. leschenaulti* (Lekagul and McNeely 1977). Males are substantially larger than females. Certain members of this genus have glandular hairs on the foreneck and the sides of the neck; these are bright tawny olive in the adult males of some species. The upper parts are usually brownish, the underparts somewhat lighter. At least two species, *R. amplexicaudatus* and *R. leschenaulti*, have a pale collar of short hairs. Otherwise, *Rousettus* tends to have a drab appearance.

Rousette bats occur in a variety of habitats, from lowlands to mountains. They roost in ancient tombs and temples, rock crevices, garden trees, and date plantations but are most common in caves. According to Lekagul and McNeely (1977), colonies of *R. leschenaulti* shift in response to the supply of fruit, and *R. amplexicaudatus* may make a foraging round trip of 40–50 km in a night. While flying in darkness, the latter species utters a high-pitched buzzing call that is the basis of a simple echolocation mechanism (Medway 1978); *R. aegyptiacus* also has been found to use echolocation (Herbert 1985). Some segments of colonies of *R. aegyptiacus* may make limited seasonal migrations (Jacobsen and Du Plessis 1976). The diet consists of fruit juices and flower nectar. In the course of obtaining nectar from different flowers, these bats carry pollen from one place to another. They have been noted at dusk flying around flowering acacia trees.

Lekagul and McNeely (1977) stated that roosting group size in *R. leschenaulti* varies from 2 or 3 to as many as 2,000 and that there is no sexual segregation. Jacobsen and Du Plessis (1976) described two roosting colonies of *R. aegyptiacus* in the eastern Transvaal, each consisting of 7,000–9,000 individuals. The bats crowded closely together in caves, maintaining bodily contact. Fighting was common and was accompanied by loud screams and hacking coughs. The genus in general tends to be noisy and restless in the daytime.

R. aegyptiacus appears to be a seasonal breeder in the Transvaal (Jacobsen and Du Plessis 1976). Mating there occurs primarily from June to September, gestation lasts about 4 months, and births take place from October to December. There is usually a single young, but occasionally twins are born. The females carry the young at first, then leave them at the roosts. By early March the young are weaned and able to fly on their own. According to Bergmans (1979*a*), pregnant females were taken in Congo in November and December; lactating females were collected there in February and March; and females give birth twice a year in Uganda, in March and September. Most births in the London Zoo have been in March, April, and May, with one young per birth the usual number and twins occurring about every fourth birth. In *R. leschenaulti* in India there are also two periods of birth, occurring about March and July–August. In an intensive study of a colony in Maharashtra State, Gopalakrishna and Choudhari (1977) found that pregnancies lasted from November to March and from March to July, with females undergoing a postpartum estrus shortly after the first birth. The gestation period was about 125 days, and there was a single young per female. Sexual maturity was attained at 5 months by females but not until 15 months by males. For *R. amplexicaudatus* in the Solomon Islands, Phillips (1968) reported that adult females were lactating in December–January but not in March–June. On Timor, however, Goodwin (1979) found most adult *R. amplexicaudatus* to be in breeding condition in March, April, and May. For the same species on Bougainville Island, McKean (1972) reported pregnant females in July and September and a lactating female in September. *R. angolensis* in Congo may breed either continuously throughout the year or biannually in April–June and

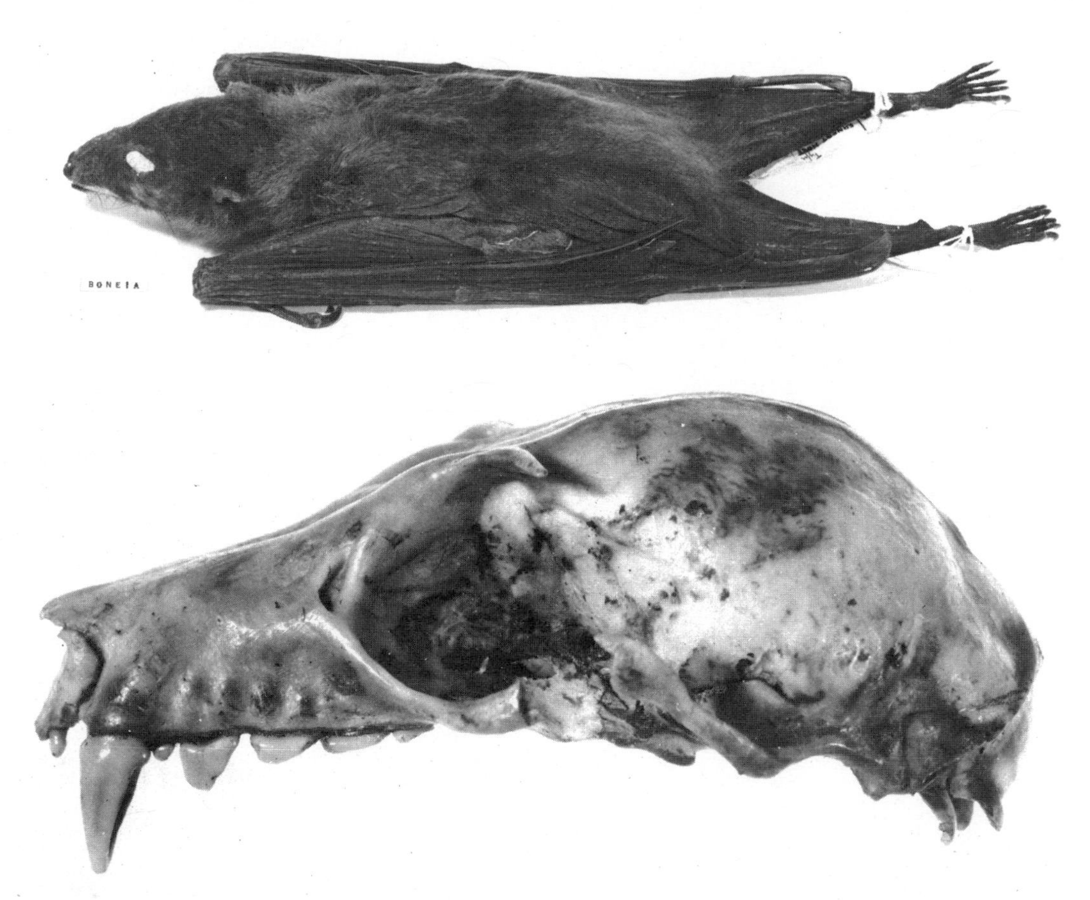

Boneia bidens, photos from British Museum (Natural History).

October–November (Bergmans 1979*a*). The known longevity record for the genus is 22 years and 11 months (Jones 1982).

CHIROPTERA; PTEROPODIDAE; **Genus BONEIA**
Jentink, 1879

The single species, *B. bidens*, is known only from northern Sulawesi between the towns of Gorontalo and Manado. The type locality is Boné (hence the generic name), a mountain range and river near Gorontalo. In their review of this species, Bergmans and Rozendaal (1988) considered *Boneia* to be a synonym of *Rousettus*, but they noted that further study might result in retention of *Boneia* as a valid subgenus.

Head and body length is about 138 mm and tail length is about 25 mm. According to Bergmans and Rozendaal (1988), forearm length is 94.3–103.5 mm and weight is 150–94 grams. General coloration is dark brown on the back, russet on the rump, and dull brownish on the undersides. Males have a mantle, lateral tufts on the neck, and long hairs on the throat. The color of these features, which are absent in females, ranges from light golden buffy to much darker brown. *Boneia* closely resembles *Rousettus* but is distinguished by having a skull with a longer occipital region, a relatively broader palate, and premaxillae separated in front. The dental characters that originally were said to distinguish *Boneia* are now known to be present in some species of *Rousettus*.

Bergmans and Rozendaal (1988) reported that *B. bidens* is found in forests and areas where forest, bush, and farm habitats intermingle. Specimens have been taken at elevations of 200–1,060 meters. *Boneia*, like *Rousettus*, probably is capable of echolocation and, also like the latter, lives gregariously in caves. Two females obtained on 17 June and 17 July each contained a single small embryo.

CHIROPTERA; PTEROPODIDAE; **Genus MYONYCTERIS**
Matschie, 1899

Little Collared Fruit Bats

There are three species (Bergmans 1976, 1980; Schlitter and McLaren 1981):

M. torquata, Sierra Leone to western Uganda, and south to northern Angola and northern Zambia;
M. relicta, southeastern Kenya to central Tanzania;
M. brachycephala, Sao Tomé Island (Gulf of Guinea).

Koopman (1975) thought that *Myonycteris* could best be treated as a subgenus of *Rousettus*, but it was retained as a full genus by Corbet and Hill (1986), Honacki, Kinman, and Koeppl (1982), and Koopman (1984*c*).

Head and body length is about 85–165 mm, tail length is

Little collared fruit bat *(Myonycteris torquata)*, photo by Merlin D. Tuttle, Bat Conservation International.

4–13 mm, and forearm length is 55–70 mm. Weight is 27–54 grams (Bergmans 1976). Coloration of the upper parts of *Myonycteris* is light brown, tawny, or dark brown, sometimes with a reddish or yellowish appearance. Males have a tawny olive or dull yellow orange ruff of hairs on the foreneck. This patch of coarse hairs, which extends onto the chest, is associated with a cutaneous glandular formation. Such ruffs and patches of hair are found in many genera of the family Pteropodidae and are always associated with skin glands.

These fruit bats occur in forest, woodland, and savannah. They apparently roost in bushes and low trees, often in direct sunlight. Nothing is known about the natural diet, but these bats have been caught near mangoes, guavas, and bananas, and captive specimens have taken soft fruits, honey, and butter (Bergmans 1976).

Studies by Thomas (1983) in the Ivory Coast indicate that *M. torquata* migrates from forest to savannah areas as the rainy season progresses in April. At this time the females are pregnant, and they return to the forests later in the wet season to give birth (about August–September). A second peak in births may occur in February–March. Most males do not return to the forests, but follow the rains into the northern part of the country and do not move south again until about October.

A study of captives from Gabon found males to be sexually active all year and females to be polyestrous; births occurred twice a year, in December–January and June. Wild females taken in Gabon were pregnant or lactating from November to March and sexually inactive in June and July. Lactating females also were collected in Cameroon in February and in the Central African Republic in May (Bergmans 1976). Three females taken in Equatorial Guinea in October had one embryo each (Jones 1971*a*). A female taken in eastern Congo in September had a half-grown young (Kingdon 1974*a*).

CHIROPTERA; PTEROPODIDAE; **Genus PTEROPUS**
Erxleben, 1777

Flying Foxes

There are 17 species groups and 59 currently recognized species (Andersen 1912; Bergmans and Rozendaal 1988; Chasen 1940; Cheke and Dahl 1981; Daniel 1975; Ellerman and Morrison-Scott 1966; Felten 1964; Felten and Kock 1972; L. S. Hall 1987; Hayman and Hill, *in* Meester and Setzer 1977; Heaney and Rabor 1982; Hill 1958, 1971*a*, 1979, 1983*b*; Hill and Beckon 1978; Klingener and Creighton 1984; Koopman 1984*a*; Laurie and Hill 1954; Lawrence 1939; Lekagul and McNeely 1977; Musser, Koopman, and Califia 1982; Phillips 1968; Ride 1970; Roberts 1977; Sanborn 1931, 1950; Sanborn and Nicholson 1950; Tate 1934; Taylor 1934; Van Deusen 1969; Wodzicki and Felten 1975, 1981):

subniger group

P. hypomelanus, known to occur on Sulawesi and otherwise only on small islands from the Bay of Bengal and the Malay Peninsula to New Guinea and the Solomons;
P. speciosus, western Mindanao, Sulu Archipelago, Besar and Mata Siri (Java Sea);
P. griseus, Luzon, Sulawesi, Timor and nearby islands, Banda Islands;
P. mearnsi, Mindanao and Basilan islands (Philippines);
P. pumilus, Palmas, Balut, Tablas, and Mindoro islands (Philippines);
P. admiralitatum, Admiralty Island north of New Guinea, Solomon Islands;
P. howensis, Lord Howe Islands (Solomons);
P. ornatus, New Caledonia (South Pacific);
P. auratus, Loyalty Islands (South Pacific);
P. dasymallus, Taiwan, Ryukyu Islands;
P. subniger, Reunion and Mauritius islands (Indian Ocean);

mariannus group

P. pelewensis, Palau Islands (western Pacific);
P. yapensis, Yap and Mackenzie islands (western Pacific);
P. ualanus, eastern Caroline Islands (western Pacific);
P. mariannus, Okinawa, Ryukyu Islands, Guam, Mariana Islands (western Pacific);
P. vanikorensis, Vanikoro Island (South Pacific);
P. tonganus, Karkar Island off northeastern New Guinea, Solomon Islands, New Hebrides, New Caledonia, Fiji Islands, Ueva Island, Tonga Islands, Samoa, Niue Island, Cook Islands;

caniceps group

P. caniceps, islands between Sulawesi and New Guinea;
P. argentatus, known only by a single specimen supposedly from Amboina Island in the Moluccas;

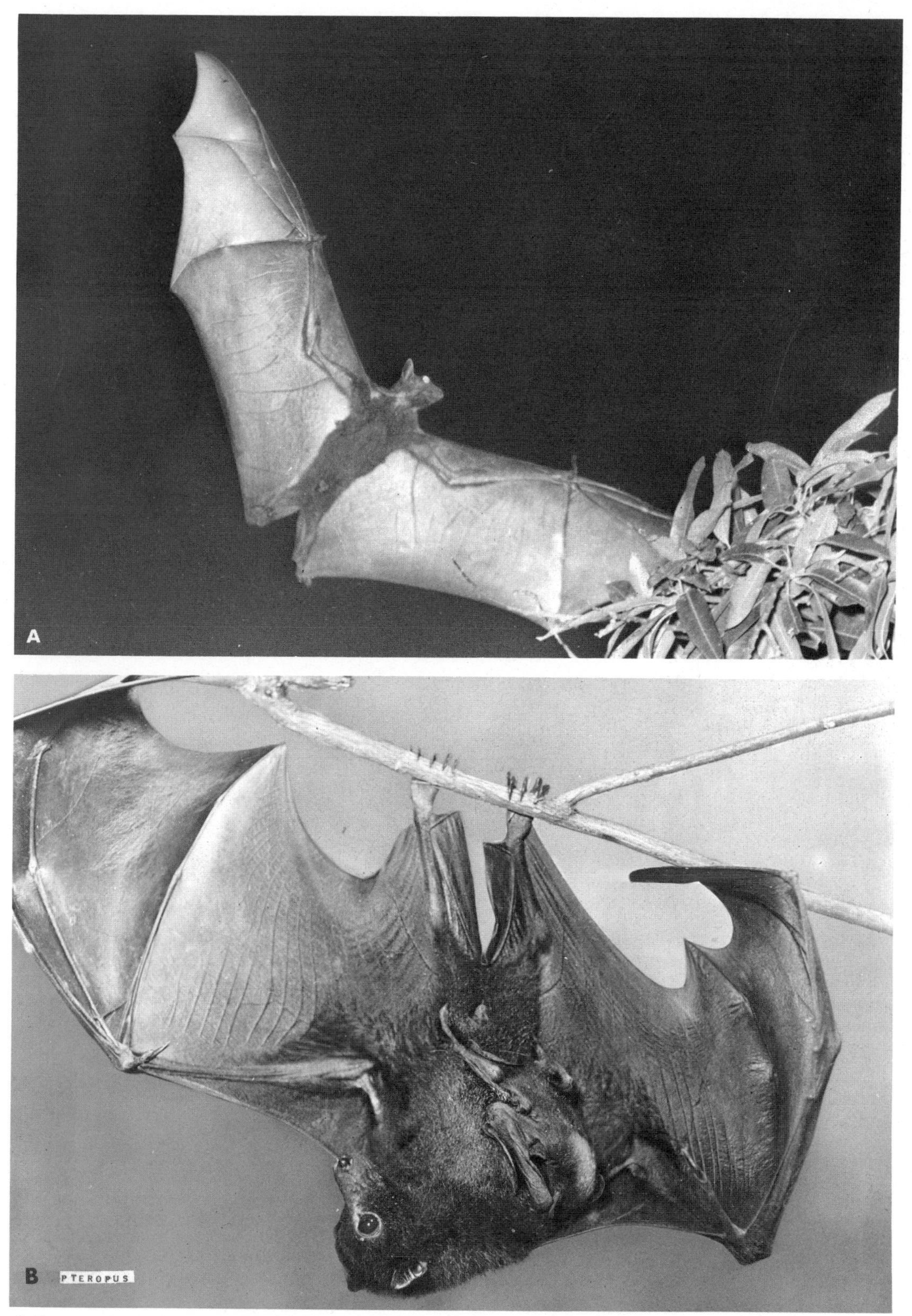

A. *Pteropus alecto,* photo by John Warham. B. *P. vampyrus,* photo by Bernhard Grzimek.

rufus group

P. rufus, Madagascar;
P. seychellensis, Comoro Islands between Africa and Madagascar, Mafia Island off east coast of Tanzania, Seychelles and Aldabra islands (Indian Ocean);
P. voeltzkowi, Pemba Island off northeast coast of Tanzania;
P. niger, Reunion and Mauritius islands (Indian Ocean);

melanotus group

P. melanotus, Andaman Islands, Nicobar Islands, Nias and Enggano islands off western Sumatra, Christmas Island (south of Java);

melanopogon group

P. melanopogon, islands between Sulawesi and New Guinea;
P. livingstonii, Johanna and Moheli islands in the Comoros;

rayneri group

P. rayneri, Solomon Islands;
P. chrysoproctus, islands between Sulawesi and New Guinea;
P. fundatus, Banks Islands in New Hebrides (South Pacific);

lombocensis group

P. lombocensis, Lombok, Flores, and Alor islands;
P. rodricensis, Rodriguez Island (Indian Ocean);
P. molossinus, Caroline Islands (western Pacific);

samoensis group

P. samoensis, Samoa and Fiji islands (South Pacific);
P. anetianus, New Hebrides Islands (South Pacific);

pselaphon group

P. vetulus, New Caledonia (South Pacific);
P. insularis, Truk and Uala islands in the Carolines (western Pacific);
P. phaeocephalus, Mortlock Island in the Carolines (western Pacific);
P. tokudae, Guam (western Pacific);
P. pselaphon, Bonin Islands (western Pacific);
P. pilosus, Palau Islands (western Pacific);
P. tuberculatus, Vanikoro Island (South Pacific);
P. nitendiensis, Santa Cruz Islands (South Pacific);
P. leucopterus, Luzon and Dinagat islands in the Philippines;

temmincki group

P. temmincki, Timor, islands between Sulawesi and New Guinea, Bismarck Archipelago;
P. personatus, Halmahera Islands;

vampyrus group

P. giganteus, Pakistan, India, Nepal, Sikkim, Bhutan, Burma, Sri Lanka, Maldive Islands (Indian Ocean);
P. lylei, southern Thailand, Cambodia, Viet Nam;
P. vampyrus, southern Thailand and Indochina, Tenasserim, Malaysia, Indonesia, Philippines;

alecto group

P. alecto, Sulawesi, Salayer Island, Bawean and Kangean islands in the Java Sea, Lombok, Sumba and Savu islands, northern and eastern Australia;

conspicillatus group

P. conspicillatus, New Guinea and nearby islands, northeastern Queensland;
P. ocularis, Ceram and Buru islands between Sulawesi and New Guinea;

neohibernicus group

P. neohibernicus, New Guinea, Bismarck Archipelago;

macrotis group

P. macrotis, southern New Guinea, Aru Islands;
P. poliocephalus, southeastern Queensland, New South Wales, Victoria;
P. pohlei, Japen Island off northwestern New Guinea;

scapulatus group

P. scapulatus, extreme southern New Guinea, northern and eastern Australia, rarely on Tasmania, one record from New Zealand;
P. woodfordi, Solomon Islands;
P. mahaganus, Santa Ysabel and Bougainville (Solomons);
P. gilliardi, New Britain Island (Bismarck Archipelago).

P. brunneus from Percy Island off northeastern Queensland, which was listed in the *subniger* group by Andersen (1912), was considered a vagrant *P. hypomelanus* by Ride (1970); however, Koopman (1984*a*) reluctantly continued to treat it as a full species. Both Corbet and Hill (1986) and Honacki, Kinman, and Koeppl (1982) included *P. pelewensis*, *P. yapensis*, and *P. ualanus* in *P. mariannus*. It is likely that many of the other names listed above also will be reduced to subspecific rank as modern systematic work progresses (Grzimek 1975).

Head and body length is 170–406 mm, forearm length is 85–228 mm, and wingspan is 610–1,700 mm; there is no tail. The largest species is *P. vampyrus*. Some reported weights are: *P. hypomelanus*, 45 grams (Lekagul and McNeely 1977); *P. macrotis*, 317–415 grams; *P. mahaganus*, 1,220–1,250 grams (McKean 1972); male *P. giganteus*, 1,300–1,600 grams; and female *P. giganteus*, 900 grams (Roberts 1977). The coloration is grayish brown or black, with the area between the shoulders often yellow or grayish yellow.

These fruit bats inhabit forests and swamps, often on small islands near coasts. Indra Kumar Sharma (pers. comm.) stated that having a large body of water nearby is essential for *P. giganteus*. By day, flying foxes usually roost in trees, and colonies may utilize the same roosting site year after year. During daylight there is much noise and motion in the roosts, and individuals sometimes fly from one place to another. At dusk the bats fly to fruit trees to feed. These bats, particularly the larger species, are strong fliers. Sterndale mentioned a *P. giganteus* that landed on a boat more than 200 miles from land, exhausted but alive, probably having been blown offshore by high winds. Flying foxes eat, rest, and digest their food for several hours while at their feeding trees; then they

Fruit bat *(Pteropus dasymallus)*, photo by Gwilym S. Jones.

return to the roosting site. In a radiotracking study of *P. giganteus*, Walton and Trowbridge (1983) found a female to leave her roost at 1800 hours, arrive at her main feeding area after 2000 hours, having stopped several times along the 12–14-km route, and then return directly to the roost at about 0400 hours, usually taking less than an hour. The principal food of *Pteropus* is fruit juices, which the bats obtain by squeezing pieces of the fruit pulp in their mouths. They swallow the juice and spit out the pulp and seeds. If the pulp is very soft, like banana, they swallow some of it. They also chew eucalyptus and probably other flowers to obtain the juices and pollen. They drink while flying to and from their feeding and sleeping locations. Some drink sea water, apparently to obtain mineral salts lacking in the plant food.

Pteropus maintains a body temperature of 33°–37° C; during cool weather, this temperature range is maintained by constant activity. These mammals have a very noticeable characteristic odor.

According to Ride (1970), Australian flying foxes congregate in great "camps" during the summer, when blossoms are abundant. Some of these roosts contain nearly 250,000 individuals at densities of 4,000–8,000 per hectare. The young are born shortly after the colonies form, and remain with the females for three to four months. Mating takes place in the camps, and soon thereafter a nongregarious phase seems to start, with segregation of the sexes. This phase culminates in a dispersed population, which persists throughout the winter. Prociv (1983) reported that a camp of 100,000 *P. scapulatus* was found near Brisbane in December, with most females associated with a male in harem groups, and some males forming their own groups. Mating occurred at this time, and births took place in April.

Lekagul and McNeely (1977) reported that *P. lyeli* also formed very large colonies, while groups of *P. vampyrus* often exceeded 100 individuals. On Niue Island, in the South Pacific, *P. tonganus* has been reported to roost singly, in pairs, or in larger groups, rarely up to 100 individuals in size (Wodzicki and Felten 1975). Neuweiler (1969) studied a colony of 800–1,000 *P. giganteus* in India and found a vertical rank order among males, each having a particular roosting place in the tree. Mating in the colony occurred from July to October and births took place mostly in March. The young were carried by the females for the first few weeks of life but subsequently were left in the tree. When the young males became independent, they separated from the colony and gathered on a neighboring tree. Gould (1977) observed individual *P. vampyrus* to defend an entire tree while feeding.

Lekagul and McNeely (1977) noted that *P. vampyrus* produced a single young in late March or early April which became independent after 2–3 months. Medway (1978) reported pregnant females of this species in the Malay Peninsula in November, December, and January. Roberts (1977) stated that the young of *P. giganteus* are born during the spring and again in the monsoon season, following an estimated gestation period of 140–50 days, and that they are carried by the mother for 6–7 weeks. McKean (1972) reported a juvenile *P. rayneri* on Bougainville Island on 21 September. Births in New Guinea and Madagascar are usually in October. On Guam, *P. mariannus* apparently is capable of producing young all year (Perez 1973). On New Caledonia, Sanborn and Nicholson (1950) found no pregnancies in *P. ornatus* from November to March but a peak in pregnancies from June to September. It is probable that in some other areas this genus does not have a definite breeding season. The known longevity record is held by a *P. giganteus* that lived for 31 years and 5 months in captivity (Jones 1982).

In many areas *Pteropus* is considered a serious pest by fruit growers. Colonies of *P. lyeli* in Thailand are said sometimes to devastate a crop overnight (Lekagul and McNeely 1977). In Australia large numbers of flying foxes have been poisoned in an effort to protect orchards and gardens (Ride 1970). Raids on bananas there are thought to have increased in recent years, as natural food supplies have disappeared through land clearing (Tidemann 1987*b*). In some areas *Pteropus* also is killed for food or medicinal purposes. Roberts (1977) noted that *P. giganteus* was much rarer in Pakistan than it had been 10–20 years earlier because orchard owners had become less tolerant, and there was increasing disruption of colonies as people sought to kill bats for their fat, thought to be a cure for rheumatism. Heaney and Heideman (1987) noted that aggregations of about 150,000 *P. vampyrus* were common in the Philippines in the 1920s but that the largest groups now seen contain no more than a few hundred individuals; the

Flying fox *(Pteropus personatus)*, photo by F. G. Rozendaal.

apparent reasons for the decline are excessive hunting for meat and deforestation.

Some of the most serious conservation problems involve species restricted to small oceanic islands. Certain of these species, such as *P. melanotus* and *P. samoensis*, are especially vulnerable because they seem not to have developed a fear of humans and tend to be diurnal (Tidemann 1987*a*). *P. tonganus*, which was still abundant on Niue in the 1920s, has declined drastically in numbers because of the destruction of its forest habitat and the introduction of shotguns (Wodzicki and Felten 1975, 1981). *P. niger* and *P. rodricensis* have been seriously affected by the clearing of forests and two resultant factors: easier access by hunters and loss of buffering protection against cyclonic winds (Cheke and Dahl 1981; Hamilton-Smith 1979). *P. niger* has disappeared from Reunion Island, and its population on Mauritius has been reduced to not more than 3,000 individuals. *P. rodricensis* may once have occurred on Mauritius and nearby Round Island but disappeared there shortly after settlement. There may have been over 1,000 on Rodriguez Island in 1955, but by 1974 only about 70 were left. Subsequent conservation efforts have contributed to an increase to about 350, and captive breeding groups also have been established, but the future of the species remains in doubt (Caroll 1984). *P. subniger*, also a resident of Reunion and Mauritius, became extinct in the mid-1800s when its last habitat, in the high forests, was destroyed; *P. livingstonii*, restricted to two small islands in the Comoros, has been reduced to only a few hundred individuals through deforestation (Cheke and Dahl 1981). Yet another Indian Ocean species, *P. seychellensis*, still occurs in several island populations that contain thousands of individuals each, but they are jeopardized by agricultural usurpation of their habitat, market hunting for tourist restaurants, and collision with and electrocution by electrical and telephone wires (Cheke and Dahl 1981; Nicoll and Racey 1981; Verschuren 1985).

On Guam and some of the other Mariana islands, *P. tokudae* and *P. mariannus* are considered delicacies and have been commercially marketed despite legal protection; they have also suffered because of human development of their habitat. The former species may already be extinct, while the latter declined in numbers from as many as 3,000 in 1957 to fewer than 50 in 1978 (Perez 1973; Wheeler and Aguon 1978). There was an increase to 850–1,000 in the early 1980s, probably as a result of migration from the nearby island of Rota, which in turn was stimulated in part by human disturbance. Numbers subsequently declined again for both islands. *P. mariannus* is more numerous in other parts of its range but is being hunted and exported to Guam. Indeed, the commercial demand on Guam may have placed *Pteropus* in jeopardy throughout the Pacific; in 1984 there was a known import of 16,250 individuals to the island, with many coming from as far away as New Guinea and Samoa (N. Payne 1986; Wiles and Payne 1986). Most of the bats taken in Samoa are thought to be *P. tonganus*, but some are *P. samoensis*. The Samoan subspecies of the latter, *P. s. samoensis*, has declined through both excessive hunting and habitat destruction (Cox 1983).

The IUCN gives the following classifications: *P. insularis*, indeterminate; *P. livingstonii*, endangered; *P. macrotis*, indeterminate; *P. mariannus*, vulnerable; *P. molossinus*, indeterminate; *P. niger*, vulnerable; *P. phaeocephalus*, indeterminate; *P. pilosus*, extinct; *P. rodricensis*, endangered; *P. samoensis*, endangered; *P. seychellensis aldabrensis*, vulnerable; *P. subniger*, extinct; *P. tokudae*, extinct; *P. tonganus*, indeterminate; and *P. voeltzkowi*, vulnerable. The following species are on appendix 1 of the CITES: *P. insularis*, *P. mariannus*, *P. molossinus*, *P. phaeocephalus*, *P. pilosus*, *P. samoensis*, and *P. tonganus*; all other species of *Pteropus* are on appendix 2. The USDI lists *P. rodricensis*, *P. tokudae*, and the Guam population of *P. mariannus* as endangered.

CHIROPTERA; PTEROPODIDAE; **Genus ACERODON** *Jourdan, 1837*

There are six species (Laurie and Hill 1954; Musser, Koopman, and Califia 1982; Taylor 1934):

A. celebensis, Sulawesi, Salayer Island, Sula Mangoli Island;
A. mackloti, Lombok, Sumbawa, Flores, Sumba, Timor, and Alor islands in the East Indies;

Acerodon jubatus, photo by Paul D. Heideman.

A. humilis, Talaud Islands between Sulawesi and Philippines;
A. jubatus, Philippines;
A. leucotis, Palawan and Busuanga islands (Philippines);
A. lucifer, Panay Island (Philippines).

Head and body length is 178–290 mm and forearm length is 125–203 mm; there is no tail. The wingspan of the largest species, *A. jubatus*, is 1.51–1.7 meters. A male of that species weighed 1,050 grams (Mudar and Allen 1986). Goodwin (1979) reported that eight specimens of *A. mackloti* weighed 450–565 grams. Coloration is variable. The forehead and sides of the head are often dark brown or black, an orange or golden yellow nape may be evident, the shoulders are usually reddish brown or chestnut, the lower back is usually darker than the area between the shoulders, and the undersides are usually dark brown or black. This genus is externally indistinguishable from *Pteropus* but differs in dental features.

In the Philippines, *Acerodon* has been found roosting in clumps of bamboo, hardwood trees, and swampy forest, usually on small offshore islands. Taylor (1934) mentioned a colony of *A. jubatus* and *Pteropus vampyrus* that covered an area of 8–10 ha. and contained some 150,000 bats. These bats flew 9–16 km into the mountains in the evening to feed on wild fruit.

On Timor, Goodwin (1979) located two colonies of *A. mackloti*, each spread out over the crowns of large fig trees on the edge of open forest. The staple food in this area was at least two species of fig. As the sun was setting the bats began to leave their roosts to make rounds of the fruit trees. They usually traveled in small, well-spaced groups of two to six individuals. If more than one bat landed in the same food tree, there would be noisy squabbling until a dominant individual drove the others off. Goodwin collected pregnant females, each with a single embryo, on Timor on 19 March and 5 May.

The species *A. lucifer* is considered extinct by the IUCN. According to Heaney and Heideman (1987), no specimens have been taken since the original series was collected on Panay in 1896. The forests, on which the species depended for roosting sites and food, have been largely destroyed by human activity. All species of *Acerodon* are on appendix 2 of the CITES.

Acerodon celebensis, photo by F. G. Rozendaal.

CHIROPTERA; PTEROPODIDAE; **Genus NEOPTERYX** *Hayman, 1946*

The single species, *N. frosti*, now is known by one specimen collected by W. J. C. Frost in 1938 or 1939 at Tamalanti in western Sulawesi, as well as by three others obtained in 1985 in the northeastern part of the island (Bergmans and Rozendaal 1988).

Head and body length of the type specimen is 105 mm, and there is no tail. According to Bergmans and Rozendaal (1988), forearm length was 104.9–110.6 mm in three adult specimens, and weight was 190 grams in an adult male and 250 grams in a pregnant female. The thick, short fur is generally tawny or brownish in color. There is a paler, woolly mantle. The muzzle is sepia with creamy white stripes on top and on each side.

The wings are attached near the midline of the back, giving the impression that the back is without fur, as the fur on the back is hidden by the naked wing membranes. The thumb has a well-developed claw, but there is no claw on the index finger. The only other fruit bats of the family Pteropodidae that lack the claw and tail are *Melonycteris woodfordi* and *Melonycteris aurantius*. *Neopteryx* also has distinctive skull and dental characters. The cheek teeth of *Neopteryx* are de-

Neopteryx frosti, photo by F. G. Rozendaal.

scribed as simplified, shortened, of about the same size, and almost equally spaced. Hayman (1946) stated that the conspicuous narrowing of the anterior portion of the palate, combined with the weakening of the teeth to such an extent (the canines both above and below are remarkably short and weak), suggests a special adaptation to a particular type of food supply.

Hayman (1946) stated that *Neopteryx* constitutes the fifth known genus in the Megachiroptera lacking a claw on the index finger. The absence of a tail distinguishes it at once from *Dobsonia, Eonycteris,* and *Notopteris,* which all lack the index claw, while *Nesonycteris,* the only fruit bat previously known to combine absence of tail with absence of index claw, has distinctive cranial and dental characters. The sharply narrowed rostrum and degenerate dentition together distinguish the skull from that of any other fruit bat. It would appear that this genus and species represents a specialized branch of the pteropine group.

The type specimen was collected at an elevation of 1,000 meters. Two others were taken in lowland primary forest. One, a pregnant female with one embryo, was caught in March.

CHIROPTERA; PTEROPODIDAE; **Genus PTERALOPEX**
Thomas, 1888

There are three species (Hill and Beckon 1978; Phillips 1968):

P. anceps, Bougainville and Choiseul islands in the Solomons;
P. atrata, Santa Ysabel and Guadalcanal islands in the Solomons;
P. acrodonta, Tavenuni Island in the Fiji Islands.

Head and body length is 255–80 mm, there is no tail, and forearm length is 116–71 mm. The fur is thick, woolly, and black. The wing membranes are sometimes mottled with white beneath. *Pteralopex* differs from *Pteropus* in having cuspidate teeth, and it can be distinguished externally from *Pteropus pselaphon* in that the wing membranes originate in the midline of the back. The original describer, Thomas, regarded *Pteralopex* as an isolated survivor from the time when the modern ancestors of the Old World fruit bats had teeth with cusps. Most Old World fruit bats do not now possess cuspidate teeth, apparently having lost them in their evolution.

P. acrodonta reportedly roosts in pairs in the fern clumps growing 6–10 meters from the ground on trunks of the larger trees in open, tall forest (Hill and Beckon 1978). An adult female *P. anceps* taken in July on Bougainville was lactating (Phillips 1968). One individual was shot at night while feeding on green coconuts.

CHIROPTERA; PTEROPODIDAE; **Genus STYLOCTENIUM**
Matschie, 1899

The single species, *S. wallacei,* is known from Sulawesi and the nearby Togian Islands (Bergmans and Rozendaal 1988; Hill 1983*b*).

Head and body length is 152–78 mm, there is no tail, forearm length is about 90–103 mm, and weight is 174–218 grams. The fur is soft and silky. The back and rump are usually light gray in color, often with a pale, reddish brown wash. The underparts are a light reddish brown. The head is reddish brown, with badgerlike white stripes and spots that enhance the appearance of this bat. There is no appreciable sexual difference in coloration.

This genus differs from *Pteropus* in dental features, the most conspicuous being the absence of the first lower incisor and the third lower molar. *Styloctenium* has been described as a "rather slightly specialized offshoot of . . . *Pteropus*" (Andersen 1912).

CHIROPTERA; PTEROPODIDAE; **Genus DOBSONIA**
Palmer, 1898

Bare-backed Fruit Bats

There are 15 species (Bergmans 1975*a,* 1978*a,* 1979*b;* Bergmans and Sarbini 1985; Boeadi and Bergmans 1987; De Jong and Bergmans 1981; Hill 1983*b;* Koopman 1979; Laurie and Hill 1954; Lidicker and Ziegler 1968; McKean 1972; Phillips 1968; Rabor 1952; Ride 1970):

D. peronii, islands from Bali to Timor in the East Indies;
D. viridis, Sulawesi, Buru, Ceram, Amboina, Banda, and Kei islands (between Sulawesi and New Guinea);
D. crenulata, Sanghir Islands northeast of Sulawesi, Halmahera Islands northwest of New Guinea;
D. beauforti, Waigeo, Biak, and Owii islands off western New Guinea;
D. praedatrix, Bismarck Archipelago;
D. inermis, Solomon Islands;
D. moluccensis, Moluccas and nearby islands, New Guinea, Bismarck Archipelago, Cape York Peninsula of northern Queensland;
D. magna, New Guinea and nearby Japen Island;

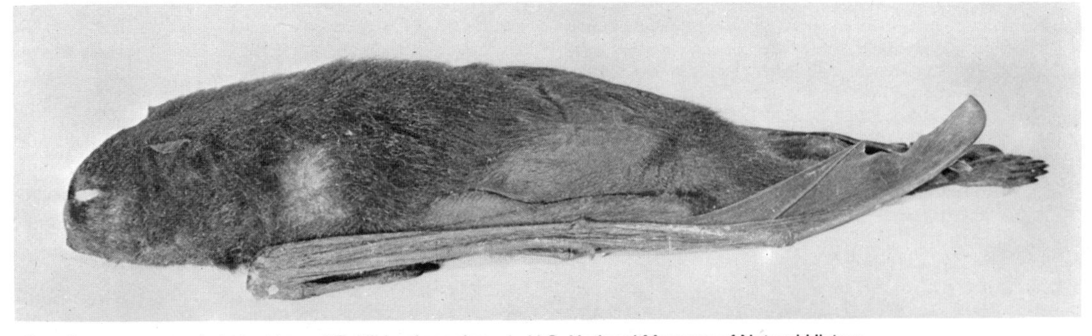

Pteralopex anceps, photo by Howard E. Uible of specimen in U.S. National Museum of Natural History.

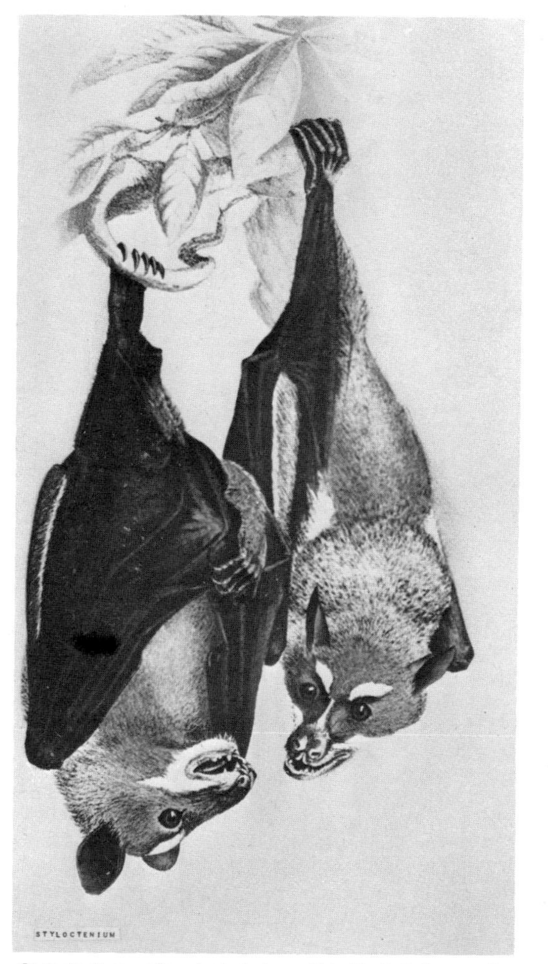

Styloctenium wallacei, photo from *Abh. Zool. Anthrop. ethn. Mus. Dresden*, "Saugetiere vom Celebes- und Philippinen-Archipel," Adolf B. Meyer.

Bare-backed fruit bat *(Dobsonia moluccensis)*, photo by P. Morris.

D. emersa, Biak and Owii islands off western New Guinea;
D. anderseni, Bismarck Archipelago;
D. pannietensis, islands southeast of New Guinea;
D. remota, Trobriand Island east of New Guinea;
D. chapmani, Negros Island (Philippines);
D. exoleta, Sulawesi, Togian Islands;
D. minor, central Sulawesi, New Guinea and nearby Japen Island.

Taylor (1934) stated that *D. peronii* had been reported from Samar Island in the eastern Philippines but thought it probable that the record either was an error or referred to *D. exoleta* or to an unknown species. Hill (1983*b*) suggested that *D. crenulata* is a subspecies of *D. viridis* and that *D. anderseni* and *D. pannietensis* should be included in *D. moluccensis*. Koopman (1982*a*) agreed with respect to *anderseni* but recognized *D. pannietensis* as a full species with *remota* as a subspecies thereof. The separation of *D. magna* from *D. moluccensis* by Bergmans and Sarbini (1985) and of *D. crenulata* from *D. viridis* by De Jong and Bergmans (1981) was not accepted by Corbet and Hill (1986).

Head and body length is about 127–242 mm, tail length is about 13–35 mm, and forearm length is 74–161 mm. Dwyer (1975*a*) reported that males of *D. moluccensis* weigh 380–500 grams, and females 325–525 grams. There is considerable variation in color, but most species are dull brown washed with olive, or grayish black.

There is no claw on the index (second) finger. The ears are pointed. The naked wing membranes are attached to the body along the spinal column in the midline of the back, as in *Pteronotus*, thus covering the hair of the body in the region that suggests the vernacular name. This form of attachment may supply a more effective wing area to aid these bats in hovering than exists for the bats whose wing membranes are attached along the side of the body.

Unlike most of the pteropodids, *Dobsonia* seems to prefer caves and tunnels rather than trees for roosting. However, it has been taken in tree roosts, and it is known to forage in areas of thick vegetation. Dwyer (1975*a*) found *D. moluccensis* in New Guinea to be abundant at an altitude of 2,700 meters. It roosted there in the twilight zone of caves and flew out by night to feed on a wide variety of fruits; it could easily be detected by its noisy flight. Bergmans (1978*a*) observed that *D. peronii* inhabited caves and trees, near the coast and inland, from sea level to about 1,200 meters. Other populations of *Dobsonia* have been noted in rock shelters on the coast at sea level and in coral limestone caves. In Australia, specimens have been collected at night while feeding on flowering bloodwood trees and native figs.

Goodwin (1979) reported that a colony of 300 *D. peronii*, composed of adults and young adults of both sexes, occupied a cave on Timor; a pregnant female was taken on this island on 20 March. Dwyer (1975*a*) reported that about 3,000 *D. mo-*

luccensis were seen roosting together but that members of this species appeared to be solitary foragers. In New Guinea *D. moluccensis* mated from April to June, and births occurred from mid-August to November. An estimated 300 *D. chapmani* were found roosting in a cave on Negros Island. A female of that species collected in May had 1 young clinging to it. Other available reproductive data include: 6 adult female *D. beauforti* taken on 25 December pregnant with 1 embryo each and another female lactating (Bergmans 1975*a*); a pregnant female *D. inermis* taken on Bougainville on 21 September; pregnant female *D. minor* collected in New Guinea in January, late May, and early June; a pregnant female *D. moluccensis* taken in New Guinea on 12 March; and juveniles of that species found in New Guinea from March to June (McKean 1972).

Bergmans (1978*a*) observed that people had destroyed over 50 percent of the forests on some of the islands inhabited by *Dobsonia* and that the bats probably had suffered accordingly. Heaney and Heideman (1987) reported that such deforestation, with consequent loss of food sources and greater access to human hunters, may have contributed to the extinction of *D. chapmani* of the Philippines. This species also declined as a result of killing and disturbance when people entered its cave roosts to mine guano.

CHIROPTERA; PTEROPODIDAE; **Genus APROTELES**
Menzies, 1977

The single species, *A. bulmerae*, was first discovered among fossil remains about 9,000–12,000 years old collected in the mountains of Chimbu Province in central Papua New Guinea (Menzies 1977). Shortly thereafter, it was realized that the same species was represented by specimens killed in 1975 in the Hindenburg Ranges of far western Papua New Guinea (Hyndman and Menzies 1980).

The mounted skin of one specimen was prepared, but lost, and the genus is known only from cranial material. Affinity to *Dobsonia* is indicated, but *Aproteles* differs in lacking lower incisors and in having less crowded, simpler teeth and a longer, more tapering rostrum.

The fossil material was found at an altitude of 1,530 meters in the refuse dump of a rock shelter. The bones are those of bats that had been killed and cooked by the human inhabitants of the shelter. Many fragments of *Aproteles* were recovered, and Menzies (1977) suggested that this genus, like *Dobsonia*, was a cave dweller, because a tree-roosting bat would hardly have been accessible in such quantity to people of the Stone Age. The specimens obtained in 1975 had been killed by a native hunter in a large cave at an altitude of 2,300 meters. In November 1977 an effort was made to locate *Aproteles* in this cave, but a local hunter had already killed or driven away nearly the entire colony. Hyndman and Menzies (1980) suggested that intensive human hunting had eliminated *Aproteles* in the central highlands of Papua New Guinea but that the genus managed to survive in sparsely populated areas farther to the west. *A. bulmerae* now is listed as endangered by the USDI.

Harpy fruit bat *(Harpyionycteris whiteheadi)*, photo by Paul D. Heideman.

CHIROPTERA; PTEROPODIDAE; **Genus HARPYIONYCTERIS**
Thomas, 1896

Harpy Fruit Bats

There are two species (Peterson and Fenton 1970):

H. whiteheadi, Mindoro, Mindanao, Negros, and Camiquin islands in the Philippines;
H. celebensis, Sulawesi.

Head and body length is 140–53 mm and forearm length is about 82–92 mm; there is no tail. Bergmans and Rozendaal (1988) recorded weights of 83–142 grams. The color is chocolate brown to dark brown above and paler below. The hind feet are very short, and the small interfemoral membrane is hidden in thick fur.

This genus differs from all other fruit bats in that the molar teeth have five or six cusps and the lower canines have three. Also, the premaxillary bones of the skull, the upper incisors, and the upper and lower canines are more strongly inclined forward than in other fruit bats. The canines cross each other almost at right angles when the jaws are closed. Oldfield Thomas, the describer, considered this genus to be an isolated group, but he thought it should be placed near *Rousettus* and *Boneia*. Knud Andersen, however, considered it a close relative of *Dobsonia*. Tate (1951*a*) associated it with the tube-nosed fruit bats. Corbet and Hill (1986) placed it in its own subfamily, the Harpyionycterinae, which in turn they placed near the Nyctimeninae, a subfamily for *Nyctimene* and *Paranyctimene*.

The type specimen was shot at dusk as it was flying around some high trees at an elevation of 1,660 meters on the island of Mindoro. Other specimens have been taken at altitudes of about 150–1,500 meters in Sulawesi and the Philippines. Peterson and Fenton (1970) speculated that the multicusp teeth of this genus might be adapted for extracting the juice of a particular type of tough-textured fruit and that the bats might not normally ingest the fruit fibers. Bergmans and Rozendaal (1988) reported the following for *H. celebensis:* pregnant females, each with a single embryo, obtained in January and September; a nursing female observed in January; and immature specimens found in January, May, June, and July.

CHIROPTERA; PTEROPODIDAE; **Genus PLEROTES**
Andersen, 1910

The single species, *P. anchietai*, is known from Angola, southeastern Zaire, and Zambia (Hayman and Hill, *in* Meester and Setzer 1977).

An adult female had a head and body length of 87 mm and a wing expanse of 343 mm. There is no external tail, and the forearm length is about 48–53 mm. The pelage is long and fine, being grayish brown above and much paler below. There is a small tuft of white hairs at the base of the ears.

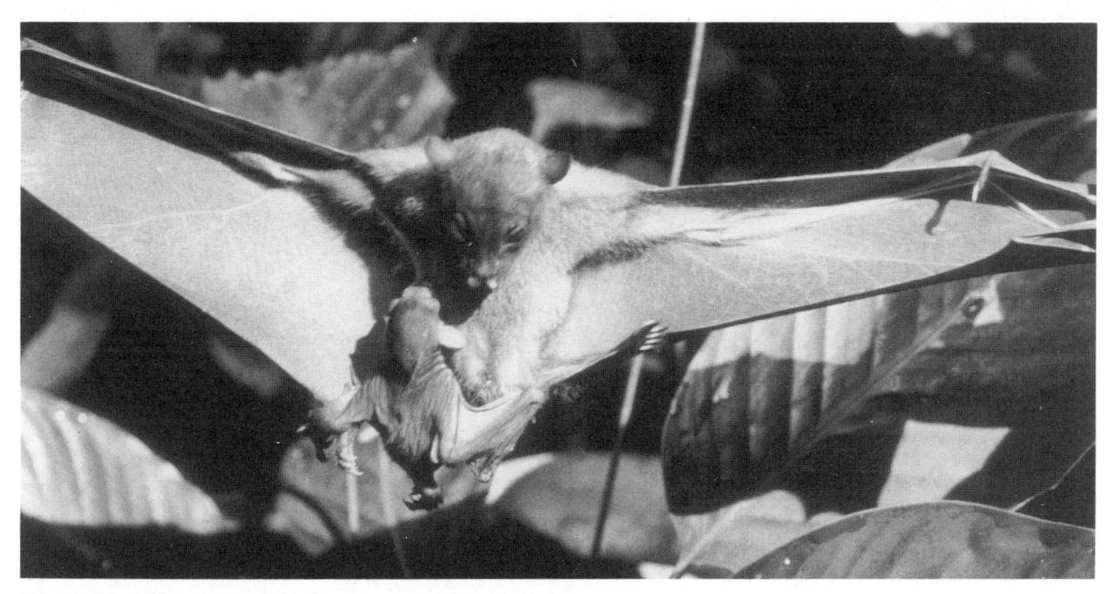

Harpy fruit bat *(Harpyionycteris celebensis)*, photo by F. G. Rozendaal.

This bat may be distinguished from all other Old World fruit bats (Pteropodidae) by the white patches at the base of the ears and the absence of a "spur," or calcar, a bone that projects from the heel to extend the interfemoral membrane. Males probably possess shoulder pouches and hair tufts. According to the original describer of this species, a cheek pouch on each side of the muzzle surrounds the eyes.

The dentition is unusually weak for bats of this group, in which weak dentition prevails. The teeth are much reduced in size, and the molars have very little structure on the surface. The dentition suggests flower and nectar feeding habits, though nothing is known specifically about the habits of *Plerotes*. It probably roosts in dense foliage during the day.

CHIROPTERA; PTEROPODIDAE; **Genus HYPSIGNATHUS**
H. Allen, 1861

Hammer-headed Fruit Bat

The single species, *H. monstrosus*, is found from Gambia to southwestern Ethiopia and south to northeastern Angola and Zambia (Hayman and Hill, *in* Meester and Setzer 1977; Koopman 1975; Largen, Kock, and Yalden 1974).

Head and body length is about 193–304 mm, there is no tail, and forearm length is 118–37 mm. The wingspan in males is as much as 907 mm. This genus has the greatest sexual dimorphism in the Chiroptera; Bradbury (1977) found that males, which averaged 420 grams, were nearly twice as heavy as females, which averaged 234 grams. The coloration is grayish brown or slaty brown. The breast is paler, and the lighter color extends up around the neck, forming a sort of collar. A white patch is present at the base of the ear. Shoulder pouches and epauletlike hair tufts are lacking in both sexes.

Male *Hypsignathus* may be recognized in flight by the large, square, truncate head. The muzzle is thick and hammer-shaped, hence the common name. Other distinctive features are enormous and pendulous lips, ruffles around the nose, a warty snout, a hairless, split chin, and highly developed voice organs in adult males. Females have a foxlike muzzle similar to that of *Epomophorus*.

In referring to this genus, Lang and Chapin (1917) commented: "In no other mammal is everything so entirely subordinated to the organs of voice." The adult male has a pair of air sacs that open into the sides of the nasopharynx and can be

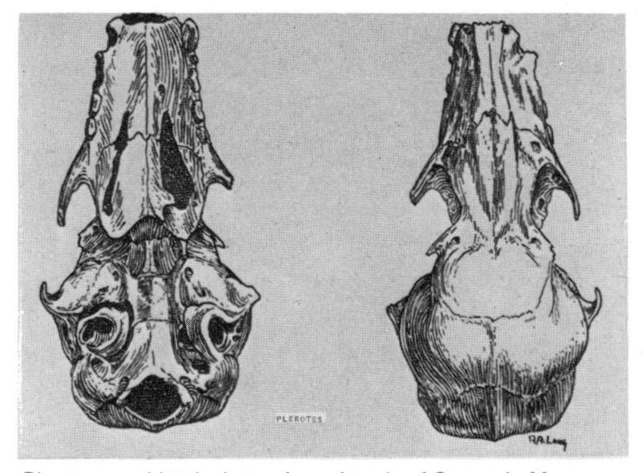

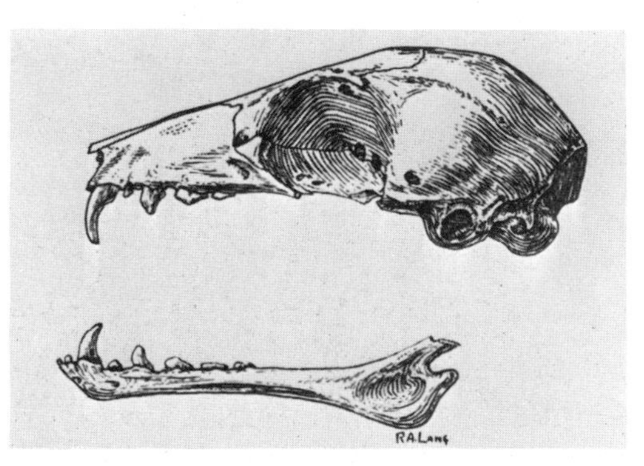

Plerotes anchietai, photos from *Annals of Carnegie Museum*.

A. Hammer-headed bats *(Hypsignathus monstrosus)*, photo by V. Aellen. B. Head; C. Mouth; D. Lips; photos by J. L. Perret.

inflated at will, as well as a great enlargement of the voice box (larynx) and vocal cords. The larynx "is nearly equal in length to one half of the vertebral column," actually filling most of the chest cavity, pushing the heart and lungs backward and sideward. The voice thus produced, a continuous croaking or quacking, is quite remarkable and probably attracts the females. The gregarious chorus reminded Lang and Chapin of "a pondful of noisy American wood-frogs, greatly magnified and transported to the treetops."

The hammer-headed bat inhabits forests, being most common in swamps, mangroves, and palms along rivers. It usually roosts in foliage but has been found in a cave. Bradbury (1977) stated that *Hypsignathus* roosted at a height of 20–30 meters during the day and would forage up to 10 km from the roost at night. With the ripening of certain fruits, this bat often seeks the high forest or native clearings to feed. It may take the juices of mangoes, soursops, and bananas. Van Deusen (1968) reported that *Hypsignathus* killed and ate tethered chickens.

In a lengthy study in Gabon, Bradbury (1977) determined that *Hypsignathus* had a "lek" mating system comparable to that of certain bovids. Twice a year, from June to August and from December to February, adult males assembled nightly at traditional sites. Each assembly numbered from 25 to 132 bats, and each bat set up a territory roughly 10 meters in diameter. The remarkable call was given from about 1830 to 2300 and 0300 to 0500 hours every night, and the males also displayed by flapping their wings. Females visited the assemblies, hovered about, and eventually settled beside one male to mate. Only 6 percent of the males accounted for 79 percent of the matings, and the relatively few successful males tended to be aggregated at certain apparently favorable locations within the overall assemblies. In comparison with the noise and activity at the nocturnal assemblies, day roosts were relatively quiet, with the bats resting alone or in small groups. At night the sexually active males would move to an assembly area, display for a while, and then fly out to forage. Examination of captives indicated that sexual maturity was attained at 6 months by females but not until 18 months by males. Wolton et al. (1982) suggested that in Liberia there is a birth peak during the middle of the rainy season in August–September and possibly another peak at the end of the rains in October–December. Kingdon (1974*a*) observed that available records suggested breeding peaks around February and July in East Africa and that the usual number of young is one, but twins had been reported.

CHIROPTERA; PTEROPODIDAE; **Genus EPOMOPS**
Gray, 1870

Epauleted Bats

There are three species (Bergmans 1979*c*, 1982; Bergmans and Jachmann 1983; Hayman and Hill, *in* Meester and Setzer 1977):

E. franqueti, Ivory Coast to southern Sudan, and south to Angola and Zambia;
E. buettikoferi, Guinea to Nigeria;

Epauleted bat *(Epomops franqueti):* Top photo from Amsterdam Zoo; Middle and bottom photos by Jan M. Haft.

E. dobsoni, Rwanda, Tanzania, Angola, southern Zaire, Zambia, Malawi, northeastern Botswana.

Head and body length is 135–80 mm, there is no tail, and forearm length is 74–102 mm. Reported weights are: male *E. franqueti,* 59–160 grams, female *E. franqueti,* 56–115 grams; male *E. buettikoferi,* 160–98 grams; and female *E. buettikoferi,* 85–132 grams (Bergmans 1975*b;* Jones 1971*a,* 1972). Coloration is tawny, brownish, or grayish, with much variation. Whitish hair tufts are present at the base of the ears. Males have two pairs of pharyngeal sacs, large shoulder pouches, and epauletlike hair tufts. These tufts, whitish or yellowish in color, are conspicuous in dried specimens, but in the live bats, as in related genera, the shoulder pouch may be so drawn in that the epaulet is almost completely concealed. Also as in the related bats, the shoulder pouches of *Epomops* are lined with a glandular membrane.

The adult males of *E. franqueti* have a bony voice box, as in *Hypsignathus,* and they emit a call that at close range sounds somewhat like "kûrnk!" or "kyûrnk!" but at a distance has a whistled, almost musical, effect. This high-pitched note can be heard throughout the night and may be emitted continuously for several minutes by a given male.

Epauleted bats are usually found in forests but also appear in open country. They roost in trees and bushes and are quite alert by day. Jones (1972) found them mainly in large trees, 4–6 meters above the ground. Their diet consists of the juice and softer parts of such fruits as guavas, bananas, and figs. The manner of feeding for *Epomops* and related genera is as follows: the expansible lips encircle the fruit; the pointed canines and premolars pierce the rind; the jaws squeeze the fruit, which is then pressed upward by the tongue against the thick palate ridges; and suction is accomplished by the large pharynx, which communicates anteriorly with the mouth by a small opening and is supported behind by the hyoid apparatus. Suction, rather than mastication, appears to be the major process in feeding.

Epauleted bats are not gregarious, usually resting alone or in groups of two or three. In Uganda *E. franqueti* has two breeding seasons, with implantation occurring in April and late September and births taking place in September and late February (Okia 1974*a*). These sequences are timed so that births occur at or shortly after the beginning of the two rainy seasons. Gestation lasts five to six months. Bergmans (1979*a*) suggested that the reproductive pattern in Congo is not much different from that found in Uganda. Gallagher and Harrison (1977) stated that breeding females of *E. franqueti* were taken in Zaire in December and January. Bergmans, Bellier, and Vissault (1974) reported the collection of possibly lactating females of *E. franqueti* and *E. buettikoferi* in the Ivory Coast in October and November.

Additional studies in the Ivory Coast by Thomas and Marshall (1984) indicate that *E. buettikoferi* has birth peaks from about February to April and from September to November. Both periods seem to be timed so that lactation coincides with the area's two rainy seasons, when fruit availability is at a maximum. Each parturition period is followed by a postpartum estrus and apparently by immediate embryonic development. The gestation period is 5–6 months, and lactation lasts 7–13 weeks. Young females can mate at an age of 6 months and give birth at 12 months. Males are sexually mature at 11 months.

CHIROPTERA; PTEROPODIDAE; **Genus EPOMOPHORUS**
Bennett, 1836

Epauleted Fruit Bats

There are six species (Bergmans 1988):

E. gambianus, Senegal and southern Mali to southern Sudan and Ethiopia, southern Zaire and Tanzania to eastern South Africa;

Epauleted fruit bat *(Epomophorus wahlbergi)*, photo by Erwin Kulzer.

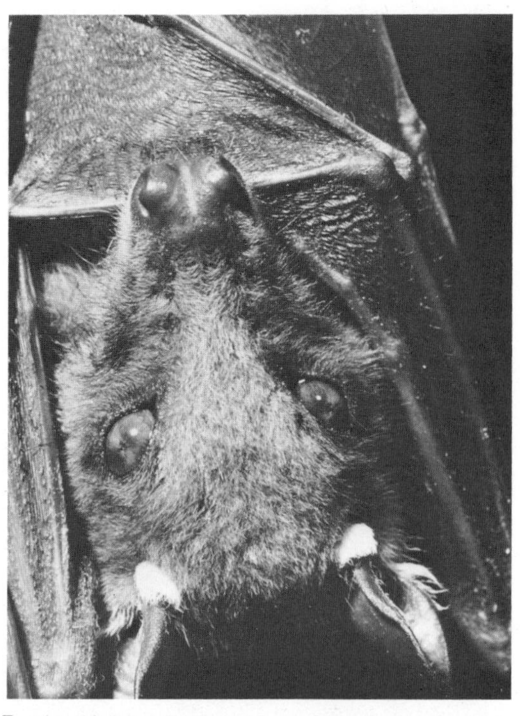

Epauleted fruit bat *(Epomophorus wahlbergi)*, photo by John Visser.

E. labiatus, northeastern Nigeria and southern Congo to northern Ethiopia and Malawi;
E. minor, southern Sudan and Ethiopia to Malawi;
E. angolensis, western Angola, northwestern Namibia;
E. wahlbergi, Cameroon and Somalia south to South Africa;
E. grandis, southern Congo, northeastern Angola.

Of several other species that often have been recognized (Corbet and Hill 1986; Hayman and Hill, *in* Meester and Setzer 1977), Bergmans (1988) regarded *E. anurus* as a synonym of the largely sympatric *E. labiatus; E. crypturus*, of southern Africa, as a subspecies of *E. gambianus; E. reii*, of Cameroon, as a synonym of *E. gambianus;* and *E. pousarguesi*, of the Central African Republic, as a subspecies of *E. gambianus*. Bergmans also transferred *E. grandis* to *Epomophorus* from the genus *Micropteropus*.

Head and body length is usually 125–250 mm, a vestigial external tail may be present beneath the interfemoral membrane, forearm length is about 60–100 mm, and the wingspan in males is about 508 mm. Weight is about 40–120 grams (Kingdon 1974*a*). These bats are grayish brown, russet, or tawny in color, with a white patch at the base of the ear in both sexes. Males have a conspicuous pair of shoulder tufts around large glandular sacs. Air sacs are present on the necks of males. Epauleted fruit bats have expansible, pendulous lips, often with peculiar folds.

These bats occur in woodland and savannah; true forest bats, they prefer the edges of forests (Kingdon 1974*a*). They roost in such places as large hollow trees, thick foliage, accumulated roots along stream banks, and below the thatch of open sheds. They often roost where there is considerable light, and sometimes groups of about half a dozen hang from the midribs of palm fronds in plain sight. They seem to be quite alert when roosting during the day. Fenton et al. (1985) found *E. wahlbergi* to switch day roosts every few days and to fly up to 4 km from these roosts to the nocturnal feeding area, females usually traveling farther than males.

The flight of one species, *E. labiatus*, has been compared to that of a raven. While flying and while feeding, members of this genus emit various squeaks, chucklings, and sharp metallic calls. These bats are likely to appear wherever fig, mango, guava, or banana trees are in fruit; some species, such as *E. wahlbergi*, make local migrations in search of ripe fruit. *E. gambianus* has been noted feeding on the nectar of the flowers of *Parkia clappertoniana* in Ghana. Control measures, in the form of poisoned fruit for bait, are sometimes utilized in areas where they feed extensively on cultivated fruit.

These fruit bats roost in small groups; there is some indication that in *E. labiatus* the females with young roost apart from the males, but about 20 *E. wahlbergi*, including young and old of both sexes, were once collected from the lower fronds of a single palm tree. According to Wickler and Seibt (1976), the latter species roosts during the day in bisexual groups of 3–100 individuals. One colony of about 100 bats in Kenya was found in the same locality every January for five years. During the breeding season, the males would leave the main roost at night and fly to another location, where they would begin to emit courting calls. The calls seemed to space out the males and attract females. While calling, the males would display the normally concealed epaulets.

Data from Congo suggest defined seasonal reproduction for *E. wahlbergi*, involving two cycles per year, with births occurring around the end of February and the beginning of September (Bergmans 1979*a*). Okia (1974*b*) found that in Uganda *E. labiatus* was a continuous but strictly cyclic breeder, with two distinct breeding seasons separated by about a month. Gestation apparently lasted about five to six months, from April to September and from October to March. Koopman, Mumford, and Heisterberg (1978) collected a lactating female *E. gambianus* in Burkina Faso in January and pregnant females in March, May, and September. Thomas and Marshall (1984) reported that in the Ivory Coast *E. gambianus* breeds twice annually, with births occurring during the rainy seasons, about April–May and October–November.

CHIROPTERA; PTEROPODIDAE; **Genus MICROPTEROPUS**
Matschie, 1899

Dwarf Epauleted Bats

There are two species (Ansell 1974; Bergmans 1979*a*; Hayman and Hill, *in* Meester and Setzer 1977; Kock 1987):

M. pusillus, Senegal to Ethiopia, south to Angola and Zambia;
M. intermedius, northeastern Angola, southern Zaire.

Another species, *M. grandis*, recently was transferred to the genus *Epomophorus* (Bergmans 1988).

Head and body length in *M. pusillus* is 67–105 mm, forearm length is about 46–65 mm, and coloration is brownish above and lighter below. Average weight in a series of specimens from Equatorial Guinea was about 20 grams for males and 22 grams for females (Jones 1972). A rudimentary external tail is present in *Micropteropus*. The pelage is moderately long, thick, and very soft.

In general appearance *Micropteropus* resembles *Epomophorus*, but the lips are not as expansible, the muzzle is shorter, and the ears are relatively shorter. Small whitish tufts of hair are present at the base of the ears, at least in *M. pusillus*, and males have shoulder pouches and epauletlike hair tufts. A collector's note on a specimen of *M. pusillus* reads: "When erected the tuft had a vibratory movement." The females of this genus lack the shoulder tufts, but female *M. pusillus* often have distinct though shallow pouches, somewhat like those of subadult males.

Most records of *M. pusillus* are from open woodlands, but the species has also been found in high forest. Specimens have been taken from between the leaves of dense bushes, usually close to the ground, and from the lower fronds of palm trees growing among vines and small trees in the bed of a stream bordering on an arid, grassy plain. This species is often exposed to direct sunlight in its roosts and is not wary, seldom flying from its shelter. It may fly if unduly disturbed, but even then it usually settles down within 10 meters of its

Dwarf epauleted bat *(Micropteropus pusillus)*, photos by Merlin D. Tuttle.

original resting place. *M. pusillus* is mainly a wanderer in search of ripe fruit. In captivity it feeds slowly and deliberately, the food being consumed by slow sucking action of the lips and mouth. The uneaten fruit pulp is dropped after fluids are removed (Jones 1972). This species also has been reported to visit the sausage tree *(Kigelia africana)* to lap nectar (Rosevear 1965).

Studies in the Ivory Coast by Thomas and Marshall (1984) indicate that *M. pusillus* has birth peaks from about March to May and from September to November. Both periods seem to be timed so that lactation coincides with the area's two rainy seasons, when fruit availability is at a maximum. Each parturition period is followed by a postpartum estrus and apparently by immediate embryonic development. The gestation period is 5–6 months, and lactation lasts 7–13 weeks. Young females can mate at 6 months and give birth at 12 months. Males are sexually mature at 11 months. Available data from Congo also suggest two reproductive cycles per year in *M. pusillus,* with births occurring there mainly in February and September (Bergmans 1979*a*). Most births of *M. pusillus* in the Garamba National Park, Zaire, apparently take place about the end of February and in November or December. Hill and Morris (1971) collected a pregnant female in Ethiopia on 30 August.

CHIROPTERA; PTEROPODIDAE; **Genus NANONYCTERIS**
Matschie, 1899

Little Flying Cow

Little flying cow *(Nanonycteris veldkampi)*, photo by Merlin D. Tuttle.

The single species, *N. veldkampi,* occurs from Guinea to the Central African Republic; earlier reports of its presence in Zaire apparently were in error (Bergmans 1982; Hayman and Hill, *in* Meester and Setzer 1977).

Head and body length is about 54–75 mm, a rudimentary tail may be present, and forearm length is 43–54 mm. A series of specimens from Nigeria weighed 19–33 grams (Happold and Happold 1978*a*). The back may be light russet and the underparts near cream buff, but much color variation exists. Both sexes have a small white patch of hair at the base of the ears. Adult males have shoulder pouches with epauletlike hair tufts. Cranial and dental features distinguish this genus from related genera. The head of *Nanonycteris* is said to have a calflike appearance.

This bat seems to favor closed forest or fringed forest within the Guinea type of open woodland (Rosevear 1965). In Ghana it was observed feeding on the nectar of the flowers of *Parkia clappertoniana. Nanonycteris* arrived at the *Parkia* trees at dusk, usually just before the departure of *Epomophorus gambianus,* which also was feeding on the flowers. *Nanonycteris* on several occasions harassed individuals of *Epomophorus* that were hanging onto the flowers; this usually resulted in *Epomophorus* releasing its position and flying off. Whereas *Epomophorus* tended to feed only in the upper parts of the trees, *Nanonycteris* tended to feed on flowers all over the tree and continued to feed in the trees in varying numbers throughout most of the night.

Nanonycteris remained on flowers from 1 to 30 seconds, considerably less than the time spent by *Epomophorus.* Baker and Harris (1957), who observed and photographed this bat, wrote:

> [*Nanonycteris*] clasps the inflorescence with its wings, holding on with its thumbs, and applies its face to the depressed ring containing the nectar. The nectar is than lapped with the narrow, rough tongue. Departure from an inflorescence is rapid and details cannot be observed easily, but from the photographs it would seem that the bat first throws itself backwards. It may then drop below the inflorescence, completing a somersault before opening its wings and flying away . . . or it may do a half-roll. . . . No evidence was obtained of any flower-eating by the bats. . . . There was . . . no sign of pollen in the stomach, . . . only nectar.

Evidently in response to food availability, *Nanonycteris* migrates northward from the forests of the Ivory Coast and onto the savannahs during the rainy season. The annual movement is nearly 750 km (Thomas 1983). In northern Ghana, Marshall and McWilliam (1982) found it only during the wet season. There it roosts alone or in small groups of well-spaced individuals. Apparently, females are polyestrous and have an extended parturition period during the early rains, about May and June. Studies by Wolton et al. (1982) in Liberia also show that *Nanonycteris* makes an annual migration; in that area there apparently is a long breeding season, from October until at least mid-March, with no marked peaks, but with the first births not occurring before mid-November.

CHIROPTERA; PTEROPODIDAE; **Genus SCOTONYCTERIS**
Matschie, 1894

There are two species (Bergmans 1973; Bergmans, Bellier, and Vissault 1974; Hayman and Hill, *in* Meester and Setzer 1977):

Scotonycteris zenkeri, photo from *Proc. Zool. Soc. London.* Inset: palate, photo from *Zoologische Beitrage,* Bonner.

S. zenkeri, Liberia to eastern Zaire, island of Bioko (Fernando Poo);
S. ophiodon, Liberia to southern Congo.

In *S. zenkeri* head and body length is 65–80 mm and forearm length is 47–55 mm. In *S. ophiodon* head and body length is 104–43 mm, and forearm length about 72–88 mm. The tail is rudimentary in this genus. A series of specimens of *S. zenkeri* from Nigeria weighed 20–27 grams (Happold and Happold 1978*a*). Liberian series weighed 18–26 grams for *S. zenkeri* and 65–72 grams for *S. ophiodon.* Coloration is about the same in both species, being russet, rust brown, or dark brown above and paler below. The small white patch at the base of the ear present in other bats that resemble *Epomophorus* is inconspicuous or absent. The species *S. zenkeri* has a white patch in front of and between the eyes and a white patch behind the eyes; *S. ophiodon* lacks the white patch behind the eyes. One specimen of *S. ophiodon* had yellow spots behind the eyes.

As determined by features of the skull and teeth, this genus closely resembles *Epomophorus* and related genera. It is very much like *Casinycteris;* in fact, it is scarcely distinguishable from that genus externally. *Scotonycteris* is best distinguished from *Casinycteris* by its normal palate, the palate in the latter genus being extremely shortened.

Both species seem to be solitary tree dwellers, but little was known about them until studies by Wolton et al. (1982) at Mount Nimba, Liberia. *S. zenkeri* was taken only at elevations below 800 meters and seemed to prefer primary closed forest to secondary growth or cultivated land. Its morphology indicated that it feeds on small, thin-skinned fruits. Collection of pregnant females in August and November and (citing Bergmans, Bellier, and Vissault 1974) of a lactating female in December suggested an extended breeding season.

Wolton et al. collected 10 specimens of *S. ophiodon* and considered the species very rare. All but one of these bats were found at elevations from 1,000 to 1,200 meters. The species seemed confined to the immediate vicinity of Mount Nimba, and there was concern that it could be adversely affected by habitat disruption. In contrast to *S. zenkeri,* which took a number of long rests during the night, *S. ophiodon* fed almost continuously. A captive ate bananas as its first choice, and then guavas and plantains. Females appeared more active and vocal than did males and emitted a high-pitched whistle through the night. Available data indicated to Wolton et al. that most females are pregnant in August and September and that there may be a second breeding at the end of the year. Indeed, there is an earlier report that a female *S. ophiodon,* collected in December in Ghana, had a youngster clinging to it.

Scotonycteris sp., photo by Norman J. Scott, Jr.

CHIROPTERA; PTEROPODIDAE; **Genus CASINYCTERIS**
Thomas, 1910

The single species, *C. argynnis,* is now known from 11 localities in southern Cameroon and Zaire (Meirte 1983).

Head and body length is about 90–95 mm, the tail is vestigial, and forearm length is about 50–63 mm. The fur is light brown in color, but the muzzle, eyelids, ears, and wings are said to be yellowish or bright orange when the bat is alive. Tufts of white hairs are present at the base of the ears, and an oblong white patch extends back between the eyes, with another present behind the eyes.

Casinycteris is remarkably similar to *Scotonycteris* in external appearance. Internally, *Casinycteris* differs from *Scotonycteris* in its extremely shortened palate and dental features. Thomas (1910*b*) wrote: "This striking bat . . . is remarkable for possessing a palate quite unlike that of other fruit-eating bats, and more recalling that found in some of the Microchiroptera. The astonishing resemblance of the type species to *Scotonycteris* . . . is also noticeable. Probably both bats bear a protective resemblance to the leaves, fresh or dry, of some local tree." Andersen (1912) also discussed the relationship between these two genera: "It might seem a little strange that these two genera, though so closely related as to be evidently modifications of one type of bat, are nevertheless inhabitants of the same faunistic area. It appears reasonable to suppose, however, that the profound difference in the posterior portion of the bony roof of the mouth must be connected either with . . . a difference in diet or a difference in the manner of feeding on the same food, so that . . . in any case there would . . . be but little competition between them."

Casinycteris argynnis, photo by J. L. Perret through V. Aellen.

Short-nosed fruit bat *(Cynopterus horsfieldi)*, photo by Lim Boo Liat.

Casinycteris evidently is restricted to the central forest block of Africa and may be locally abundant (Meirte 1983). One specimen was collected while hanging alone about three meters from the ground in the dense foliage of a bush.

CHIROPTERA; PTEROPODIDAE; **Genus CYNOPTERUS**
F. Cuvier, 1825

Short-nosed Fruit Bats, or Dog-faced Fruit Bats

There are six species (Ellerman and Morrison-Scott 1966; Hill 1983*b*; Kitchener and Foley 1985; Lekagul and McNeely 1977; Medway 1978; Paradiso 1971; Roberts 1977; Taylor 1934):

C. brachyotis, Sri Lanka, Andaman and Nicobar islands, southern Burma, Thailand, southern China, Indochina, Malay Peninsula, Sumatra and nearby islands, Java, Kangean Islands, Borneo and many associated islands, Bali, Sulawesi, Philippines;
C. minor, southeastern Sulawesi;
C. archipelagus, Polillo Island (Philippines);
C. horsfieldi, southern Thailand, Malay Peninsula, Sumatra, Java, Borneo;
C. sphinx, probably southeastern Pakistan, India and Sri Lanka to Indochina, southern China, Hainan, Malay Peninsula, Sumatra and nearby islands, Java, possibly Borneo, Bali, Serasan Island;
C. titthaecheilus, Sumatra, Krakatoa, Sebesi, Nias, Java, Bali, Lombok, Timor.

Head and body length is 70–127 mm, tail length is 6–15 mm, forearm length is 55–92 mm, wingspan is 305–457 mm, and adult weight is about 30–100 grams. The fur is dense and variable in color, but usually some shade of olive brown. These bats have prominent, almost tubular nostrils, and the upper lip is divided by a deep vertical groove. Their vocalizations in captivity have been described as loud, squeaking cries of two distinct notes.

These bats are found in forests and open country, from sea level to 1,850 meters. In Sri Lanka they prefer to roost in the folded leaves of the talipot palm. Groups of 6–12 may roost in one palm or in a group of palms. The older males seem to be solitary. This is the only Old World bat known to make shelters: *Cynopterus* occasionally bites off the "center seed string" in the fruit clusters of the kitul palm, thus leaving a hollow in which to hang. Caves, deserted mines, and under the eaves of houses are common roosting sites; hollow trees are used only rarely.

In flight, short-nosed fruit bats are usually seen among bushes and low trees. They may travel 97–113 km in one night to feed on such fruits as palms, figs, guavas, plantains, mangoes, and chinaberries, as well as on such flowers as those of the sausage tree *(Oroxylum indicum)*. Local abundance is often associated with the fruiting of such trees, so that they occasionally damage fruit plantations. These bats seem to subsist mainly on the juice of fruits, rather than the pulp. By carrying off the fruit of the date palm *(Phoenix sylvestris)* and eating it elsewhere and disseminating the seeds through its droppings, *C. sphinx* becomes an agent of seed dispersal for this plant.

Medway (1978) stated that pregnant females of *C. brachyotis* and *C. horsfieldi* had been collected throughout the year in the Malay Peninsula and that breeding was apparently nonseasonal. A single young was produced and was carried by the female during the early part of its life. According to Lekagul and McNeely (1977), breeding in Thailand was also nonseasonal, but most pregnancies in *C. brachyotis* occurred from March to June, with peaks also in January and September. Most pregnancies in *C. sphinx* reportedly were found in February and June, and gestation in that species was 115–25 days. Observations by Sreenivasan, Bhat, and Geevarghese (1974) suggested a well-defined breeding season for *C. sphinx* in Mysore, India. Two periods of pregnancy occurred from December or January to August, with a three-month gestation period. None of the pregnant females collected in this study had more than one embryo. Sandhu (1984) found births in this species to take place in February–March and again in June–July, with the second gestation period overlapping the first lactation period. Gestation lasted about 120 days. The young weighed about 11 grams at birth, were weaned 40–45 days later, and were carried by the mother for 45–50 days. Sexual maturity was attained by females at 5–6 months and by males at 15–20 months.

In certain areas, such as northern Thailand, dog-faced fruit bats are caught and sold in the markets for medicinal use. Some Chinese are said to eat these bats, because they believe them to be "strength-giving."

CHIROPTERA; PTEROPODIDAE; **Genus MEGAEROPS** *Peters, 1865*

There are four species (Ellerman and Morrison-Scott 1966; Hill 1983*b*; Hill and Boeadi 1978; Lekagul and McNeely 1977; Taylor 1934; Yenbutra and Felten 1983):

M. niphanae, Darjeeling (northeastern India), Thailand, southern Indochina;
M. ecaudatus, southern Thailand, Malay Peninsula, Sumatra, Borneo;
M. kusnotoi, Java;
M. wetmorei, southern Philippines.

Yenbutra and Felten (1983) described *M. niphanae* on the basis of specimens from Thailand. Meanwhile, Hill (1983*b*) reported, with question, the presence of the previously described *M. ecaudatus* in northeastern India. Both authorities suggested that the specimens from each of the involved areas, as well as from Indochina, showed morphological affinity. Corbet and Hill (1986) subsequently listed the range of the two species as shown above, though with question.

Head and body length is 76–102 mm and forearm length is about 49–59 mm. *M. wetmorei* has a rudimentary tail, whereas the other three species lack a tail. Lekagul and McNeely (1977) reported weights of 20–38 grams for *M. ecaudatus*. Coloration in *M. ecaudatus* is yellowish brown above; the nape of the neck is pale gray, contrasting with the brown of the back; and the breast and belly are pale silvery gray. *M. wetmorei* has a silvery to ashen gray head, the ashen gray back is slightly lighter than the head, and the individual hairs have pale brown tips. The coloration beneath is slightly lighter than the back. In *M. kosnotoi* the dorsal pelage is grayish brown and the ventral surface is a slightly paler gray.

The members of this genus are similar to *Cynopterus*, but the braincase is deeper. There is only one pair of incisors in the lower jaw, whereas *Cynopterus* has two pairs; the second upper incisor is reduced in length. The base of the thumb is partially enveloped in the wing membrane, so that it folds inward when the wing is folded.

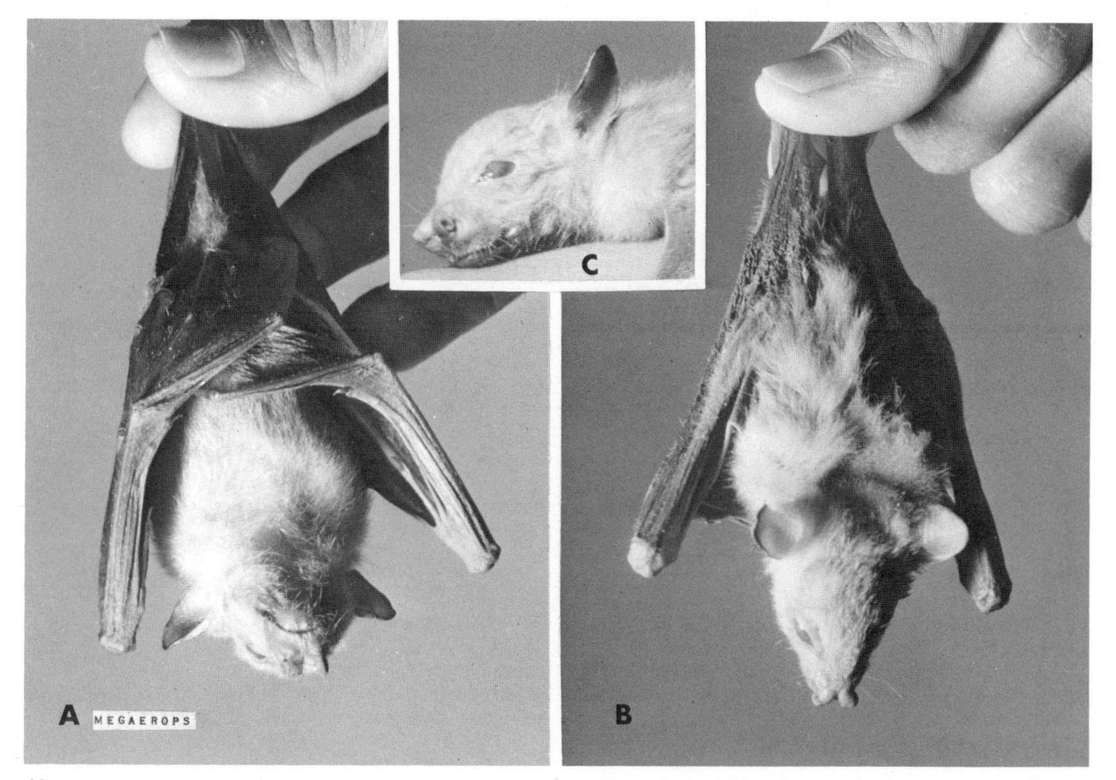

Megaerops ecaudatus: A. Ventral view; B. Dorsal view; C. Head showing tube nose. Photos by Lim Boo Liat.

In Thailand *M. ecaudatus* has been found both in mountains and on plains, in evergreen and dry evergreen forest, ranging from 500 to 3,000 meters (Lekagul and McNeely 1977). In the Malay Peninsula this species occurs in forest and open country and in lowlands and mountainous areas. Observations in the Malay Peninsula suggest an extended breeding season from about February to June (Medway 1978).

CHIROPTERA; PTEROPODIDAE; **Genus PTENOCHIRUS**
Peters, 1861

There are two species (Heaney and Rabor 1982; Taylor 1934; Yoshiyuki 1979):

P. minor, Mindanao, Dinagat, and Palawan islands (Philippines);
P. jagori, Philippines.

In *P. minor* head and body length is 86–101 mm, tail length is 9–15 mm, and forearm length is 68–77 mm (Yoshiyuki 1979). in *P. jagori* head and body length is about 105 mm, tail length is about 15 mm, and forearm length is 80.0–86.5 mm (Taylor 1934). Coloration is dark brown above and paler below. These bats resemble *Cynopterus* but differ from that genus in having only one pair of lower incisors instead of two and much reduced second upper incisors.

Several specimens have been collected in small caves. Another was found in the same coconut grove as a specimen of *Cynopterus brachyotis*, hanging in much the same manner from the underside of a large frond. Two *Ptenochirus* collected on Negros Island were feeding on kapok fruit *(Ceiba pentandra)*. Taylor (1934) thought *P. jagori* to be a solitary species.

Mudar and Allen (1986) found *P. jagori* to be the most abundant bat in their study area in northeastern Luzon. Although collected in a variety of habitats, it was most common over rivers, streams, and ponds. Pregnant and lactating females w .re taken in May, August, and October.

CHIROPTERA; PTEROPODIDAE; **Genus DYACOPTERUS**
Andersen, 1912

The single species, *D. spadiceus*, is known from a few specimens collected in the Malay Peninsula, Sumatra, Borneo, and Luzon (Hill 1961*a*; Kock 1969*a*; Peterson 1969; Start 1972*a*, 1975). Corbet and Hill (1986) listed *D. brooksi*, of Sumatra, as a separate species.

Head and body length is 107–52 mm, tail length is about 13–18 mm, forearm length is 76–89 mm, and reported weight is 70–100 grams. The coloration is blackish on the face, yellowish on the shoulders, brown on the back and sides, and dull whitish on the chest and belly. There is a small opening just behind the orbit of the eye. In this respect, *Dyacopterus* resembles the genera *Cynopterus, Ptenochirus*, and *Megaerops*.

Dyacopterus has been found roosting in a tree trunk in a forest and has been netted in a paddy field (Medway 1978). It also has been caught near caves (Payne, Francis, and Phillipps 1985).

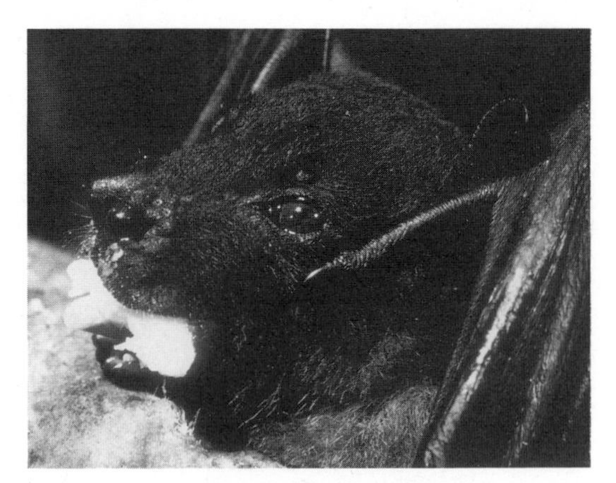

Ptenochirus jagori, photo by Paul D. Heideman.

CHIROPTERA; PTEROPODIDAE; **Genus CHIRONAX**
Andersen, 1912

Black-capped Fruit Bat

The single species, *C. melanocephalus*, occurs in the Malay Peninsula, Sumatra, Java, Nias Island, Borneo, and Sulawesi (Chasen 1940; Hill 1974*b*; Hill and Francis 1984; Lekagul and McNeely 1977).

Head and body length is about 65–70 mm, there is no external tail, forearm length is 40–50 mm, and weight is 12–17 grams (Lekagul and McNeely 1977; Payne, Francis, and Phillipps 1985). The head is black, the back is gray brown to reddish brown, the undersides are pale buff, and the wings are not spotted.

Chironax closely resembles *Balionycteris* but differs in dental characters, the nonspotting of the wings, and smaller size. *Chironax* differs from *Cynopterus* in several features, the most conspicuous being the absence of a tail and the absence of a small opening just behind the orbit. In the absence of a postorbital foramen *Chironax* resembles the genera *Balionycteris, Thoopterus, Penthetor*, and *Sphaerias*.

Chironax is usually found at elevations above 600 meters. Groups of about two to eight individuals have been found during the daytime, resting on the undersides of tree ferns several meters above the ground, in virgin forests in Java. These individuals seemed to have been feeding mainly on wild figs *(Ficus)*. In the Malay Peninsula pregnant females have been recorded only from January to March, suggesting that breeding there is restricted to the early months of the year (Medway 1978). Pregnant females, each with a single embryo, were collected in Sulawesi in March and April (Bergmans and Rozendaal 1988).

CHIROPTERA; PTEROPODIDAE; **Genus THOOPTERUS**
Matschie, 1899

Short-nosed Fruit Bat

The single species, *T. nigrescens*, now is known by the type specimen from Morotai Island in the North Moluccas and by many more recently collected specimens from Sulawesi (Bergmans and Rozendaal 1988; Hill 1983*b*). It has been reported from Luzon in the Philippines, but its occurrence there is doubtful.

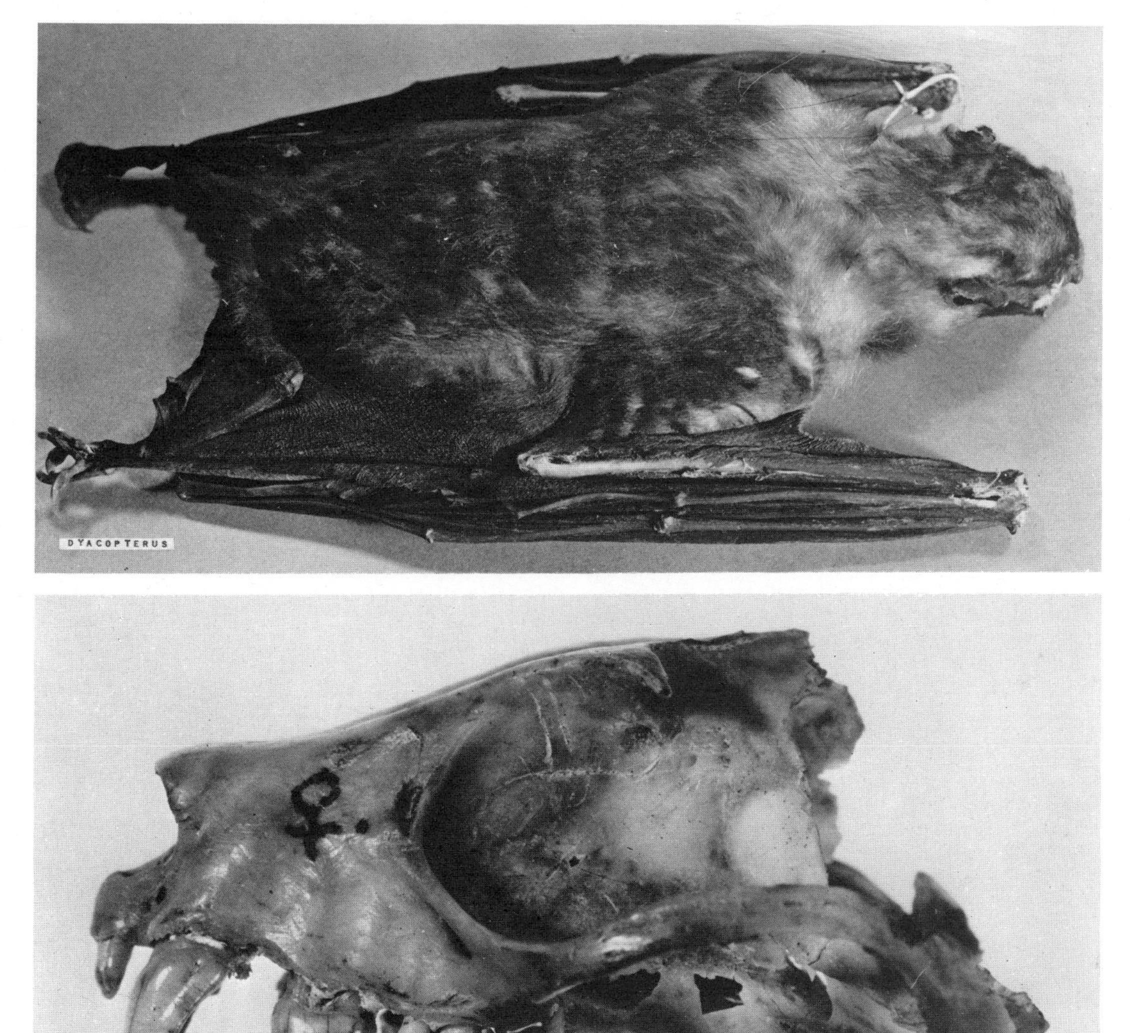

Dyacopterus spadiceus, photos from British Museum (Natural History).

Head and body length has not been recorded, and the tail is a mere spicule. Forearm length is about 70–82 mm and weight is 67–99 grams (Bergmans and Rozendaal 1988). The coloration of the head and body is grayish brown. *Thoopterus* lacks postorbital foramina. It possesses grooved upper canines without ridges at their bases; the fourth premolar and the first molar are greatly increased in breadth. The membranes attached to the hind feet are inserted on the second toe.

The species *T. nigrescens* was formerly placed in *Cynopterus* but was removed from that genus because of structural peculiarities. Hill (1983*b*) suggested affinity between *Thoopterus*, *Penthetor*, and *Latidens*. *Thoopterus* is almost identical to *Latidens* in external form and color and has a similarly long, strong rostrum. Dentally, however, *Thoopterus* approaches *Penthetor*. The three genera are easily distinguished by number of incisor teeth, *Thoopterus* having 8, *Penthetor* 6, and *Latidens* 4.

According to Bergmans and Rozendaal (1988), *Thoopterus* has been collected in forests at elevations of 50–1,780 meters and evidently uses communal roosts. Pregnant females, each with a single embryo, were taken in January. Subadults were taken in January, March, October, and December.

CHIROPTERA; PTEROPODIDAE; **Genus SPHAERIAS**
Miller, 1906

Mountain Fruit Bat

The single species, *S. blanfordi*, is now known from Tibet, southwestern China, northern India, Nepal, eastern Burma, and northern Thailand (Lekagul and McNeely 1977; Wang Sung, *in* Honacki, Kinman, and Koeppl 1982).

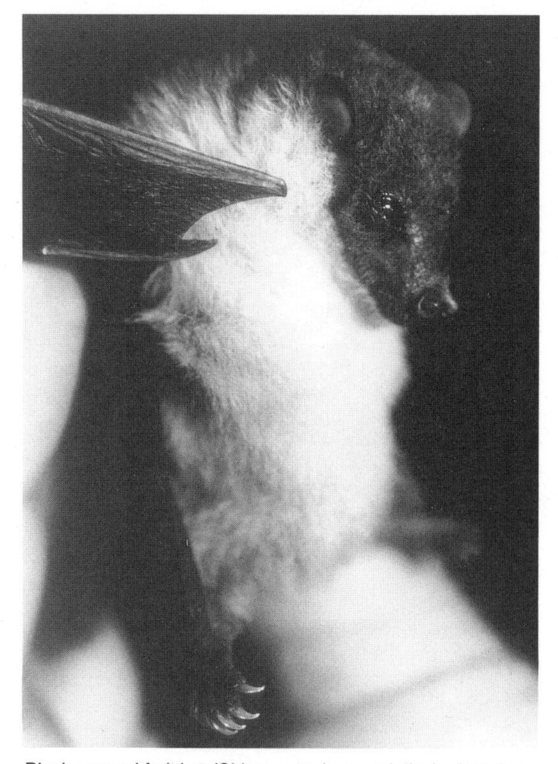

Black-capped fruit bat *(Chironax melanocephalus)*, photo by F. G. Rozendaal.

Short-nosed fruit bat *(Thoopterus nigrescens)*, photo by F. G. Rozendaal.

Mountain fruit bat *(Sphaerias blanfordi)*, photo by B. Elizabeth Horner and Mary Taylor of specimen in Harvard Museum of Comparative Zoology.

Head and body length is about 64–80 mm, there is no external tail, and forearm length is about 50–52 mm. The color is dull grayish brown above and paler below. The absence of a tail is a character exhibited by several other supposedly related genera, but *Sphaerias* is unique in having the interfemoral membrane reduced to a narrow rim along the femur and the upper part of the tibia.

This bat seems to be confined to montane forest from 800 to 2,700 meters. Specimens from Thailand were collected in a pine and oak forest (Lekagul and McNeely 1977).

CHIROPTERA; PTEROPODIDAE; **Genus BALIONYCTERIS**
Matschie, 1899

Spotted-winged Fruit Bat

The single species, *B. maculata*, occurs in the Malay Peninsula, the Rhio Archipelago near Singapore, and Borneo (Lekagul and McNeely 1977).

Head and body length is about 50–66 mm, there is no external tail, and forearm length is 39–43 mm. Reported weight is 9.5–14.5 grams (Lim, Shin, and Muul 1972; Medway 1978). The color is sooty brown above and grayish below; the head is blackish, and the dark brown wings are marked with yellow spots. The only other fruit bats with well-defined yellow spots on the wings are the tube-nosed bats, *Nyctimene* and *Paranyctimene*. *Balionycteris* closely resembles the black-capped fruit bat, *Chironax*, in structural features.

The color pattern is suggestive of foliage and tree-dwelling habits. In the Malay Peninsula this bat is locally common in the forests of the lowlands and foothills. It is not normally a cave dweller but has been recorded from a cave in Sabah. In Selangor small groups of adults and young have been found roosting in the crowns of palms and in clumps of ferns epiphytic on forest trees (Medway 1978).

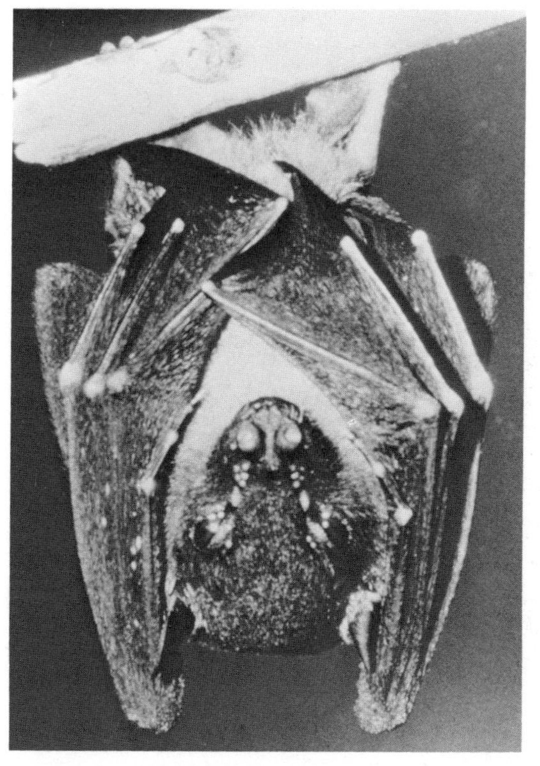

Spotted-winged fruit bat *(Balionycteris maculata)*, photo by Yong Hoi-Sen.

CHIROPTERA; PTEROPODIDAE; **Genus AETHALOPS**
Thomas, 1923

Pygmy Fruit Bat

The single species, *A. alecto*, occurs in the Malay Peninsula, Sumatra, western Java, and Borneo (Hill 1983*b;* Medway 1977, 1978).

This is about the smallest of the Old World fruit bats. Head and body length is 65–73 mm, there is no external tail, and forearm length is 42–52 mm. One specimen weighed 19.3 grams (Medway 1978). The coloration is often black or one of various shades of dark gray.

This genus is characterized externally by the furred, narrow interfemoral membrane, the absence of a tail, the minute calcar bone on the foot, and the small ears. Boeadi and Hill (1986) stated that the inner upper incisors of *Aethalops* are less robust than the outer pair, whereas the reverse is true in the related genera *Balionycteris* and *Chironax*. *Aethalops* also differs from *Chironax* in having two and not four lower incisors, and from *Balionycteris* in lacking a small last upper molar.

In the Malay Peninsula this bat is uncommon, known only from forests above about 1,000 meters in elevation. Pregnant females have been recorded from February to June (Medway 1978).

CHIROPTERA; PTEROPODIDAE; **Genus PENTHETOR**
Andersen, 1912

Dusky Fruit Bat

The single species, *P. lucasi*, inhabits the Malay Peninsula, the Rhio Archipelago near Singapore, and Borneo (Medway 1978).

Head and body length is about 114 mm, tail length is 8–10 mm, forearm length is 57–67 mm, and weight is 30–55 grams. The fur is coarse and smoky brown in color. This fruit bat may be recognized by the combination of one pair of lower incisors and the extreme thinness of the tail.

Andersen (1912) stated that *Penthetor* is, without doubt, the Indo-Malayan representative of the Austro-Malayan *Thoopterus*. He thought that its principal claim to stand as a genus distinct from *Thoopterus* is the absence of the first incisor on the lower jaw and the shortening of the second incisor on the upper jaw. Also, the toothrows extend farther backward; the incisors are more sharply needle-pointed; and the premolars and molars, though similar in outline and characters, are very similar to those of *Thoopterus* but lack all trace of the surface cusps so conspicuous in that genus. The tail is not reduced to a mere spicule, the insertion of the membranes on the hind feet is different, the digits are considerably shorter, and the tibia is unusually long.

The following account is based primarily on Medway (1978). In the Malay Peninsula the dusky fruit bat is irregularly distributed in lowland and hill forests. It normally roosts in caves, rock shelters, or crannies between large boulders, a habit that limits its distribution. It emerges imme-

Pygmy fruit bat *(Aethalops alecto)*, photo by B. Elizabeth Horner and Mary Taylor of specimen in Harvard Museum of Comparative Zoology.

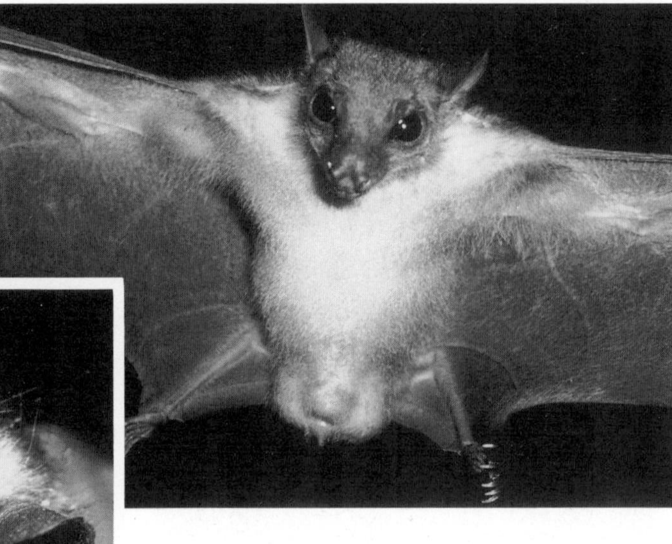

Dusky fruit bat *(Penthetor lucasii)*, photos by Klaus-Gerhard Heller.

diately after dusk to feed at the nearest fruit plantation. Seeds and husks of many kinds of fruit often litter the cave floor under roosts, indicating that fruits are carried back to the cave to be eaten. This bat appears to be gregarious. Reproduction in the Malay Peninsula apparently is seasonal, as a large proportion of females examined in September were pregnant, a few were pregnant in June, but none were pregnant in January, February, March, or July. Start (1972*b*) reported that a female taken in Sarawak in May was lactating and that one taken in January had an advanced embryo.

CHIROPTERA; PTEROPODIDAE; **Genus HAPLONYCTERIS**
Lawrence, 1939

Fischer's Pygmy Fruit Bat

The single species, *H. fischeri*, was long known only by a single specimen collected in 1937 on the slopes of Mount Halcon, Mindoro, Philippine Islands. Since the 1970s numerous additional specimens have been taken on Luzon, Mindanao, Palawan, Negros, Leyte, Biliran, Bohol, and Dinagat islands (Heaney, Heideman, and Mudar 1981; Heaney and Rabor 1982; Heideman 1988; Mudar and Allen 1986).

A series of 27 specimens from northeastern Luzon averaged 75.3 mm in total length and 52.1 mm in forearm length (Mudar and Allen 1986). A series of 15 specimens from Negros averaged 17.9 grams in weight (Heaney, Heideman, and Mudar 1981). There is no tail. The thumb is long, about 25 mm in length, and the hind foot is short, about 13 mm in length. The coloration is given by the describer as "cinnamon brown" on the back, pale "mummy brown" in the shoulder region, darker in the head region, and "wood brown" beneath, with a slight tinge of silver along the midline of the belly.

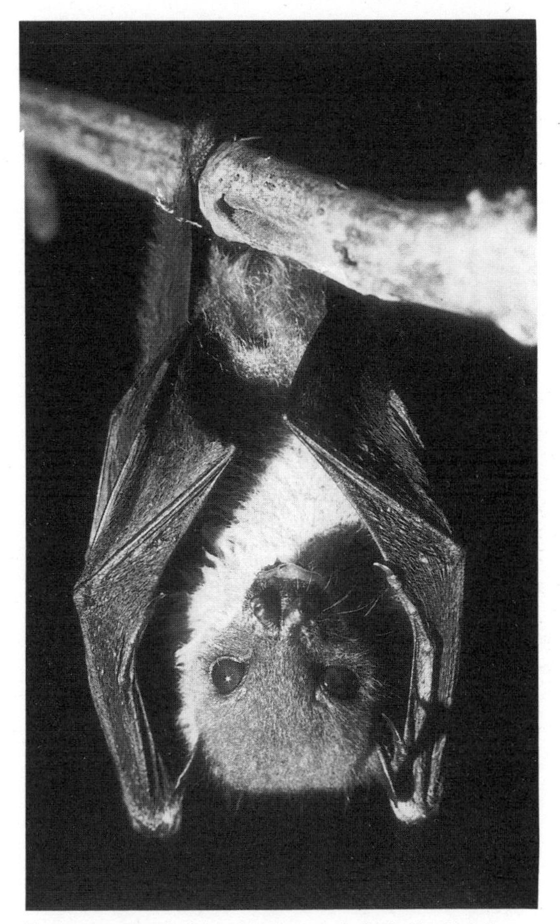

Fischer's pygmy fruit bat *(Haplonycteris fischeri)*, photo by Paul D. Heideman.

None of the other genera similar to *Cynopterus* has such a relatively long thumb. One pair of incisors, one pair of canines, and four pairs of premolars and molars also set *Haplonycteris* apart from related genera. The describer stated: "*Haplonycteris* belongs in the group of small bats that resemble *Cynopterus* in which the postorbital foramen has become obliterated. In the reduction of the tooth formula with the strengthening of the remaining teeth, it appears to be the most highly evolved of this group. The unusually developed cusps and transverse ridges also indicate a greater degree of differentiation" (Lawrence 1939).

Mudar and Allen (1986) found *Haplonycteris* to exhibit a marked preference for forest habitat and often to be the only bat caught in nets set away from water or agricultural fields. Heideman (1988) reported that this bat has an annual reproductive pattern that seems associated with seasonal patterns of abundance of the fruit upon which it feeds. Births are highly synchronous within a given population, occurring in May or June on Negros Island but up to 3 months later in other areas. Although mating may occur throughout the year, synchronization of births is achieved by a postimplantation delay in embryonic development. Females also undergo a postpartum estrus 1–3 weeks after giving birth, and most become pregnant at this time. The period of delay lasts up to 8 months and is followed (starting about March on Negros) by 3 months of rapid embryonic growth. A single young is born, and lactation lasts about 10 weeks. Most females that are born in June on Negros become pregnant themselves in October or November, at 5 or 6 months.

CHIROPTERA; PTEROPODIDAE; **Genus OTOPTEROPUS**
Kock, 1969

The single species, *O. cartilagonodus*, is known only from seven specimens taken in Mountain and Isabella provinces in northern Luzon, Philippines (Kock 1969*c*; Mudar and Allen 1986).

This is a very small, long-haired, and tailless flying fox related to *Cynopterus*. Head and body length is 55–73 mm, forearm length is about 46–48 mm, and two specimens weighed 17 and 21 grams. The coloration of the back is dark blackish brown; the belly is lighter, with more gray coloration. The hairs on the middle of the breast and belly are tipped with white. The eyes in *Otopteropus* are very large. The most noteworthy character of the genus is the ears, the edges of which are marked with reddish thickenings. On the front edge of the ear the thickening is less broad than it is thick; on the hind edge is a well-marked lobe. The naked nose has tubular nostrils. Although there is no tail, a cartilaginous calcarium is present. The uropatagium is thickly furred on the upper side and on the distal edge of the underside. The tibia are thickly furred, but the feet and toes only thinly so. The wing membranes have a reticulated pattern. The dental formula for this genus is: (i 1/1, c 1/1, pm 3/3, m 1/1) × 2 = 24.

The few known specimens have been taken in moun-

Otopteropus cartilagonodus, photo by Lawrence R. Heaney.

tainous areas, and the long, thick hair suggests adaptation for higher altitudes. Kock (1969*c*) wrote that the big eyes and small hearing apparatus of this genus, as well as the relatively narrow teeth and weak lower jaw, indicate a very species-specific kind of life for this bat. Mudar and Allen (1986) collected a pregnant female on 30 April.

CHIROPTERA; PTEROPODIDAE; **Genus ALIONYCTERIS**
Kock, 1969

The single species, *A. paucidentata*, is known only by seven specimens from Mount Katanglad, Bukidnon Province, Mindanao, Philippines (Kock 1969*b*).

Head and body length was 64 mm in each of three specimens, and forearm length was 43.7–45.6 mm in five. The fur is brownish black in color, and on the dorsal side there are numerous soft guard hairs that are about double the length of the underfur. The blackish ears are naked, their edges being slightly thickened, as is usual in related genera. The rear legs, feet, and toes are thickly haired, as is the proximal third of the underarms.

Alionycteris is a very small, long-haired flying fox related to *Cynopterus*. It has long thumbs, no external tail, no interfemoral membrane, and no postorbital foramina. The divided naked nose is composed of two 4-mm-long cylinders that stand free on the end. The dental formula, which distinguishes this genus from all other known genera, is: (i 1/1, c 1/1, pm 3/3, m 1/2) × 2 = 26.

Kock (1969*b*) wrote that the various dental and cranial characters suggest that this bat consumes very soft food. The thick hair covering of the body suggests that *Alionycteris* is adapted for life at higher elevations in the mountains.

CHIROPTERA; PTEROPODIDAE; **Genus LATIDENS**
Thonglongya, 1972

The single species, *L. salimalii*, is known only by one specimen collected from High Wavy Mountains, Madura district, South India, at an altitude of 750 meters. The specimen is in the collections of the Bombay Natural History Society, Bombay, India.

This is a medium-sized bat, similar in general appearance to *Cynopterus* but without an external tail. The length of the forearm in the single specimen is 67.5 mm. The fur of the head is blackish brown with a light grayish base, darker then that of the other parts. The body fur is dense and long, light brown in color, about 5 mm long on the midback. The fur of the underparts, including chin and throat, is shorter and thinner than that of the upper parts and is light gray brown in color. The ear membrane is thin and oval, with no white rim. The wing membrane, brownish and rather thin, starts at the first foot-toe and has no white along the fingers, as in *Cynopterus*. The index claw is present.

Latidens possesses only one pair of upper and lower incisors, a character known among the Megachiroptera in *Dobsonia, Haplonycteris,* and *Harpyionycteris*. However, *Harpyionycteris* differs from *Latidens* in possessing strongly proclivous upper incisors and upper and lower canines. *Latidens* is readily separated from the other genera mentioned above in having the cheek teeth 4–4/5–5, whereas in *Dobsonia* they are 5–5/6–6 and in *Haplonycteris* 4–4/4–4.

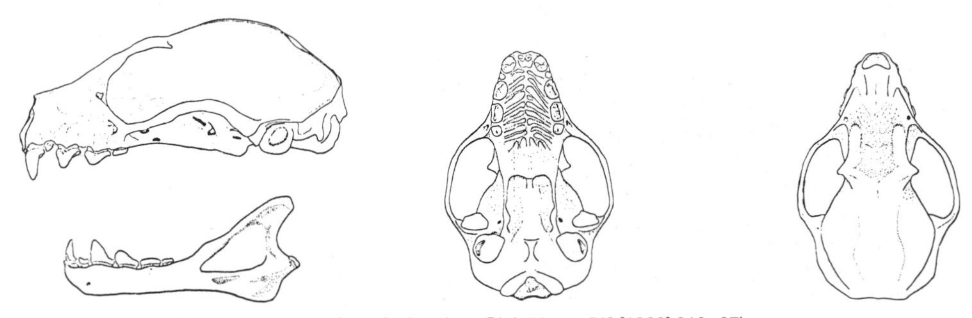

Skull of *Alionycteris paucidentata*, after Kock (*Senckenbergiana Biol.* 50, no. 5/6 [1969]:319–27).

CHIROPTERA; PTEROPODIDAE; **Genus NYCTIMENE**
Borkhausen, 1797

Tube-nosed Fruit Bats

There are 14 species (Heaney and Peterson 1984; Hill 1983*b*; Koopman 1979, 1982*a*, 1984*a*; Laurie and Hill 1954; Phillips 1968; Ride 1970; Smith and Hood 1983; Troughton 1931):

N. minutus, Sulawesi, Amboina Islands;
N. albiventer, Halmahera Islands, New Guinea and nearby islands;
N. draconilla, New Guinea;
N. robinsoni, eastern Queensland;
N. cephalotes, Sulawesi, Timor, Moluccas, western New Guinea and nearby Numfoor and Umboi islands;
N. vizcaccia, Bismarck Archipelago, Solomon Islands;
N. masalai, New Ireland Island in the Bismarck Archipelago;
N. rabori, Negros Island (Philippines);
N. malaitensis, Malaita Island (Solomons);
N. major, Bismarck Archipelago, Solomon Islands, Trobriand Islands, Louisiade and D'Entrecasteaux archipelagoes;
N. sanctacrucis, Santa Cruz Islands (South Pacific);
N. cyclotis, New Guinea;
N. aello, Halmahera Islands, New Guinea;
N. celaeno, western and northwestern New Guinea.

Head and body length is 75–136 mm, tail length is 15–28 mm, and forearm length is 50–86 mm. One adult weighed 42 grams. McKean (1972) reported that two *N. albiventer* weighed 25 and 27 grams. The coloration is usually buffy gray above, but on some species it is buffy or creamy. The underparts are usually paler. A dark brown spinal stripe is usually present, and the wings, forearms, and ear membranes are speckled with yellow. The only other fruit bats with contrasting yellow spots are *Paranyctimene* and *Balionycteris.*

Tube-nosed fruit bat *(Nyctimene rabori),* photo by Lawrence R. Heaney.

Tube-nosed bat *(Nyctimene major),* photo by L. J. Brass.

The common name refers to the tubular nostrils that project from the upper surface of the muzzle to a length of about 6 mm. This genus is distinguished from *Paranyctimene,* the only other genus of bats with which it could be confused, by dental features.

Nyctimene has been taken singly in foliage and on tree trunks, but the mottled and broken color pattern increases its chance of concealment. In its resting position, the animal hangs freely, wrapped in the wing membranes. When these bats begin calling or when there is a disturbance, the nostrils are stretched outward and moved with a slight trembling motion. Possibly the peculiar shape of the nostrils serves some purpose in giving ultrasonic sounds for orientation. While in flight, these bats emit a high, whistling note.

McKean (1972) reported the collection of *Nyctimene* in the following habitats: *N. aello,* primary and secondary rainforest, and swamp forest; *N. albiventer,* primary and secondary rainforest, primary montane forest, sago swamp, and the vicinity of native gardens; and *N. cyclotis,* primary rainforest. Altitude ranged from sea level to 1,650 meters.

Vestjens and Hall (1977) reported that the stomachs of three *N. albiventer* contained insects but that most other stomachs had fruit or vegetable matter. Earlier observations indicated that captive individuals preferred soft, juicy fruit and would not take insects that were offered. To eat, the animal hangs horizontally or obliquely in a fruit bush, with its thumbs inserted into the fruit. The lips are turned up on the fruit, and pieces are bitten out with the lower jaw, the upper teeth merely aiding the lips in supporting the lower jaw. The bitten pieces are shoved toward the breast and belly by the muzzle, then masticated, mainly for the juice. A fringe of fleshy lobes on the inner edge of the lips seems to aid the bat. Captives that fed on guavas were not observed to bite into the inside of the fruit, and the nostrils did not come in contact with the fruit at any time. The pulp of young coconuts is also eaten.

McKean and Price (1967) reported that two females of *N. robinsoni* taken in late August in Queensland contained single embryos. McKean (1972) reported the following reproductive information: pregnant females of *N. aello* collected in New Guinea in January and February; a lactating female *N. albiventer* taken on Bougainville in July; pregnant females of *N. albiventer*, each with one embryo, taken in Papua New Guinea in January, July, and August; and lactating females of *N. albiventer* taken in Papua New Guinea in February, May, and August.

The IUCN classifies *N. rabori*, which has lost most of its forest habitat to human activity, as endangered. Heaney and Heideman (1987) noted that it still survives in a few forest fragments and that there is some hope that these areas can be protected.

CHIROPTERA; PTEROPODIDAE; **Genus PARANYCTIMENE**
Tate, 1942

Lesser Tube-nosed Fruit Bat

The single species, *P. raptor*, is now known to occur in Papua New Guinea (Laurie and Hill 1954; McKean 1972) but probably is widespread throughout New Guinea (Koopman 1982*a*).

Head and body length is 77–80 mm, tail length is 13–20 mm, and forearm length is 47–55 mm. McKean (1972) reported that one individual weighed 21.5 grams. The coloration is grayish brown above and dull yellowish buff below. There is a greenish cast to the membranes and the fur around the nostrils. A dorsal spinal stripe is lacking. The spotted color pattern and the tubular nostrils are similar to those of the genus *Nyctimene*.

Tate stated that this bat can be distinguished from *Nyctimene* by the high molar and premolar cusps, the extreme height and slenderness of the upper and lower canines and premolars, the elongation of the postdental palate, and the absence of the dorsal stripe. Both upper and lower canines are daggerlike, with apparent seizing or grappling functions. As in *Nyctimene*, the lower incisors are absent.

McKean (1972) reported the collection of 27 specimens in primary and secondary forest, at altitudes ranging from sea level to 260 meters. Of these specimens, 7 were pregnant females collected in January, February, and May, each with one embryo. The type specimen of *P. raptor*, an adult female taken in August, was carrying a nursing young.

CHIROPTERA; PTEROPODIDAE; **Genus EONYCTERIS**
Dobson, 1873

Dawn Bats

There are four species (Bhat, Sreenivasan, and Jacob 1980; Ellerman and Morrison-Scott 1966; Hill 1983*b*; Laurie and Hill 1954; Lekagul and McNeely 1977; Medway 1977; Rozendaal 1984; Taylor 1934):

E. spelaea, northern India, Andaman Islands, Burma, Thailand, Indochina, Malaysia, Indonesia, Philippines;

Lesser tube-nosed fruit bat *(Paranyctimene raptor)*, photo by Tim Flannery.

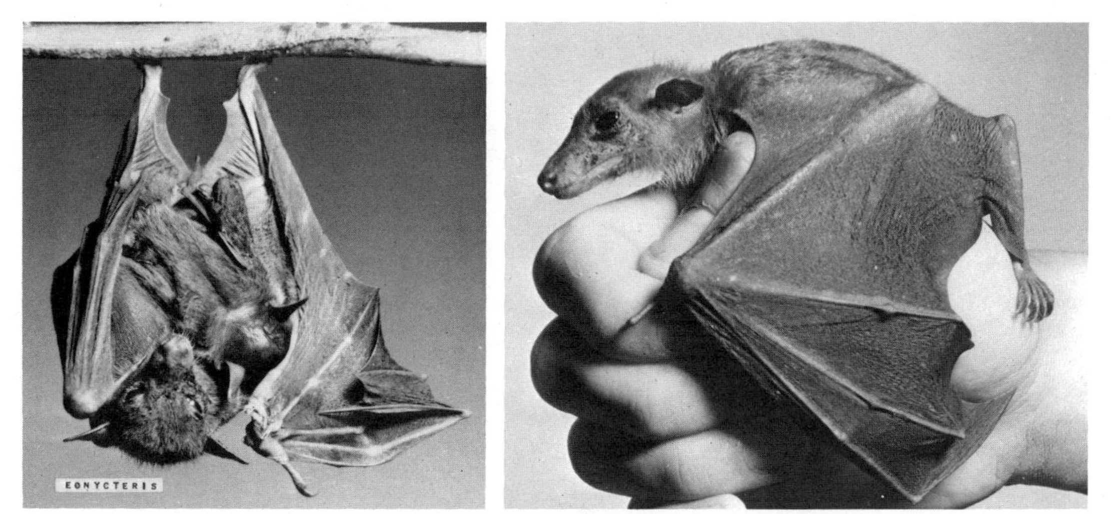

Dawn bats *(Eonycteris spelaea)*, young bat clinging to its mother, photos by Lim Boo Liat.

E. robusta, Luzon;
E. major, throughout Borneo;
E. rosenbergi, known only by three specimens from northern Sulawesi.

Head and body length is about 85–125 mm, tail length is about 12–33 mm, and forearm length is 60–81 mm. Weight of *E. spelaea* is 55–82 grams in males and 35–78 grams in females (Beck and Lim 1973; Bhat, Sreenivasan, and Jacob 1980). The upper parts are brownish, usually dark brown, and the underparts are paler. The sides of the neck in females tend to be thinly haired, whereas the same area in males is covered with a "ruff" of long hairs slightly darker in color than the hairs of the head and chest.

In *Eonycteris* and five other genera of fruit bats in the subfamily Macroglossinae, the tongue is long, slender, and protrusible, with brushlike projections for picking up nectar and pollen, and the snout is long and slender. The cheek teeth are small, barely showing above the gums. The macroglossine type of tongue, in combination with the length of the tail, is characteristic of *Eonycteris.* In all other genera of macroglossine bats, except *Notopteris,* the tail is reduced to a spicule or is absent; in *Notopteris* it is as long as the forearm. The absence of a claw on the index finger is also an aid in distinguishing *Eonycteris. E. robusta* has a kidney-shaped gland on either side of the anus in both sexes.

Dawn bats occur in a variety of habitats, including forests and cultivated areas. They usually roost in caves, but *E. major* also has been taken in hollow trees in Borneo (Medway 1977). The natural food seems to consist mainly of pollen and nectar (Medway 1978).

E. spelaea has been noted darting in and out among agave flowers, which are remarkable for their long and projecting stamens. The bats alighted occasionally for a few minutes. Their preference for this particular type of agave was so marked "that neither shooting nor artificial light would frighten them away." The stomach of one of these bats contained only pollen. D. P. Erdbrink stated that in his experience the feeding habits of *Eonycteris* are so similar to those of *Nanonycteris* that the description of the latter given by Baker and Harris (1957) and quoted in the account of *Nanonycteris* could apply to *Eonycteris* verbatim.

These bats are gregarious. Several hundred individuals of *E. robusta* were found clinging to the ceiling of a cave in Luzon. Bhat, Sreenivasan, and Jacob (1980) found that roosting colonies were divided into clusters that were segregated by sex. Beck and Lim (1973) reported that colonies of *E. spelaea* in Malaysia range from a few dozen individuals in shallow limestone shelters to tens of thousands in the larger massifs. In a study at one of the major roosting caves in the Malay Peninsula, these authorities found that at any one time, more than 50 percent of the adult females were either pregnant or lactating or both. The females were polyestrous, and successive pregnancies could begin in the late stages of lactation. The gestation period seemed to be slightly longer than 6 months, possibly as long as 200 days. The usual number of young was one, but two fetuses were found in rare instances. Shortly after parturition, the young took hold of a nipple and remained firmly attached for 4–6 weeks as the female flew about. After this period the young could detach readily, and they made short flights on their own, but weaning did not occur until at least 3 months. Sexual maturity was attained sometime after the first year of life, in males possibly not until after the second year.

CHIROPTERA; PTEROPODIDAE; **Genus MEGALOGLOSSUS**
Pagenstecher, 1885

African Long-tongued Fruit Bat

The single species, *M. woermanni,* occurs from Guinea to Uganda and southern Zaire (Bergmans and Van Bree 1972; Hayman and Hill, *in* Meester and Setzer 1977).

Head and body length is 60–82 mm, the tail is a mere spicule or is absent, and forearm length is 37–50 mm. Jones (1971*a*) reported a weight of 8.4–15.6 grams in nine females and 13.2 grams in one male. Coloration is dark brown above and lighter brown below, with an indistinct dark longitudinal stripe from the middle of the crown to the nape. Adult males have creamy or buff-colored hairs on the foreneck and sides of the neck.

This is the only genus of the subfamily Macroglossinae with the tail reduced or absent and with the fifth metacarpal bone of the hand shorter than the third. Another set of diagnostic characters is the long, slender, and protrusible tongue with brushlike projections, the clawed index finger, and the fifth hand bone shorter than the third.

African long-tongued fruit bat *(Megaloglossus woermanni)*, photos by Merlin D. Tuttle, Bat Conservation International.

Wolton et al. (1982) collected this bat at all altitudes in both primary and secondary vegetation. It has been found roosting in plantain leaves, in the foliage of shrubs, and in a native hut. It feeds on pollen and nectar (Rosevear 1965). Bergmans and Van Bree (1972) reported the apparently unusual event of two individuals of this genus being collected in a cave. Kingdon (1974*a*) stated that a lactating female was taken in mid-December and that two lactating females and a flying juvenile were taken in Uganda in late February. Jones (1971*a*) reported the collection of a lactating female in April and a pregnant female in September in Equatorial Guinea, and two pregnant females in August in Ghana. In Congo, pregnant females were collected in March, and nursing young in February and May (Bergmans 1979*a*). In Liberia pregnant or lactating females were taken in July and August (Wolton et al. 1982). Czekala and Benirschke (1974) reported a pregnant female with two fetuses but indicated that normally a single young was produced.

CHIROPTERA; PTEROPODIDAE; **Genus MACROGLOSSUS**
F. Cuvier, 1824

Long-tongued Fruit Bats

There are three species (Chasen 1940; R. E. Goodwin 1979; Hill 1983*b;* Koopman 1979, 1984*a;* Laurie and Hill 1954; Lekagul and McNeely 1977; Ride 1970; Taylor 1934):

M. sobrinus, eastern India to Indochina, Malay Peninsula, Sumatra, Nias Island, Sipora, Krakatoa, Java, Bali;
M. minimus, southern Thailand and Viet Nam, Malay Peninsula, Java, Kangean Islands, Borneo, Natuna Islands, southern Philippines, Sulawesi and nearby islands, Molucca Islands, New Guinea and nearby islands, Bismarck Archipelago, Solomon Islands, Timor, northern Australia;
M. fructivorus, southern Mindanao.

Koopman (1982*a*) recognized *M. lagochilus,* found from Sulawesi to the Solomons and northern Australia, as a species

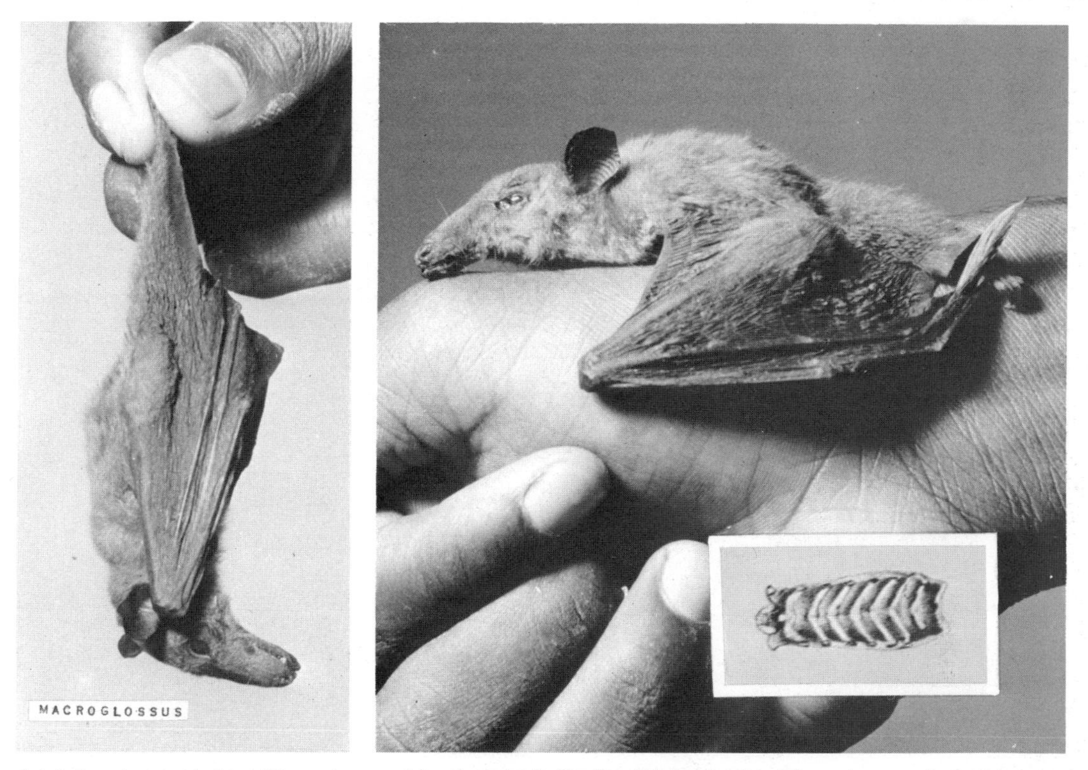

Asiatic long-tongued fruit bat *(Macroglossus minimus)*, photos by Lim Boo Liat. Inset: palate (*Macroglossus* sp.), photo from *Jahrb. Hamburgischen, Wissenschaft. Anstalten.*

distinct from *M. minimus.* Kitchener and Foley (1985) reported that measurements of specimens from Bali overlap both *M. sobrinus* and *M. minimus.*

Head and body length is about 60–85 mm, the tail is rudimentary, and forearm length is 36–49 mm. Lekagul and McNeely (1977) gave weights of 18.5–23 grams for *M. sobrinus* and 12–18 grams for *M. minimus.* Coloration is russet brown above and paler below, with a more or less definite longitudinal stripe of darker brown from the crown to the nape. Males lack the neck tufts present in adult males of *Eonycteris* and *Megaloglossus. Macroglossus* is distinguished from the other genera of the subfamily Macroglossinae by cranial and dental features. The call has been described as a piercing high shriek.

According to Lekagul and McNeely (1977), *M. sobrinus* is found in evergreen forest from sea level to 2,000 meters, while *M. minimus* is usually confined to coastal areas, especially in the vicinity of mangroves. Although long-tongued fruit bats shelter under the branches of trees and under roofs, the preferred daytime retreat seems to be in the rolled leaves of hemp and banana plants. On the island of Bali an individual of *M. sobrinus* was found in a pisang leaf. Like many of the smaller fruit bats, *Macroglossus* is usually solitary, and this reduces the chance of discovery. Five men looked in hemp plants for one day in the Philippines and found only one *Macroglossus.* Lekagul and McNeely (1977), however, mentioned that *M. sobrinus* has been found roosting in groups of 5–10 individuals. As food, *Macroglossus* seems to prefer the pollen and nectar of the cultivated jambu *(Eugenia)* and century plants *(Agave).* The juices of soft fruits are also taken.

Two female *M. minimus* collected in August on Buru Island had embryos almost ready to be born. Phillips (1968) reported that lactating female *M. minimus* were collected in the Solomons in January, March, and December. Lim, Shin, and Muul (1972) reported that a female *M. minimus* captured in Sarawak in mid-June had a single embryo and that pregnant females of this species had been found in the Malay

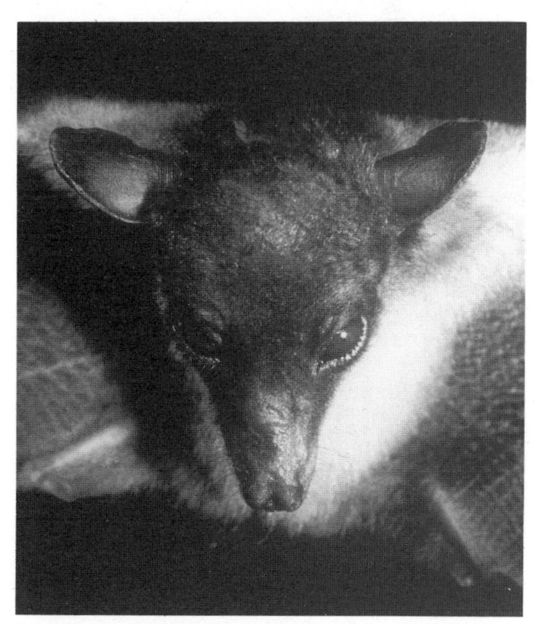

Long-tongued fruit bat *(Macroglossus minimus)*, photo by Paul D. Heideman.

Peninsula in April and June. McKean (1972) found pregnant females of *M. minimus,* each with a single embryo, in every month of the year during which any females were collected in New Guinea and the Solomons. Smith and Hood (1981) thought this species to breed throughout the year in the Bismarck Archipelago.

CHIROPTERA; PTEROPODIDAE; **Genus SYCONYCTERIS**
Matschie, 1899

Blossom Bats

There are three species (Hill 1983*b;* Koopman 1982*a,* 1984*a;* Laurie and Hill 1954; Lidicker and Ziegler 1968; McKean 1972; Ride 1970; Rozendaal 1984; Ziegler 1982*a*):

S. australis, Buru, Ceram, Kei and Amboina islands, New Guinea, Bismarck Archipelago, Louisiade and D'Entrecasteaux archipelagoes, Trobriand Islands, eastern Queensland and New South Wales;
S. hobbit, northeastern Papua New Guinea;
S. carolinae, Halmahera Island in the northern Moluccas.

Head and body length is about 50–72 mm, the tail is a mere spicule, and forearm length is 38–50 mm in *S. australis* and *S. hobbit* but 60 mm in the much larger *S. carolinae.* McKean (1972) gave the weight of *S. australis* as 11.5–25.0 grams, and Ziegler (1982*a*) listed weights of 15.1–15.7 grams for *S. hobbit.* The coloration is reddish brown or grayish brown to dark brown above and lighter below. Males do not have neck tufts.

Syconycteris australis, photo by Stanley Breeden.

As in the other bats of the subfamily Macroglossinae, the tongue is long, slender, and protrusible, with brushlike projections that pick up nectar and pollen. This genus is distinguished by dental features, especially the large size of the upper incisors and the great difference in size between the first and second lower incisors.

McKean (1972) collected *S. australis* in a variety of habitats covering almost all forest and woodland formations occurring in Papua New Guinea and at altitudes ranging from sea level to 2,860 meters. Ziegler (1982*a*) found *S. hobbit* only in montane moss forest at elevations above 2,200 meters. Bats of this genus feed mainly on nectar and pollen. One *S. australis* was caught on a fruit that it was devouring. In New South Wales this species may form "camps," like those of *Pteropus,* for a period of two to four weeks in October (Ride 1970). McKean (1972) found pregnant or lactating females of *S. australis* in New Guinea during most months from January to September. Each pregnant female contained only one embryo.

CHIROPTERA; PTEROPODIDAE; **Genus MELONYCTERIS**
Dobson, 1877

There are three species (Laurie and Hill 1954; Phillips 1968):

M. melanops, Papua New Guinea, Bismarck Archipelago;
M. aurantius, Florida and Choiseul islands (Solomons);
M. woodfordi, Solomon Islands.

Melonycteris melanops. Inset: muzzle. Photos from *Proc. Zool. Soc. London.*

Melonycteris woodfordi, photo from *Proc. Zool. Soc. London.*

The latter two species have sometimes been placed in the separate genus or subgenus *Nesonycteris* Thomas, 1887, but Phillips (1968) synonymized this name with *Melonycteris.*

Head and body length is about 77–106 mm, there is no tail, and forearm length is about 42–65 mm. In *M. melanops* the head is dark brown, the back is pale to golden cinnamon, and the underparts are dark brown, almost black. In *M. woodfordi* and *M. aurantius* the head and nape are pale russet, the back is orange to russet cinnamon, and the underparts are grayish brown, paler than the back. All three species have long, slender, protrusible tongues with bristlelike papillae.

M. melanops may be distinguished from all other fruit bats of the family Pteropodidae by its tongue and by the color of the underparts, which are darker than the back. This species normally has a claw on the index (second) finger and has four lower incisors. *M. woodfordi* and *M. aurantius* differ from all other Old World fruit bats by usually lacking a claw on the index finger and by possessing five pairs of upper cheek teeth, six pairs of lower cheek teeth, two pairs of upper incisors, and one pair of lower incisors. None of the species of *Melonycteris* has the neck tufts that are present on adult males of *Eonycteris* and *Megaloglossus.*

These bats are assumed to feed on nectar, pollen, and the juices of soft fruits. Phillips (1966) reported that the type specimen of *M. aurantius,* an adult female, was lactating when obtained in October. McKean (1972) reported the collection of pregnant *M. woodfordi* in July and September.

CHIROPTERA; PTEROPODIDAE; **Genus NOTOPTERIS**
Gray, 1859

Long-tailed Fruit Bat

The single species, *N. macdonaldi,* occurs on the New Hebrides Islands, the Fiji Islands, and New Caledonia (Hill 1983*b*). It also has been reported from Ponapé in the Carolines, but Corbet and Hill (1986) listed this record with question.

Head and body length is about 100 mm, tail length is about 60 mm, and forearm length is 60–72 mm. Nelson and Hamilton-Smith (1982) reported an average weight of 51 grams for 12 females and 61 grams for two males. *Notopteris* is the only member of the family Pteropodidae with a long,

Long-tailed fruit bat *(Notopteris macdonaldi)*, photos by Anthony Healey.

free tail. Coloration is olive brown to dark brown, and neck tufts are lacking.

Notopteris resembles *Dobsonia* in that it lacks a claw on the index finger and has the wing membranes attached along the midline of the back, thus covering the fur on the back of the body. The tongue is like those of other bats in the subfamily Macroglossinae, that is, long, slender, protrusible, and with bristlelike papillae.

This bat frequently roosts in caves, though natives on New Caledonia have seen them roosting in hollow trees. About 300 long-tailed fruit bats have been observed in a cave on New Caledonia, resting in clusters of 5–25 individuals. Their roosting habits are similar to those of some of the Microchiroptera, especially *Tadarida brasiliensis.* The diet is assumed to comprise nectar, pollen, and the juices of fruits. A captive female was kept alive for three days on the juices from canned pears, peaches, and sugar water; it refused solid food. Death was attributed partly to a lack of food.

The sexes apparently do not segregate at any time. On New Caledonia, pregnant or lactating females have been collected in July and December, and young have been noted in August, December, and January; females evidently are polyestrous (Nelson and Hamilton-Smith 1982). As in some other bats, the youngster clings to the underparts of its mother with mouth, wing hook, and claws.

CHIROPTERA; **Family RHINOPOMATIDAE; Genus RHINOPOMA**
E. Geoffroy St.-Hilaire, 1818

Mouse-tailed Bats, or Long-tailed Bats

The single known genus, *Rhinopoma*, contains three species (Aggundey and Schlitter 1984; Aulagnier and Destre 1985; Hill 1977*a*; Koch-Weser 1984; Kock and Felten 1980; Lekagul and McNeely 1977; Nader and Kock 1983*b*):

R. microphyllum, Morocco, Mauritania, Senegal, Burkina Faso, Nigeria, Egypt, Sudan, Israel and Saudi Arabia to India, Sumatra;
R. hardwickei, Morocco and Mauritania to Egypt and Kenya, Israel and South Yemen to Afghanistan and Burma, Socotra Island, one record from Thailand;
R. muscatellum, Oman, Iran, southern Afghanistan, Pakistan, possibly Ethiopia.

Head and body length is 53–90 mm, tail length is 43–75 mm, forearm length is 45–75 mm, and adult weight is about 6–14 grams. The soft fur is lacking on the face, rump, and posterior portion of the abdomen. The coloration is grayish brown or dark brown above and usually paler below.

The bats of this genus are the only ones in the suborder Microchiroptera with a tail nearly as long as the head and body combined. The tail membrane is extremely short and

Mouse-tailed bat *(Rhinopoma hardwickei)*, photo by P. Morris.

narrow. The large ears, connected by a band of skin across the forehead, extend beyond the nostrils when the ears are laid forward. A well-developed tragus is present. The snout bears a small, rounded nose leaf. The valvular nostrils appear as slits that open forward and can be closed at will. The nasal chambers are swollen laterally, the muzzle having a distinct ridgelike outgrowth of skin. The nasal bones are expanded laterally and vertically, and the frontal bones are depressed, forming a concavity in the forehead. The teeth are of the normal insectivorous kind. The dental formula is: (i 1/2, c 1/1, pm 1/2, m 3/3) $\times$ 2 = 28.

Mouse-tailed bats are usually found in treeless and arid regions. They roost in caves, rock clefts, wells, pyramids, palaces, and houses. *R. microphyllum* has inhabited certain pyramids in Egypt for 3,000 years or more. These bats often hang by the thumbs as well as the feet. *R. muscatellum* has a characteristic and unusual flight; it rises and falls, much like some small birds. Usually flying at least six to nine meters above the ground, this species travels by a series of glides, some of great length, and occasionally it flutters.

Fat, sometimes equaling the normal weight of the individual, accumulates in the fall beneath the naked skin, especially in the abdominal region; this fat is absorbed during the winter. *R. microphyllum*, at least, does not hibernate, but it does remain in a torpor during the winter and does not move about in search of food. During this period the bat is able to survive for several weeks in captivity without food and water. The resorption of the accumulated fat in the winter suggests that torpidity is probably an adaptation to circumvent the scarcity of insect food during the cold season (Anand Kumar 1965).

These bats sometimes roost in colonies numbering many thousands of individuals. *R. hardwickei*, however, may roost alone or in small groups of 4–10. Several such groups may form a large, loose group of 80–100 individuals. The smaller groups may be sexually segregated for at least part of the year (Lekagul and McNeely 1977).

The reproductive pattern of *R. microphyllum* in Jodhpur, India, can be related to its seasonal movement, which occurs just before and after its winter torpor. The females are monestrous, mating in March and giving birth to a single young in July–August. Gestation lasts at least 123 days. The young are weaned after 6–8 weeks and attain sexual maturity in their second year of life (Anand Kumar 1965). The following reproductive data for *R. hardwickei* were summarized by Qumsiyeh and Jones (1986): young are born in June and July in India, females pregnant with single embryos have been taken in Egypt and Sudan in late March, and lactating females have been taken in the same countries in August.

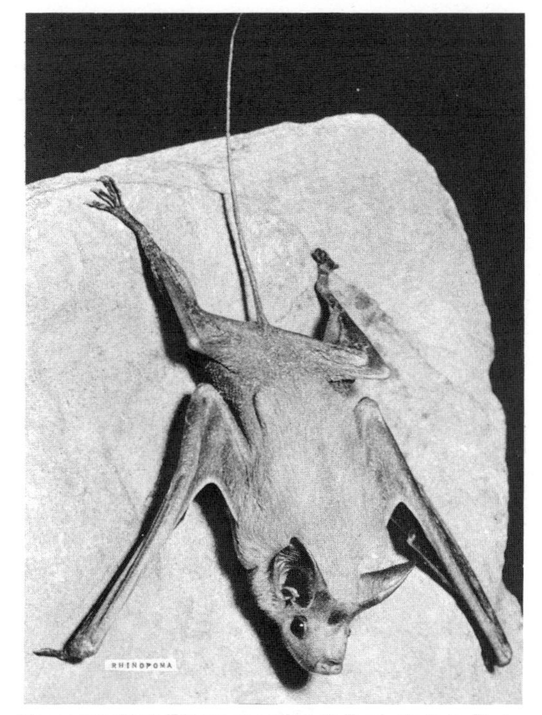

Mouse-tailed bat *(Rhinopoma microphyllum)*, photo by Erwin Kulzer.

No fossils referable to the Rhinopomatidae have been found.

CHIROPTERA; **Family EMBALLONURIDAE**

Sac-winged Bats, Sheath-tailed Bats, and Ghost Bats

This family of 12 Recent genera and 48 species is widely distributed in the tropical and subtropical regions of the world. The sequence of genera presented here follows that of Robbins and Sarich (1988), who recognized two subfamilies, Taphozoinae, with *Taphozous* and *Saccolaimus*, and Emballonurinae, with the remaining genera. A third subfamily, Diclidurinae, comprising the genera *Cyttarops* and *Diclidurus*, sometimes is recognized.

The size ranges from small to relatively large. Head and body length is 37–157 mm, tail length is 6–36 mm, and forearm length is 37–97 mm. Adults weigh 5–105 grams. Most members of this family are gray, brown, or black in color, but *Diclidurus*, the ghost bat, has an unusual color for bats, being white or white mixed with gray. In the subgenus *Peronymus* of *Peropteryx* the wings are white from the forearm outward. In the proboscis bat, *Rhynchonycteris*, the upper surface of the forearm is dotted with tufts of hair.

The face and lips are smooth. A nose leaf is not present. The ears are often united across the top of the head, and a tragus is present. Many forms have glandular wing sacs that open on the upper surface of the wing. These glands secrete a red substance of strong odor. The sacs can be seen by holding the bat with its head up and its belly toward the light and gently extending its wings; if present, they appear near the shoulder. The wing sacs are larger and better developed in the males; perhaps the odor of their secretion attracts the op-

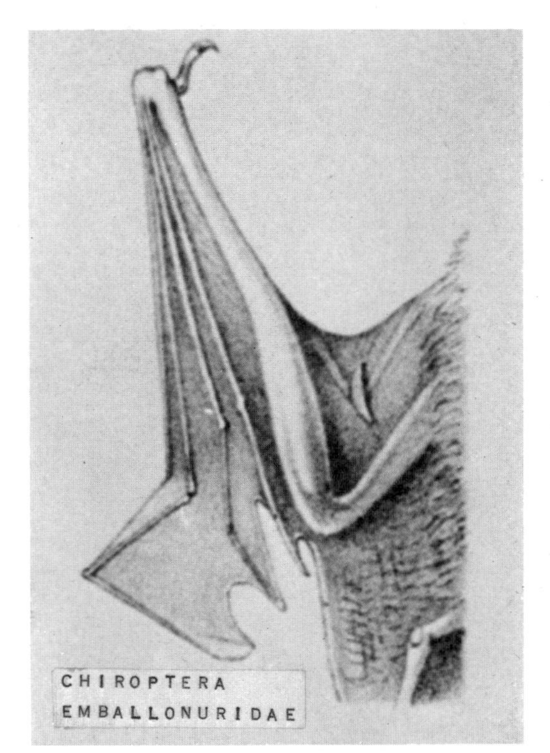

Wing of emballonurid bat, showing the glandular wing sac that is present in many members of the family Emballonuridae, photo from *Biologia Centrali-Americana, Mammalia*, E. R. Alston.

posite sex. In the genus *Taphozous* a glandular mass is located between the lower jaws. The common name "sheath-tailed bats" refers to the nature of the tail attachment. In the members of this family the tail pierces the tail membrane, and its tip appears completely free on the upper surface of the membrane, the base of the tail being loosely enclosed in the membrane. Thus, in flight the tail membrane can be lengthened by stretching the hind limbs, the membrane then slipping quite easily over the tail vertebrae. In this way, by pulling in or moving out their hind legs, these bats can "set their sail." Most forms can steer and turn exceptionally well. The hind legs are slender, and the foot is normal.

The molar teeth have a **W** pattern of cusps and ridges. The dental formula varies. In *Emballonura* it is (i 2/3, c 1/1, pm 2/2, m 3/3) $\times$ 2 = 34; in the American genera and *Coleura* it is (i 1/3, c 1/1, pm 2/2, m 3/3) $\times$ 2 = 32; and in *Taphozous* it is (i 1/2, c 1/1, pm 2/2, m 3/3) $\times$ 2 = 30.

Shelters include rocky crevices, caves, ruins, houses, trees, curled leaves, and hollow logs. These bats are agile in their retreats, scrambling about with considerable dexterity and often clinging to vertical walls or crawling into crevices. At rest they often fold the tips of the wings back on their upper surface. Some forms roost in large colonies; others assemble in groups of about 10–40 individuals. These groups seem to operate as units in most of their activities. Other forms are generally solitary. Some species occasionally shelter with other species. When they do, the species usually remain separate. In *Rhynchonycteris* the females roost apart from the males when the young are born, and different shelters are used by adult male and female *Balantiopteryx plicata* during the summer; most of the other forms seem to remain together all year. Some species of *Taphozous* breed throughout the year, but most members of this family probably have a fairly well-defined breeding season. Sheath-tailed bats find objects in the dark by echolocation, but they may also guide themselves by vision. Some species are rapid fliers. They feed on insects and occasionally on fruit.

The geological range of this family is middle Eocene to early Miocene in Europe, early Miocene to Recent in Africa, Pleistocene to Recent in South America, and Recent in other parts of the range (Koopman 1984*c*).

CHIROPTERA; EMBALLONURIDAE; **Genus TAPHOZOUS**
E. Geoffroy St.-Hilaire, 1813

Tomb Bats

There are 2 subgenera and 13 species (Aggundey and Schlitter 1984; Chasen 1940; Corbet 1978; Ellerman and Morrison-Scott 1966; R. E. Goodwin 1979; Hayman and Hill, *in* Meester and Setzer 1977; Hill 1983*b*; Kitchener 1980*a*; Kock 1974*b*; Koopman 1975, 1982*a*, 1984*a*; Laurie and Hill 1954; Lawrence 1939; Lekagul and McNeely 1977; McKean and Friend 1979; Ride 1970; Roberts 1977; Taylor 1934):

subgenus *Taphozous* E. Geoffroy St.-Hilaire, 1813

T. mauritianus, savannah regions throughout Africa south of the Sahara, Madagascar, Mauritius, Reunion, Assumption Island, Aldabra Island;
T. hildegardeae, coastal and central Kenya, northeastern Tanzania;
T. perforatus, Senegal to India, and south to Zimbabwe;
T. longimanus, India and Sri Lanka to southern Indochina and Malay Peninsula, Sumatra, Java, Borneo, Flores;
T. melanopogon, India to Indochina, Yunnan (southern China), Malay Peninsula, Sumatra, Java, Borneo, Sulawesi, Savu, Sumbawa, Timor, possibly Lombok;
T. theobaldi, central India, eastern Burma, Thailand, Indochina;
T. philippinensis, Philippines;
T. georgianus, Western Australia, Northern Territory, Queensland;
T. hilli, Western Australia, Northern Territory;
T. kapalgensis, northern Northern Territory;
T. australis, northeastern Queensland, islands of Torres Strait, one probably accidental or erroneous record from southeastern New Guinea;

subgenus *Liponycteris* Thomas, 1922

T. hamiltoni, southern Sudan, Kenya;
T. nudiventris, Mauritania to Egypt and Tanzania, Palestine and Arabian Peninsula to Burma and Malay Peninsula.

Saccolaimus (see account thereof) sometimes is considered a subgenus of *Taphozous.* Honacki, Kinman, and Koeppl (1982) suggested that *T. philippinensis* may be a subspecies of *T. melanopogon,* and Corbet and Hill (1986) did not list *T. philippinensis* at all.

Head and body length is about 62–100 mm, tail length is 20–35 mm, forearm length is 50–75 mm, and weight is 10–50 grams. The upper parts are grayish or various shades of brown, sometimes with a reddish or cinnamon cast. Some species have whitish spots on the body. The underparts are creamy, pale brown, or white. Males of the species *T. longimanus* are usually cinnamon brown in color, whereas most females of this species are dark gray.

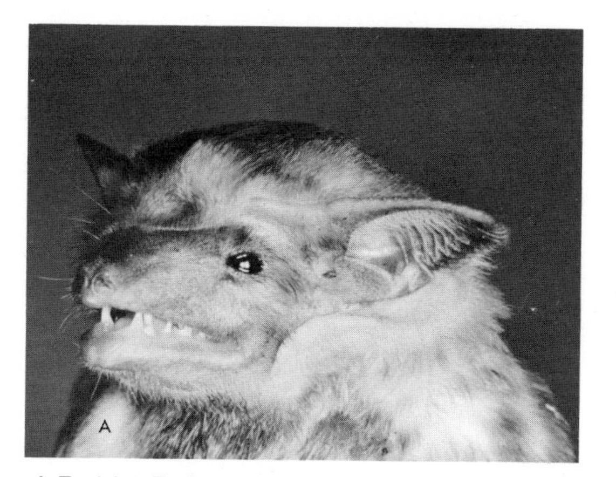

A. Tomb bat *(Taphozous hildegardeae)*. B. *T. mauritianus*. Photos by David Pye.

A wing pocket or pouch is present in all species. Most species also have a glandular sac in the lower throat. This sac is more highly developed in the males than in the females and may be entirely absent in the latter. The species *T. melanopogon* lacks this sac and instead has small pores that open into the area where the sac would be.

Tomb bats are found in a variety of habitats, including rainforest, open woodland, and fairly arid country. They roost in tombs, old buildings, rock crevices, caverns in rocky deserts, shallow caves, cliffs along seashores, and trees. *T. longimanus* is often found in the top of coconut palms. Tomb bats are quite agile and active in crawling around crevices and rock walls and sometimes cling on vertical walls. They are numerous in certain large tombs. Hibernation is suggested by the seasonal accumulation of fat.

An almost inaudible "tic-tic-tic" may be uttered when roosting, whereas the call when wounded and on the ground is shrill and piercing. Some species emit a loud cry in flight. Tomb bats are strong fliers. They may begin feeding before dusk at heights of 60–90 meters, coming down to lower levels as the evening progresses. The diet consists of flying insects. In a study in India, Subbaraj and Chandrashekaran (1977) found that *T. melanopogon* emerged from its roost at about the same time every day, regardless of the time of sunset, so that the bats began flying at sunset in some months and well after dark in others. Kingdon (1974*a*) wrote that *T. mauritianus* has three hours of intensive activity after sunset and then alternates long rests with short flights. According to Roberts (1977), *T. nudiventris* makes seasonal migrations, with colonies reoccupying a summer roost in a matter of a few days.

Lekagul and McNeely (1977) reported the following roosting group sizes in Thailand: *T. longimanus*, 2–20; *T. melanopogon*, 150–4,000; and *T. theobaldi*, in the thousands. Kingdon (1974*a*) wrote that *T. nudiventris* shelters in groups of 200–1,000 and forms all-female groups during late pregnancy and lactation, and Roberts (1977) mentioned one colony of over 2,000. While roosting, each individual *T. melanopogon* occupies a definite vertical territory, and males and females may be in separate areas. The mating season of this species is January–February, but in *T. longimanus* there may be continuous breeding, with each female undergoing more than one pregnancy per year (Lekagul and McNeely 1977). In a study in India, Gopalakrishna (1955) found *T. longimanus* to breed all year, with each female having a rapid succession of pregnancies and producing one young at a time. In Pakistan one colony of *T. nudiventris* reportedly arrives at its summer roost every year in the first week of March, and

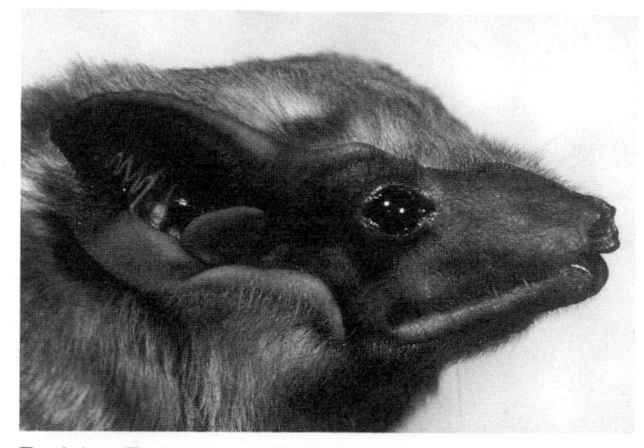

Tomb bat *(Taphozous philippinensis)*, photo by Paul D. Heideman.

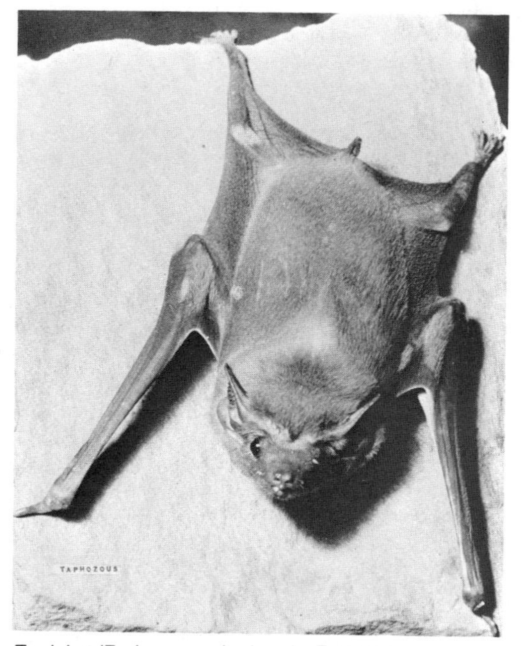

Tomb bat (*Taphozous* sp.), photo by Erwin Kulzer.

each female gives birth to a single young in mid-April. The young are carried by the females until the age of eight weeks (Roberts 1977). The birth seasons of *T. mauritianus* in northeastern Zaire are April–May and November–December (Kingdon 1974*a*). In studies in Western Australia, Kitchener (1973, 1976) found *T. georgianus* to be monestrous, with an anestrous period from mid-fall to midwinter. A single young was produced from late November to late April, following a gestation period of about four months.

CHIROPTERA; EMBALLONURIDAE; **Genus SACCOLAIMUS**
Temminck, 1838

There are 5 species (Chasen 1940; Corbet 1978; Ellerman and Morrison-Scott 1966; Feiler 1980; R. E. Goodwin 1979; Hayman and Hill, *in* Meester and Setzer 1977; Hill 1983*b;* Koopman 1975, 1982*a*, 1984*a;* Laurie and Hill 1954; Lawrence 1939; Lekagul and McNeely 1977; Ride 1970; Taylor 1934):

S. peli, Liberia to western Kenya and northeastern Angola;
S. saccolaimus, India, Sri Lanka, Bangladesh, southern Burma, Malay Peninsula, Sumatra, Java, Borneo, Sulawesi, Timor, New Guinea, Solomon Islands, northwestern Northern Territory, northeastern Queensland;
S. pluto, Philippines;
S. flaviventris, throughout Australia except Tasmania;
S. mixtus, southern and eastern New Guinea, Cape York Peninsula of northern Queensland.

Saccolaimus was considered only a subgenus of *Taphozous* by Corbet and Hill (1986), but the two were regarded as generically distinct by Barghoorn (1977), Honacki, Kinman, and Koeppl (1982), Koopman (1979, 1984*a*, 1984*b*), and Robbins and Sarich (1988).

In the largest species, *S. peli*, of Africa, head and body length is 110–57 mm, tail length is 27–36 mm, forearm length is 84–97 mm, and weight is 92–105 mm (Kingdon 1974*a*). In the three species that occur in Australia, head and body length is 72–100 mm, tail length is 20–35 mm, forearm length is 62–80 mm, and weight is 30–60 grams (Hall, *in* Strahan 1983; Richards, *in* Strahan 1983). The upper parts vary in color from reddish brown to jet black; the underparts usually are paler, sometimes entirely white. In *S. saccolaimus* the pelage is irregularly flecked with patches of white and the posterior portion of the back is naked. There is a large pouch under the chin in both sexes. From *Taphozous*, *Saccolaimus* differs in lacking a wing pouch and in having a deep groove on the lower lip (Lekagul and McNeely 1977). *Saccolaimus* also is considered generically distinct on the basis of numerous cranial characters, especially in that the auditory bullae are completely ossified (Barghoorn 1977).

According to Kingdon (1974*a*), *S. peli* shelters by day in deep forests and comes out at night to forage along the margins of the forest and in clearings and river valleys. Its flight is acrobatic and is accompanied by loud calls. It pursues moths and beetles. *S. saccolaimus* is found in a variety of woodland and forest habitats, and roosts in hollow trees, caves, old tombs, buildings, and openings between boulders (Hall, *in* Strahan 1983). *S. flaviventris* may migrate to warmer areas for the winter (Richards, *in* Strahan 1983).

S. flaviventris is usually solitary but occasionally forms colonies of fewer than 10 individuals (Richards, *in* Strahan 1983). Chimimba and Kitchener (1987) reported *S. flaviventris* to begin mating in August and to give birth to a single young from December to mid-March. Groups of *S. saccolaimus* in Thailand contain five or six individuals (Lekagul and McNeely 1977). According to Hall (*in* Strahan 1983), this species is gregarious but does not form tight clusters; females are lactating during the wet season in Australia and produce a single young. Pregnant female *S. peli* have been reported in June and December in eastern Zaire (Kingdon 1974*a*).

CHIROPTERA; EMBALLONURIDAE; **Genus EMBALLONURA**
Temminck, 1838

Old World Sheath-tailed Bats

There are 10 species (Bruner and Pratt 1979; Ellerman and Morrison-Scott 1966; Hayman and Hill, *in* Meester and Setzer 1977; Hill 1956, 1971*c*, 1983*b*, 1985*a*; Hill and Beckon

Papuan sheath-tailed bat *(Saccolaimus mixtus)*, photo by B. G. Thomson / National Photographic Index of Australian Wildlife.

1978; Laurie and Hill 1954; Lekagul and McNeely 1977; Lemke 1986; McKean 1972; Medway 1977; Smith and Hood 1981; Tate and Archbold 1939; Taylor 1934):

- *E. atrata,* Madagascar;
- *E. monticola,* Malay Peninsula (including southern Burma and Thailand), Sumatra, Java, Borneo, Sulawesi, and small nearby islands;
- *E. alecto,* Borneo, Sulawesi, Moluccas, Philippines;
- *E. nigrescens,* New Guinea and other islands from Sulawesi to Solomons;
- *E. semicaudata,* New Hebrides, Palau, Mariana, Fiji, and Samoa islands;
- *E. sulcata,* Caroline Islands (western Pacific);
- *E. beccarii,* New Guinea, Bismarck Archipelago, Kei Islands, Trobriand Islands;
- *E. raffrayana,* New Guinea, Ceram, Solomon Islands;
- *E. dianae,* Papua New Guinea, New Ireland in the Bismarck Archipelago, Rennell and Malaita islands in the Solomons;
- *E. furax,* southwestern New Guinea, Papua New Guinea, Bismarck Archipelago.

Head and body length is about 38–62 mm, tail length is 10–20 mm, and forearm length is 30–53 mm. Lekagul and McNeely (1977) gave the weight of *E. monticola* as 30–40 grams. Coloration is rich brown to dark brown above and paler below. *Emballonura* is the only genus in the family with two pairs of upper incisors. Wing sacs are not present.

E. monticola is a forest species; it has been seen flying in dense shade during the day (Lekagul and McNeely 1977). This species roosts in a variety of rather exposed sites, including inside hollow logs, under overhanging earth banks, between boulders, and in rock shelters; in caves it rarely penetrates beyond the twilight zone, and it may occupy brightly lit sectors (Medway 1978). *E. nigrescens* has been collected in secondary forest, freshwater mangrove, and village environs; it roosts under the leaves of broad-leaved trees and in the roofs of native huts (McKean 1972). This species flies before dusk in the highest levels of the rainforest. The usual diet of

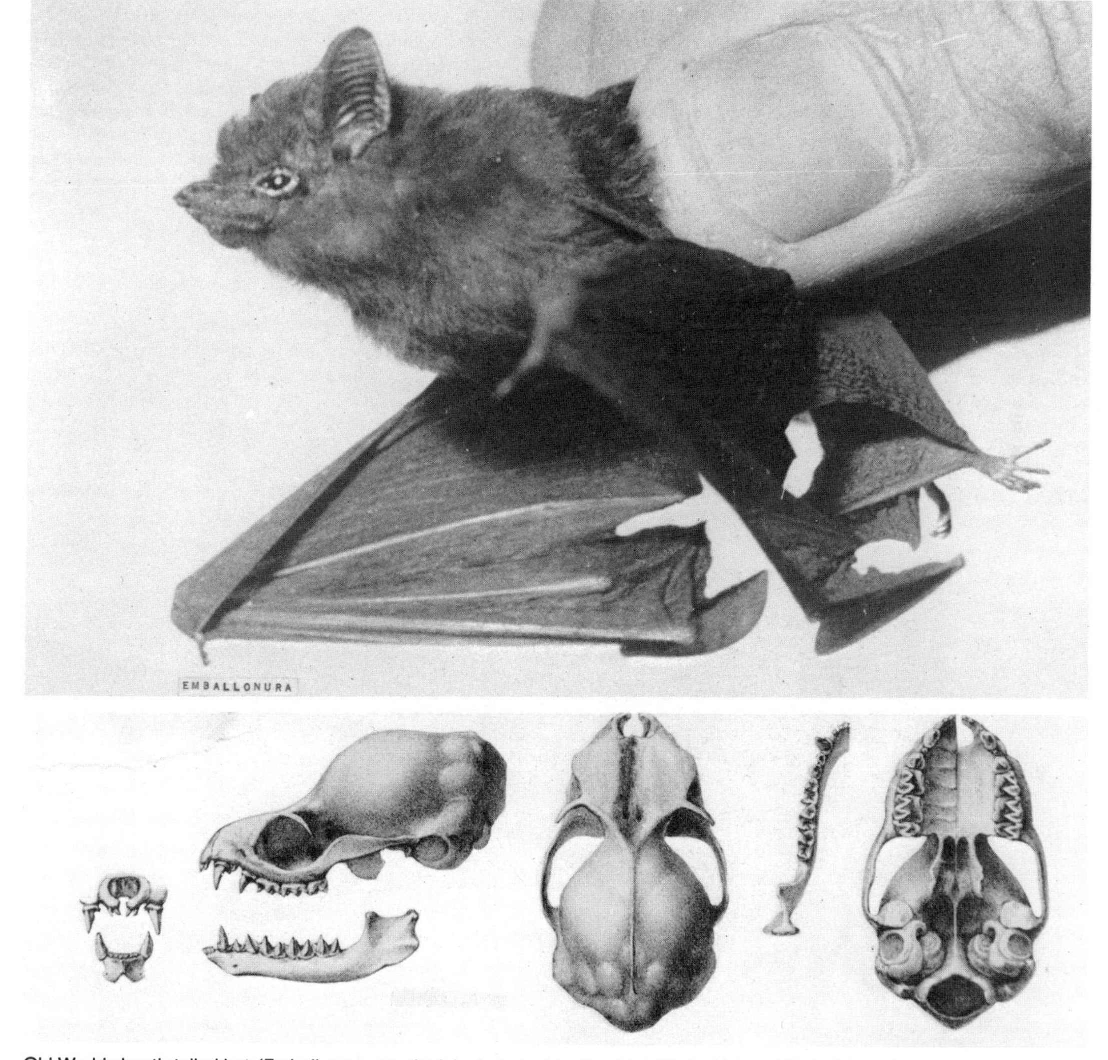

Old World sheath-tailed bat *(Emballonura monticola),* photo by Lim Boo Liat. Skull photos of *Emballonura* from *Die Fledermäuse des Berliner Museums für Naturkunde,* Wilhelm K. Peters.

Old World sheath-tailed bat *(Emballonura monticola)*, photo by Klaus-Gerhard Heller.

Emballonura is insects, but fruit is eaten occasionally. Vestjens and Hall (1977) found 44 stomachs of *E. nigrescens* to contain mostly wingless ants, suggesting that the bats had fed on or near the ground or on trees, or possibly at their roosting site.

E. monticola is moderately gregarious, usually associating in groups of 2–20 (Medway 1978), but one colony of 100–150 was found in a cave (Lekagul and McNeely 1977). Pregnant female *E. monticola,* bearing one embryo each, were found in the Malay Peninsula in February, March, October, and November (Medway 1978). Pregnant female *E. nigrescens* were collected in New Guinea in February, May, June, and July (McKean 1972).

Bruner and Pratt (1979) stated that the subspecies *E. semicaudata rotensis,* of Rota Island in the Marianas, had not been seen for many years and might be extinct. Perez (1972) suggested that this same bat may once have been abundant on Guam but had become extremely rare. Lemke (1986) obtained reports that *E. semicaudata rotensis* was present in the Marianas in some numbers through the 1960s but subsequently had declined, perhaps in association with the swiftlets that occupied the same caves. During intensive surveys of the Marianas in 1983–84 he located only two groups of about four bats each, both on Aguijan Island. This subspecies may thus rank as one of the world's most critically endangered mammals.

CHIROPTERA; EMBALLONURIDAE; **Genus COLEURA**
Peters, 1867

African and Arabian Sheath-tailed Bats

There are two species (Hayman and Hill, *in* Meester and Setzer 1977):

C. afra, southwestern Arabian Peninsula, Guinea-Bissau to Somalia, and south to Angola and Mozambique;
C. seychellensis, Seychelles Islands.

Head and body length is 55–65 mm, tail length is 12–20 mm, and forearm length is 45–56 mm. Nicoll and Suttie (1982) reported weight of *C. seychellensis* to average 10.2 grams in adult males and 11.1 grams in parous females. Coloration is reddish brown, dark brown, or sooty brown above and somewhat paler below. This genus resembles *Emballonura* but has only one pair of upper incisors, as opposed to two pairs, and there are differences in the skulls. *Coleura* does not have wing sacs.

Coleura roosts in caves and houses, generally in crevices and cracks. Bats of this genus usually do not roost upside down; instead, they crawl into a cranny or press their underside flat against a stone wall. *C. afra* has been found in the same cave with *Triaenops persicus* and *Asellia tridens.* Thousands of bats have been found in a cave partially filled with water during high tide on the rocky coast of the Indian Ocean. Large colonies of *C. afra* and *Taphozous hildegardeae* were in the front of this cave, and smaller colonies of *Triaenops afer* and *Hipposideros caffer* were in the deeper passages. The species were segregated. The behavior of *Coleura* and *Taphozous* was much the same in this cave; they rested flat against the walls, individually or closely together, and only when approached within about 50 cm did they fly away.

African sheath-tailed bats *(Coleura afra)*, photo by Bruce J. Hayward.

They would not leave the cave during the day, even when noise was used to try to drive them out. Members of the genus *Coleura* sometimes become pests when large numbers successfully colonize a house. Overlapping tiles or corrugated iron sheets are suitable resting places. The diet is mainly insects.

McWilliam (1987*a*) analyzed the population structure of *C. afra* roosting in coral caves along the coast of Kenya. Colonies consisted of up to 50,000 individuals or more, but each bat had a precise roosting place to which it returned. Colonies were divided into clusters, each with about 20 individuals, comprising mainly a harem of adult females and their young, plus a single adult male. Solitary bats, mainly bachelor males, roosted on the periphery of the clusters. Females were found to be polyestrous and to give birth twice a year, primarily in April and November. These periods evidently were timed so that lactation would continue for about three months during the rainy seasons, when availability of insects was at a maximum. Many of the young females remained in their natal clusters for at least a year, and since they attained sexual maturity during this period, each cluster tended to be a permanently related unit. Pregnant female *C. afra* also have been collected in December in Tanzania and in April in South Yemen. In Sudan breeding has been found to be markedly seasonal, with births occurring in October, at the end of the rains. Only one young is produced at a time (Kingdon 1974*a*).

Little was known of *C. seychellensis* until studies by Nicoll and Suttie (1982). Colonies roosted in caves and evidently were divided into harem groups like those of *C. afra*. Births occurred during the November–December rainy season over three years, though the presence of young in one year suggested that females are polyestrous. This species once was abundant but has declined drastically, perhaps because of environmental disruption and predation by introduced barn owls. Only two occupied caves were located during searches on three islands. The IUCN now classifies *C. seychellensis* as endangered.

CHIROPTERA; EMBALLONURIDAE; **Genus RHYNCHONYCTERIS**
Peters, 1867

Proboscis Bat, or Sharp-nosed Bat

The single species, *R. naso*, occurs from southern Mexico to northern Peru and central Brazil and on the island of Trinidad (Cabrera 1957; Hall 1981).

Head and body length is 37–43 mm, tail length is about 12 mm, and forearm length is 35–41 mm. Four males and three females from Trinidad weighed 2.1–4.3 grams. The whitish-tipped hairs and chocolate brown underfur impart an unusual grizzled yellowish gray color. Two curved lines, white or gray in color, are usually present on the lower back and rump. The skull is very small, with the rostrum elongate and the frontal area sharply depressed. This genus may be identified by the small tufts of grayish hair on the forearms, the elongate and pointed muzzle, and the absence of wing sacs.

This bat is generally found in forests, usually along waterways. During the day it roosts adjacent to the water on the branches or boles of trees, on rocks, or on the sides of cliffs (Bradbury and Emmons 1974; Husson 1978). It has also been discovered in the curled leaves of the heliconia, or false banana plant. Because of the color pattern, groups of *Rhynchonycteris*, when roosting, resemble patches of li-

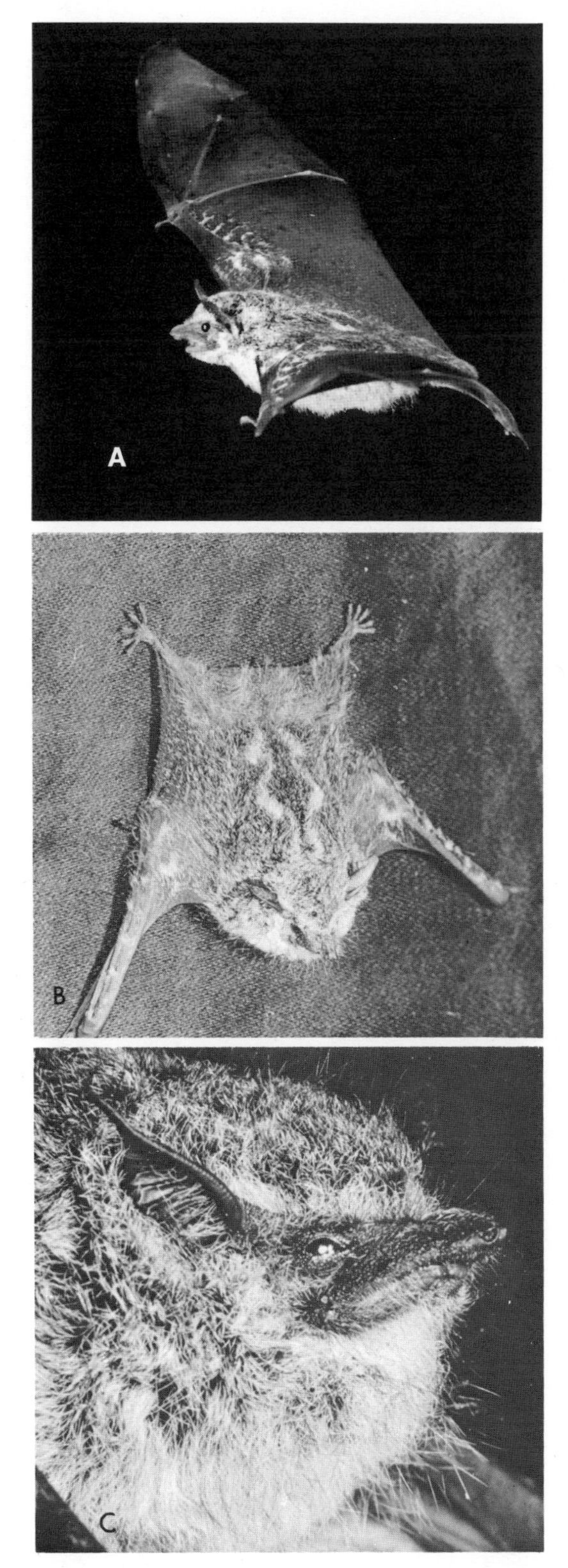

Proboscis bat *(Rhynchonycteris naso)*: A. Photo by J. Scott Altenbach; B & C. Photos by David Pye.

chens, and their habit of slightly curving the head and nose backward increases the resemblance to the curled edge of a lichen. The latter position, however, may not be assumed until the bat is threatened. Proboscis bats usually reassemble at the home perch within 20 minutes after being scattered by a disturbance. Groups of roosting *Rhynchonycteris* have also been reported to bear a remarkable resemblance to cockroaches (Dalquest 1957). There is usually a definite space between each roosting bat, and in common with *Saccopteryx* and *Peropteryx*, *Rhynchonycteris* folds its wings at about a 45° angle to the body when inactive (Goodwin and Greenhall 1961). Husson (1978) stated that groups of 8–10 hang downward in a vertical row on tree branches, with each bat straight above the other, about 10 cm apart.

In studies in Trinidad and Costa Rica, Bradbury and Vehrencamp (1977) found *Rhynchonycteris* to forage almost entirely over water, up to a height of 3 meters. The foraging range averaged 1.1 ha., and population density in the study area was 7.6/ha. The diet of the proboscis bat consists of small insects.

Colonies of *Rhynchonycteris* roost together and seem to associate as a unit in most of their activities. Bradbury and Vehrencamp (1977) found colony size to range from 3 to 45 individuals and to average 5–11, depending on area. Each colony used three to six roosting sites, among which it moved at intervals. In some colonies there appeared to be a dominant male which resided within the colony more constantly than other adult males. The dominant male typically foraged alone at the boundaries of the colony's foraging range and chased intruding conspecifics. The reproductive females of the colony, and their young when capable, foraged together in a small area. Other males and nonreproductive females foraged in solitary "beats." At the age of two to four months, the young of both sexes dispersed to adjacent colonies. In Costa Rica births did not occur during the early dry season (January–March) but were recorded in April, June, July, and October. Bradbury and Emmons (1974) recorded births in Trinidad in July and August and found, contrary to earlier reports, that females with young did not roost away from the main colony. Graham (1987) found pregnant females in Peru in May, July, August, and September, and Webster and Jones (1984*a*) reported one taken in Ecuador on 12 June.

Shaggy-haired bat *(Centronycteris maximiliani):* Top, photo from *Die Fledermäuse des Berliner Museums für Naturkunde*, Wilhelm K. Peters; Bottom, photo by Richard K. LaVal.

CHIROPTERA; EMBALLONURIDAE; **Genus CENTRONYCTERIS**
Gray, 1838

Shaggy-haired Bat

The single species, *C. maximiliani*, occurs from southern Mexico to Peru and Brazil (Cabrera 1957; Hall 1981; Tuttle 1970).

Least sac-winged bat *(Balantiopteryx plicata)*, photo by Lloyd G. Ingles. Inset showing wing sac, photo from . . . *American Bats of the Subfamily Emballonurinae*, Colin C. Sanborn.

Head and body length is about 50–62 mm, tail length is 18–23 mm, and forearm length is 43–48 mm. Coloration of the long, soft hair is raw umber or tawny. The hairs in front of the eyes and those on the interfemoral membrane are sometimes reddish in color. There are no stripes.

This genus resembles *Saccopteryx* but is more slender in body form, and as far as is known, there is no wing sac. Also, the lower border of the orbit of the skull projects only slightly, so that the toothrow is visible from above and is not hidden as in *Saccopteryx*.

Starrett and Casebeer (1968) observed that the rarity of specimens of this genus suggests that it requires relatively heavy forest in which to live. Its very slow and highly maneuverable, floppy flight makes it well adapted to hunting insects among trees and in natural and artificial pathways and clearings. Baker and Jones (1975) observed several individuals in flight, traveling extremely slowly and in a straight path along the right of way of a telegraph line in a second-growth forest. *Centronycteris* has also been collected from holes in the boles of trees. A pregnant female with a single fetus was taken in Nicaragua on 26 May (Greenbaum and Jones 1978). In Costa Rica two pregnant females were collected on 15 May and a subadult was taken on 15 September (LaVal 1977).

CHIROPTERA; EMBALLONURIDAE; **Genus BALANTIOPTERYX**
Peters, 1867

Least Sac-winged Bats

There are three species (Cabrera 1957; Hall 1981):

B. plicata, western and southern Mexico, including southern Baja California, to Costa Rica;
B. io, southern Mexico to Belize and Guatemala;
B. infusca, Ecuador.

Head and body length is about 48–55 mm, tail length is about 12–21 mm, forearm length is 35–49 mm, and adult weight is about 4–9 grams. The upper parts are dark chestnut brown, dark brown, or dark gray, and the underparts are usually paler. There are no dorsal stripes. The wing sac is in the center of the membrane between the upper arm and forearm. This genus can be distinguished from the other bats of the family Emballonuridae that have wing sacs by the greatly inflated rostrum.

In Jalisco, Mexico, *B. plicata* is common in many areas and has a known altitudinal range of sea level to about 1,300 meters (Watkins, Jones, and Genoways 1972). *Balantiopter-*

A. White-lined bat *(Saccopteryx bilineata)*, photo by Rexford Lord. B & C. Two-lined bats *(S. leptura)*, photos by Francis X. Williams, California Academy of Sciences, through Robert T. Orr.

yx roosts in caves and culverts and under boulders, often near lakes and rivers. Individuals have been found in limestone caves in the crevices at the tops of stalactites, but these bats usually hang in exposed places in fairly well illuminated spots, with appreciable space between each other. Starrett and Casebeer (1968) found one colony of *B. plicata* in a series of fissures in the face of a sea cliff that was accessible only at low tide, and another in the shadow of a huge rafter near the mouth of a partially collapsed manganese mine. Least sac-winged bats are among the first bats to appear in the evening, often coming out before sunset. Their flight is erratic but relatively slow, and the diet is insects.

These bats are gregarious. Approximately 300 *B. io* were found clinging in a random manner to the ceiling of a cave in late July and early August in Oaxaca, Mexico. Pregnant females of this species, each with a single embryo, have been taken from March to July (Arroyo-Cabrales and Jones 1988*b*). Bradbury and Vehrencamp (1977) investigated one cave in Costa Rica that contained 1,500–2,000 *B. plicata*. These bats in turn were divided into separate colonies of 50–200 each. The bats foraged alone or in groups, and no evidence of territoriality was observed. There were usually more males than females in a colony, but some males moved away during the rainy season. There was a single highly synchronized birth season starting in late June, and females were capable of reproduction in their first year of life. In Mexico, pregnant female *B. plicata* have been reported from May to August, and lactating females were found in September (Jones, Choate, and Cadena 1972; Watkins, Jones, and Genoways 1972). The gestation period is about 4.5 months; the single young weighs about 2 grams at birth and is precocial. It is carried by the mother for its first week of life, can fly by itself at 2 weeks, and is fully weaned at about 9 weeks (Arroyo-Cabrales and Jones 1988*a*).

CHIROPTERA; EMBALLONURIDAE; **Genus SACCOPTERYX**
Illiger, 1811

White-lined Bats

There are four species (Anderson, Koopman, and Creighton 1982; Cabrera 1957; Hall 1981; Koopman 1982*a*):

- *S. bilineata*, southern Mexico to Bolivia and southeastern Brazil, Trinidad;
- *S. leptura*, southern Mexico to central Brazil and Bolivia, Trinidad;
- *S. canescens*, northern Colombia to the Guianas and central Brazil;
- *S. gymnura*, northeastern Brazil.

Head and body length is 37–55 mm, tail length is 12–20 mm, forearm length is 35–52 mm, and adult weight is 3–11 grams. According to Bradbury and Emmons (1974), female *S. bilineata* are larger; average weight on Trinidad was 7.5 grams in males and 8.5 grams in nonpregnant females. Coloration is dark brown above and paler brown below. As in *Rhynchonycteris*, two whitish longitudinal lines are present on the back, but *Saccopteryx* does not have tufts of hair on the forearm as does the former genus. A pouchlike gland is present in the membrane that extends from the shoulder to the wrist. This is a scent gland that can be opened by muscular action. It is well developed in males but vestigial or absent in females and may serve to attract the opposite sex.

In Venezuela, Handley (1976) found *S. bilineata* and *S. leptura* mainly in forests but *S. canescens* mainly in open areas. Most specimens of each species were collected near streams or in other moist places. These bats roost in less shaded locations than do many other bats. In southern Mexico they have been found under concrete bridges, in culverts under highways, on the trunks of tall trees such as wild figs *(Ficus)*, inside hollow trees, and at or near the entrances of limestone caves. Tuttle (1970) found *S. bilineata* to be one of the first bats to fly after sundown. In studies in Costa Rica and Trinidad, Bradbury and Vehrencamp (1977) found roosting *S. leptura* to prefer the exposed boles of large trees in riparian forest and *S. bilineata* to roost mostly in hollow trees or the buttress cavities of large trees. Each colony of *S. bilineata* had a series of sites where it moved seasonally to forage for insects. Population density for *S. leptura* was 2.5/ha. in Costa Rica and 17.6/ha. in Trinidad, and respective foraging ranges averaged 0.14 ha. and 1.4 ha. For *S. bilineata*, population density averaged about 0.7/ha., and each colony used an annual foraging area of about 6–18 ha.

Detailed investigations in Costa Rica and Trinidad have revealed marked differences in the social structure of *S. bilineata* and *S. leptura* (Bradbury and Emmons 1974; Bradbury and Vehrencamp 1977). Social groups of *S. bilineata* are composed of a single adult male and a harem of 1–8 females. A number of such groups may be found in a single tree, and together they form a colony of 40–50 individuals, but each male actively defends an area of about 1–3 sq meters of vertical roost and performs elaborate visual and vocal displays to attract a harem of females. There may also be lone adult males in a colony that seek to form their own harems. If a male with a harem is approached by another male, there will be high-frequency barking and rarely striking with wings but no noticeable injuries. Such territorial defense and competition for females goes on throughout the year. There is much movement of females between harems, though the females of an established harem appear aggressive toward newcomers. Individual bats forage alone but adjacent to other colony members. The foraging sites of a colony are divided into exclusive territories defended by harem males against other males. Each of these group territories is in turn divided into territories for each female, from which she chases outside females. When a female changes harems at a roost site, she also changes her foraging site.

In *S. leptura* group size is about one to five in Trinidad and two to nine in Costa Rica. This species appears to have monogamous bonds, and the most common kind of group is simply an adult male and female, but other combinations also occur. Adult males do not defend roosting territories, are not aggressive toward one another, and do not perform rituals to attract females. Unlike *S. bilineata*, in which a harem occupies a single roost for a lengthy period, *S. leptura* roams among a number of roosts. Group composition changes frequently, and both sexes may shift location. *S. leptura* forages either in groups or alone, but even if alone, the foraging areas are adjacent and form a group territory that is actively defended. In Costa Rica the young of *S. leptura* are born at the beginning of the rainy season in May and again in November. In Trinidad births occur only in May. The females carry their young for 10–15 days, after which the young can fly by themselves. The young are weaned at about 2.5 months.

In Trinidad the reproductive season of *S. bilineata* is remarkably synchronized, with each female producing a single young from late May to mid-June, just after the rainy season. The females never leave the young at the roosts, but carry them during nightly foraging, or possibly drop them off in another tree for the night. The young can fly by themselves at about 2 weeks. By 10–12 weeks weaning has occurred, and at this time nearly all young females disperse to other sites. The young males, however, tend to remain near the parents and appear to await the opportunity to take over a harem or split off some females for themselves.

In addition to the above information provided by Bradbury and Emmons (1974) and Bradbury and Vehrencamp (1977), records of pregnant female *S. bilineata* have originated in Jalisco, Mexico, in April (Watkins, Jones, and Genoways 1972); the Yucatan Peninsula in February (Jones, Smith, and Genoways 1973); Guatemala and Belize in March, April, and May (Rick 1968); El Salvador in March and Costa Rica in March, April, May, September, and December (LaVal and Fitch 1977); Panama in January, February, March, and April but not in July (Fleming, Hooper, and Wilson 1972); and Peru in June, July, and August (Tuttle 1970).

CHIROPTERA; EMBALLONURIDAE; **Genus CORMURA**
Peters, 1867

Wagner's Sac-winged Bat

The single species, *C. brevirostris*, is found from Nicaragua to Peru, the Guianas, and central Brazil (Cabrera 1957; Hall 1981; Koopman 1982*b*).

Head and body length is 50–60 mm, tail length is 6–12 mm, and forearm length is 43–50 mm. The upper parts are either deep blackish brown or reddish brown, and the underparts are paler. A long sac-shaped pouch containing a scent gland is in the small triangular membrane that is in front of the upper arm and forearm. The open end of the sac is at the end nearest the wrist.

Cormura resembles *Peropteryx* in the absence of striping,

Wagner's sac-winged bat *(Cormura brevirostris)*, photos by Merlin D. Tuttle.

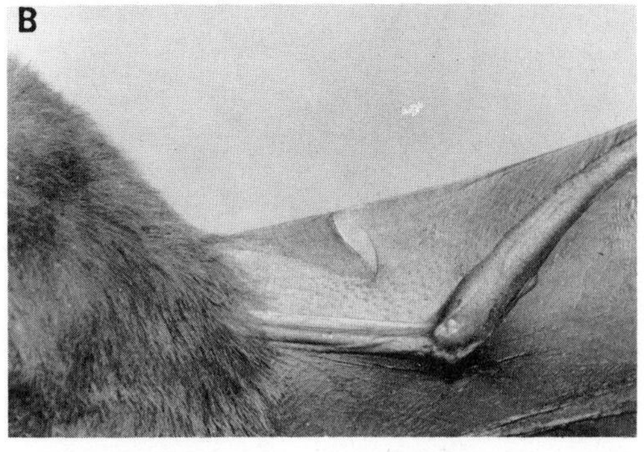

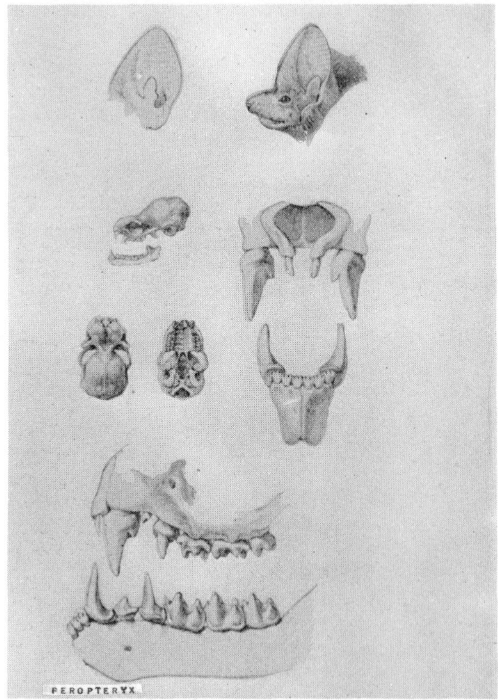

Doglike bat *(Peropteryx macrotis):* A. Photo by P. Morris; B. Photo by Merlin D. Tuttle. Drawings of ear, head, and cranial details from *Die Fledermäuse des Berliner Museums für Naturkunde,* Wilhelm K. Peters.

in coloration, and in size, but it can be distinguished from that genus by its shorter feet and more robust structure and by the differently placed attachments of the wing membrane to the feet, i.e., from the ankle in *Cormura* and from the femur in *Peropteryx.*

One Wagner's sac-winged bat was collected in a hollow fallen tree, and another was taken under the projecting end of a fallen tree. Fleming, Hooper, and Wilson (1972) reported that the stomachs of nine specimens contained 100 percent insect food. These authors also indicated that pregnant females had been found in Panama in April and May but not in June, July, September, and October.

CHIROPTERA; EMBALLONURIDAE; **Genus PEROPTERYX**
Peters, 1867

Doglike Bats

There are two subgenera and three species (Anderson, Koopman, and Creighton 1982; Cabrera 1957; Goodwin and Greenhall 1961; Hall 1981; Koopman 1978*a;* Lemke et al. 1982; Myers, White, and Stallings 1983; Tuttle 1970):

subgenus *Peropteryx* Peters, 1867

P. macrotis, southern Mexico to Paraguay and central Brazil, Grenada, Trinidad and Tobago;
P. kappleri, southern Mexico to Peru and southern Brazil;

subgenus *Peronymus* Peters, 1868

P. leucopterus, eastern Colombia, Venezuela, the Guianas, eastern Peru, Amazonian Brazil.

Peronymus was recognized as a full genus by Baker, Genoways, and Seyfarth (1981) and Corbet and Hill (1986) but not by Cabrera (1957), Honacki, Kinman, and Koeppl (1982), Koopman (1978*a,* 1984*b*), or Robbins and Sarich (1988). Handley (1976) considered *P. trinitatus* of Trinidad and Venezuela to be a species distinct from *P. macrotis.*

Head and body length is about 45–55 mm, tail length is about 12–18 mm, and forearm length is 38–54 mm. Bradbury and Vehrencamp (1977) reported that the weight of *P. kappleri* averaged 9 grams for females and 11 grams for males. Coloration of the upper parts is usually some shade of brown, and the underparts are paler. In *P. leucopterus* the wings are white from the forearm outward.

The ears of *P. macrotis* and *P. kappleri* are separate at the base, but in *P. leucopterus* they are joined by a low membrane. *Peropteryx* differs from *Saccopteryx* in the absence of dorsal stripes and in the position and size of the wing sacs. In *Peropteryx* the wing sacs are near the upper edge of the antebrachial membrane and open outward.

Doglike bats occur in forests, swamps, savannahs, and cultivated areas (Handley 1976). They usually roost in shallow caves or rock crevices where light can enter, but they have been taken from the dark recesses of ruins in the Yucatan and from underground passages of ruins in Chiapas, Mexico. A few have been found in dead trees, and several have been found hanging from palm thatching at the entrance to a ruin. *Peropteryx* has been discovered with *Saccopteryx,* roosting under the arch of a natural bridge. As might be expected of bats that roost in exposed locations, they are alert in the daytime. They sometimes hang from a horizontal surface but often suspend themselves from a vertical surface by spreading out the forearms and feet. These bats, in common with *Saccopteryx* and *Rhynchonycteris*, fold their wings at about a 45° angle to the body when roosting.

In a study in Costa Rica, Bradbury and Vehrencamp (1977) found colonies of *P. kappleri* roosting in fallen logs and the cavities of large trees, usually about one meter above the ground. Population density was about 0.6/ha. Colonies averaged 4.3 (1–6) individuals, and usually several adults of each sex were present. There was no harem formation and no territoriality. Births began in May and possibly continued until October. LaVal (1977) reported a pregnant female *P. kappleri* in April in Costa Rica. Pregnant females of *P. leucopterus* have been found in March, April, May, and June.

In east-central Brazil, Willig (1983) found *P. macrotis* roosting in small aggregations of up to 10 individuals. Each group contained only 1 adult male, thus suggesting the maintenance of harems. Reproductive data for *P. macrotis* include: in the Yucatan, 14 of 16 females taken on 18 April were pregnant, each with a single embryo, and a lactating female was taken in August (Jones, Smith, and Genoways 1973); in Guatemala, pregnant females with one embryo were taken from February to April, and a lactating female was collected in July (Jones 1966; Rick 1968); in southwestern Colombia, 15 of 33 adult females collected in July were pregnant (Arata and Vaughan 1970); and in Peru, pregnant females were taken in August (Tuttle 1970).

CHIROPTERA; EMBALLONURIDAE; **Genus**
CYTTAROPS
Thomas, 1913

The single species, *C. alecto*, occurs from Nicaragua to northeastern Brazil and the Guianas (Cabrera 1957; Hall 1981).

Head and body length is about 50–55 mm, tail length is about 20–25 mm, and forearm length is about 45–47 mm. The coloration is a uniform dull smoky gray. The most distinctive feature of this bat is the tragus of the ear; the lower half of the outer margin consists of a large, angular structure that is unique among bats. A wing sac is not present, and apparently there are no glands associated with the tail.

The type specimen was collected in a garden. A series of specimens from Costa Rica was collected during the day as the bats hung in groups of one to four under the fronds of coconut palms located near buildings. The groups were composed of mixed sexes and ages. Insect remains were found in the digestive tracts of some of these specimens (Starrett 1972).

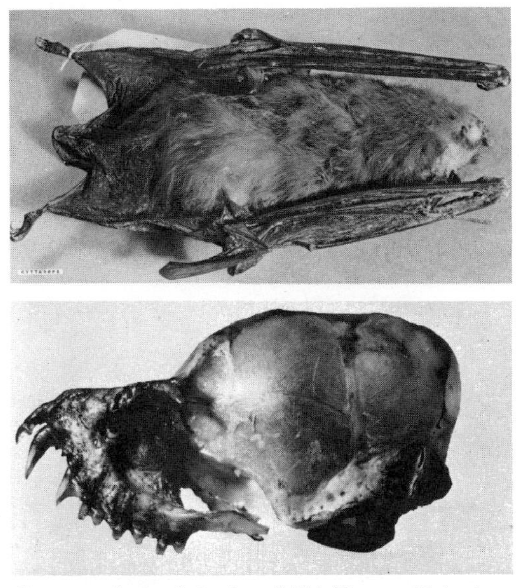

Cyttarops alecto, photos from British Museum (Natural History).

CHIROPTERA; EMBALLONURIDAE; **Genus**
DICLIDURUS
Wied-Neuwied, 1820

Ghost Bats, or White Bats

There are four species (Cabrera 1957; Ceballos and Medellín L. 1988; Goodwin and Greenhall 1961; Hall 1981; Handley 1976; Husson 1978; Jones, Swanepoel, and Carter 1977; Koopman 1982*b*; Ojasti and Linares 1971; Tuttle 1970):

D. albus, southern Mexico to Panama, Colombia, Ecuador, eastern Peru, Venezuela, Surinam, northern and eastern Brazil, Trinidad;
D. ingens, southeastern Colombia, southern Venezuela;
D. scutatus, Venezuela, the Guianas, eastern Peru, northern Brazil;
D. isabellus, southern Venezuela, Amazonian Brazil.

The last of the above species long was known from only a

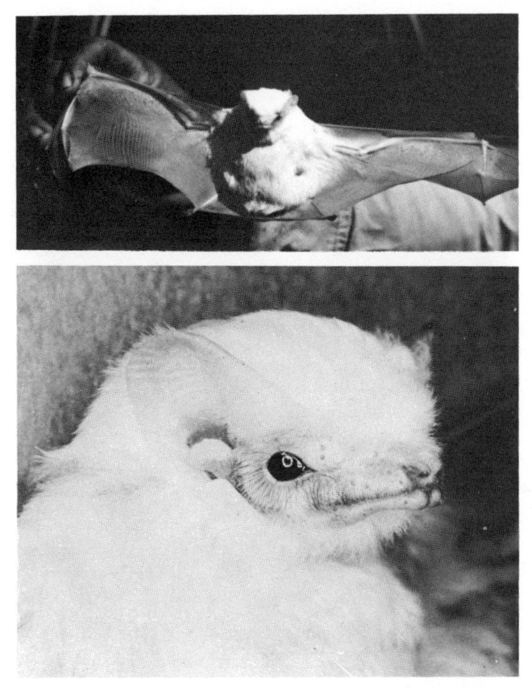

White bat *(Diclidurus albus)*: Top, photo by Cornelio Sanchez Hernandez; Bottom, photo by Merlin D. Tuttle.

single specimen and was placed in the separate genus *Depanycteris* Thomas, 1920. Handley (1976), however, who collected 28 specimens of this species in Venezuela, placed it in the genus *Diclidurus. D. virgo* of South America sometimes has been treated as a species distinct from *D. albus.*

Head and body length is 50–80 mm, tail length is 12–25 mm, and forearm length is 45–73 mm. Coates-Estrada and Estrada (1985) reported a weight of 32.8 grams for one specimen. The coloration is unusual for bats, generally being white or white mixed with gray. The hairs of the upper parts are slate-colored at the base and white or gray for most of their length. The hairs of the underparts are slate-colored on the basal half and white on the outer half. The wings are also white. A wing sac is not present. These are the only white bats except for *Ectophylla alba* and occasional albinos. In the type specimen of *D. isabellus* the ventral surface of the body is dull buffy white, but the upper parts are pale brown.

The eyes are large and the ears are short and rounded. The tail is shorter than the tail membrane and protrudes upward through it. Near where the tail penetrates the membrane, there appears to be either a glandular structure or a small pouch. The short thumb has only a rudimentary claw and is partly concealed in the wing membrane.

Handley (1976) collected most specimens of *D. ingens, D. albus,* and *D. scutatus* in open yards, but he obtained *D. isabellus* only in evergreen forest; all species were found predominantly near streams or in other moist places. According to Ceballos and Medellín L. (1988), *D. albus* has been taken in tropical rainforests, tropical dry-deciduous and semideciduous forests, coconut plantations, and disturbed vegetation; it roosts by day under the leaves of palms and may undertake local seasonal or migratory movements. Starrett and Casebeer (1968) reported the collection of *D. albus* in Costa Rica, both in the lowlands and at altitudes of 1,100–1,500 meters. Some specimens were taken as they pursued moths and other insects. The call of *D. albus* was reported to be a "unique musical twittering not heard in other Costa Rican bats." Jones (1966) observed about 100 *D. albus* in the floodlights of an oil rig in Guatemala; the bats foraged at heights of about 3–135 meters.

Ceballos and Medellín L. (1988) stated that *D. albus* is solitary for most of the year but that early in the breeding season up to four individuals, usually a male and several females, may be found roosting within 5–10 cm of each other. Mating probably occurs in January and February, and pregnant females, each with a single embryo, have been taken from January to June.

CHIROPTERA; **Family CRASEONYCTERIDAE; Genus CRASEONYCTERIS**
Hill, 1974

Kitti's Hog-nosed Bat

The single known genus and species, *Craseonycteris thonglongyai,* has thus far been found only at Sai Yoke, Kanchanaburi Province, western Thailand (Hill 1974*d;* Lekagul and McNeely 1977).

About the size of a large bumblebee, this bat may be the world's smallest mammal. Head and body length is 29–33 mm, there is no tail, forearm length is 22–26 mm, and weight is about 2 grams. There seem to be two color phases, one with the upper parts brown to reddish and the other with the upper parts distinctly gray. The underparts are paler; the wings and interfemoral membrane are dark. The ears are relatively large, and the eyes are small and largely concealed by fur.

Distinctive external characters include the very small size, lack of a tail (though there are two caudal vertebrae), lack of a calcar, presence of a large interfemoral membrane, and a prominent glandular swelling at the base of the underside of the throat in males. The muzzle is rather piglike, slightly swollen around the nostrils and chin. The wide, slightly

Kitti's hog-nosed bat *(Craseonycteris thonglongyai)*, photo by Jeffrey A. McNeeley.

crescent-shaped nostrils open directly in the face of the nose pad and are separated by a relatively wide septum, which broadens to form a small nose pad. The large ears reach beyond the tip of the muzzle when laid forward. The tragus is a little less than half the length of the ear, basally narrow, then broadly expanding to reach its greatest width about halfway along its length. The wing is relatively long and wide, with a long tip adapted for hovering flight. The membrane in front of the forearm and upper arm (propatagium) is broad, the thumb is short with a well-developed claw, and the hind foot is long, narrow, and slender. The penis is relatively large. Females have one pair of pectoral and one pair of pubic nipples, the latter probably being vestigial.

The skull is very small, with a slightly inflated, globose braincase. There are no lambdoidal crests, postorbital processes, or supraoccipital ridges, but a rather prominent sagittal crest occurs in both sexes. The zygomata are slender but complete; the bullae are relatively large and flattened on the inner face, and the palate is short and wide. The dental formula is: (i 1/2, c 1/1, pm 1/2, m 3/3) × 2 = 28. The upper incisors are relatively large and have strong cingula at the base. The lower incisors are long and narrow, about equal in size, and tricuspid, the central cusp being the largest.

The Craseonycteridae are considered to be related to the Rhinopomatidae and the Emballonuridae but to differ sufficiently to warrant familial status. *Craseonycteris* most nearly resembles *Rhinopoma,* the sole genus of the family Rhinopomatidae, but differs in having nostrils that are not valvular or slitlike, large ears not joined anteriorly by a band of integument, a tragus that is widest in the middle, no tail, a more inflated braincase, and relatively larger incisors.

This bat has been collected in January, March, October, and December at Sai Yoke, a formerly forested but now completely cleared area riddled with limestone caves. *Craseonycteris* was found only in small caves, far from the entrance, in the most remote caverns. Approximately 10–15 individuals were seen roosting high on a wall, well separated from one another. Further observations indicated that this bat begins to fly just at dusk and that its usual hunting pattern is to fly around the tops of bamboo clumps and teak trees. It may glean on foliage, as well as take small insects on the wing. According to Hill and Smith (1981), the stomach contents of one specimen consisted of the fragments of small insects and a spider. *C. thonglongyai* is classified as rare by the IUCN and as endangered by the USDI.

CHIROPTERA; **Family NYCTERIDAE; Genus NYCTERIS**
G. Cuvier and E. Geoffroy St.-Hilaire, 1795

Slit-faced Bats, or Hollow-faced Bats

The single known genus, *Nycteris,* contains 13 species (Bergmans and Van Bree 1986; Ellerman and Morrison-Scott 1966; Hayman and Hill, *in* Meester and Setzer 1977; Koch-Weser 1984; Koopman 1975; Koopman, Mumford, and Heisterberg 1978; Medway 1977; Nader and Kock 1983*a;* Rautenbach, Fenton, and Braack 1985; Van Cakenberghe and De Vree 1985):

N. arge, Sierra Leone to southern Zaire and western Kenya;
N. nana, Ivory Coast to southern Zaire and western Kenya;
N. major, Ivory Coast to eastern and southern Zaire;

Slit-faced bat *(Nycteris thebaica),* photo by Merlin D. Tuttle, Bat Conservation International.

N. intermedia, Ivory Coast to western Tanzania and Angola;
N. tragata, Malay Peninsula (including Tenasserim), Sumatra, Borneo;
N. javanica, Java, Bali, Kangean Islands;
N. grandis, Guinea to Uganda and Zimbabwe;
N. hispida, most of Africa south of the Sahara;
N. macrotis, Gambia to Sudan and Somalia, and south to Angola and Botswana, Madagascar;
N. woodi, Cameroon, Somalia to Transvaal;
N. thebaica, open country throughout most of Africa, Palestine, Arabian Peninsula, island of Corfu (Greece);
N. gambiensis, Senegal to Burkina Faso and Benin;
N. vinsoni, Mozambique.

An old report of *N. javanica* from Timor was not accepted by Bergmans and Van Bree (1986) or Goodwin (1979). *N. madagascariensis* of Madagascar is sometimes considered a full species and sometimes a synonym of *N. thebaica,* but Van Cakenberghe and De Vree (1985) indicated that it is a synonym of *N. macrotis.*

Head and body length is 40–93 mm, tail length is about 43–75 mm, forearm length is 32–60 mm, and adult weight is 10–30 grams. The pelage is long and rather loose. The coloration is from rich brown or russet to pale brown or grayish. *N. macrotis* exhibits considerable orange color variation in Zaire, and an albino specimen of *N. nana* has been found.

The muzzle is divided by a longitudinal furrow, with the nostrils in the anterior end of this groove. This furrow is margined and concealed by nose leaves and expands posteriorly into a deep pit on the forehead. The interorbital region of the skull is deeply concave. The lower lip has a granular surface at the tip. The large ears are longer than the head and united at their bases by a low membrane. A small tragus is

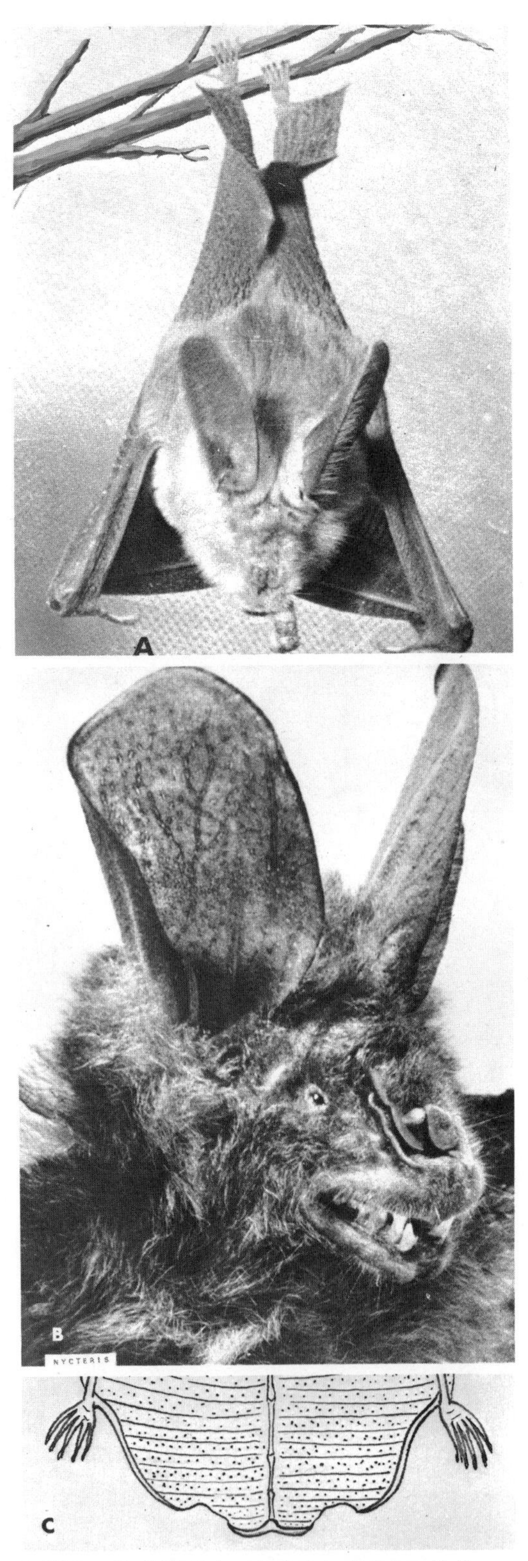

A. Slit-faced bat (*Nycteris* sp.), photo by V. Aellen. B. *N. thebaica*, photo by Erwin Kulzer. C. *N. javanica* showing terminal cartilage, drawing by William J. Schaldach, Jr., of specimen in U.S. National Museum of Natural History.

present. The wings are broad. The long tail is completely enclosed within the large interfemoral membrane and terminates in a T-shaped tip, which serves as a support for the free edge of the tail membrane. This is a distinctive character that is unique among mammals.

The upper incisors closely resemble the lower incisors in size and form. The dental formula is: (i 2/3, c 1/1, pm 1/2, m 3/3) $\times$ 2 = 32.

Most species inhabit woodland savannah or dry country, but some, including *N. grandis* and *N. javanica*, occupy dense forest (Hayman and Hill, *in* Meester and Setzer 1977; Lekagul and McNeely 1977; Rosevear 1965). Roosting sites include hollow trees, dense foliage, rocky outcrops, caves, buildings, ruins, culverts, abandoned wells, and even porcupine and aardvark burrows. These bats feed on a variety of arthropods, including moths, butterflies, other insects, spiders, and sun spiders. Scorpions form a staple part of the diet of *N. thebaica*. Most species flutter around trees and bushes, but they also come into lighted verandas and rooms in search of insects. Captive *N. grandis* have captured and consumed other bats and frogs (Fenton, Gaudet, and Leonard 1983), and this species is now known to regularly feed on fish, frogs, birds, and bats in the wild (Fenton, Thomas, and Sasseen 1981).

Most species shelter alone, in pairs, or in small family groups, but *N. hispida* occasionally gathers in groups of about 20, and *N. thebaica* and *N. gambiensis* are even more gregarious. Smithers (1971) reported a colony of 500–600 *N. thebaica* in one cavern. He also observed that the young of this species were born in Botswana during the warm, wet summer months from about September to February. Bernard (1982) found this same species to mate in early June in South Africa; gestation then lasted five months and lactation another two months. Kingdon (1974*a*) wrote that thousands of *N. thebaica* had been reported roosting together in caves in South Africa. Other data summarized by Kingdon include a pregnant female *N. macrotis* found in Uganda in late December; a lactating female *N. macrotis* found in Tanzania in November; and embryos or young of *N. grandis* found in Gabon in April, August, and November. Pregnant female *N. thebaica*, with one embryo each, have been recorded from Egypt in April (Gaisler, Madkour, and Pelikan 1972) and from Zambia in August (Ansell 1960). In South Africa *N. thebaica* gives birth in November and December (Herselman and Norton 1985). *N. javanica* breeds throughout the year, and females mate again shortly after giving birth (Lekagul and McNeely 1977). Young *N. nana* nurse from 45 to 60 days.

No fossils referable to this family have been found.

CHIROPTERA; **Family MEGADERMATIDAE**

False Vampire Bats and Yellow-winged Bats

This family of four Recent genera and five species occurs in Africa, south-central and southeastern Asia, the East Indies, and Australia.

Head and body length is 65–140 mm, an external tail is lacking, and forearm length is 50–115 mm. The Australian member of this family, *Macroderma gigas*, is the largest of the bats other than some of those in the family Pteropodidae and the New World species *Vampyrum spectrum*; it is known as the Australian giant false vampire.

These bats are noted for their large, basally united ears (largest in the genus *Lavia*), divided tragus, and long, erect

African yellow-winged bat *(Lavia frons)*, photos by David Pye.

nose leaf. The genera differ from each other in size and in dental and skull features.

Upper incisors are lacking. The upper canines project noticeably forward, the shaft having a large secondary cusp. The dental formula in the subgenera *Megaderma* and *Lyroderma* of the genus *Megaderma* is (i 0/2, c 1/1, pm 2/2, m 3/3) × 2 = 28, but in the genera *Cardioderma*, *Macroderma*, and *Lavia* it is (i 0/2, c 1/1, pm 1/2, m 3/3) × 2 = 26.

Megaderma, *Cardioderma*, and *Macroderma* roost in caves, rock crevices, buildings, and hollow trees and are often the sole occupant of their retreat, but *Lavia* generally shelters in trees and bushes. These bats shelter singly or in groups of as many as 50–100 individuals. Although not particularly rapid fliers, they are skillful and dexterous in flight. *Lavia* is often active during the day, but the other members of this family hunt only at night. These bats, and others with a nose leaf, possibly fly with the mouth closed, emitting high-frequency sounds through their nostrils.

The common name "false vampire bat" reflects the old but erroneous belief that these bats feed on blood, as do the true vampire bats (Desmodontinae) of the New World. The diet of *Megaderma* and *Macroderma* consists of insects and a variety of small vertebrates, and *Lavia* feeds mainly, if not entirely, on insects. Although *Megaderma lyra* may first consume the blood and then eat the flesh of an animal it has captured, no member of this family preys on another animal solely for its blood.

The geological range of this family is early Oligocene to early Pliocene in Europe, early Oligocene to Recent in Africa, Pleistocene to Recent in Asia, and middle Miocene to Recent in Australia (Hand 1985; Koopman 1984*c*; Koopman and Jones 1970).

CHIROPTERA; MEGADERMATIDAE; **Genus MEGADERMA**
E. Geoffroy St.-Hilaire, 1810

Asian False Vampire Bats

There are two subgenera and two species (Ellerman and Morrison-Scott 1966; Roberts 1977):

subgenus *Megaderma* E. Geoffroy St.-Hilaire, 1810

M. spasma, India and Sri Lanka to Indochina, Malaysia, Indonesia, Philippines;

subgenus *Lyroderma* Peters, 1872

M. lyra, eastern Pakistan and Sri Lanka to southeastern China and northern Malay Peninsula.

Head and body length is about 65–95 mm, there is no external tail, and forearm length is 50–75 mm. Lekagul and McNeely (1977) listed weights of 23–28 grams for *M. spasma* and 40–60 grams for *M. lyra*. *M. spasma* is smoky bluish gray above and brownish gray below. *M. lyra* is grayish brown above and whitish gray below.

These bats roost in groups in caves, pits, buildings, and hollow trees and are usually the sole occupant of their retreat. *M. spasma* generally inhabits wetter areas than does *M. lyra* (Lekagul and McNeely 1977). *M. lyra* is carnivorous, feeding on insects, spiders, and small vertebrates such as other bats, rodents, birds, frogs, and fish. Captured prey is carried to the

Asian false vampire bat *(Megaderma spasma)*, photo by Paul D. Heideman.

roost to be eaten, and vertebrate remains may be found beneath the roost. *M. spasma* is less carnivorous, apparently favoring grasshoppers and moths, but it also carries prey to a feeding roost (Lekagul and McNeely 1977). Both species often feed among trees and undergrowth, usually flying within three meters of the ground, and they will enter houses to pick lizards and insects off the wall.

Both species usually roost in groups of about 3–30, but a seasonal colony of 1,500–2,000 *M. lyra* has been reported in India. The sexes of *M. spasma* live together year round, whereas the adult males and females of *M. lyra* appear to segregate when birth is imminent. Mating in India takes place from about November to January, gestation lasts 150–60 days, and the young are born from April to June, before the onset of the rains. There is usually one young, occasionally two. The young grow rapidly, are initially carried by the females, and are suckled for 2–3 months. In *M. lyra* the males are sexually mature at 15 months, and the females at 19 months.

CHIROPTERA; MEGADERMATIDAE; **Genus CARDIODERMA**
Peters, 1873

African False Vampire Bat

The single species, *C. cor*, is found in East Africa from northeastern Ethiopia and southern Sudan to northern Zambia. Although it is sometimes considered a subgenus of *Megaderma*, most recent authors have given *Cardioderma* full generic rank (Hayman and Hill, *in* Meester and Setzer 1977; Kingdon 1974*a*; Koopman 1975).

Head and body length is 70–77 mm, there is no tail, forearm length is 54–59 mm, and adult weight is 21–35 grams. This bat is uniformly blue gray in color. From *Megaderma*, *Cardioderma* differs in having a narrower frontal shield, only one upper premolar, and no reduction of the third upper molar (Lekagul and McNeely 1977).

Most of what is known about this genus stems from a detailed study by Vaughan (1976) in bushland in southern Kenya. The bat was found to roost in hollow baobab trees, in groups of up to 80 individuals. At night each bat utilized an exclusive foraging area; two such areas measured 0.10 ha. and 0.11 ha. in mid-April, during the wet season, and two others measured 0.55 ha. and 1.01 ha. in late August, during the dry season, when food was less abundant. During the night, individuals spent considerable time perching in low vegetation, listening for terrestrial prey. Flights to capture prey and to move between perches were brief, usually lasting less than five seconds and covering less than 25 meters. Food was generally captured on the ground and consisted mainly of

African false vampire bat *(Cardioderma cor)*, photo by Erwin Kulzer.

African false vampire bat *(Cardioderma cor)*, photo by D. W. Yalden.

large, ground-dwelling beetles, but centipedes and scorpions, and rarely small bats, were also taken. During part of the rainy season, leaf gleaning and aerial capture of insects were important. During the March–April wet season, individuals spent considerable time "singing" and establishing their exclusive foraging areas. The "song" consisted of four to nine high-intensity pulses, each with a sharp chip like that given by some passerine birds. There was also a less frequent, more variable, and louder flight call. Females gave birth twice each year, during the rainy periods, in March or early April and in November. More than one young per female was never observed. Kingdon (1974*a*) indicated that the breeding season may be more extended in some areas and stated that gestation lasts about three months. Additional studies in Kenya have suggested the existence of long-lasting male-female pairs and territories (McWilliam 1987*b*).

CHIROPTERA; MEGADERMATIDAE; **Genus MACRODERMA**
Miller, 1906

Australian Giant False Vampire Bat, or Ghost Bat

The single species, *M. gigas,* may be found throughout that part of Australia north of 29° S (Ride 1970). Quaternary fossil remains indicate that the same species formerly occupied much of southern Australia (Molnar, Hall, and Mahoney 1984). Based on habitat suitability and reported observations, Filewood (1983) stated that there is little doubt that *M. gigas* also occurs in New Guinea.

This is among the largest of the bats of the suborder Mi-

Australian giant false vampire bat *(Macroderma gigas)*, photo by Stanley Breeden.

crochiroptera. Head and body length is 100–140 mm, there is no external tail, forearm length is 105–15 mm, and wingspan is about 600 mm. Hudson and Wilson (1986) listed weight as 130–70 grams. The hairs of the back are usually white with pale grayish brown tips that impart an ashy gray effect, and the head, ears, nose leaf, membranes, and underparts are generally whitish. Several specimens from Queensland were deep brownish drab above and plumbeous below, with darkened skin. The vernacular name "ghost bat" refers to the usual pale coloration.

The ghost bat occurs in tropical areas, roosting in caves, rock crevices, and old mine shafts. It emerges from its shelter after sunset and flies a straight, smooth course. Tidemann et al. (1985) found it to fly an average of 1.9 km from the roost to the foraging area. There it would hang from a tree at heights of up to 3 meters and watch for prey. *Macroderma* is carnivorous and drops on small mammals from above, envelops them with its wings, and kills by biting the head and neck. The most common prey in the wild is house mice, while native rodents, small marsupials, other bats, birds, reptiles, and insects are also taken (Douglas 1967). The prey is carried to a high point or back to the main roost to be consumed, and uneaten remains often accumulate under such sites. Individuals of other species of bats have been found in a cave occupied by the ghost bat; these smaller bats, *Taphozous georgianus* and *Pipistrellus pumilus,* were wedged deeply into crevices, possibly to avoid attacks by the ghost bat.

Macroderma roosts alone or in small groups. Ride (1970) stated that field observations of this genus in Western Australia showed that females give birth to a single young about November in southern areas and about September in the far north. It seems likely that at these times the females congregate without the males in maternity colonies, while the males gather in all-male colonies. In a study of a cave colony of about 150 individuals in central Queensland, however, Toop (1985) found some males to always be present with the females. Mating occurred in April; the group then dispersed for the cooler months and came together again prior to the birth season, which occurred from mid-October to late November. The young were initially carried by the mothers and then left at the roost. They began flying on their own at 7 weeks, weaning was complete by March, and both sexes reached reproductive maturity in their second year of life.

On the basis of numerous remains and guano deposits, it can be supposed that *Macroderma* formerly was common in the southern part of Australia, where it does not now occur. It has been suggested that the absence of *Macroderma* from this region may be related to increasingly arid conditions. Such aridity would deplete the insect life and so result in a reduction in the numbers of the insectivorous bats upon which the ghost bat feeds. For some years it was thought that the ghost bat was extremely rare even where it did survive, but such a view seems to have developed because of the secretive habits of this bat. Recent field studies have shown *Macroderma* still to have a wide distribution and to be relatively common in some areas (Ride 1970). Nonetheless, quarrying operations and other human activity jeopardize some of the remaining colonies, and the IUCN classifies the ghost bat as vulnerable.

CHIROPTERA; MEGADERMATIDAE; **Genus LAVIA**
Gray, 1838

African Yellow-winged Bat

The single species, *L. frons*, occurs from Gambia to Ethiopia and south to northern Zambia (Hayman and Hill, *in* Meester and Setzer 1977).

Head and body length is 58–80 mm, there is no external tail, and forearm length is 49–63 mm. Kingdon (1974*a*) gave a weight of 28–36 grams. The color of the body is blue gray, pearl gray, or slaty gray; the lower back is sometimes brownish or olive green; and the underparts occasionally are yellow. The ears and wings are reddish yellow.

This bat inhabits forests and open country, generally being found around swamps, lakes, and rivers where trees and bushes fringe the water and afford roosting sites. It prefers to roost where it can observe its surroundings, hence usually in an environment where the herbaceous growth is not too dense. On the savannahs of Kenya, Vaughan and Vaughan (1986) found *Lavia* to roost in acacia trees at heights of 5–10 meters above the ground. It also has been found in the cavity of a tree and in buildings. This bat is alert and often active during the day. It sometimes betrays its presence by the motion of its long, sensitive ears. Vaughan and Vaughan reported that the head is held up almost constantly in daylight, the body revolves, and the eyes were never seen to be closed. When disturbed, it generally flies away for a considerable distance, even on a bright day.

Unlike other members of this family, *Lavia* is not known to feed on small vertebrates. The diet is mainly, if not entirely, insects. Vaughan and Vaughan (1986) suggested that *Lavia* is closely associated with acacia trees, which attract many insects during periods of flowering. The method of feeding has been compared to that of a flycatcher: the bat often hangs from a branch at night, constantly scanning the area until an insect is detected, and then swoops down to it and returns to the same or another perch to consume the catch. This bat has been flushed from grass, where it may have been feeding. It can dodge about between bushes and trees with considerable skill.

Vaughan and Vaughan (1986) found *Lavia* to live in permanent, territorial pairs. This lifestyle may be an adaptation to environmentally produced stress resulting from the recurring periods of low food availability when acacias are not in flower. Monogamy is necessary to allow the female to concentrate her energy on raising young. The members of a pair roost together, less than a meter apart, and interact at night through flying rituals. They maintain an exclusive foraging territory of 0.60–0.95 ha. The male regularly patrols this area and drives away intruding conspecifics. The roosting site is usually near the edge of the territory, proximal to, but more than 20 meters from, the closest neighboring pair. A single young is born to the pair in early April, at the start of the long rainy season. Vaughan and Vaughan (1987) added that the young clings tenaciously to its mother for several weeks, even when she forages. The young then is left at the roost for about a week and begins flying alone in early May. Weaning occurs 20 days after the first solo flight, when the young is about 55 days old, but it continues to share its parents' territory and roost for at least another 30 days.

Kingdon (1974*a*) provided the following additional reproductive data: in eastern Zaire, pregnancies are most nu-

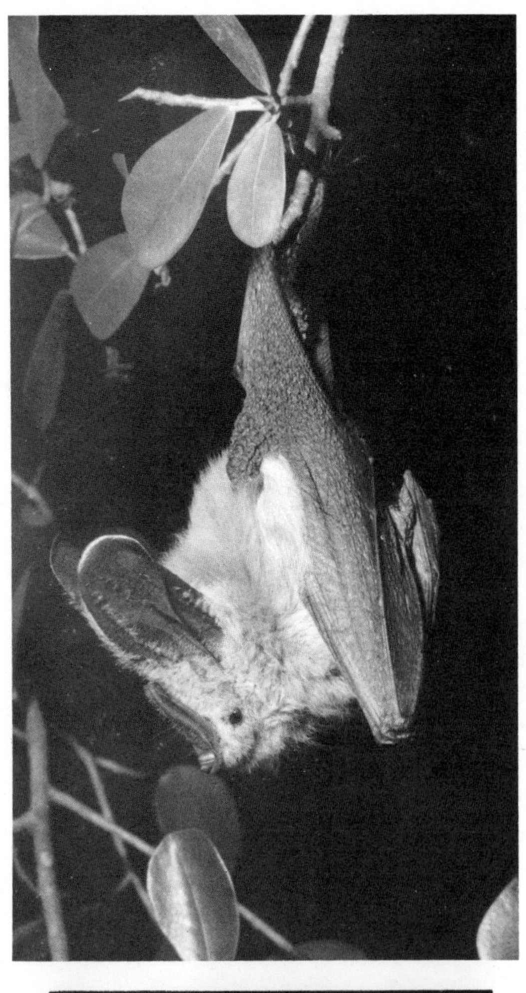

African yellow-winged bat *(Lavia frons)*, photo by Bruce J. Hayward. Inset: skull, photo by P. F. Wright of specimen in U.S. National Museum of Natural History.

merous from January to April, but there are scattered records at other times; in Garamba National Park, in extreme northeastern Zaire, there appears to be a defined breeding season, with births starting in April; and gestation lasts about three months. Koopman, Mumford, and Heisterberg (1978) reported the collection of pregnant females in Burkina Faso, each with a single embryo, in March, May, and September.

CHIROPTERA; **Family RHINOLOPHIDAE; Genus RHINOLOPHUS**
Lacépède, 1799

Horseshoe Bats

The single Recent genus, *Rhinolophus*, contains 69 species (Aellen 1973; Aggundey and Schlitter 1984; Allen 1939*b*; Bergmans and Van Bree 1986; Brosset 1984; Chasen 1940; Corbet 1978; DeBlase 1980; Ellerman and Morrison-Scott 1966; R. E. Goodwin 1979; Hayman and Hill, *in* Meester and Setzer 1977; Herselman and Norton 1985; Hill 1962, 1982*a*, 1983*b*, 1986; Hill, Harrison, and Jones 1988; Hill and Schlitter 1982; Hill and Thonglongya 1972; Hill and Yoshiyuki 1980; Koopman 1975, 1979, 1982*a*; Laurie and Hill 1954; Lawrence 1939; Lekagul and McNeely 1977; McFarlane and Blood 1986; McKean 1972; Medway 1977, 1978; Qumsiyeh 1985; Ride 1970; Taylor 1934):

R. ferrumequinum, the entire southern Palaearctic region from Great Britain and Morocco to Afghanistan and Japan;
R. clivosus, central and southwestern Asia, most of the open country of Africa from Algeria and Cameroon eastward, Liberia;
R. bocharicus, Azerbaijan, northeastern Iran, Turkmen and Uzbek S.S.R., northern Afghanistan;
R. darlingi, Nigeria and Namibia to Tanzania and Transvaal;
R. silvestris, Gabon, Congo;
R. deckeni, eastern Kenya and Tanzania, Zanzibar and Pemba islands;
R. landeri, Gambia to Somalia, and south to Angola and eastern Transvaal;
R. guineensis, Senegal, Sierra Leone, Guinea;
R. alcyone, forest zone from Senegal to northeastern Zaire;
R. affinis, northern India to southern China and Indochina, Malay Peninsula, Sumatra, Java, Kangean Islands, Borneo, Flores;

A. Lesser horseshoe bat *(Rhinolophus hipposideros)* carrying young in flight, photo by Ronald Thompson. B. Greater horseshoe bat *(R. ferrumequinum)*, photo by Liselotte Dorfmüller. C. Lesser horseshoe bat *(R. hipposideros)* in sleeping position, photo by Eric J. Hosking.

R. rouxi, India and Sri Lanka to southeastern China;
R. thomasi, Yunnan (southern China), Burma, Thailand, Indochina;
R. malayanus, Thailand, Indochina, Malay Peninsula;
R. robinsoni, Thailand, Malay Peninsula and islands along east coast;
R. stheno, southern Thailand, Malay Peninsula, Sumatra, Java;
R. borneensis, Cambodia, Borneo and nearby islands, Java;
R. celebensis, Java, Sulawesi;
R. madurensis, Madura and Kangean Islands northeast of Java, Timor;
R. importunus, Java;
R. nereis, Siantan and Bunguran islands between Malay Peninsula and Borneo;
R. simplex, Lombok and Sumbawa islands east of Java;
R. megaphyllus, Papua New Guinea, Bismarck Archipelago, St. Aignan's Island in Louisiade Archipelago, eastern Australia;
R. keyensis, islands between Sulawesi and New Guinea;
R. virgo, Philippines;
R. anderseni, Philippines;
R. hipposideros, British Isles to Arabian Peninsula and Central Asia, Morocco to Sudan;
R. euryale, Mediterranean region of southern Europe and northern Africa, east to Iran and Turkmen S.S.R.;
R. mehelyi, southern Europe, Morocco to Iran;
R. blasii, Italy to Afghanistan, Morocco to Ethiopia and Transvaal;
R. capensis, southern Africa;
R. simulator, Guinea, northern Nigeria, Cameroon, Sudan to South Africa;
R. denti, Guinea, southern Africa;
R. swinnyi, southern Zaire and Tanzania to South Africa;
R. adami, Congo;
R. acuminatus, southern Thailand and Cambodia, Malay Peninsula, Sumatra, Nias and Enggano islands, Java, Bali, Lombok, Borneo, Palawan;
R. lepidus, Afghanistan and India to southern China and Malay Peninsula, Sumatra;
R. pusillus, northern India to southern China and Malay Peninsula, Hainan, Sumatra, Java, Borneo;
R. cornutus, Japan, Ryukyu Islands, possibly eastern China;
R. subbadius, northern India, Nepal, Assam, Burma, northern Viet Nam;
R. monoceros, Taiwan;
R. cognatus, Andaman Islands;
R. imaizumi, Iriomote Island (southern Ryukyu Islands);
R. refulgens, Malay Peninsula and nearby islands, Sumatra;
R. rex, Sichuan (south-central China);
R. macrotis, northern India, Nepal, southern China, Thailand, Indochina, Malay Peninsula, Sumatra, Philippines;
R. marshalli, known only by a single specimen from southern Thailand;
R. paradoxolophus, known only by one specimen from northern Viet Nam and another from northeastern Thailand;
R. trifoliatus, northern India, Tenasserim, southern Thailand, Malaysia, Indonesia;
R. luctus, India and Sri Lanka to Thailand, southeastern China, Taiwan, Hainan, Malaysia, Indonesia;
R. pearsoni, northern India to southeastern China and Thailand;
R. yunanensis, Yunnan (southern China), eastern Assam, northern Burma, Thailand;
R. philippinensis, Philippines, Borneo, Sulawesi, Timor, Kei Islands southwest of New Guinea, northeastern Queensland;
R. hirsutus, known only by a single specimen from Guimaras Island (Philippines);
R. sedulus, Malay Peninsula, Borneo;
R. euryotis, Sulawesi, Molucca Islands, Aru Islands, New Guinea, Bismarck Archipelago;
R. creaghi, Borneo, Madura Island northeast of Java;
R. canuti, Java, Timor;
R. coelophyllus, Burma, Thailand, Malay Peninsula;
R. shameli, Burma, Thailand, Cambodia, Malay Peninsula;
R. inops, Mindanao;
R. rufus, Luzon, Mindanao;
R. subrufus, Luzon, Mindoro, Negros, Mindanao;
R. arcuatus, Sumatra, Borneo, Philippines, Sulawesi, Buru Island, New Guinea;
R. maclaudi, Guinea;
R. ruwenzorii, Ruwenzori region of eastern Zaire and western Uganda;
R. hilli, Ruwenzori region of eastern Zaire, western Uganda, and western Rwanda;
R. hildebrandti, East Africa from southern Sudan and Somalia to Transvaal;
R. fumigatus, Africa south of the Sahara;
R. eloquens, southern Sudan, Kenya, Uganda.

The organization of the above list, and the maintenance of the Rhinolophidae as a completely separate family from the Hipposideridae, is based in part on information presented by Lekagul and McNeely (1977), which in turn stems partly from unpublished work by J. E. Hill and the late Kitti Thonglongya. Koopman's (1975) arrangement of African species of *Rhinolophus* would indicate some differences from the above sequence. The species *R. paradoxolophus* was formerly placed in the separate genus *Rhinomegalophus* Bourret, 1951, but is now considered a species of *Rhinolophus* (Hill 1972*a*; Thonglongya 1973). The species named above as *R. hilli* may be part of *R. ruwenzorii*, which in turn may be only a subspecies of *R. maclaudi* (Baeten, Van Cakenberghe, and De Vree 1984). *R. madurensis* was listed, with some question, as a subspecies of *R. celebensis* by Hill (1983*b*). Bergmans and Rozendaal (1982) described *R. tatar* from northeastern Sulawesi, but Hill (1983*b*) considered it a subspecies of *R. euryotis*.

Head and body length is about 35–110 mm, tail length is 15–56 mm, and forearm length is 30–75 mm. The weight of *R. hipposideros*, one of the smaller species, ranges from 4 to 10 grams; *R. ferrumequinum*, a larger species, weighs from 16.5 to 28 grams. Color varies greatly, ranging from reddish brown to deep black above and paler below.

These bats have a peculiar, complex, nose-leaf expansion of the skin surrounding the nostrils. It consists of three parts. The lower part, which is horseshoe-shaped, covers the upper lip, surrounds the nostrils, and has a central notch in the lower edge. Above the nostrils, the appendage is a pointed, erect structure, the lancet, attached only by its base. Both the horseshoe and the lancet are flattened from front to back. The sella, located between the horseshoe and the lancet, is flattened from side to side; it is connected at its base by means of folds and ridges. The shape and arrangement of the nose leaf varies from species to species. These bats generally fly with their mouth closed and emit ultrasonic sounds through the nostrils. The sounds thus emitted may be oriented with the aid of the nose leaf.

The ears are large and lack a tragus. Two teatlike processes not connected with a mammary gland, known as dummy teats, are found on the abdomens of females in addition to the

Horseshoe bat *(Rhinolophus ferrumequinum):* Left photo, showing the complex nose leaf of the Rhinolophidae, by Walter Wissenbach; Right photo by P. Rödl.

two functional mammae on the chest. An infant horseshoe bat may grasp the dummy teats of its mother while she carries it during flight.

Young horseshoe bats shed milk teeth before birth. The teeth of adults exhibit the normal cuspidate pattern found in insectivorous bats. The dental formula is: (i 1/2, c 1/1, pm 2/3, m 3/3) $\times$ 2 = 32. The nasal region of the skull is considerably expanded. All the toes have three bones, except the first, which has two; the Hipposideridae, in contrast, have only two bones in each toe. The eyes of horseshoe bats are quite small, and the field of vision seems to be partly obstructed by the large nose leaf, so sight is probably of little importance.

These bats, roosting where they can hang freely, do not close their wings alongside their body as do most bats, but wrap them around the body. The small bare patch on the back at the base of the tail is covered by the upturned tail and membrane; the bat is thus completely enclosed in its flight membranes and resembles the pod of a fruit or the cocoon of an enormous insect. When the bat is at rest, the basal axis of the head makes a right angle with the vertebral column, so that it looks in the direction of its ventral surfaces.

The wings are broad with rounded ends. Horseshoe bats generally have a fluttering, butterflylike, or hovering flight. Their relatively short tails and small tail membranes are not large enough to form a pouch for holding insects. When a large insect is caught in flight, it may be tucked into the wing membrane under the arm while the bat manipulates it with its mouth. Horseshoe bats sometimes alight with large prey in order to eat more easily.

Horseshoe bats occur throughout the temperate and tropical zones of the Old World, being found in a great variety of forested and nonforested habitats, at both high and low altitudes. They roost in caves, buildings, foliage, and hollow trees. The species living in temperate regions hibernate during the winter in retreats other than their summer roosts, but they awaken readily and change their hibernating sites occasionally, sometimes flying 1,500 meters or more to a new place. The body temperature of *R. ferrumequinum* has been recorded in Berlin as from 8° C in hibernation to 40° C in periods of normal activity.

These bats begin feeding on insects and spiders later in the evening than most bats and often return to the roost to eat their catch. They usually hunt within six meters of the ground, and will also feed on the ground. Like many bats, they generally have regular feeding territories or hunting areas.

Most species roost in moderate-sized groups, but some are solitary. In some species the sexes live together all year, whereas in others the females form maternity colonies. Lekagul and McNeely (1977) wrote that *R. lepidus* occurs in groups of 3–4 to 400 and, unlike most species of *Rhinolophus,* forms roosting clusters. The clusters are segregated by sex during the birth season. Individuals of this species may establish well-defined foraging territories near their roosts. In general, individuals of *Rhinolophus* are solitary hunters, while those of *Hipposideros* forage in small groups.

In some species, including those that hibernate, mating occurs during the fall, but there is a delay in ovulation, so that actual fertilization does not occur until the spring. In other species mating and fertilization occur in the spring. Menzies (1973) stated that both *R. hipposideros* in Europe and *R. megaphyllus* in Australia undergo delayed fertilization. In *R. landeri* in Nigeria mating takes place in November, no development occurs in December–January, actual gestation takes place in February–April, births occur in late April and early May, and lactation goes on through June. In general for *Rhinolophus,* gestation takes about 7 weeks, a single young is produced in late spring, and sexual maturity is attained by 2 years of age (Lekagul and McNeely 1977). Medway (1978)

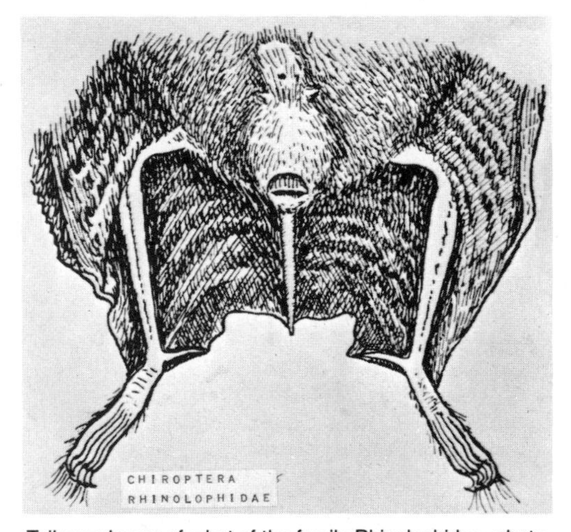

Tail membrane of a bat of the family Rhinolophidae, photo from *Exploration du Parc Albert National, Mamm.*

reported pregnant females of several species of *Rhinolophus* in the Malay Peninsula from February to April but also stated that pregnancies in *R. trifoliatus* had been recorded in March, May, June, September, October, and November. Sreenivasan, Bhat, and Geevarghese (1973) studied a colony of 2,000 *R. rouxi* in India and found a short, well-defined breeding season. Pregnant females were present only from January to April, lactation lasted from April to June, and there was arrested reproductive activity from June to November. Lekagul and McNeely (1977) stated that longevity in *Rhinolophus* seldom exceeds 6–7 years, but there are several records of remarkably long life spans. The known record is that of an individual *R. ferrumequinum* that was captured in France in 1982, 29 years after initial banding, and then seen again in 1983 (Caubere, Gaucher, and Julien 1984).

R. ferrumequinum was considered vulnerable by Smit and Van Wijngaarden (1981), endangered in Europe by Stebbings (1982), and endangered throughout the world by Stebbings and Griffith (1986). This species is declining rapidly because of disturbance of its roosts in caves and buildings, vandalism, habitat modifications resulting in loss of large insect prey, and increasing use of insecticides that may be absorbed by the bats. In England during the last century numbers have fallen by over 98 percent, to about 2,200 individuals. Similar problems have befallen *R. hipposideros, R. blasii, R. euryale,* and *R. mehelyi,* which also were listed as endangered in Europe by Stebbings and Griffith (1986). A slow recovery of *R. euryale* has been noticed since the most dangerous pesticides were banned in the 1980s (Brosset et al. 1988).

The geological range of this family is middle Eocene to Recent in Europe, and Recent in other regions now occupied.

CHIROPTERA; **Family HIPPOSIDERIDAE**

Old World Leaf-nosed Bats

This family of 9 Recent genera and 63 species inhabits tropical and subtropical regions in Africa and southern Asia, east to the Philippine Islands, the Solomon Islands, and Australia. Some authors (e.g., Corbet 1978; Ellerman and Morrison-Scott 1966; Koopman 1984*c*) consider the Hipposideridae to be only a subfamily of the Rhinolophidae. Other authors (e.g., Corbet and Hill 1986; Hayman and Hill, *in* Meester and Setzer 1977; Lekagul and McNeely 1977) maintain both groups as distinct families, and this procedure is followed here. The Hipposideridae differ from the Rhinolophidae in the form of the nose leaf, the foot structure, the absence of the lower small premolar, and the structure of the shoulder and hip girdles.

The muzzle has an elaborate leaflike outgrowth of skin. This nose leaf consists of an anterior horseshoe-shaped part, sometimes with smaller accessory leaflets, and an erect transverse leaf, corresponding to the lancet in the nose leaf of the Rhinolophidae, usually divided into three cell-like parts, the apices of which may be produced into points. It lacks a sella, which is the median projection of the nose leaf in the Rhinolophidae.

Head and body length is 28–110 mm. An external tail is absent in some forms but may be as much as 60 mm in length in others; when present, the tail is enclosed in the tail membrane. Forearm length is 30–110 mm. *H. commersoni gigas,* with a forearm length of 110 mm, is one of the largest bats in the suborder Microchiroptera. In several species, the sexes differ in body size, size of nose leaves, and color of fur. The ears are well developed or short and united across the forehead in some genera. A tragus is lacking. In these bats each toe has only two bones, whereas in the Rhinolophidae each toe consists of three bones, except the first toe, which has two bones. The dental formula is: (i 1/2, c 1/1, pm 1–2/2, m 3/3) × 2 = 28 or 30. The nasal region of the skull is only slightly inflated.

These bats utilize caves, underground chambers made by people, buildings, and hollow trees as retreats. *Hipposideros fulvus* has been found in burrows of the large porcupine (*Hystrix*) in Africa. Some species associate in groups of hundreds, others in small groups, and some forms roost singly. Sometimes they associate with other species of bats. Some forms of *Hipposideros* hibernate; *Asellia* may hibernate; and there is a record of apparent hibernation for *Coelops.*

Horseshoe bats catch most of their insect prey in flight. As in other insectivorous bats, the jaws are worked with a slightly side-to-side movement as well as up and down, so that the cusps of the upper and lower teeth sweep past each other in a shearing action that cuts up the hard parts of insects. These bats possibly fly with the mouth closed, emitting ultrasonic sounds through the nostrils. *Asellia tridens* emits far-reaching, beamed pulses, like the members of the family Rhinolophidae, as well as shorter pulses, like the members of the family Vespertilionidae.

The geological range of this family is Eocene to Oligocene, and perhaps Miocene, in Europe, and Recent over the present range.

CHIROPTERA; HIPPOSIDERIDAE; **Genus HIPPOSIDEROS**
Gray, 1831

Old World Leaf-nosed Bats

There are 51 species (Aggundey and Schlitter 1984; Bergmans and Van Bree 1986; Brosset 1984; Chasen 1940; Cranbrook 1984; Ellerman and Morrison-Scott 1966; R. E. Goodwin 1979; Gould 1978*a*; Happold 1987; Hayman and Hill, *in* Meester and Setzer 1977; Hill 1963*a*, 1983*b*, 1985*a*; Hill and Francis 1984; Hill and Morris 1971; Hill and Yenbutra 1984; Hill, Zubaid, and Davison 1985, 1986; Jenkins and Hill 1981; Largen, Kock, and Yalden 1974; Laurie and Hill 1954; Lekagul and McNeely 1977; Nader 1982; Rautenbach, Schlitter, and Braack 1984; Roberts 1977; Schlitter et al. 1986; Smith and Hill 1981):

H. megalotis, Ethiopia, Kenya;
H. bicolor, southern Thailand, Malay Peninsula, Teratau and Tioman islands, Sumatra, Java, Borneo, Philippines, Timor;
H. pomona, India and Sri Lanka to southern China and Indochina, Hainan, Malay Peninsula;
H. macrobullatus, Kangean Islands, Sulawesi, Ceram;
H. fulvus, Afghanistan, Pakistan, India, Sri Lanka;
H. ater, India and Sri Lanka, East Indies including Philippines and New Guinea, northern Australia;
H. cineraceus, Pakistan to northern Viet Nam, Malay Peninsula, Borneo, Kangean Islands northeast of Java;
H. nequam, known only by a single specimen from the Malay Peninsula;
H. calcaratus, New Guinea, Japen Island, Bismarck Archipelago, Solomon Islands;
H. maggietaylorae, New Guinea, Bismarck Archipelago;
H. coronatus, northeastern Mindanao;
H. jonesi, Guinea to Nigeria;
H. ridleyi, Malay Peninsula, Singapore, Borneo;
H. dyacorum, Borneo;
H. sabanus, Malay Peninsula, Sumatra, Borneo;
H. obscurus, Philippines;
H. marisae, Guinea, Liberia, Ivory Coast;
H. halophyllus, Thailand;
H. pygmaeus, Philippines;
H. galeritus, India, Sri Lanka, Burma, Malay Peninsula, Sumatra, Java, Borneo, and many small nearby islands;
H. cervinus, Malay Peninsula, Sumatra, Java, Borneo, Kangean Islands, Philippines, Sulawesi, New Guinea, Solomon Islands, New Hebrides, Cape York Peninsula of northern Queensland;

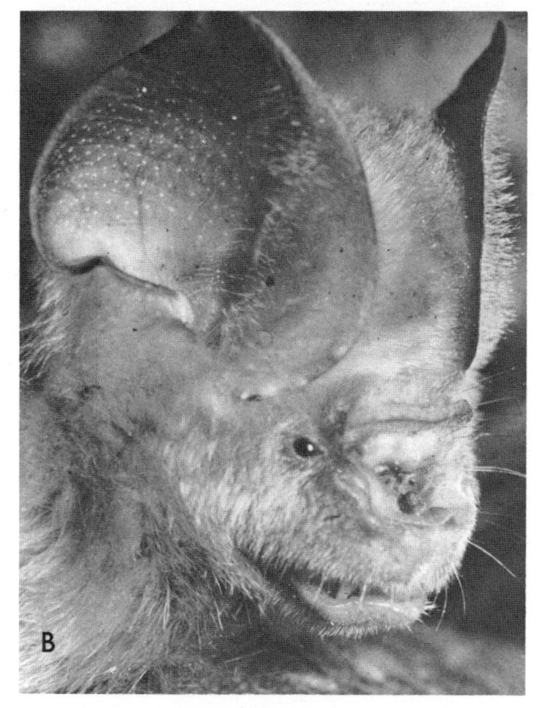

A. Head of *Hipposideros armiger terasensis* showing construction of nose leaf in the family Hipposideridae, photo by Robert E. Kuntz. B. Head of *H. ruber*, photo by P. Morris.

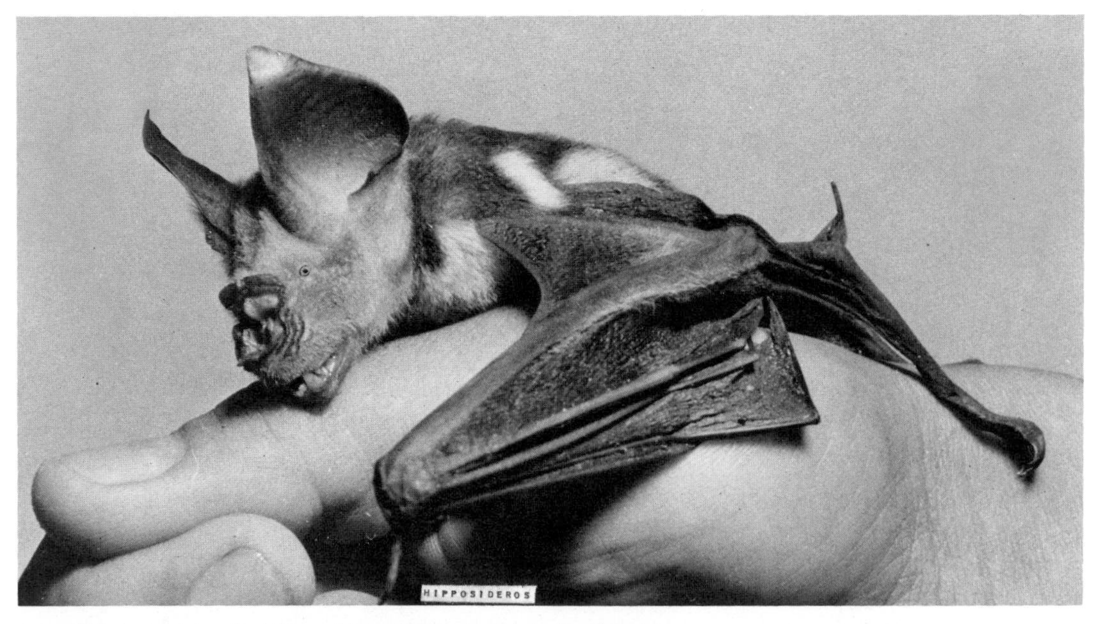

Large Malayan leaf-nosed bat *(Hipposideros diadema)*, photo by Lim Boo Liat.

H. breviceps, North Pagi Island west of Sumatra;
H. curtus, Cameroon, island of Bioko (Fernando Poo);
H. fuliginosus, Ghana to Cameroon, eastern Zaire, Ethiopia;
H. caffer, southwestern Arabian Peninsula, most of Africa except central forested region;
H. ruber, Senegal to Ethiopia, and south to Angola and Zambia;
H. lamottei, known only from Mount Nimba area of Guinea;
H. beatus, Sierra Leone to northern Zaire and southwestern Sudan;
H. coxi, known only by a single specimen from Sarawak (Borneo);
H. papua, Misori Island off northwestern New Guinea;
H. cyclops, Guinea-Bissau to southern Sudan and Kenya;
H. camerunensis, Cameroon, eastern Zaire, western Kenya;
H. muscinus, Papua New Guinea;
H. wollastoni, southwestern New Guinea, western Papua New Guinea;
H. semoni, eastern New Guinea, northern Queensland;
H. corynophyllus, known only by holotype from western Papua New Guinea;
H. stenotis, northern Australia;
H. pratti, southern China, northern Viet Nam;
H. lylei, eastern Burma, western Thailand, Malay Peninsula;
H. armiger, Nepal and Assam to southern China and Indochina, Malay Peninsula, Taiwan, Hainan;
H. turpis, southern Ryukyu Islands, extreme southern Thailand;
H. abae, Guinea-Bissau to southern Sudan and Uganda;
H. larvatus, eastern Burma, Thailand, Indochina, Hainan, Malay Peninsula, Sumatra, Java, Kangean Islands, Borneo;
H. lekaguli, southern Thailand, Malay Peninsula;
H. speoris, India, Sri Lanka;
H. lankadiva, India, Sri Lanka;
H. schistaceus, Bellary (southern India);
H. diadema, Nicobar Islands, southern Burma and Thailand, Indochina, Malay Peninsula, Sumatra, Java, Kangean Islands, Timor, Philippines, New Guinea, Solomon Islands, northeastern Queensland;
H. dinops, Sulawesi, Peleng Island, Solomon Islands;
H. inexpectatus, northern Sulawesi;
H. commersoni, Gambia to Somalia, south to Namibia and Transvaal, Madagascar.

Koopman (1982*a*) regarded *H. cervinus* as a subspecies of *H. galeritus*.

Head and body length is about 35–110 mm, tail length is 18–70 mm, and forearm length is about 33–105 mm. Medway (1978) listed the following weights: *H. bicolor*, 8–10 grams; *H. cineraceus*, 7–8 grams; *H. galeritus*, 6–7 grams; *H. lylei*, 33–47 grams; *H. armiger*, 40–60 grams; *H. larvatus*, 14–20 grams; and *H. diadema*, 34–50 grams. Kingdon (1974*a*) gave weights of 30–39 grams for *H. cyclops* and 74–180 grams for *H. commersoni*. Coloration of the upper parts varies from reddish to some shade of brown, and the underparts are paler. Some species have two color phases, reddish and gray.

These bats are distinguished from those of the genus *Rhinolophus* by features of the nose leaf and ear, characters of the teeth, the greater posterior width of the skull, and the characters separating the families Hipposideridae and Rhinolophidae, as explained in the accounts thereof. Many species of *Hipposideros* have a sac behind the nose leaf that can be everted at will; this sac secretes a waxy substance and is found chiefly in the males.

Members of this genus usually roost in hollow trees, caves, and buildings. *H. fulvus* has been found in burrows of the large crested porcupine *(Hystrix)* in Africa. Some species are gregarious; others associate in small family groups or at times may be alone. The nose leaf and ears often twitch while these bats are hanging. Some species hibernate.

These bats fly lower than most bats and catch insects such as beetles, termites, and cockroaches; cicadas apparently form an important food source for some species. *H. armiger* appears to locate these insects by their calls. *H. commersoni* in Africa eats the beetle larvae that are inside wild fig fruits

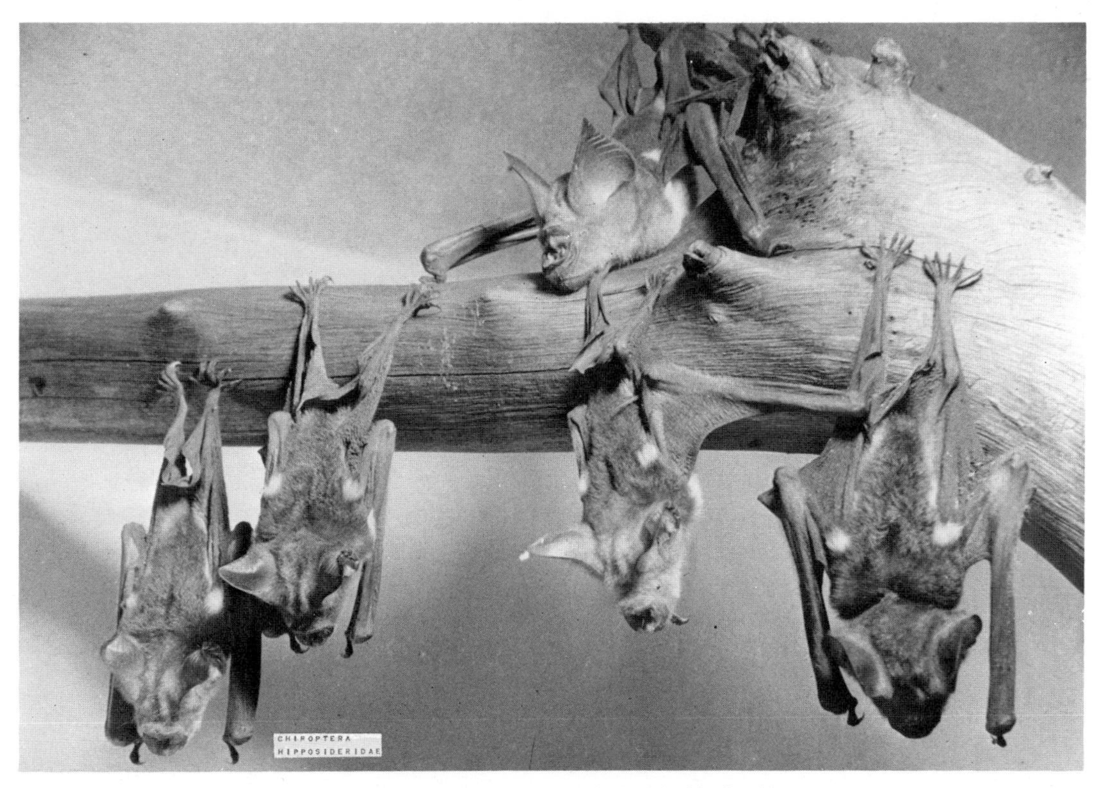

A small group of large Malay leaf-nosed bats *(Hipposideros diadema)*, photo by Lim Boo Liat.

and incidentally consumes some fig pulp. *Hipposideros* often returns to its roost to eat its catch and seems to have certain feeding territories that it regularly patrols.

The species *H. caffer* may emit a long, drawn-out whistle when handled. A female of another species returned several times to its captured young in response to its sharp twitter.

Medway (1978) indicated that investigated species of *Hipposideros* in the Malay Peninsula are gregarious, sometimes gathering in large colonies. Female *H. diadema* are known to congregate during March and April, when each gives birth to one young. *H. bicolor* and *H. cineraceus* have marked seasonal reproduction in this area, with mating in October, births in April, and the young flying by June. *H. armiger* apparently has two breeding seasons, pregnant females having been recorded in February–May and September–October. Gould (1978*a*) added that the newly rediscovered *H. ridleyi* in the Malay Peninsula gave birth during April. Lim, Shin, and Muul (1972) found pregnant female *H. galeritus*, each with one embryo, in Sarawak in mid-June. McKean (1972) reported a pregnant female of this species in New Guinea in September.

Roberts (1977) wrote that in Pakistan *H. fulvus* gathers in colonies of only 10–20, but Madhavan, Patil, and Gopalakrishna (1978) reported that colonies in India usually numbered 50–100. In India this species is a strictly seasonal breeder; mating occurs in mid-November, gestation lasts 150–60 days, and births take place in late April and early May. A single young is produced and is carried by the female for 20–22 days. Sexual maturity is attained at 18 or 19 months. A banded female *H. fulvus* was recovered at the age of at least 12 years; this individual was pregnant with a well-developed embryo.

Kingdon (1974*a*) stated that *H. cyclops* of Africa does not assemble in groups of over 12 and that usually the groups comprise only females. This species has a well-defined breeding season in eastern Zaire, with pregnant females having been recorded from January to March and young animals found in late April and May.

Ansell (1960) reported pregnant female *H. caffer* and *H. commersoni* in Zambia in August and October. *H. commersoni* roosts in large groups, sometimes numbering hundreds of bats. *H. caffer* roosts in still larger colonies, 500,000 reportedly having been found in one cave in Gabon. Bell (1987) suggested that the groups are divided into harems, one of which contained 7 adult females, each with a single young, plus 1 adult male. In Gabon *H. caffer* has a well-defined breeding season, with births occurring synchronously but with some colonies giving birth in March and others in October. Records suggest that in tropical populations of this species births generally occur about April north of the equator and about October to the south, perhaps because of a separate origin for the populations involved. Mutere (1970) reported that in *H. caffer* in Uganda, pregnancies were noted only from December to March, and birth of the single young took place just before the peak rains, when the most insect food was available. Menzies (1973) found that in Nigeria *H. caffer* mates in November and implantation is delayed for about two months. Births occur in late April and early May, and lactation continues through June. Studies in South Africa (Bernard and Meester 1982) suggest that the period of delayed embryonic development increases with distance from the tropics. Females of *H. caffer* there were found to be monestrous; mating took place in late April (early winter), and births occurred in December. The total gestation period was about 220 days, or approximately 100 days longer than in the tropics.

The species *H. ridleyi* is classified as indeterminate by the IUCN and as endangered by the USDI. It survives only in small patches of lowland peat forest that have been heavily logged. *H. papua* is designated rare by the IUCN.

CHIROPTERA; HIPPOSIDERIDAE; **Genus ASELLIA**
Gray, 1838

Trident Leaf-nosed Bats

There are two species (Corbet 1978; Hayman and Hill, *in* Meester and Setzer 1977):

A. tridens, Morocco and Senegal to Egypt and Ethiopia, Zanzibar, Arabian Peninsula to Pakistan;
A. patrizii, Ethiopia.

Head and body length is about 46–62 mm, tail length is 18–27 mm, and forearm length is 45–52 mm. Gaisler, Madkour, and Pelikan (1972) stated that weight in *A. tridens* is 6–10 grams. The coloration is variable. Pale buffy gray and orange brown color phases occur in some areas; buff-colored individuals have been noted in Egypt; and *Asellia* with a pale yellow silky sheen have been collected in Iraq.

The nose leaf consists of three parts: an anterior horseshoe-shaped part, a central triangular part, and a posterior tridentate part. The nostrils are located in the anterior part, and there is a frontal sac behind the nose leaf. The ears are large and nearly naked.

Gaisler, Madkour, and Pelikan (1972) reported *A. tridens* to be among the most desert-adapted bats; they found this species at several oases in Egypt. Caves and artificial structures are often utilized as roosts. In Egypt *A. tridens* has also been seen on the inner walls of wells and underground channels at oases. Two individuals of this species were found roosting with *Pipistrellus kuhli* under the corrugated iron roof of a shed in Iraq in June, when the temperature inside the shed was probably at least 38° C. These bats were active at this temperature and flew about inside the shed when disturbed. *Asellia* has been found in a cave by the sea, associated with *Coleura* and *Triaenops*. According to Qumsiyeh (1985), *A. tridens* in Iraq disperses to hibernating quarters (cellars and tombs) from mid-September to mid-November and returns to summer roosts in April.

Hundreds of *Asellia* were found roosting in an underground tunnel in Oman. In the evening, these bats flew from the tunnel in groups of about 20, flying swiftly and close to the ground to nearby palm groves to feed. In another area trident leaf-nosed bats have been observed flying within a few meters of the ground across almost a mile of barren desert to a feeding area.

DeBlase (1980) discovered a colony of about 5,000 *A. tridens* in a cave in Iran. Gaisler, Madkour, and Pelikan (1972) reported that the largest colony of *A. tridens* that they saw, consisting of about 500 bats, was in a dark room in a ruined temple. They also stated that 13 females taken in Egypt in April–May were pregnant with one embryo each. In South Yemen pregnant females have been found in April. Qumsiyeh (1985) reported colonies of 300–1,000 or more in Egypt and stated that in Iraq births begin in early June, gestation is assumed to be 9–10 weeks, and lactation lasts 40 days.

CHIROPTERA; HIPPOSIDERIDAE; **Genus ANTHOPS**
Thomas, 1888

Flower-faced Bat

The single species, *A. ornatus,* is known only from the Solomon Islands. Oldfield Thomas had six specimens when he named and described this genus and species. These specimens are in the British Museum of Natural History. The U.S. National Museum of Natural History and the Field Museum of Natural History also have specimens.

Head and body length is about 50 mm, the tail is very

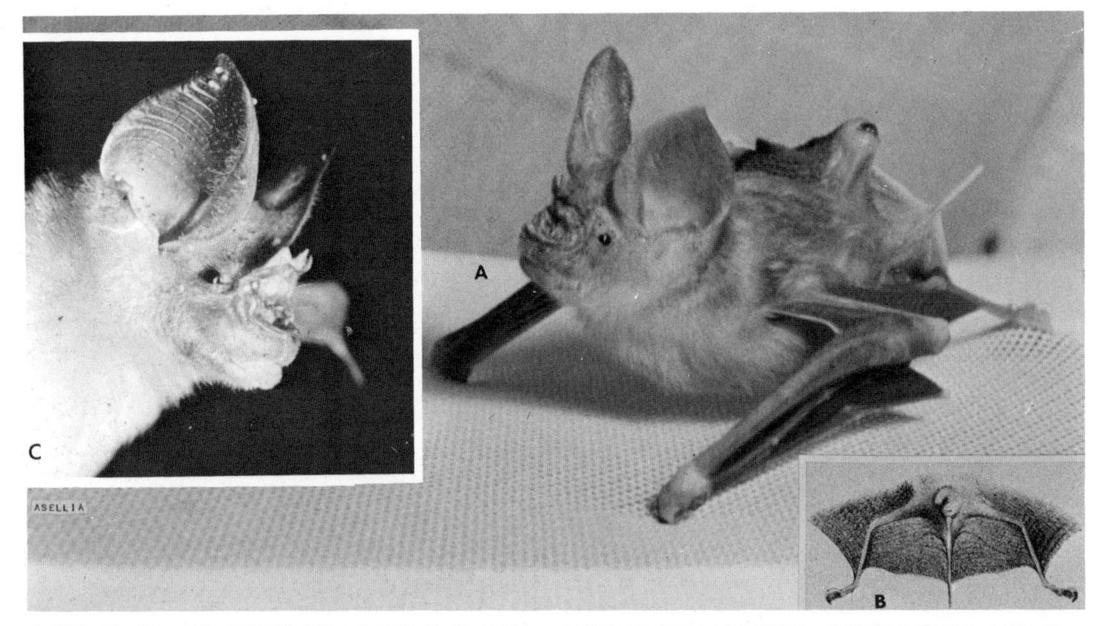

A. Trident leaf-nosed bat *(Asellia tridens)*, photo by Erwin Kulzer. B. Tail membrane of *A. tridens*, photo from *Zoology of Egypt, Mammalia*, John Anderson. C. Face of *A. patrizii*, photo by D. W. Yalden.

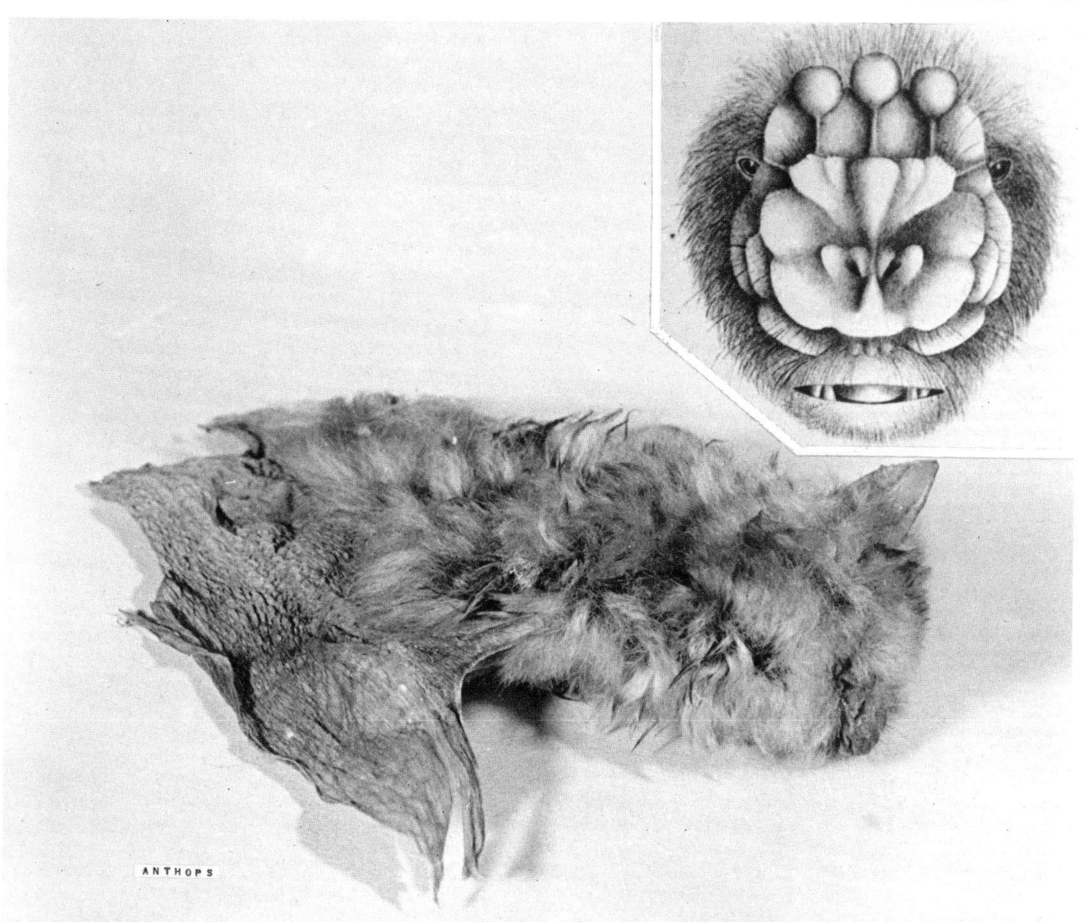

Flower-faced bat *(Anthops ornatus)*, photo by Howard E. Uible of specimen in U.S. National Museum of Natural History. Inset photo from *Proc. Zool. Soc. London.*

short, and forearm length is about 48–51 mm. The pelage is long, soft, and silky, and the coloration is grizzled grayish buff.

This genus is similar to *Hipposideros,* differing from it in the greatly reduced tail and in skull and dental features. The tail in *Anthops* is less than half as long as the femur. The horseshoe-shaped part of the nose leaf has two lateral leaflets, the inner one being quite small and the outer one large. Traces of a small frontal sac are present.

CHIROPTERA; HIPPOSIDERIDAE; **Genus ASELLISCUS**
Tate, 1941

Tate's Trident-nosed Bats

There are two species (Laurie and Hill 1954; Lekagul and McNeely 1977; McKean 1972; Schlitter, Williams, and Hill 1983; Zubaid 1988):

A. stoliczkanus, extreme southeastern China, northern Indochina, eastern Burma, western Thailand, Penang and Tioman islands off the coast of the Malay Peninsula;

A. tricuspidatus, New Guinea, islands from the Moluccas to the New Hebrides.

Head and body length is about 38–45 mm, tail length is 20–40 mm, and forearm length is 35–45 mm. McKean (1972) reported a weight of 3.5–4.0 grams for *A. tricuspidatus* and a

Tate's trident-nosed bat (*Aselliscus* sp.), photo by Boonsong Lekagul.

weight of 6–8 grams for *A. stoliczkanus*. The coloration of *A. tricuspidatus* is bright brown above and buffy brown below. *A. stoliczkanus* has been described as both brownish and sooty, so perhaps there are two color phases in this species.

The upper margin of the transverse nose leaf is divided into three points, and two lateral leaflets margin the horseshoe. No frontal sac is present in either sex. The tail extends beyond the membrane, as in *Asellia*. *Aselliscus* is best distinguished by features of the skull and teeth.

Available records indicate that these bats are cave dwellers. Smith and Hood (1981) encountered *A. tricuspidatus* in great abundance in caves and tunnels on New Britain and New Ireland islands. Individuals there were always found hanging singly and evenly spaced (30–40 cm apart) within discrete groups that usually contained 40–50 bats. An especially large colony of several hundred individuals was seen in a limestone cave, in which the bats hung from the tips or sides of stalactites. Females were seen carrying a single young. McKean (1972) reported that 62 specimens of *A. tricuspidatus* were taken while roosting in limestone caves in New Guinea, at altitudes of 98–260 meters. Of 26 females collected there on 1 July, 12 were pregnant with one embryo each; the other 14 apparently were immature. Phillips (1967) reported that 4 female *A. stoliczkanus* taken in early June in Laos were carrying young. Topal (1974) collected pregnant or lactating females of the same species in Viet Nam during May.

CHIROPTERA; HIPPOSIDERIDAE; **Genus RHINONYCTERIS**
Gray, 1847

Golden Horseshoe Bat

The single species, *R. aurantius*, occurs from the Pilbara and Kimberley regions of northern Western Australia, through the northern part of the Northern Territory, and along the Gulf of Carpentaria into northwestern Queensland (Churchill, Helman, and Hall 1987). The generic name sometimes is spelled *Rhinonicteris*.

Head and body length is 45–53 mm, tail length is 24–28 mm, forearm length is 47–50 mm, and weight is 8–10 grams (Jolly, *in* Strahan 1983). The most common color pattern is bright orange upper parts and paler underparts, but there is much variation. Specimens have ranged from dark rufous brown, through orange, dark lemon, and pale lemon, to white (Churchill, Helman, and Hall 1987). The fur is fine and silky.

According to Hill (1982*b*), *Rhinonycteris*, *Triaenops*, and *Cloeotis* form a group of genera characterized principally by a number of common features of the nose leaf. All have a straplike projection extending forward from the internarial region over the anterior leaf to its edge, and all have a strongly cellular posterior leaf. None of the remaining six genera of the Hipposideridae have the anterior median straplike process.

According to F. W. Jones (1923–25),

> the crown of the head is well raised above the face, the face itself being almost entirely occupied by the nose-leaf. The ears are short; laid forward they cover the eye; they are sharply pointed at the tip. The inner margin of the auricle sweeps forward with a fairly uniform convexity to the tip; the outer margin starts straight, or slightly concave, from the tip and then sweeps round, boldly convex, joining the side of the head behind and slightly below the eye. The inner margins are widely separated by the crown of the head. The nose-leaf is large and complex; it consists of two parts, a lower part which forms a horse-shoe ring below the nostrils, and an upper part which surmounts the nostrils. The lower part consists of two bilateral leaves, notched in the middle line below, and there joined by a projecting shelf, which springs from the interspace between the nostrils. From the inner margin of each of the leaves a small curved projection passes along the outer margin of each nostril. The complex upper portion of the nose-leaf surmounts the nostrils like a crown, a curious middle-line process jutting out just above the nostrils. Above this process the crown-like portion of the leaf is extensively honeycombed and sculptured.

According to Churchill, Helman, and Hall (1987), the golden horseshoe bat is found in a variety of habitats, including grassland, open forest, dense palm forest, and mangroves. It usually roosts in caves and mines, often together with *Macroderma*. Hot (29°–30° C), humid (95–100 percent relative humidity) sites are preferred. *Rhinonycteris* evidently is unable to lower its body temperature and does not enter torpor but becomes lethargic if exposed to ambient temperatures under 20° C. It emerges about 0.5–1.5 hours after sunset and has been observed to forage for insects 1–3 meters above the vegetation. Its flight is relatively faster than that of other hipposiderids. Specimens also have been collected in and around buildings. Vestjens and Hall (1977) reported that three stomachs of *Rhinonycteris* collected in the Northern Territory contained only moths, and five stomachs contained moths and traces of other insects.

Churchill, Helman, and Hall (1987) listed roost sites con-

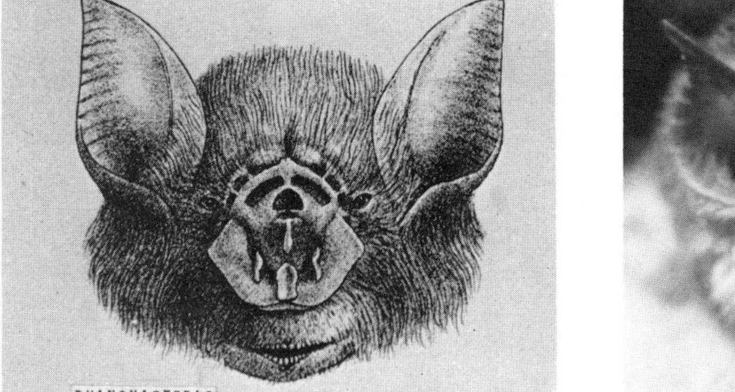

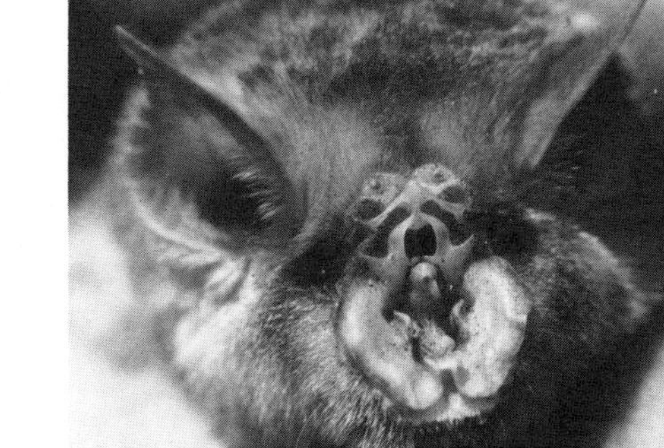

Golden horseshoe bat *(Rhinonycteris aurantius):* Left, photo from *Catalogue of the Chiroptera in the Collection of the British Museum*, G. E. Dobson; Right, photo by Simon Jolly.

Golden horseshoe bat *(Rhinonycteris aurantius)*, photo by G. B. Baker / National Photographic Index of Australian Wildlife.

taining from 1 to 5,000, commonly several hundred, bats. Spacing between roosting individuals is at least 15 cm. Breeding is thought to occur during the wet season, from October to April. The largest known colony, at Cutta Cutta Cave in the Northern Territory, was thought to have about 5,000 bats in 1966. Subsequently, the entrance of the cave was covered with a steel mesh grill intended to keep people out but to allow the bats to pass through. It was not successful, and only a few hundred bats survived. Following removal of the grill, numbers have fluctuated up to about 2,000, and other means have been taken to prevent tourists from entering the part of the cave used by the bats.

CHIROPTERA; HIPPOSIDERIDAE; **Genus TRIAENOPS**
Dobson, 1871

Triple Nose-leaf Bats

There are two species (Corbet 1978; Hayman and Hill, *in* Meester and Setzer 1977; Hill 1982*b*; Kock and Felten 1980):

T. furculus, western and northern Madagascar, Aldabra and Cosmoledo islands;
T. persicus, southwestern Iran, southern Pakistan, Oman, South Yemen, possibly Egypt, East Africa from Ethiopia to Mozambique, Congo, eastern Madagascar.

Honacki, Kinman, and Koeppl (1982) listed *T. rufus* and *T. humbloti* of eastern Madagascar as species separate from *T. persicus.*

Head and body length is 35–62 mm, tail length is 20–34 mm, forearm length is 45–55 mm, and adult weight is usually 8–15 grams. *T. persicus* exhibits great diversity in coloration, through grays, browns, and reds; these may be correlated with age. Individuals in some areas are pale buff, almost white. *Triaenops* resembles *Hipposideros* to some extent but differs from that genus in features of the nose leaf, as described above in the account of *Rhinonycteris. Triaenops* evidently is closely related to *Rhinonycteris* but differs in having a tridentate posterior section of the nose leaf such as is also found in *Cloeotis.* Cranially, *Triaenops* differs from both genera in its raised and inflated rostrum and relatively much larger cochleae but resembles *Rhinonycteris* in the structure of the anteorbital region and zygoma and in the curiously thickened premaxillae (Hill 1982*b*).

T. persicus has been found in a cave by the sea in South Yemen, associated with *Coleura afra* and *Asellia tridens.* This species was collected in Oman while flying around the opening of an underground water tunnel. It apparently roosted in the tunnel in small numbers, along with hundreds of *Asellia tridens.* The *Triaenops* emerged quite early in the evening, while there was still light, and flew low over the ground. Thousands of *T. persicus* and *Coleura afra* were

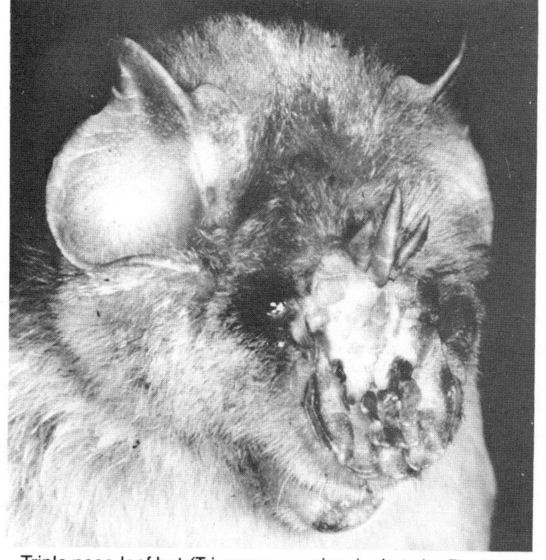

Triple nose-leaf bat *(Triaenops persicus)*, photo by David Pye.

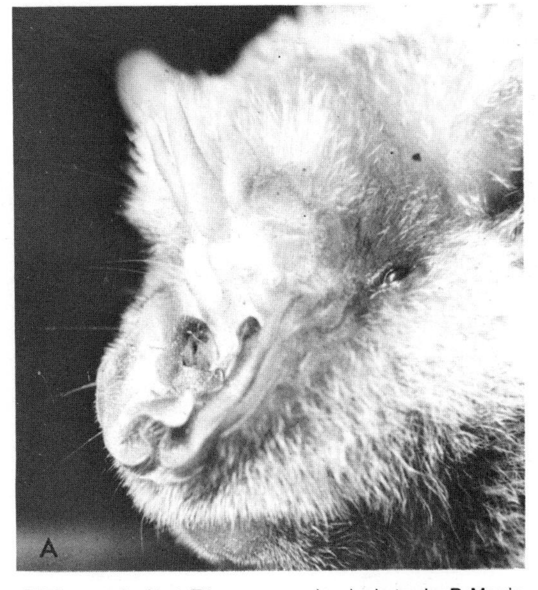

Triple nose-leaf bat *(Triaenops persicus)*, photos by P. Morris.

found in a cave with many passages and chambers at different levels. The *T. persicus* in this cave assembled in groups just behind the entrance before flying out to feed on insects. At least some of the young of *T. persicus* are born in January; 24 males and 6 females, 2 pregnant, were taken from the Amboina Caves in Tanzania, only a few meters above sea level, in December.

CHIROPTERA; HIPPOSIDERIDAE; **Genus CLOEOTIS**
Thomas, 1901

African Trident-nosed Bat

The single species, *C. percivali*, occurs from southern Zaire and Kenya to the Transvaal in South Africa (Hayman and Hill, *in* Meester and Setzer 1977).

Head and body length is 33–50 mm, tail length is 22–33 mm, and forearm length is 30–36 mm. Ansell (1986) listed weights of 3.8–5.9 grams. Two subspecies are recognized: *C. p. percivali* and *C. p. australis*. The type specimen of *C. p. percivali* is colored as follows: bright buffy face, grayish crown, grayish brown back, and smoky brown wings and membranes, with the hairs on the abdomen having slaty gray bases and yellowish white tips. *C. p. australis* is usually buffy brown or dark brown above and yellowish below.

In this genus, the well-developed tail is longer than the femur. Three pointed processes, tridentlike, are present at the back of the nose leaf. The ears are rounded and short, scarcely projecting above the level of the hair. *Cloeotis* resembles *Triaenops* in that its nose leaf has a tridentate posterior portion but differs in such cranial characters as its larger braincase and lower rostrum (Hill 1982*b*).

In South Africa this bat has been found in large numbers in caves with narrow entrances. Records from Zimbabwe and Botswana also indicate that *Cloeotis* is a cave dweller and that the diet is insectivorous (Smithers 1971). In Zambia, Whitaker and Black (1976) found this bat to feed almost entirely on adult Lepidoptera. Colonies may contain hundreds of individuals, which roost in tight clusters. Pregnant females, each with a single embryo, were collected in Zambia and Zimbabwe in October (Ansell 1986; Smithers 1971, 1983).

CHIROPTERA; HIPPOSIDERIDAE; **Genus COELOPS**
Blyth, 1848

There are two species (Cranbrook 1984; Hill 1972*b*, 1983*b*; Lekagul and McNeely 1977; Taylor 1934):

C. frithi, eastern India to southeastern China and northern Indochina, Taiwan, Malay Peninsula, Java, Bali;
C. robinsoni, known by two specimens from the Malay Peninsula, one from Teratau Island off southern

African trident-nosed bat *(Cloeotis percivali)*, photo by Herbert Lang through J. Meester.

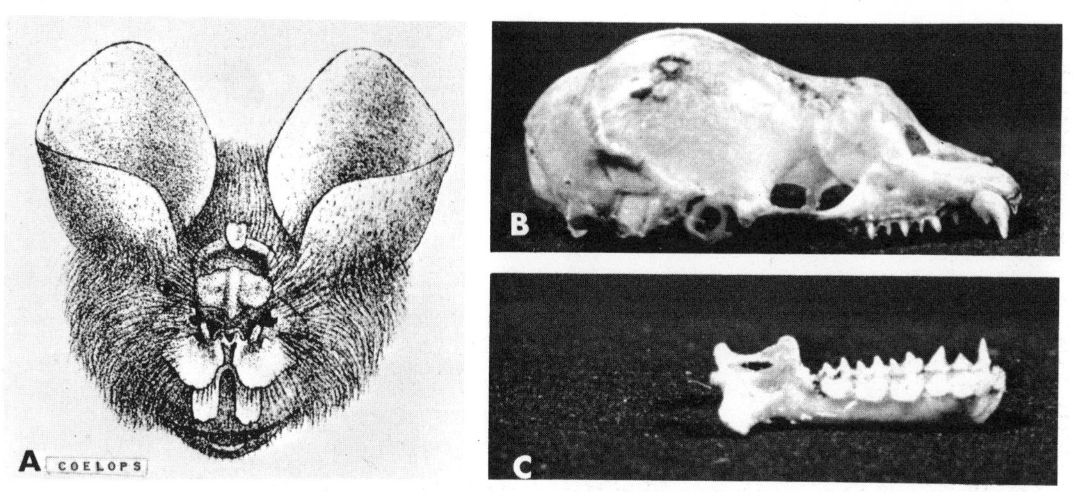

A. *Coelops frithi*, photo from *Catalogue of the Chiroptera in the Collection of the British Museum*, G. E. Dobson. B & C. *C. robinsoni*, photos by P. F. Wright of specimen in U.S. National Museum of Natural History.

Thailand, one from Borneo, and one from Mindoro Island in the Philippines.

A third specific name, *C. hirsuta*, has sometimes been applied to the specimen from the Philippines and one other from the Malay Peninsula. Hill (1972*b*), however, suggested that *C. hirsuta* is conspecific with *C. robinsoni*, and Medway (1978) did not use the name *C. hirsuta*.

Head and body length is 28–50 mm, the tail is extremely short or absent, and forearm length is 33–47 mm. Lekagul and McNeely (1977) listed weights of 7–9 grams for *C. frithi* and 6–7 grams for *C. robinsoni*. The coloration above is bright brown, ashy brown, dusky, or blackish, and the underparts are brownish or ashy. Lateral leaflets are lacking on the nose leaf, or they are obscured by dense, stiff hairs. A frontal sac is present in some individuals. The ears are short and rounded.

The type specimen of *C. frithi sinicus* was taken in Sichuan from a "warm-air cave in which it was evidently hibernating." On Taiwan, Gwilym S. Jones (pers. comm.) found that *C. frithi* was caught in old Japanese pillboxes as well as in caves; it was "not abundant, nor was it particularly rare." In Java, *C. frithi* inhabits forests, sheltering during the day in hollow trees in groups of 16 individuals or fewer. The following are Javan records: 2 females, each pregnant with 1 embryo, taken in January and a female and single young obtained in March.

CHIROPTERA; HIPPOSIDERIDAE; **Genus PARACOELOPS**
Dorst, 1947

The single species, *P. megalotis*, is known only from the type specimen, which was collected at Vinh, Annam, Viet Nam, by M. David Beaulieu. This specimen, an adult male, was obtained in 1945 and is now in the Paris Museum. The generic name refers to the funnel-shaped ears.

Measurements of the type are as follows: head and body length, 45 mm; forearm length, 42 mm; and ear length, 30 mm. There is no tail. This bat weighed 7 grams. The hairs are quite long, particularly on the upper parts. The interfemoral membrane is dark brown and lacks hair. The top of the head is bright golden yellow, in sharp contrast with the brownish back. The underparts are light beige, the bases of the hairs are yellow, and the ears are pale brown.

This medium-sized bat is characterized by large ears, a nose leaf, and the absence of a tail. The ears are separate and rounded at the top. The nose leaf is horseshoe-shaped and surmounted by a rounded leaf with radial striations. The tail membrane, 30 mm long, is not indented by attachment to the end of the tail and is supported by the long heel bones, the calcanea.

CHIROPTERA; **Family MORMOOPIDAE**

Moustached Bats, Naked-backed Bats, or Ghost-faced Bats

This family of two Recent genera and eight species is found in the New World, from southern Arizona and southern Texas south through Mexico, Central America, and South America

Paracoelops megalotis, photo by F. Petter.

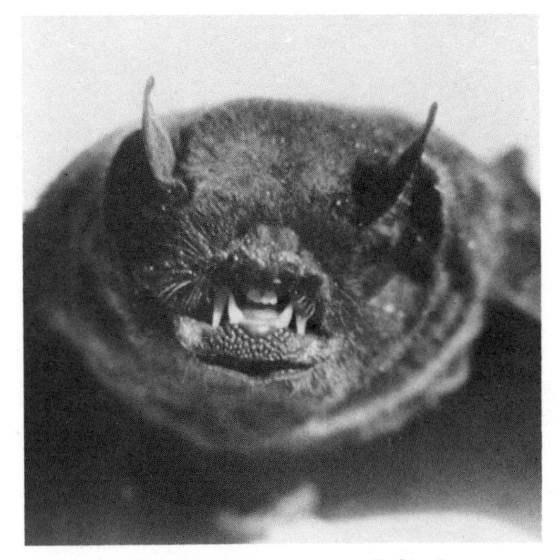

Pteronotus gymnonotus, photo by John P. O'Neill.

to the Mato Grosso of Brazil. These bats are generally restricted to tropical habitats below 3,000 meters. Until recently, this group was usually considered a subfamily of the Phyllostomidae, with the name Chilonycterinae. Some authorities, such as Hall (1981) and Husson (1978), still give the mormoopids only subfamilial rank, but Jones and Carter (1976), in their comprehensive review of the Phyllostomidae, excluded the Mormoopidae. The increasing acceptance of the latter group as a distinct family is based largely on the revisionary work of Smith (1972), and much of the following familial account is abstracted from his report.

Bats of this family range in size from the small *Pteronotus quadridens*, with a forearm length of 35–40 mm, to the large *P. parnellii*, with a forearm measuring 54–65 mm. Mormoopids differ from the phyllostomid bats in that they do not possess a nose leaf. Instead, the lips have been expanded and ornamented with various flaps and folds that form a "funnel" into the oral cavity when the mouth is opened. Short, bristlelike hairs surround this funnel and may act to direct airflow toward the scooplike mouth. The nostrils have been incorporated into the expanded upper lip; above and between them are various bumps and ridges that form a sort of nasal plate. The eyes in the Mormoopidae are small and inconspicuous, in contrast to the large and prominent eyes of the Phyllostomidae.

The wing membrane is variously attached to the side of the body. In several species the membranes meet and fuse on the dorsal midline, giving these species a naked-backed appearance. The attachment is somewhat lower in the other species. A tail is always present in this family, and the latter half protrudes from the tail membrane.

The pelage in this family is short, fine, and densely distributed over the body. It was formerly thought that these bats occurred in two distinct color phases, but Smith (1972) has shown that these phases are actually seasonal variations. It is noteworthy that naked-backed bats possess a thick, long pelage beneath the dorsally fused wing membranes.

The tragus of the Mormoopidae is different from that of any other group of bats. The tragus varies in the different genera from a seemingly simple lanceolate structure to one with a secondary fold of skin that lies at a right angle to the main longitudinal axis of the structure. This secondary fold is barely more than a pocketlike structure in the cranial edge of the tragus in *Pteronotus parnellii* and is best developed in the genus *Mormoops*. In addition to the above external characters, the family differs from the Phyllostomidae in a number of cranial and postcranial skeletal characters.

In a study of the forelimbs of the Mormoopidae and Phyllostomidae, Vaughan and Bateman (1970) found the mormoopids to be characterized by specializations furthering reduction of the weight of the wing. Such adaptations were considered to be associated with maneuverable, rapid flight and the ability to remain continuously in the air for long periods. In contrast, the wings of phyllostomids were found to be less well adapted for efficient flight and to retain muscular patterns allowing food handling and clambering in vegetation.

Mormoopids are gregarious cave dwellers, sometimes roosting in very large colonies. They are exclusively insect eaters. The geological range is Pleistocene to Recent in North America and the West Indies and Recent in South America (Koopman 1984*c*).

CHIROPTERA; MORMOOPIDAE; **Genus PTERONOTUS**
Gray, 1838

Naked-backed Bats, Moustached Bats, or Leaf-lipped Bats

There are three subgenera and six species (Hall 1981; Koopman 1982*b*; Smith 1972):

subgenus *Phyllodia* Gray, 1843

P. parnellii, Sonora and Tamaulipas (Mexico) to Mato Grosso region of central Brazil, Greater Antilles, Trinidad;

subgenus *Chilonycteris* Gray, 1839

P. macleayii, Cuba, Isle of Pines, Jamaica;
P. quadridens, Cuba, Jamaica, Hispaniola, Puerto Rico;
P. personatus, Sonora and Tamaulipas (Mexico) to central Brazil, Trinidad;

subgenus *Pteronotus* Gray, 1838

P. davyi, Sonora and Nuevo Leon (Mexico) to northern Peru, Lesser Antilles, Trinidad;
P. gymnonotus, southern Mexico to Peru and Mato Grosso region of central Brazil.

P. gymnonotus was formerly known as *P. suapurensis* (see J. D. Smith 1977). *Chilonycteris* is sometimes recognized as a full genus that includes *P. parnellii* as well as the three species indicated above (see, for example, Husson 1978).

Head and body length is 40–77 mm, tail length is 15–30 mm, and forearm length is 35–65 mm. Adults of *P. parnellii* usually weigh 10–20 grams; two female *P. davyi* weighed 7.5–10 grams. Coloration is variable, often being light or dark brown, grayish brown, or ochraceous orange, with the underparts usually paler.

The lower lip of *Pteronotus* has platelike outgrowths and small papillae. A nose leaf is not present, and the tail is well developed. The wings in *P. davyi* and *P. gymnonotus* attach along the midback and cover the fur of the back, giving the appearance that the back is naked. These two species are known as naked-backed bats. In the other species, the wings are attached along the sides of the body and the back is furred.

These bats occupy a variety of habitats. In Jalisco, Mexico, three species of *Pteronotus* were collected along waterways lined with dense vegetation, at altitudes from sea level to

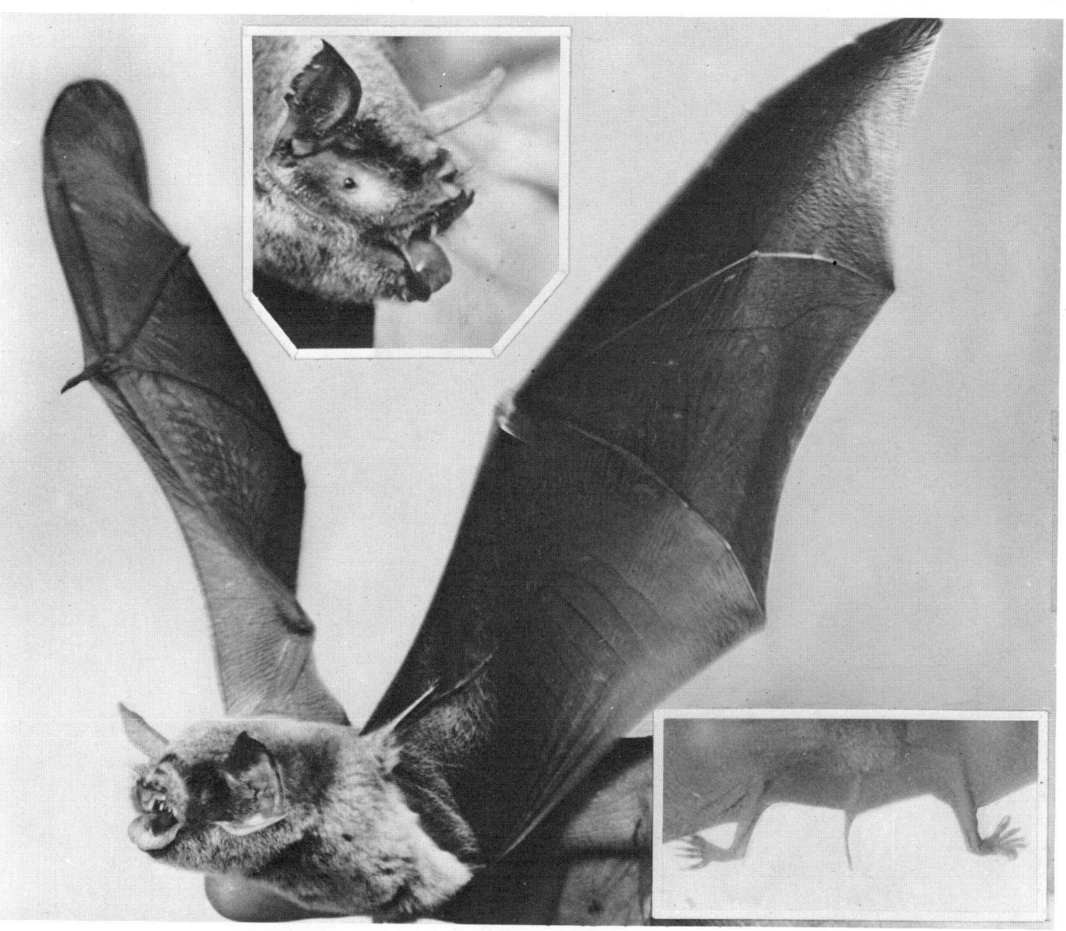

Mustache bat *(Pteronotus parnellii)*. Insets: head, photo by Ernest P. Walker; tail, photo by Harold Drysdale, Trinidad Regional Virus Laboratory, through A. M. Greenhall.

about 1,700 meters (Watkins, Jones, and Genoways 1972). In Venezuela, *P. davyi* and *P. gymnonotus* were most often taken in dry, open areas, while *P. parnellii* was found mainly in moist places within forests (Handley 1976). Although bats of this genus have been found in houses, they roost primarily in caves and tunnels. Some species may also shelter in hollows of plants, such as thorny bamboo. When hanging in caves, these bats seek the darker recesses, seldom near the entrances. They generally hang singly rather than in compact masses. *P. personatus* has been taken from a cave, where it preferred to lie flat on horizontal and nearly horizontal surfaces. The voice of *Pteronotus* has been described as a sibilant, birdlike chirp.

Bateman and Vaughan (1974) studied a combined group of *P. parnellii, P. personatus, P. davyi*, and *Mormoops megalophylla* that inhabited a cavern system in Sinaloa, Mexico. The bats began flying shortly after sunset; some returned as early as 1.5 hours after departure, but most appeared to remain away from the roost for 5–7 hours. The flyways to foraging grounds were at least 3.5 km, and probably several times that length for some individuals. It was estimated that these bats consumed 1,902–3,805 kg of insects each night. The total number of bats in the cavern system was estimated as 400,000–800,000. Additional investigation indicated that male *P. parnellii* roosted separately from the females for much of the year.

LaVal and Fitch (1977) determined that in Costa Rica *P. parnellii* was seasonally monestrous, with pregnant females being found from January to May but not from July to December. The following data also indicate seasonal reproduction in this genus: in Sinaloa, Mexico, pregnancies in *P. parnellii, P. davyi*, and *P. personatus* in May and June and lactating *P. parnellii* in July (Jones, Choate, and Cadena 1972); in Jalisco, Mexico, pregnant female *P. parnellii* and *P. davyi* in May and June but none from July to October (Watkins, Jones, and Genoways 1972); in Michoacán, female *P. parnellii* pregnant in March but not in February, September, or December (Hernandez et al. 1985); in the Yucatan, most female *P. parnellii* pregnant in February but none pregnant in July, August, and January (Jones, Smith, and Genoways 1973); in Guatemala, most female *P. parnellii, P. davyi*, and *P. personatus* pregnant in March but none pregnant in January (Jones 1966) and a lactating female *P. parnellii* taken in late July (Rick 1968); in Nicaragua, pregnant female *P. parnellii* in February and March, pregnant *P. davyi* and *P. gymnonotus* in April and May, no pregnancies from June to August, and a lactating female *P. parnellii* in April (Jones, Smith, and Turner 1971); in Haiti, a pregnant female *P. quadridens* in February (Klingener, Genoways, and Baker 1978); on Margarita Island, Venezuela, a pregnant female *P. parnellii* in July but two not pregnant in November (Smith and Genoways 1974); and in Surinam, pregnant females in July (Genoways and Williams 1979*b*). All pregnancy records refer to a single embryo per female.

Naked-backed bats *(Pteronotus davyi)*, photo by Ernest P. Walker.

CHIROPTERA; MORMOOPIDAE; **Genus MORMOOPS**
Leach, 1821

Leaf-chinned Bats

There are two species (Graham and Barkley 1984; Hall 1981; Smith 1972):

M. blainvillii, Cuba, Jamaica, Hispaniola, Puerto Rico, Mona Island, a single pre-Columbian specimen from Exuma Island (Bahamas);
M. megalophylla, southern Arizona and Texas to Honduras, southern Baja California, Caribbean coast of South America and adjacent islands, northern Ecuador, northwestern Peru.

Hall (1981), for technical reasons of nomenclature, used the name *Aello* Leach, 1821 for this genus and the name *Aello cuvieri* for the species *Mormoops blainvillii.*

Head and body length is about 50–73 mm, tail length is 18–31 mm, forearm length is 45–61 mm, and adult weight is usually 12–18 grams. Coloration in *M. megalophylla* is reddish brown or other shades of brown, buff, and cinnamon. In the pale phase, *M. blainvillii* is light brown above and buffy below, whereas in the dark phase the upper parts are dark brown and the underparts are ochraceous tawny.

The lower lip has fleshy peglike projections, the chin has leaflike projections, and there are long stiff hairs at the sides of the mouth, so that it is almost hidden. The upturned nose is short and has grooves, ridges, and pits. The tail is well developed. The braincase is greatly deepened, its floor being so elevated that the lower border of the foramen magnum is above the level of the rostrum. The tongue is protruded while drinking, but the water taken up by the foliation of the lips is sucked in, with the head raised, an action resembling chewing.

M. megalophylla occupies diverse habitats from desert scrub to tropical forest, and roosts in caves, mines, tunnels, and rarely buildings (Barbour and Davis 1969). In southwestern Texas, Easterla (1973) found this species to prefer the hot lowlands. Handley (1976) collected most specimens in Venezuela in moist, forested areas. In Jalisco, Mexico, this species ranges in altitude from sea level to about 2,000 meters (Watkins, Jones, and Genoways 1972). One observer noted that at Alamos, Sonora, Mexico, *M. megalophylla* emerged from retreats and began flight about 10 minutes before full dark. The bats seemed most active at about 2300 hours and

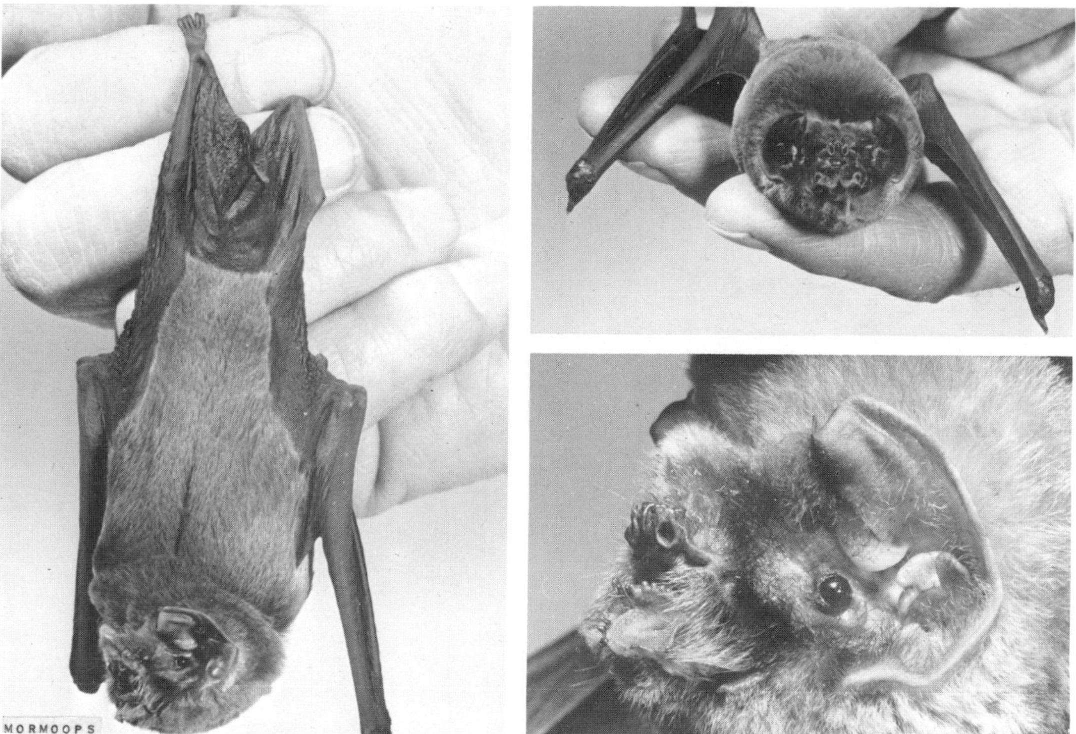

Leaf-chinned bat *(Mormoops megalophylla)*, showing the remarkable folds of skin on the chin and the peculiar folding of the ear and the shape of the nose, photos by Ernest P. Walker.

again at about midnight. Leaf-chinned bats feed later in the evening than do most bats that hunt insects. They seem to become less active in the winter, but they do not hibernate. *M. megalophylla* appears to hunt just above the ground. It searches for insects over land and water.

Leaf-chinned bat *(Mormoops megalophylla)* in flight, photo by Ernest P. Walker.

Goodwin (1970) stated that *M. blainvillii* has an extraordinarily swift flight, even in the narrow passageways of caves. Its wings often produce a characteristic humming sound. It penetrates farther into the depths of caves than does any other Jamaican bat.

M. megalophylla is colonial, 500,000 having been found in one cave in Nuevo Leon, but it does not roost in clusters (Barbour and Davis 1969). Easterla (1973) reported up to 4,000 per cave in southwestern Texas. In the latter area pregnant females, with one embryo each, were taken in June and lactating females were found from June to August. A pregnant female with a single embryo was collected in the Yucatan in February (Jones, Smith, and Genoways 1973). Pregnant females also have been taken in Coahuila in March, in Veracruz in April, in Sonora in May, and in Arizona in June.

CHIROPTERA; **Family NOCTILIONIDAE; Genus NOCTILIO**
Linnaeus, 1766

Bulldog Bats, or Fisherman Bats

The single known genus, *Noctilio*, contains two species (Baker, Genoways, and Patton 1978; Buden 1985; Cabrera 1957; Davis 1976*b*; Dolan and Carter 1979; Hall 1981; Hershkovitz 1975*a*; Hood and Jones 1984; Polaco 1987):

N. leporinus, Sinaloa and southern Veracruz (Mexico) to northern Argentina and southeastern Brazil, Greater and Lesser Antilles, Bahamas;

N. albiventris, southern Mexico to northern Argentina.

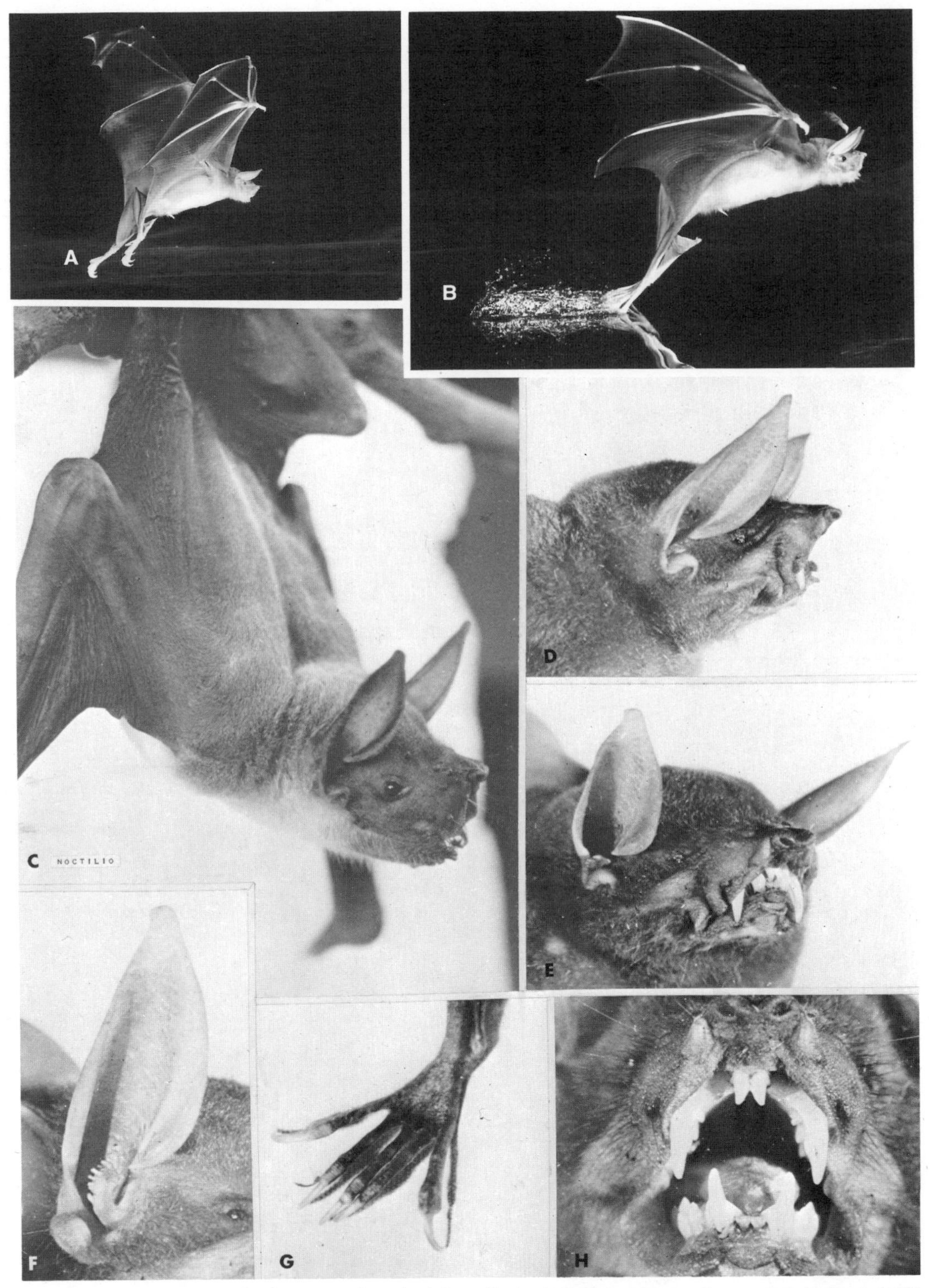

Fisherman bats *(Noctilio);* A & B *N. leporinus,* photos by J. Scott Altenbach. C. *N. leporinus mexicanus,* photo by Bernardo Villa. D–H, *N. leporinus leporinus,* photo by Harold Drysdale, Trinidad Regional Virus Laboratory, through A. M. Greenhall.

N. albiventris often has been referred to, incorrectly, as *N. labialis.*

In the larger species, *N. leporinus,* head and body length is 98–132 mm and forearm length is 70–92 mm. Klingener, Genoways, and Baker (1978) reported that a male weighed 78 grams, and a female 60 grams. In *N. albiventris* head and body length is 57–85 mm and forearm length is 54–70 mm; and Hood and Pitocchelli (1983) noted adult weights of 18–44 grams. The upper parts of *N. leporinus* are bright orange rufous in the males and gray or dull brown in the females. In *N. albiventris* the upper parts are grayish brown to yellowish or bright rufous, many males being bright rufous and many females dull brown to drab. Both species usually have a paler middorsal line and paler underparts.

The pointed muzzle and the nose lack excrescences. The lips are full and appear swollen. The upper lips are smooth but are divided by a vertical fold of skin under the nostrils, forming a "harelip" or hood over the mouth; the edges of the lower lips are smooth, but there is a lip pad or wart at the middle and a fold of skin under the pad, in addition to other semicircular folds of skin on the chin. The cheeks are elastic and can be greatly extended. The nostrils open forward and down, with the somewhat tubular nose projecting slightly beyond the lips. There is no nose leaf. The large, slender, pointed ears are separate, and a tragus, with a serrated outer margin, is present. The tail is well developed, being more than half as long as the thigh bone, and extends to about the middle of the well-developed tail membrane. It appears close to the upper surface of the membrane and perforates the surface, its tip being free. The calcar, or heel extension, is large and bony, particularly in *N. leporinus.* This species has unusually long hind legs and large hind feet with well-developed claws. The limbs of *N. albiventris* are not so modified for fishing. The hind legs are almost completely free from the wing membrane and are relatively short, particularly the femur. Both species have quite narrow and long wings in comparison with other families of the Microchiroptera.

The dental formula is: (i 2/1, c 1/1, pm 1/2, m 3/3) × 2 = 28. Of the two pairs of upper incisors, the first or middle pair is larger and nearly conceals the smaller second pair, located directly behind the first incisors. The molars exhibit a **W** pattern of cusps and ridges.

Handley (1976) found both species in a variety of vegetative conditions in Venezuela, but always near streams or in other moist places. *N. leporinus* usually roosts in rock clefts and fissures, dark caves, and hollow trees but has also been taken in buildings; *N. albiventris* generally shelters in hollow trees, foliage, and buildings. The former species has been found in sea caves with *Mormoops, Carollia,* and *Glossophaga,* and *N. albiventris* has been found in the same hollow trees as *Molossus major.* A roost of *Noctilio* is often detected by the unusually strong and penetrating musky odor of these bats.

Noctilio is one of only three genera of bats now known to catch and eat fish; the others are *Myotis (M. vivesi, M. adversus)* and *Megaderma. N. leporinus* fishes over the sea, at the edge of the surf, in large rivers, and in freshwater ponds. It generally feeds at dusk and during the night but has been seen in late afternoon flight over water in the company of pelicans. Presumably the bat catches small fish disturbed by the pelicans. Groups of *N. leporinus* move in zigzag flight, chirping and skimming the surface of the water. Small, surface-swimming fish (25–76 mm long) are caught with the sharp, long claws of the feet and are lifted quickly to the bat's mouth. The fish are either eaten in flight or carried to a roost to be eaten while the bat is resting. The bat does not scoop fish into its tail membrane, as once was claimed, but may transfer fish to the membrane after capture. Each of two captive individuals of this species was estimated to consume 30–40 fish each night. It evidently uses echolocation to find fish and determine their swimming velocity (Wenstrup and Suthers 1984). *N. leporinus* also uses its feet to catch insects (Novick and Dale 1971). Its diet includes aquatic crustaceans and such insects as winged ants, crickets, scarab beetles, and stinkbugs. The flight of *N. leporinus* is rather stiff-winged and not particularly rapid, but powerful. When knocked into water, it swims well, using its wings as oars.

Although *N. albiventris* had not previously been reported to feed on fish, Suthers and Fattu (1973) induced captive specimens to catch floating insects and pieces of fish on the surface of a pool, by dipping their feet into the water in a manner like that of *N. leporinus;* these specimens thrived on a diet of fish. Fleming, Hooper, and Wilson (1972) reported that 28 stomachs of *N. albiventris* from Central America contained 100 percent insect food. Brown, Brown, and Grinnell (1983) found *N. albiventris* to employ echolocation to capture insects from the surface of the water.

In a semiarid region of northeastern Brazil, Willig (1983, 1985*b*) found *N. leporinus* to roost by day in groups of up to 30 individuals and to forage by night in groups of 5–15. Reproductively, the species exhibited clear seasonal monestry in this region, with the later stages of gestation and the period of lactation corresponding with the wet season and greatest availability of fish and insects. Pregnancies occurred from September until January, and lactation was first seen in November and continued until April.

Additional data on social structure and reproduction include the following: in Sinaloa, Mexico, 3 of 7 female *N. leporinus* collected in mid-June were lactating (Jones, Choate, and Cadena 1972); in Jalisco, Mexico, a lactating female *N. leporinus* was taken in January, and a female pregnant with 1 embryo was taken in April (Watkins, Jones, and Genoways 1972); in the Yucatan, a colony of *N. leporinus* collected on 7 July included 3 adult males, 7 lactating females, and 6 young (Jones, Smith, and Genoways 1973); in Nicaragua, most female *N. albiventris* collected in April were pregnant with 1 embryo, as were 2 female *N. leporinus* taken in March, and 1 female *N. leporinus* collected in June was lactating (Jones, Smith, and Turner 1971); in Costa Rica, pregnant female *N. leporinus* were taken in February, April, and August (LaVal and Fitch 1977); in Panama, *N. albiventris* was reported to mate in late November or December and to give birth in late April or early May (Anderson and Wimsatt 1963); on Montserrat in the Lesser Antilles, 2 of 4 females collected in late July were lactating (Jones and Baker 1979); in Peru, pregnant females of both species were collected in July, and *N. albiventris* was reported to forage in groups of 8–15 (Tuttle 1970); and captive infant *N. albiventris* were nursed for almost three months (Brown, Brown, and Grinnell 1983). According to Jones (1982), a captive *N. leporinus* lived for 11 years and 6 months.

Martin (1972) stated that specimens of *N. leporinus* had been found in possibly late Pleistocene sites in Puerto Rico and Cuba. Otherwise, this family lacks a fossil history.

CHIROPTERA; **Family PHYLLOSTOMIDAE**

American Leaf-nosed Bats

This family of 48 Recent genera and 148 species is found in the tropical and subtropical regions of the New World, from the southwestern United States and the West Indies to northern Argentina. For years there has been confusion about the spelling of the name of this family, with some authors using

Saussure's long-nosed bat (*Leptonycteris* sp.) hovering at a spike of *Agave* flowers while obtaining nectar and pollen, photo by Bruce Hayward.

Phyllostomatidae and others writing Phyllostomidae. In a detailed analysis of the Greek and Latin origins of the name, Handley (1980*a*) determined that the proper spelling is Phyllostomidae. The sequence of genera presented here follows primarily that of Jones and Carter (1976, 1979), who recognized the following six subfamilies:

Subfamily Phyllostominae

Micronycteris	*Tonatia*	*Trachops*
Macrotus	*Mimon*	*Chrotopterus*
Lonchorhina	*Phyllostomus*	*Vampyrum*
Macrophyllum	*Phylloderma*	

Subfamily Glossophaginae

Glossophaga	*Anoura*	*Choeroniscus*
Monophyllus	*Scleronycteris*	*Choeronycteris*
Leptonycteris	*Lichonycteris*	*Musonycteris*
Lonchophylla	*Hylonycteris*	
Lionycteris	*Platalina*	

Subfamily Carolliinae

Carollia	*Rhinophylla*

Subfamily Stenodermatinae

Sturnira	*Ectophylla*	*Pygoderma*
Uroderma	*Artibeus*	*Ametrida*
Vampyrops	*Ardops*	*Sphaeronycteris*
Vampyrodes	*Phyllops*	*Centurio*
Vampyressa	*Ariteus*	
Chiroderma	*Stenoderma*	

Subfamily Phyllonycterinae

Brachyphylla	*Erophylla*	*Phyllonycteris*

Subfamily Desmodontinae

Desmodus	*Diaemus*	*Diphylla*

The recognition of the Desmodontinae (vampire bats) as a phyllostomid subfamily rather than a full family now has received general acceptance (Corbet and Hill 1986; Honacki, Kinman, and Koeppl 1982; Jones and Carter 1976, 1979; Koopman 1984*c*). Hall (1981) did state that "the sanguinivorous habits and corresponding morphological adaptations . . . as well as other obvious differences, contrast so sharply with those of other bats that the vampires are here accorded family rank." However, Baker, Honeycutt, and Bass (1988) concluded that "the affiliation of vampire bats within the Phyllostomidae is clear from immunological, morphological, and chromosomal data. To consider the vampires as a separate family obscures the evolution of sanguivory and constructs an unnatural classification." Pine and Ruschi (1976) listed a number of unconfirmed reports suggesting that certain phyllostomids, other than the Desmodontinae, occasionally feed on blood.

Other problems of systematic relationships are more controversial. Based primarily on morphological features of those phyllostomids specialized for feeding on nectar, Griffiths (1982) proposed (1) moving the genera *Lonchophylla, Lionycteris,* and *Platalina* from the Glossophaginae to a new subfamily, the Lonchophyllinae; (2) recognizing close affinity between the remaining glossophagines and the subfamily Phyllonycterinae; (3) placing *Brachyphylla* in its own subfamily, the Brachyphyllinae; and (4) recognizing that the genera *Glossophaga*, *Monophyllus*, and *Lichonycteris* form a systematic division of the Glossophaginae. These recommendations generally were accepted by Koopman (1984*b*), but they have also come under criticism (Haiduk and Baker 1982; Smith and Hood 1984). Two recent systematic checklists (Corbet and Hill 1986; Jones, Arroyo-Cabrales, and Owen 1988) did not follow Griffiths's proposals, and Griffiths (1985) found that molar morphology actually indicated that *Brachyphylla* is in the same systematic group as *Phyllonycteris* and *Erophylla*. On the basis of studies using albumin immunology, Honeycutt and Sarich (1987) did suggest that certain species of the Glossophaginae are closely related to the Phyllonycterinae (which they called the Brachyphyllinae).

Phyllostomids range in size from small to the largest of the American bats, *Vampyrum spectrum*. Head and body length is 40 to approximately 135 mm; the tail is absent or 4–55 mm in length; and forearm length is 31 to about 105 mm. When an external tail is lacking, the free edge of the tail membrane is deeply and broadly notched. The fur is variable in color. One species *(Ectophylla alba)* is whitish, and a number of species have two color phases. Several species in the subfamily Stenodermatinae have white spinal stripes and/or white facial stripes.

A nose leaf is usually present, but it is reduced or absent in a few genera. When it is present, it is not as complexly developed as in the families of Old World leaf-nosed bats. Those genera in which the nose leaf is reduced or absent often have platelike outgrowths on the lower lip. The ears are connected by a band of tissue across the top of the head in some species and are variable in form, but usually they are rather narrow and tend to be pointed. In some genera they are greatly elongated. A tragus is present and is variously thickened or notched. The males of *Phyllostomus* have a glandular throat sac. In the subfamilies Glossophaginae and Phyllonycterinae, the snout is elongate and the tongue is long, highly extensible, and covered with bristlelike papillae, much as in the Old World fruit bats, Pteropodidae (Macroglossinae). In most members of the subfamily Phyllostominae, and in some of the Glossophaginae, the dental formula is: (i 2/2, c 1/1, pm 2/3, m 3/3) × 2 = 34. Most other genera have 26–32 teeth.

In the subfamily Desmodontinae, the short and conical muzzle lacks a true nose leaf, having instead naked pads bearing U-shaped grooves at the tip that have been likened to a nose leaf. The ears are rather small, and the tail membrane is short. All the long bones of the wing and leg are deeply grooved for the accommodation of muscles. The dental formulas for the three genera of vampire bats are: *Desmodus*, (i 1/2, c 1/1, pm 2/3, m 0/0) × 2 = 20; *Diaemus*, (i 1/2, c 1/1, pm 1/2, m 2/1) × 2 = 22; and *Diphylla*, (i 2/2, c 1/1, pm 1/2, m 2/2) × 2 = 26. The incisors and the canines are specialized for cutting, being sicklelike or shearlike, with their cutting edges forming a V. The cheek teeth are greatly reduced, and all traces of crushing surfaces are absent. Other anatomical indications of the liquid diet are the short esophagus and the slender caecumlike stomach.

Vampire bats seem to prefer retreats of almost complete darkness, sheltering mostly in caves but also in old wells, mine shafts and tunnels, hollow trees, and buildings. Single individuals, small groups, or colonies of thousands utilize a specific roost. About 20 other species of bats have been found in the same retreats as *Desmodus* but not closely associated with it. Vampire bats are shy and agile in their roosts; they can walk rapidly, using their feet and thumbs, on horizontal or vertical surfaces. These bats do not leave their shelters until after dark and are most active before midnight. They forage low over the ground, flying fairly straight courses. Vampire bats, like those New World bats known to feed on

Long-tongued bat *(Glossophaga soricina)* hovering at a cluster of flowers of a banana while obtaining nectar and pollen, photo by Bernardo Villa.

fruit, emit pulses having only about one-thousandth of the sound energy of those used by bats that feed on flying insects or fish.

Some other species of the Phyllostomidae also roost in dark areas, whereas still others shelter in fairly well-illuminated places. Some members of this family are gregarious, others associate in small groups, and some roost singly. American leaf-nosed bats are often associated with other species of bats in their retreats, which may be caves, culverts, trees, buildings, or animal burrows. Several species form their own shelters by modifying leaves. It had been thought that the bats of this family were true homeotherms, never exhibiting torpor; however, laboratory studies have demonstrated the ability of three phyllostomids to estivate: *Glossophaga soricina*, *Vampyrops helleri*, and *Carollia perspicillata* (Rasweiler 1973).

Indirect evidence indicates that those species found in the southwestern United States migrate to a warmer climate for the winter. American leaf-nosed bats fly with their mouth closed, transmitting sounds through their nostrils. Except in the insectivorous forms, the soft impulses emitted by the members of this family are a complex and changing mixture of high frequencies of short duration.

Most forms feed on insects, fruit, nectar, and pollen; *Phyllostomus hastatus*, *Chrotopterus*, and *Vampyrum* are carnivorous. Some species aid in seed dissemination, and those that seek nectar, pollen, or insects from flowers pollinate the flowers. The scientific names of certain genera suggest that they drink blood, but their food habits are actually much like those of most other phyllostomids.

The Desmodontinae are the true vampires, feeding, so far as is known, only on fresh blood. Their method is to alight on or near the prospective victim and walk or climb onto it. Vampires attack in areas without hair or feathers or where the hair is scant. These include the naked skin around the anus and vagina, the ears and neck of cattle, and the wattles and combs of chickens. When a suitable area is found, the bat makes a quick shallow bite with its very sharp teeth that cuts away a small thin piece of skin. This operation is practically

A. Vampire bat *(Desmodus rotundus)* lapping blood from a saucer. (The bat's left forearm is broken near the elbow.) B. The face of a vampire bat *(D. rotundus)*, showing the extremely sharp incisor teeth, which are used to make the small incisions for obtaining blood. Photos by Ernest P. Walker.

painless and usually does not disturb the sleeping or quiet animal or human being. The bats do not bite deeply or struggle with a victim.

The wound thus produced is 3–6 mm wide, 5–10 mm long, and 1–5 mm deep. The tongue of the bat is then applied to the wound. The lower and lateral surfaces of the tongue have grooves that function like capillary tubes, through which the blood flows. The upper surface of the tongue remains free of blood. The tongue is protruded and retracted slowly and may produce a partial vacuum in the mouth cavity and in this manner aid in moving the flow of blood into the mouth. The blood sometimes flows from the wound for as long as eight hours, as the saliva of the bat may delay coagulation of the blood on the wound. *Desmodus* sometimes consumes so much blood at a meal that it is barely able to fly. The feeding period may not be more than 30 minutes in length.

Practically any warm-blooded animal that is quiet may be attacked by a vampire bat. Domestic animals often are victims because of their accessibility, but vampires seldom bite dogs, since dogs can hear sounds of higher frequency than some of the larger mammals are able to hear and so may detect the bat's approach. Stock raising in some tropical areas is uneconomical because of attacks by vampires. These bats bite sleeping humans, but not often.

The quantity of blood lost is usually not great, though it may be smeared extensively and present a startling appearance. The real danger of vampire bites lies in the diseases and infections that may result: these animals can transmit rabies and the livestock disease murrina, and the open wounds may become infected with bacteria and parasitic insect larvae, such as screw worms. The danger of infection is probably greatest when the flow of blood from the wound is not particularly rapid and the vampire, as a result, actually licks the wound. The bats may die of the diseases they transmit.

Some species of the Phyllostomidae form maternity colonies; in others the sexes remain together throughout the year. There is one, rarely two, young per birth. Breeding usually is soon followed by fertilization and development of the young.

The geological range of this family is Miocene and Pleistocene to Recent in South America, and Pleistocene to Recent in the West Indies and North America.

CHIROPTERA; PHYLLOSTOMIDAE; **Genus MICRONYCTERIS**
Gray, 1866

Little Big-eared Bats

There are 7 subgenera and 10 species (Anderson, Koopman, and Creighton 1982; Carter et al. 1981; Davis 1976*a;* Genoways and Williams 1986; Jones and Carter 1976, 1979; McCarthy 1987; McCarthy and Blake 1987; Trajano 1982; Williams and Genoways 1980*a*):

subgenus *Micronycteris* Gray, 1866

M. megalotis, Jalisco and Tamaulipas (Mexico) to Bolivia and southern Brazil, Grenada, Trinidad;
M. schmidtorum, Yucatan Peninsula of Mexico to northwestern South America;
M. minuta, Nicaragua to Bolivia and Brazil, Trinidad;

subgenus *Xenoctenes* Miller, 1907

M. hirsuta, Honduras to Peru and Surinam, Trinidad;

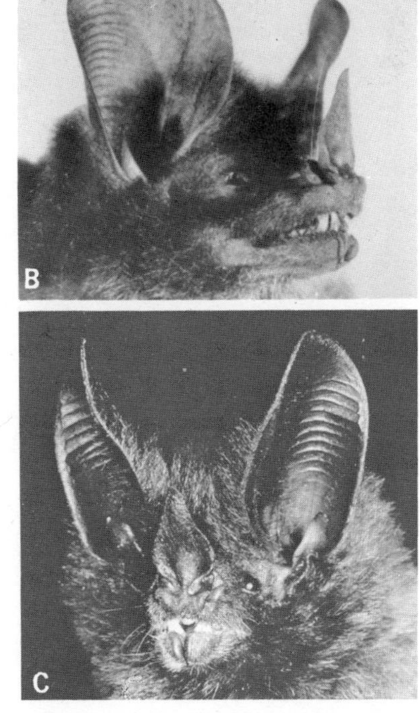

Little big-eared bat *(Micronycteris megalotis):* A. Photo by Bruce J. Hayward; B. Photo by Harold Drysdale, Trinidad Regional Virus Laboratory, through A. M. Greenhall; C. Photo by David Pye.

subgenus *Lampronycteris* Sanborn, 1949

M. brachyotis, southern Mexico to Amazonian Brazil, Trinidad;

subgenus *Neonycteris* Sanborn, 1949

M. pusilla, eastern Colombia, northern Brazil, and possibly adjacent regions of South America;

subgenus *Trinycteris* Sanborn, 1949

M. nicefori, Belize to northeastern Peru and northern Brazil, Trinidad;

subgenus *Glyphonycteris* Thomas, 1896

M. sylvestris, Nayarit and Veracruz (Mexico) to eastern Peru and southeastern Brazil, Trinidad;
M. behni, Peru, central Brazil;

subgenus *Barticonycteris* Hill, 1964

M. daviesi, known only from Costa Rica, Panama, Guyana, Surinam, Amazonian Peru, and Brazil.

Barticonycteris was treated as a full genus by Corbet and Hill (1986). In contrast, Genoways and Williams (1986) considered *Barticonycteris* a synonym of *Micronycteris* and put the species *M. daviesi* in the subgenus *Glyphonycteris.*

Head and body length is 42–69 mm, tail length is about 10–14 mm, forearm length is 31–57 mm, and adult weight is usually 4–16 grams. Coloration is usually some shade of brown above and brownish or buffy below. The upper parts in *M. megalotis* are dark brown tinged with russet, or dark brown without the russet tinge, some specimens exhibiting intermediate colors. *M. minuta* is orange rufous or brown. In *M. brachyotis* the upper parts are olive brown, the throat is yellowish or orangish, and the chest and belly are tawny olive.

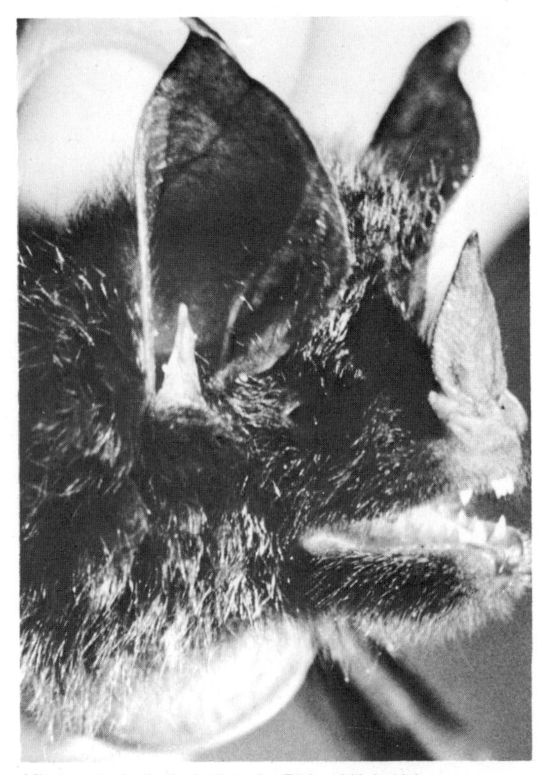

Micronycteris daviesi, photo by Richard K. LaVal.

A prominent nose leaf is present. In some species the ears are connected by a high notched band; in others they are not. In the bats of this genus the tail extends to the middle of the interfemoral membrane, and the middle lower premolar is not reduced. *Barticonycteris* is distinguished from the other subgenera by large size and the presence of a single pair of upper incisors so expanded basally as to fill the intercanine space completely and to extend posteriorly to a line joining the upper canines.

Little big-eared bats occupy a variety of habitats, including desert scrub and tropical forest. Handley (1976) found most species in Venezuela in moist places within forests, but many specimens were also taken in dry, open areas, and *M. nicefori* was found predominantly in dry forest. In Venezuela the most common roosting sites were holes in trees and hollow trees and logs. These bats are also known to roost in caves and crevices, animal burrows, culverts, buildings, and ancient ruins. They often roost where there is some light. Available data, summarized by Gardner (1977), indicate that *Micronycteris* is primarily insectivorous but that a variety of fruits are also consumed, their importance probably varying seasonally.

These bats roost alone or in small groups, usually not more than 20 being found together. Rick (1968) reported the discovery of a colony of *M. brachyotis* in a small chamber in Mayan ruins in Guatemala on 10 July. Present were a single adult male, 9 adult females, and 5 nursing young. Of the 7 adult females that were examined, 6 were lactating and 1 was pregnant with a single embryo. LaVal and Fitch (1977) found pregnant female *M. megalotis* in Costa Rica in April, a lactating female in May, subadults in August and September, and only nonpregnant females from October to February. Wilson (1979) summarized reproductive information for *Micronycteris,* showing that overall for the genus pregnant and lactating females have been found from February to August. These data suggest the following possible patterns: *M. megalotis,* seasonal breeding, with females pregnant in the northern part of the range during the beginning of the rainy season and there being perhaps two annual breeding cycles per female in the southern part of the range; *M. minuta,* breeding initiated at the beginning of the rainy season; *M. hirsuta,* a bimodal reproductive pattern in Trinidad; and *M. sylvestris,* breeding late in the rainy season in the north and early in the rainy season in the south.

Medellín L. et al. (1983) discovered what was perhaps the largest colony of *Micronycteris* on record. In 1978 they found a group of over 300 *M. brachyotis* in a coastal cave in southern Veracruz, Mexico. Observations over the next few years suggested that females are polyestrous. Unfortunately, the colony declined in numbers during the period and had disappeared entirely by July 1981. This loss was attributed to human habitat disturbances of the surrounding forest and to deliberate killing by persons who mistakenly believed the bats to belong to a vampire species.

CHIROPTERA; PHYLLOSTOMIDAE; **Genus MACROTUS**
Gray, 1843

Big-eared Bats

There are two species (Davis and Baker 1974; Jones and Carter 1976):

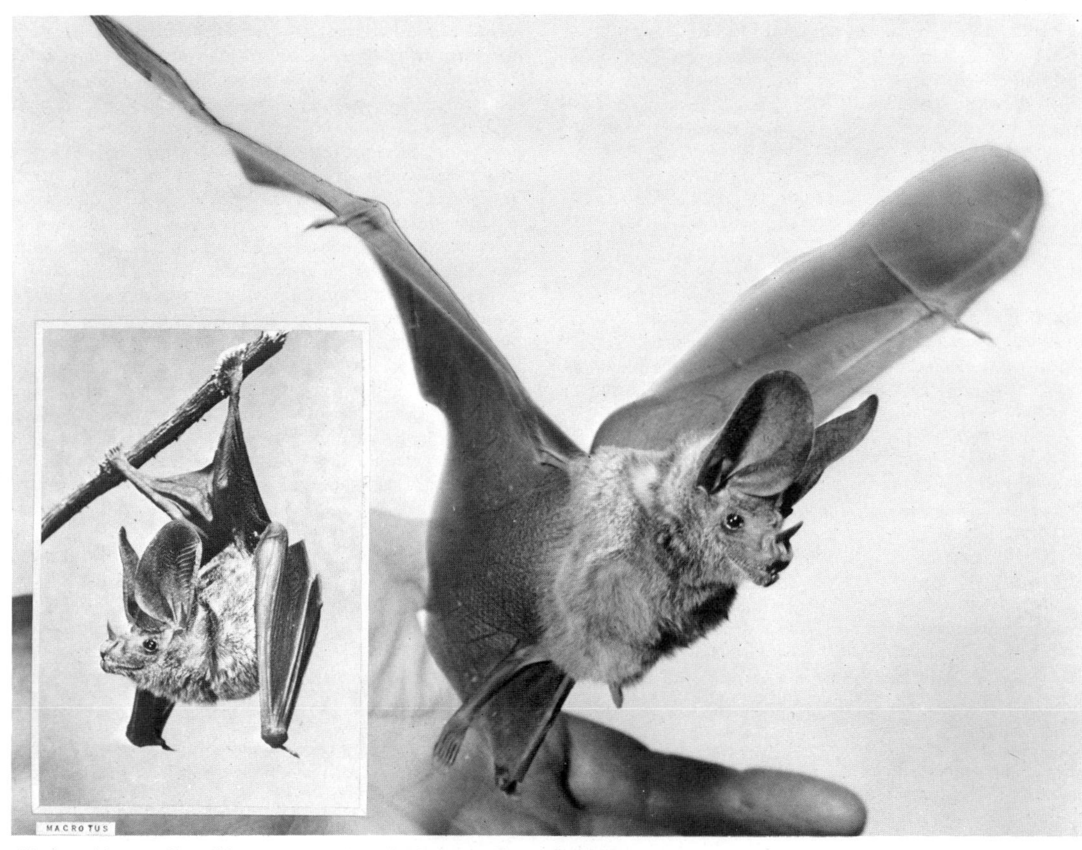

Mexican big-eared bat *(Macrotus waterhousii)*, photos by Ernest P. Walker.

M. waterhousii, western and central Mexico to the Yucatan and Guatemala, Greater Antilles, Bahamas;
M. californicus, southwestern United States, northwestern Mexico including Baja California.

Hall (1981) did not consider *M. californicus* to be specifically distinct from *M. waterhousii.*

Head and body length is 50–69 mm, tail length is 35–41 mm, forearm length is 45–58 mm, and adult weight is usually 12–20 grams. The upper parts are brownish or grayish, and the underparts are brown or buff, generally with a silvery or whitish wash.

This genus can be distinguished externally from *Micronycteris* by the larger ears and the longer tail. The ears of *Macrotus* are united, and the tail extends slightly beyond the interfemoral membrane. The nose leaf is erect and lanceolate, shaped somewhat like an arrowhead.

Big-eared bats usually frequent arid lowlands, but *M. waterhousii* also occurs in more humid locations in the West Indies. In Jalisco, Mexico, Watkins, Jones, and Genoways (1972) found *M. waterhousii* mainly in relatively arid areas where subtropical vegetation predominated. Caves are a favorite roosting site, but this genus also occupies mine tunnels and buildings. Complete darkness is not required, and the bats are often found within 10–30 meters of the entrance of a cave or in partially lighted buildings. *Macrotus* emerges later than most bats, usually 90–120 minutes after sunset (Anderson 1969). Unlike most bats native to the United States, *Macrotus* cannot crawl, but it is among the most agile and alert in flight (Barbour and Davis 1969). Some individuals of *M. californicus* winter north of the Mexican border, though most probably migrate to warmer regions. Hoffmeister (1986) reported that colonies in most caverns in southern Arizona are about the same size in summer and winter. Three individuals of *M. californicus* were found in a semidormant condition in March in northern Arizona, but the species does not experience deep hibernation. Although they can feed on the ground, the members of this genus feed mainly in flight and then hang to digest their catch. The diet consists of insects and fruits, including those of cacti.

These bats are gregarious, colonies of dozens or hundreds being common in some areas. Studies of a colony of *M. californicus* in southern Arizona showed both sexes to be present in March and April, but during the summer the females segregated in maternity colonies and the males dispersed in small groups. From August to October the sexes reassociated, but during the winter only males were present. Ovulation, insemination, and fertilization occurred in September and October. The embryo then grew slowly until March in a process known as delayed development. Subsequent development was more rapid, and births took place from May to early July after a total pregnancy of about 8 months. Normally there was a single young, rarely two. Weaning took place at about 1 month. Young females could breed during the first autumn after their birth, but males were not sexually mature until the following year. Maximum life expectancy was estimated to be more than 10 years (Anderson 1969; Barbour and Davis 1969). Other reproductive studies, as summarized by Wilson (1979), have found pregnant or lactating *M. californicus* from March to July in California and northwestern Mexico. Pregnant or lactating *M. waterhousii* have been taken in Mexico, Cuba, and the

Bahamas from February to July, and lactating females have been found in Jamaica in December. Klingener, Genoways, and Baker (1978) reported that five of six adult females collected in southern Haiti on 16 May were pregnant and that no pregnancies were found in January and August.

CHIROPTERA; PHYLLOSTOMIDAE; **Genus LONCHORHINA**
Tomes, 1863

Sword-nosed Bats

There are four species (Anderson, Koopman, and Creighton 1982; Hernandez-Camacho and Cadena 1978; Jones and Carter 1976; Ochoa G. and Ibáñez 1982; Trajano 1982):

L. orinocensis, southern Venezuela, southeastern Colombia;
L. fernandezi, known only from the type locality in southern Venezuela;
L. aurita, southern Mexico to Bolivia and southern Brazil, Trinidad, doubtfully New Providence in the Bahamas;
L. marinkellei, known only from the type locality at Durania in Amazonian Colombia.

Head and body length is 51–74 mm, tail length is 32–69 mm, and forearm length is 41–59 mm. The following weights were reported for *L. aurita* from Trinidad: a female 14.5 grams; a pregnant female 16.6 grams; and an immature male 11.2 grams. The fur is usually light reddish brown.

The long, sharply pointed nose leaf is about as long as the large, separate ears. The lower lip is grooved in front, with a raised cushion on either side. The tail extends to the edge of the interfemoral membrane.

Handley (1976) found most *L. aurita* in Venezuela in moist places within forests, while nearly all *L. orinocensis* were caught emerging from hot dry roosts in large rocks on prairies. *L. aurita* seems to roost mainly in caves and tunnels. Some 500 were found in a tunnel in a dense Panama forest. They were hanging in clusters toward the back of the tunnel, while bats of another genus, *Carollia,* occupied the front of the tunnel in about equal numbers. *L. aurita* has also been found hanging singly and concealed in large colonies of other species in a water canal. Another large colony was found in a highway culvert in Oaxaca, Mexico, again associated with large numbers of *Carollia.*

Wilson (1979) stated that the breeding season in *L. aurita* evidently is correlated with the beginning of the rainy season. Pregnant females of this species have been found in southern Mexico, Central America, and Trinidad from February to April. Hernandez-Camacho and Cadena (1978) reported that

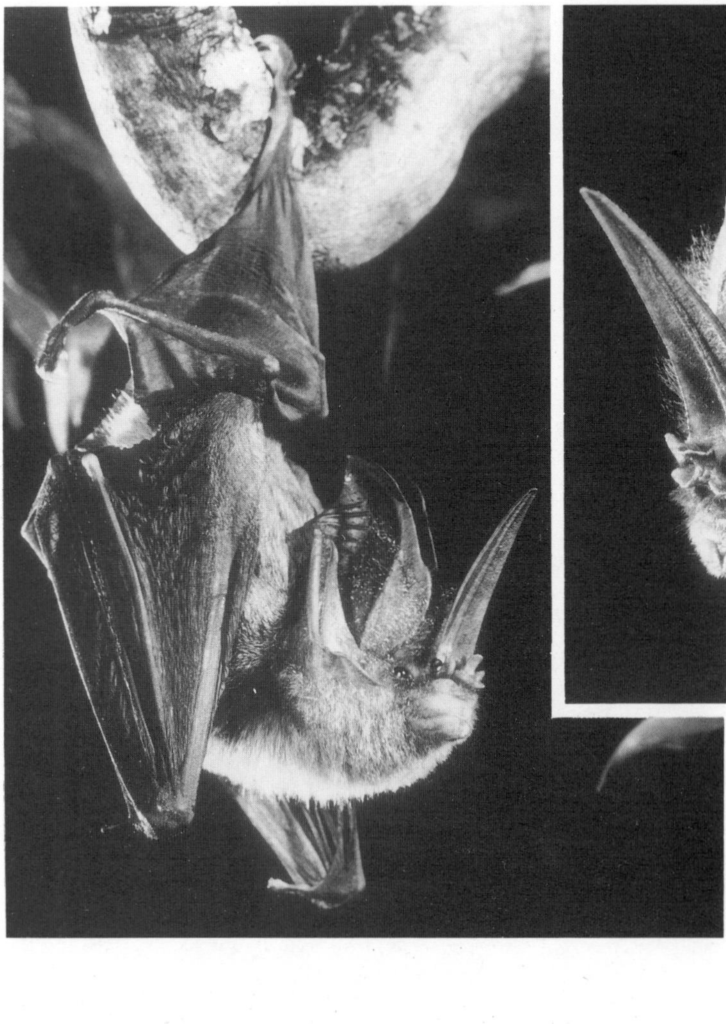

Sword-nosed bat *(Lonchorhina aurita),* photos by Merlin D. Tuttle, Bat Conservation International.

the holotype of *L. marinkellei*, an adult female, was pregnant with one embryo when collected on 8 August.

CHIROPTERA; PHYLLOSTOMIDAE; **Genus MACROPHYLLUM**
Gray, 1838

Long-legged Bat

The single species, *M. macrophyllum*, occurs from southern Mexico to Peru, northern Argentina, and southeastern Brazil (Harrison 1975*a*; Jones and Carter 1976).

Head and body length is 43–62 mm, tail length is about 37 mm, forearm length is about 34–45 mm, and adult weight is usually 6–9 grams. The coloration is sooty brown above and somewhat paler below. This bat can be distinguished externally by the long hind limbs, the long tail enclosed within the broad interfemoral membrane, the large and broad nose leaf, and the slender body form.

Handley (1976) collected most specimens in Venezuela in moist areas within forests; all roosting individuals were found in culverts. Other reports also indicate roosting in culverts, caves, sea caves, irrigation tunnels, and abandoned buildings (Harrison 1975*a*). The diet is mainly insectivorous and may include some aquatic life (Gardner 1977). Seymour and Dickerman (1982) reported groups of up to 59 individuals in culverts in a coastal marsh of Guatemala. Females usually were outnumbered by males and frequently were entirely

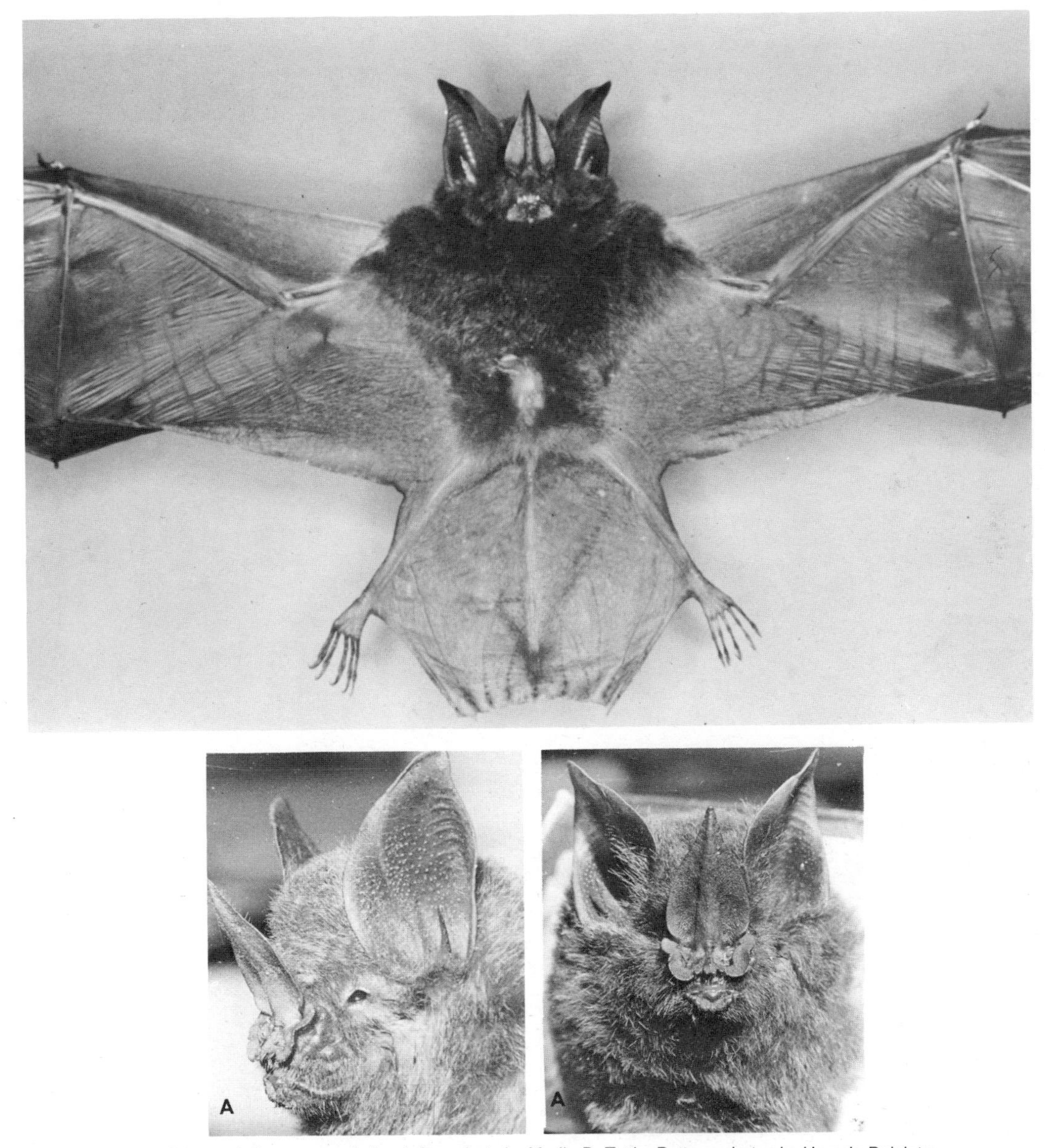

Long-legged bat *(Macrophyllum macrophyllum)*: Top, photo by Merlin D. Tuttle; Bottom, photos by Horacio Delpietro.

absent, suggesting alternate, undiscovered roosts. Pregnant females were taken in this area in both the wet and dry seasons. Additional pregnant females have been collected in El Salvador in October, in Costa Rica in March and May, and in French Guiana in October and November (Wilson 1979).

CHIROPTERA; PHYLLOSTOMIDAE; **Genus TONATIA**
Gray, 1827

Round-eared Bats

There are six currently recognized species (Davis and Carter 1978; Gardner 1976; Genoways and Williams 1979*b*, 1980, 1984; Hall 1981; Jones and Carter 1976, 1979; McCarthy 1982*b*, 1987; McCarthy, Cadena G., and Lemke 1983; McCarthy and Handley 1987; McCarthy, Robertson, and Mitchell 1988; Marques and Oren 1987; Medellín L. 1983; Myers, White, and Stallings 1983; Ojasti and Naranjo 1974):

T. bidens, southern Mexico and Belize to Paraguay and eastern Brazil, Trinidad;
T. carrikeri, Colombia, Venezuela, Surinam, northern Brazil, Peru, Bolivia;
T. brasiliense, southern Mexico to Amazonian Peru and eastern Brazil, Trinidad;
T. silvicola, eastern Honduras to Bolivia and northern Argentina;
T. evotis, southeastern Mexico to Caribbean versant of Honduras;
T. schulzi, Guyana, central Surinam, central French Guiana, northern Brazil.

Head and body length is 42–82 mm, tail length is 12–21 mm, forearm length is 32–60 mm, and adult weight is approximately 9–30 grams. The coloration is ochraceous to dark brown above and paler below. The underparts of *T. carrikeri* are white.

The ears are rounded and large, as long as the head or longer. Those of some species are separate, while those of others are united at the base. The tail extends to about the middle of the interfemoral membrane.

In Venezuela, Handley (1976) collected four species of *Tonatia,* mostly in moist areas within forests, and he reported *T. silvicola* to roost in hollow trees and termite nests. The species *T. carrikeri* and *T. brasiliense* also have been reported to roost in hollowed-out termite nests, both abandoned ones and those still in use by the insects. The diet of *Tonatia* consists mainly of insects, taken in flight and gleaned, and may also include a variety of fruits (Gardner 1977).

Fenton and Kunz (1977) stated that *T. bidens* and *T. silvicola* apparently roost alone or in small groups. Tuttle (1970) collected four colonies of *T. silvicola,* each comprising 6–10 individuals, from hollow termite nests. The type series of *T. carrikeri,* comprising 2 males and 5 females, some mature and some not yet fully grown, was collected in December from a hollowed-out termite nest hanging from a vine about 5 meters above the ground (Allen 1911). Wilson (1979) suggested that *T. bidens* and *T. brasiliense* might breed twice a year and that *T. silvicola* apparently gives birth during the early half of the rainy season. Data summarized by him include: pregnant *T. bidens* in January, February, May, July, and August; pregnant *T. brasiliense* in February, April, and July, and lactation in August; and pregnant *T. silvicola* in January, March, July, and August.

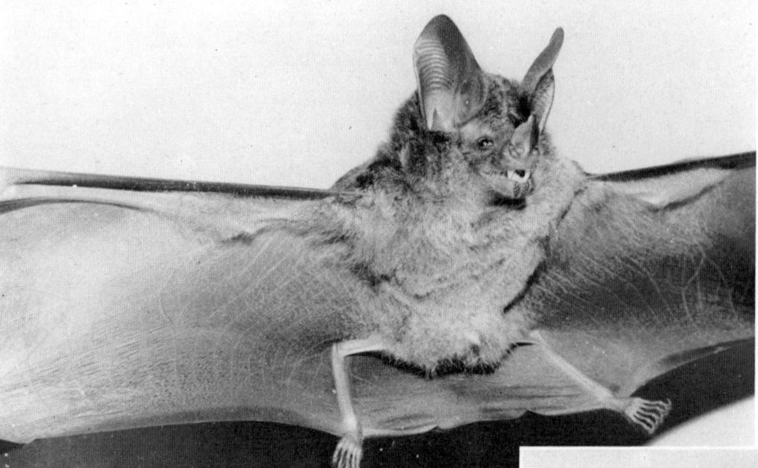

Round-eared bat *(Tonatia bidens),* photos by Merlin D. Tuttle. Inset: skull *(T. bidens),* photo by P. F. Wright of specimen in U.S. National Museum of Natural History.

CHIROPTERA; PHYLLOSTOMIDAE; **Genus MIMON**
Gray, 1847

Gray's Spear-nosed Bats

There are two species (Anderson, Koopman, and Creighton 1982; Carter et al. 1981; Hall 1981; Jones and Carter 1976, 1979; Koopman 1978*a*):

M. bennettii, southern Mexico to northern Colombia, eastern South America from Guyana and Surinam to southeastern Brazil;
M. crenulatum, southern Mexico to Bolivia and the Mato Grosso region of central Brazil, Trinidad.

Anthorhina Lydekker, 1891, sometimes is used as a genus or subgenus to include *M. crenulatum* (Husson 1978). *Mimon koepckeae*, from Huanhuachayo, Peru, was named as a full species by Gardner and Patton (1972) but was reduced to subspecific rank by Koopman (1978*a*). *Mimon cozumelae*, from southern Mexico to northern Colombia, has sometimes been considered a species distinct from *M. bennettii*.

Head and body length is approximately 50–75 mm, tail length is 10–25 mm, and forearm length is 48 to about 57 mm. A female *M. bennettii* weighed 22.9 grams and two males averaged 21.5 grams (Valdez and LaVal 1971). The coloration in *M. bennettii* is uniformly pale brownish except for whitish patches behind the ears. The fur in this species is long and woolly. In *M. crenulatum* the upper parts are bright mahogany brown or blackish brown in the fresh pelage; with age these colors become obscured with yellow, orange, and red. This species has a whitish or yellow orange patch behind the ear and a pale-colored spinal line; both the patch and the line are sometimes indistinct or even absent. The underparts are whitish to rusty, occasionally grayish. The fur in *M. crenulatum*, at least in the typical subspecies, is medium long and lax.

Gray's spear-nosed bat *(Mimon crenulatum)*, photo by Merlin D. Tuttle, Bat Conservation International.

The ears are separate. In *M. bennettii* the chin has a broad naked area divided by a longitudinal groove; in *M. crenulatum* the lower lip has a V-shaped notch in front bordered by wartlike protuberances. This genus may be distinguished from *Chrotopterus* by its smaller size, and from *Phyllostomus* by its pointed, rather than rounded, ears.

In Venezuela, Handley (1976) found most specimens of *M. crenulatum* near streams or in other moist places within forests, and some were roosting in hollow trees. This species has also been obtained from hollow, decayed tree stumps in wooded areas of Panama and Ecuador. *M. bennettii* seems to prefer to roost in dark, damp caves below the level of the ground but has also been taken in highway culverts in Oaxaca in the company of *Carollia, Glossophaga, Trachops*, and *Desmodus*. This species and *Phyllostomus discolor* have been noted side by side in the same cave. The diet of *Mimon* probably consists of a variety of arthropods and fruits (Gardner 1977).

M. bennettii does not roost in large groups, the usual number in a given retreat being two to four. Pregnant and lactating females of this species have been taken in southern Mexico and Central America from March to August, and apparently a single young is produced at the beginning of the rainy season in that region (LaVal and Fitch 1977; Wilson 1979). Pregnant female *M. crenulatum* have been taken in southern Mexico in February, in Costa Rica in April, in Venezuela in March, in Surinam in July (Genoways and Williams 1979*b*), and in Peru in July.

CHIROPTERA; PHYLLOSTOMIDAE; **Genus PHYLLOSTOMUS**
Lacépède, 1799

Spear-nosed Bats

There are four species (Anderson, Koopman, and Creighton 1982; Baker, Dunn, and Nelson 1988; Carter et al. 1981; Jones and Carter 1976; Koopman 1982*b*; Myers and Wetzel 1983):

P. discolor, southern Mexico to northern Argentina, Trinidad;
P. hastatus, Honduras to Paraguay and southeastern Brazil, Trinidad;
P. elongatus, east of the Andes in South America from Colombia to Bolivia and southeastern Brazil;
P. latifolius, southeastern Colombia, Guyana, Surinam, northern Brazil.

P. hastatus is one of the larger American bats, with a head and body length of about 100–130 mm, forearm length of about 83–95 mm, wingspread of about 457 mm, and adult weight of about 50–100 grams. In the smallest species, *P. discolor*, head and body length is about 75 mm, forearm length is 55–65 mm, and weight is about 20–40 grams. The length of the tail in *Phyllostomus* ranges from 10 to about 25 mm. The coloration is dark brown or blackish brown, grayish, reddish brown, or chestnut brown above and somewhat paler below.

Distinguishing features include the robust form, the simple, well-developed nose leaf, the widely separated ears, the short tail, and the heavy skull. The glandular throat sac, well developed in the males, is rudimentary in the females. The lower lip has a V-shaped groove edged with wartlike protuberances.

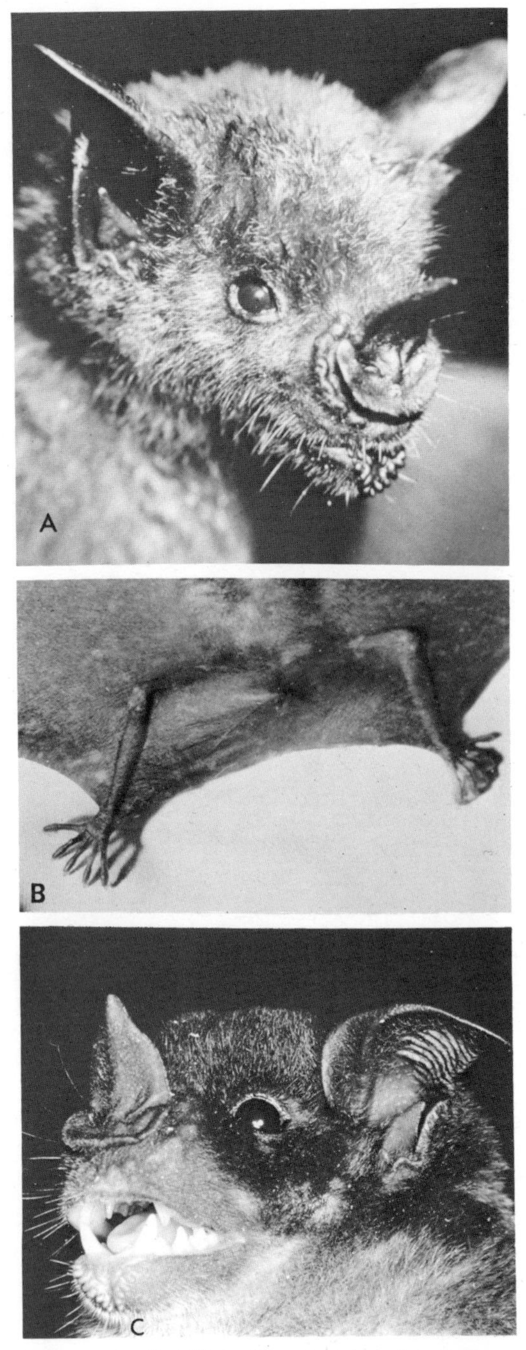

A. Spear-nosed bat *(Phyllostomus hastatus)*, photo by P. Morris. B. *P. d. discolor*, photo by Harold Drysdale, Trinidad Regional Virus Laboratory, through A. M. Greenhall. C. *P. hastatus*, photo by David Pye.

In Venezuela, Handley (1976) found *P. discolor, P. hastatus,* and *P. elongatus* to occur mostly near streams and in other moist places, but many specimens were taken in dry areas. About half of the specimens were collected in forests, and about half in more open areas. Roosts included caves, culverts, hollow trees, holes in trees, and buildings. In Peru, Tuttle (1970) found roosting colonies of *P. elongatus* in large hollow trees and those of *P. hastatus* in hollow trees, termite nests, caves, and thatched roofs. Both species were netted most often near gardens where bananas were grown. A radiotracking study in Trinidad determined that after *P. hastatus* emerged from its roosting caves in the evening, it flew to feeding sites some 1–5 km away (Fenton and Kunz 1977). Data summarized by Gardner (1977) indicated that bats of the genus *Phyllostomus* are omnivorous. *P. hastatus* preys on small vertebrates as well as insects, while the diet of *P. discolor* consists mainly of fruit, pollen, nectar, and insects caught in flowers. The diets of *P. elongatus* and *P. latifolius* are not known as well but probably include flower parts, fruits, insects, and small vertebrates such as anoles and geckos gleaned from vegetation. Fleming, Hooper, and Wilson (1972) found stomachs of *P. discolor* and *P. hastatus,* taken wild in Central America, to contain almost entirely insect food. Captive *P. discolor* are said to feed on fruit and to refuse meat. Captive *P. hastatus* have eaten fruit and small vertebrates; one such captive fell on a bat of the genus *Molossus,* crushed it with bites to the back and head, and later ate it.

Spear-nosed bats are gregarious; several thousand *P. hastatus* were found in a cave in Panama, and in Trinidad numerous individuals of this species fly together as a group from roost to feeding grounds. In Peru, Tuttle (1970) found colonies of *P. elongatus* to number 7–15 and those of *P. hastatus* to number 10–100 or more. More detailed studies (cited by Fenton and Kunz 1977) have determined that colonies of *P. hastatus* are divided into harems, with up to 30 adult females per dominant male, along with juveniles and nonharem males. Dominant males defend their harems against other males. *P. discolor* also occurs in harems, but the females number only 1–12 per dominant male and tend to be more nomadic.

In a radiotracking study of cave-dwelling colonies of *P. hastatus* in Trinidad, McCracken and Bradbury (1981) found harem clusters to contain an average of 18 females each and to be highly stable; some individuals roosted together for years. The single harem male also sometimes remained for several years, but turnover of male residency was frequent and did not disrupt female social structure. Bachelor male groups, averaging about 17 bats each, were present and were less stable. Resident males repelled intruding males in quick, violent fights but generally ignored female movements. Each individual in the colony had a separate foraging area; those belonging to the females of a particular cluster adjoined and were sometimes shared, but females of different harems foraged well apart from one another. Harem males also hunted separately. Mating in the colonies took place from about October to February, and there was a synchronized birth period in April and May. The young weighed about 13 grams at birth, were carried for several days, and were then left at the roost while the mothers foraged. They could fly within the cave at about 6 weeks and went out on their own at 2 months. Juveniles of both sexes dispersed after several months and were not recruited into their parental social units. Young females from different clusters formed new stable harems.

Additional reproductive data summarized by Wilson (1979) indicate that *P. discolor* may be an acyclic or continuous breeder in some areas, though possibly it is monestrous in Costa Rica. For *P. hastatus* the data could support either a monestrous pattern (in Nicaragua, Panama, and Trinidad) or a polyestrous one (Colombia); the reproductive strategy possibly varies geographically. In both species pregnant or lactating females have been recorded for most months of the year. One record not listed by Wilson is of a pregnant *P. discolor* collected in Peru in the period 31 October to 8 November (Davis and Dixon 1976). The smaller amount of data for *P. elongatus* (pregnancies in Colombia in June, in Peru in July and August) indicate that breeding occurs in the middle of the rainy season.

CHIROPTERA; PHYLLOSTOMIDAE; **Genus PHYLLODERMA**
Peters, 1865

Peters's Spear-nosed Bat

The single species, *P. stenops*, occurs from southern Mexico to Bolivia and southeastern Brazil (Barquez and Ojeda 1979; Hall 1981; Jones and Carter 1976, 1979; Trajano 1982). Based on an analysis of genetic data from protein variation, Baker, Dunn, and Nelson (1988) proposed that this species be transferred to the genus *Phyllostomus*.

Head and body length is 85–120 mm, tail length is about 16–22 mm, and forearm length is 67–80 mm. LaVal (1977) reported a weight of 71 grams for one female. The coloration is dark brown above and buffy to pale brown below. The wing membranes are described as blackish brown with whitish tips in the subspecies *P. s. septentrionalis* and dark brown in *P. s. stenops*.

This bat generally resembles *Phyllostomus* but may be readily distinguished by the two-lobed middle upper incisors, the narrow lower molars, and the presence of a small third lower premolar. A male specimen in the British Museum of Natural History appears to have a glandular throat sac.

In Venezuela, Handley (1976) collected all specimens in lowlands near streams and in other moist places; about half were taken in forests and about half in more open areas, such as pastures, orchards, and marshes. In Peru, Gardner (1976) collected a specimen in a small clearing in a cloud forest at 2,600 meters. In Costa Rica, LaVal (1977) captured a female adjacent to a small forest stream. The feces of this specimen consisted mostly of the large seeds of one species of the family Annonaceae, while in captivity this specimen eagerly ate bananas and drank sugar water with a long, extensible tongue. Another individual was caught in Brazil while eating the larvae and pupae from an active nest of a social wasp (Jeanne 1970). LaVal's (1977) specimen, taken on 9 February, was pregnant with one large embryo.

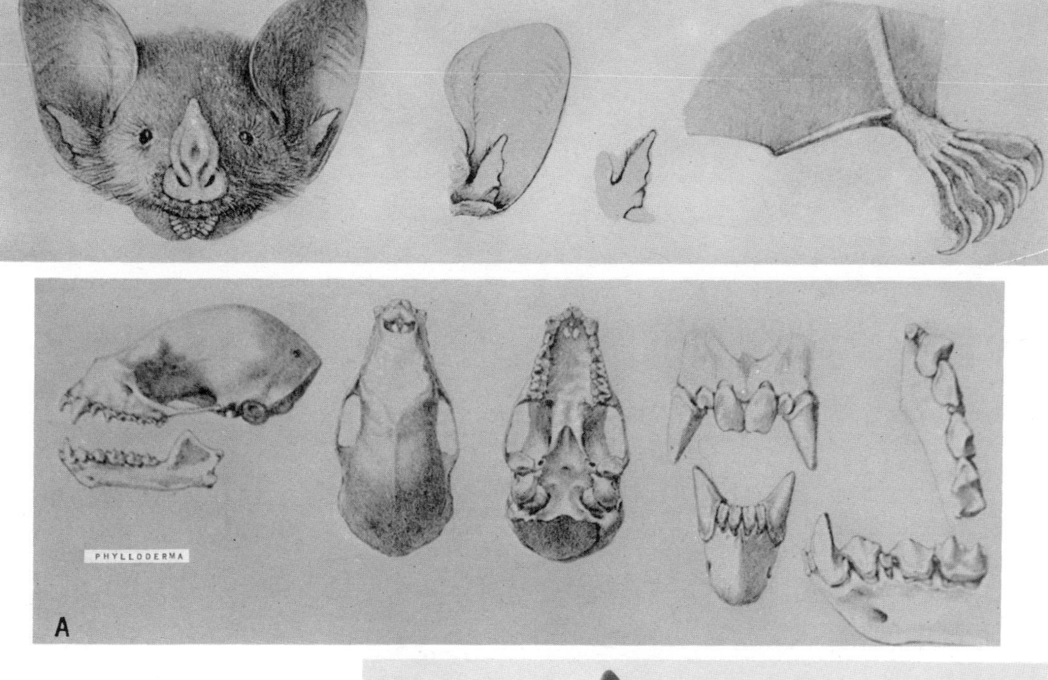

Peters's spear-nosed bat *(Phylloderma stenops):* A. Photo from *Die Fledermäuse des Berliner Museums für Naturkunde*, Wilhelm K. Peters; B. Photo by David Pye.

CHIROPTERA; PHYLLOSTOMIDAE; **Genus TRACHOPS**
Gray, 1847

Frog-eating Bat

The single species, *T. cirrhosus,* occurs from southern Mexico to Bolivia and southern Brazil (Jones and Carter 1976) and also on Trinidad (Carter et al. 1981).

Head and body length is 76–88 mm, tail length is 12–21 mm, and forearm length is 57–64 mm. LaVal and Fitch (1977) reported an average weight of 32.3 grams for a series of specimens from Costa Rica. The upper parts are dark reddish brown, cinnamon brown, or somewhat darker; the underparts are dull brownish washed with gray.

This genus can be distinguished by the wart-studded lips and the small size and peculiar position of the second lower premolar. The ear is longer than the head. The tail is much shorter than the femur and projects from the upper surface of the interfemoral membrane.

In Venezuela this bat was collected mostly in humid lowland forest and was found to roost in hollow trees (Handley 1976). Other reported roosting sites include caves, a Mayan building (Rick 1968), culverts, and an abandoned railroad tunnel (Starrett and Casebeer 1968). Data summarized by Gardner (1977) indicate that the diet of this bat consists of insects, small vertebrates such as lizards, and possibly some fruit. Ryan, Tuttle, and Barclay (1983) showed that *Trachops* locates frogs and distinguishes frog species by listening for and analyzing the frogs' calls. Ryan and Tuttle (1983) added that such an ability allows the bat to safely discriminate between poisonous and palatable prey.

Colonies of six individuals of both sexes have been discovered in hollow trees on Trinidad. Pregnant or lactating females have been collected in southern Mexico and Central America in February, March, April, May, August, and December; on Trinidad in March; and in Peru in July. These data suggest an extended breeding season, or that the season is geographically variable (Wilson 1979).

CHIROPTERA; PHYLLOSTOMIDAE; **Genus CHROTOPTERUS**
Peters, 1865

Peters's Woolly False Vampire Bat

The single species, *C. auritus,* ranges from southern Mexico to Paraguay and northern Argentina (Jones and Carter 1976).

Head and body length is about 100–112 mm, tail length is 7–17 mm, and forearm length is 75–87 mm. Rick (1968) listed weights of 72.7 grams for a male and 90.5 grams for a female. The long, soft hair is dark brown on the upper parts and grayish brown on the underparts.

Frog-eating bat *(Trachops cirrhosus)*, photos by Merlin D. Tuttle, Bat Conservation International.

Peters's woolly false vampire bat *(Chrotopterus auritus)*, photo by Louise Emmons.

On the front of the neck there is an opening into a glandular pocket, apparently somewhat like that on some other bats. The lips and chin are smooth except for a small wart in the center of the lower lip and a narrow elevation on either side. The ears are large, ovate, and separate. The tail is practically absent. In *Chrotopterus* there are four upper and two lower incisors, whereas *Vampyrum* has four upper and four lower incisors.

In Venezuela this bat occurs in forested lowlands, usually near streams or in other moist places (Handley 1976). Reported roosting sites include: caves in the Yucatan (Jones, Smith, and Genoways 1973), a Mayan ruin in Guatemala (Rick 1968), and a hollow tree in Costa Rica (Starrett and Casebeer 1968). The diet includes insects, fruit, and apparently a substantial proportion of small vertebrates such as other bats, opossums *(Marmosa)*, mice, birds, lizards, and frogs (Gardner 1977; Sazima 1978).

Colonies contain 1–7, usually 3–5, individuals (Medellín L. 1989). Pregnant females have been taken in southern Mexico in April and July and in Argentina in July; a lactating female was found in the Yucatan in July (Wilson 1979). Taddei (1976) found reproductive activity in southeastern Brazil only in the second half of the year; one female, captured pregnant and maintained in isolation, gave birth to a single young after 99 days.

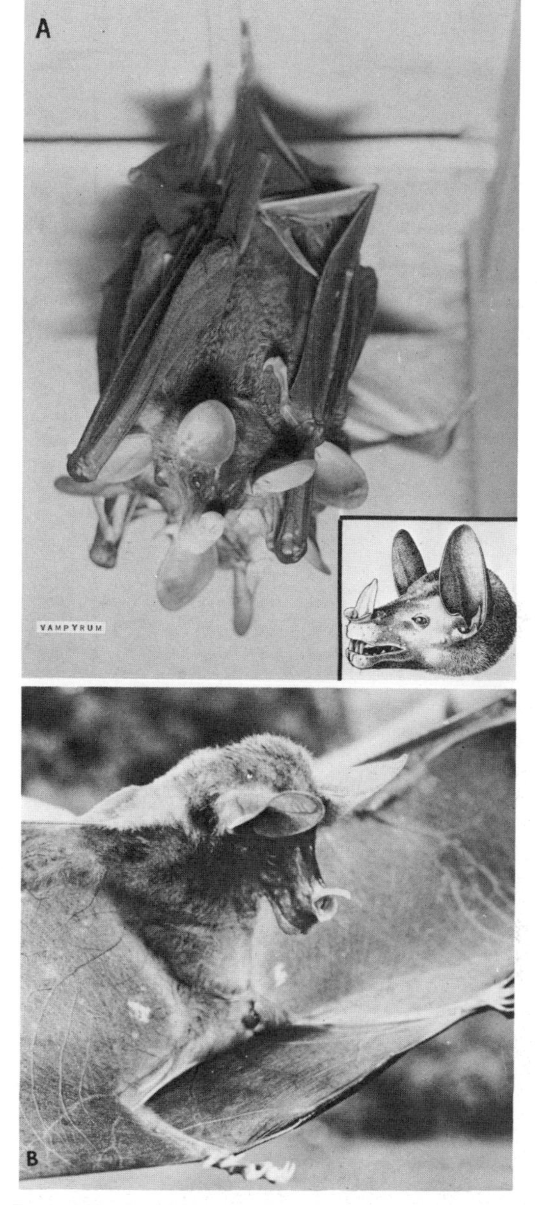

Tropical American false vampire bats *(Vampyrum spectrum)*: A. Photo from New York Zoological Society; inset photo from *Zool. Verhandel.*, "The Bats of Suriname," A. M. Husson; B. Photo by Richard K. LaVal.

CHIROPTERA; PHYLLOSTOMIDAE; **Genus VAMPYRUM**
Rafinesque, 1815

Linnaeus's False Vampire Bat, or Spectral Vampire

The single species, *V. spectrum*, ranges from southern Mexico to Peru and central Brazil and on Trinidad; a single specimen from Jamaica probably represents an accidental occurrence (Carter et al. 1981; Hall 1981; Jones and Carter 1976).

This is the largest New World bat; head and body length is about 125–35 mm, there is no tail, forearm length is about 100–108 mm, and the wingspan is usually 762–914 mm, but some measure as much as 1,016 mm. Adults weigh approximately 145–90 grams. The coloration is reddish brown above and slightly paler below. This bat is readily distinguished from *Chrotopterus* by its larger size, absence of a tail, and four upper and four lower incisors, instead of four upper and two lower incisors. As in *Chrotopterus*, the chin is smooth.

In Venezuela this bat occurs in lowlands and foothills; Handley (1976) collected all specimens in moist areas, with 40 percent being taken in evergreen forest, 40 percent in yards, and 20 percent in swamps. On Trinidad groups of not more than five individuals have been found roosting in hollow tree trunks, and on the upper Amazon scores were observed flying out of a church in the twilight. In Costa Rica a radiotracked individual hunted over an area of 3.2 ha. and spent most of its time in deciduous woodland, secondary growth, and forest edge (Vehrencamp, Stiles, and Bradbury 1977). *Vampyrum* was once thought to be a true vampire, but actually it is carnivorous and does not drink blood. According to Gardner (1977), the diet consists of birds, bats, rodents, and possibly some fruit and insects. A lactating female was taken on Trinidad in May, and reproductively inactive individuals were found in Costa Rica in February, July, August, and October (LaVal and Fitch 1977; Wilson 1979).

Greenhall (1968), reporting on a breeding pair that he successfully kept in captivity for five years, said that the bats readily adapted to captivity and became tame and gentle. They fed on white mice, birds, and raw meat, and drank water. The standard daily meal consisted of two adult white mice per bat. Young chicks and small pigeons were also relished. Raw fruit, such as bananas, mangoes, papayas, and citrus, was offered, but it was never eaten. When living mice were presented, the bats would crawl stealthily down the wall to within a few centimeters of the floor and wait for a mouse to pass underneath. Then a bat would drop on the mouse and kill it. With the mouse clamped tightly between its jaws, the bat would hitch itself backward up the wall to its roost to eat. The bat would hold and steady the mouse with the claws of its thumbs and thoroughly masticate it from head to heel, excluding the tail, which it discarded. Greenhall stated that the bats were careful, slow stalkers that rarely missed their aim. The head of a bird or rodent was carefully seized near the snout or beak, and the large canine teeth were quickly sunk into the skull. The captive *Vampyrum* mated and gave birth to a single young in late June, but Greenhall was not able to ascertain the gestation period. The female took good care of the offspring, and the male was also solicitous of his family. Sometimes he wrapped his huge wings around both mother and young, and frequently around the mother alone.

CHIROPTERA; PHYLLOSTOMIDAE; **Genus GLOSSOPHAGA**
E. Geoffroy St.-Hilaire, 1818

There are five species (Gardner 1986; Hall 1981; Jones and Carter 1976; Webster and Handley 1986; Webster and Jones 1980*b*, 1982, 1983, 1984*b*, 1984*c*, 1987):

- *G. soricina*, northern Mexico to Paraguay and northern Argentina, Jamaica, Bahamas;
- *G. leachii*, central Mexico to central Costa Rica;
- *G. commissarisi*, west-central Mexico, southern Mexico to eastern Peru and western Brazil;
- *G. morenoi*, southern Mexico;
- *G. longirostris*, Colombia, Venezuela and nearby Caribbean islands, Guyana, Trinidad, Lesser Antilles.

According to Gardner (1986), *G. morenoi* was mistakenly given the name *G. mexicana* by Webster and Jones (1980*b*).

Head and body length is 48–65 mm, tail length is about 7 mm, and forearm length is 32–42 mm. LaVal and Fitch (1977) listed average weights of 10.5 grams for *G. soricina*

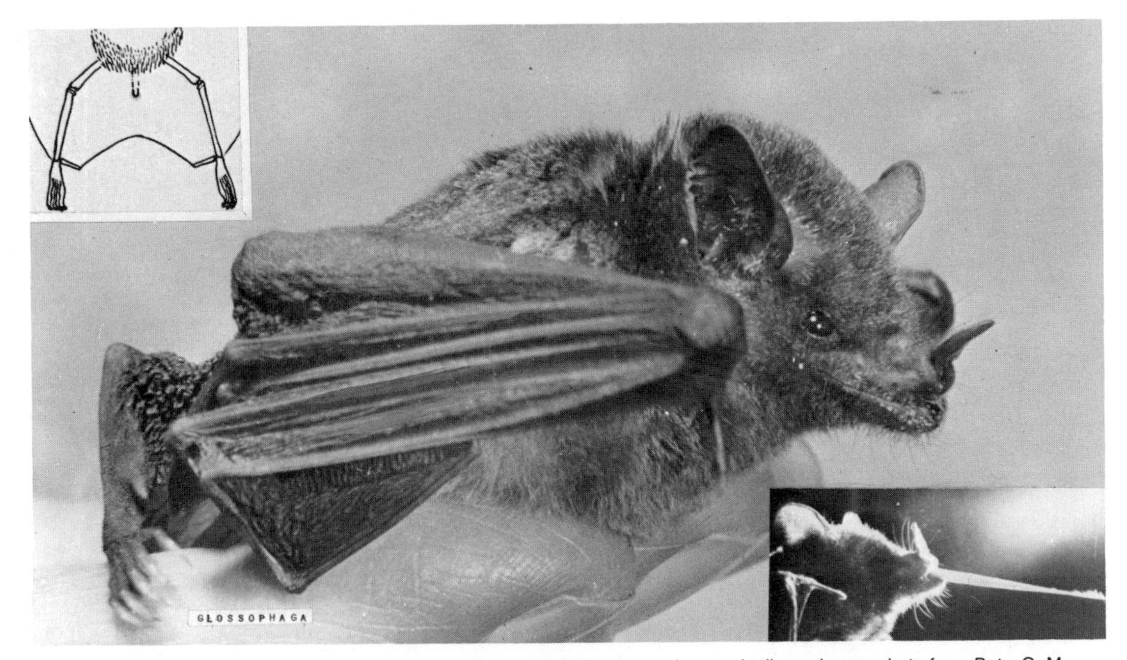

Long-tongued bat *(Glossophaga soricina)*, photo by Ernest P. Walker. Insets: legs and tail membrane, photo from *Bats*, G. M. Allen; head with tongue extended, photo by Stefan Vogel.

and 9.3 grams for *G. commissarisi.* The coloration varies from dark brown through lighter brown to reddish brown.

Glossophaga belongs to the long-snouted, extensible-tongued glossophagine group of bats. In this group the tongue is covered with bristlelike papillae, and the cheek teeth are narrow and elongate.

In Venezuela, Handley (1976) found *G. longirostris* mostly in arid lowlands and llanos, while *G. soricina* was taken mainly in moist, open areas. Roosting sites included caves, buildings, rock crevices, and hollow trees. In Jalisco, Mexico, Watkins, Jones, and Genoways (1972) found *G. commissarisi* in forested hills and mountains at altitudes of about 400–1,600 meters. Information summarized by Gardner (1977) indicates that the diet of *Glossophaga* consists of insects, fruit, pollen, nectar, and flower parts.

It once was assumed that *G. soricina* ate mainly fruit and probably nectar. Two species of Calabash tree are known to be visited by this species. In large flight cages, however, insects were actively hunted and readily eaten. In fact, insects were the favorite food of captives observed over a 14-month period in El Salvador. Before the rains began, honey water was almost preferred to insects, but it was much less important at other times. Captive *G. soricina* will hover and drink sugared or honeyed liquid from a flat dish. While in hovering flight, they will also drink honey water with rapid tongue movements from freely hanging petri dishes. The El Salvador captives also consumed fruit juices and pulp throughout the period of observation.

Lemke (1984) found certain *G. soricina* to defend feeding territories of about 3–10 sq meters around nectar-rich agave flowers. Maternity colonies of several hundred females and their young have been found in midsummer in San Luis Potosi, but the sexes have been found together in midsummer in Guerrero. In Veracruz, Hall and Dalquest (1963) found colonies of *G. soricina* numbering about 1,000 individuals, including both males and females. Wilson (1979) listed records of pregnant *G. soricina* for every month of the year and noted that the data suggest that this species is polyestrous in most areas. Willig (1985*b*) reported seasonal, bimodal polyestry in Brazil. Normally there is a single young, but one set of twins was reported in Chiapas, Mexico. According to Wilson (1979), *G. longirostris* apparently breeds during the rainy season, with pregnant or lactating females having been found from February to September. For *G. commissarisi* he listed pregnancies in January, February, April, July, and September and stated that the data are not inconsistent with a pattern of bimodal polyestry. LaVal and Fitch (1977) reported *G. commissarisi* to be seasonally polyestrous in Costa Rica and recorded pregnant females in February, March, and October. Webster and Jones (1984*b*, 1985) reported collection of a pregnant female *G. morenoi* in March and a lactating female in May; and collection of pregnant female *G. leachii* in February, April, June, July, August, September, and November and lactating females in February, March, June, and November. Jones (1982) reported that a specimen of *G. soricina* was still living after 10 years in captivity.

CHIROPTERA; PHYLLOSTOMIDAE; **Genus MONOPHYLLUS**
Leach, 1821

There are two species (Hall 1981; Jones and Carter 1976):

M. redmani, Cuba, Jamaica, Hispaniola, Puerto Rico, southern Bahamas;
M. plethodon, Lesser Antilles from Anguilla to Barbados, known from subfossils in Puerto Rico.

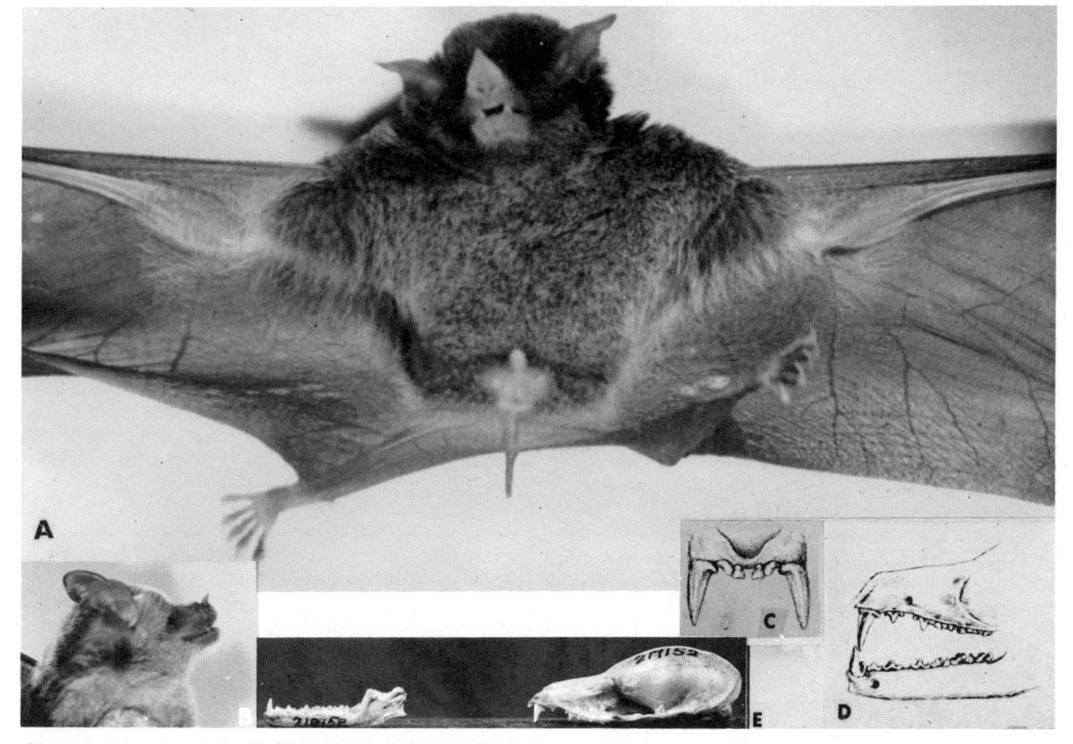

Monophyllus redmani: A & B. Photos by J. R. Tamsitt; C & D. Photos from *Proc. Zool. Soc. London;* E. Photo by P. F. Wright of specimen in U.S. National Museum of Natural History.

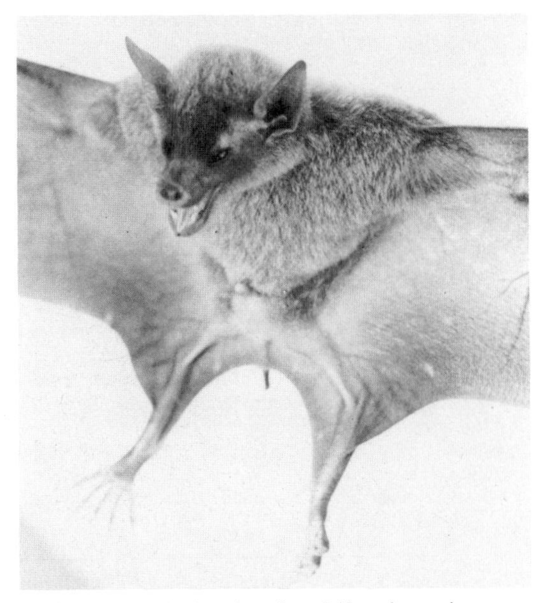

Monophyllus plethodon, photo from J. Knox Jones, Jr.

Head and body length is about 50–80 mm, tail length is about 4–16 mm, and forearm length is about 35–45 mm. Klingener, Genoways, and Baker (1978) gave a weight range of 8–13 grams for *M. redmani* in Haiti. Homan and Jones (1975*b*) reported weights of 12.5–17.2 grams for *M. plethodon* on Dominica. Coloration is various shades of brown, sometimes washed with gray underneath.

This genus is distinguished from *Glossophaga* and related bats mainly by dental features. The muzzle is elongate; the long tongue is supplied with papillae; the cheek teeth are narrow and elongate; and the tail usually projects for about half its length beyond the edge of the narrow interfemoral membrane.

M. redmani evidently roosts mainly in caves, and *M. plethodon* has also been collected in caves (Homan and Jones 1975*a*, 1975*b*). Goodwin (1970) observed that *M. redmani* prefers large, deep caves with high humidity. H. E. Anthony mentioned a unique metallic buzzing sound suggestive of the droning flight of a huge beetle that is probably made by *M. redmani* when disturbed and flying about. The flight of this species has been described as strong but not particularly erratic or angular. Various authors have indicated that *Monophyllus* feeds on soft fruit or nectar and possibly also insects, but there are no firm data on food habits (Homan and Jones 1975*a*).

M. redmani has been found roosting alone and in groups; clusters may be formed and apparently are sometimes segregated by sex. Pregnancies in this species have been recorded in December, January, February, and May. Pregnant female *M. plethodon* have been found on Dominica in March and April, and a lactating female was taken on Guadeloupe in late July. Both species are known to produce only one young at a time (Baker, Genoways, and Patton 1978; Klingener, Genoways, and Baker 1978; Homan and Jones 1975*a*, 1975*b*; Wilson 1979).

CHIROPTERA; PHYLLOSTOMIDAE; **Genus LEPTONYCTERIS**
Lydekker, 1891

Saussure's Long-nosed Bats

There are three species (Jones and Carter 1976):

L. nivalis, southern Texas to Guatemala;
L. sanborni, southern Arizona and New Mexico to El Salvador;
L. curasoae, islands of Aruba, Bonaire, and Curacao, and adjacent mainland in Colombia and Venezuela.

Hall (1981) and some other recent authors have used the name *L. yerbabuenae* for the species *L. sanborni*.

Head and body length is 70–95 mm, the tail is minute and appears to be lacking but actually consists of three vertebrae, and forearm length is 46–57 mm. The weight is usually 18–30 grams. These bats are usually reddish brown or sooty brown above and cinnamon or brown below.

This genus is characterized by dental features: the third molar is not present, and the lower incisors are usually present. In the only other genus of bats lacking the third molar *(Lichonycteris)*, the lower incisors are absent. The muzzle is elongate; the long tongue is supplied with papillae; and the cheek teeth are elongate and slender. The interfemoral membrane is very narrow.

According to Barbour and Davis (1969), *L. nivalis* favors high pine-oak country, while *L. sanborni* occurs in lowland desert scrub. Handley (1976) collected *L. curasoae* mainly in dry thorn forest. All species roost in caves, tunnels, and buildings. They emerge relatively late in the evening to feed. Populations of *L. nivalis* and *L. sanborni* in the southwestern United States apparently migrate to Mexico for the winter. They may return year after year to the same locality for the summer (Barbour and Davis 1969; Fenton and Kunz 1977; Hayward and Cockrum 1971). The diet consists of nectar, pollen, fruit, and insects. The insects may be ingested accidentally while feeding on pollen and nectar. *L. nivalis* is known to visit the flowers of *Malvaviscus*, flowers and fruits of cacti, and perhaps the flowers of jimson weed *(Datura)*. The bats probably use their long muzzle to reach the spineless parts of the cactus fruits, their canine teeth to tear the skin, and their tongue to lap up the juices. The tongue of *L. nivalis* can be extended to 76 mm. This manner of feeding on fruits may be utilized by other bats as well. The diet of *Leptonycteris* often results in the accumulation of yellow or red droppings on the floors of roosts.

These bats may occur in groups of 10,000 or more. During the spring, females of *L. sanborni* in the northern part of their range form maternity colonies numbering into the thousands (Barbour and Davis 1969). Apparently, adult male *L. nivalis* also separate from the females in the summer and do not occupy the northern part of the range of the species (Hensley and Wilkins 1988). Smith and Genoways (1974) reported that in July a colony of *L. curasoae* on Margarita Island, Venezuela, consisted of approximately 4,000 females nursing nearly full-grown young. In November the colony contained no young, but pregnant females and adult males in breeding condition were then present. The young of *L. nivalis* apparently are born during the summer in Texas and Mexico, while pregnant female *L. sanborni* have been found in Mexico in February, March, April, July, September, and November (Wilson 1979).

According to Arita and Wilson (1987), *L. sanborni* and *L. nivalis* are vital in the pollination of certain kinds of cacti and agaves. As the bats seek the nectar from the flowers of these

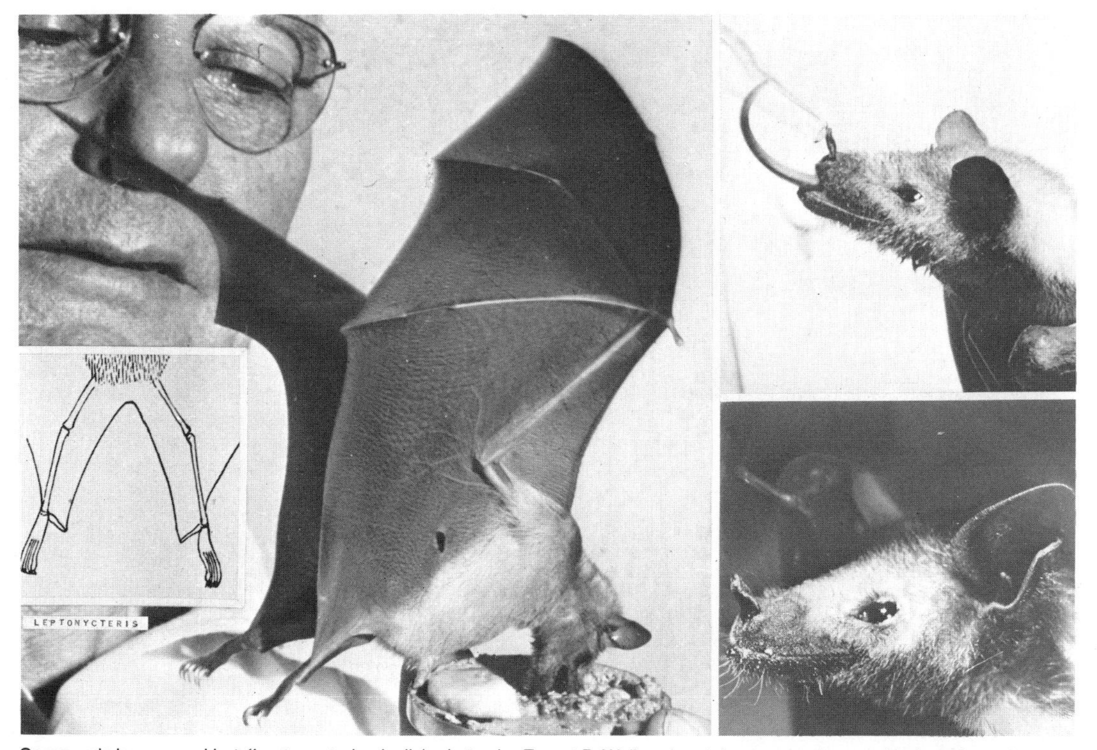

Saussure's long-nosed bat *(Leptonycteris nivalis)*, photos by Ernest P. Walker. Inset: *Leptonycteris* sp., with the calcars that extend the membrane at the heel clearly shown, photo from *Bats*, G. M. Allen.

plants, their fur becomes coated with pollen, which then is transferred to other flowers visited. The clearing of natural agaves may have resulted in greatly reduced bat populations, a situation that in turn jeopardizes the survival of some of the agaves. Colonies of *Leptonycteris* also have been destroyed deliberately by people who mistakenly believe the bats to belong to a vampire species. Such problems have been responsible for the elimination of most of the major known colonies in Mexico and for a drastic reduction in the number of these bats that migrate into the southwestern United States. Both *L. sanborni* and *L. nivalis* are now classified as vulnerable by the IUCN and as endangered by the USDI.

CHIROPTERA; PHYLLOSTOMIDAE; **Genus LONCHOPHYLLA**
Thomas, 1903

There are eight species (Alberico 1987; Gardner 1976; Hill 1980*a*; Jones and Carter 1976; Sazima, Vizotto, and Taddei 1978; Taddei, Vizotto, and Sazima 1983; Webster and Jones 1984*a*):

L. hesperia, Peru;
L. mordax, Ecuador, Bolivia, Brazil;
L. dekeyseri, east-central Brazil;
L. concava, Costa Rica, Panama, Colombia;
L. robusta, Nicaragua, Costa Rica, Panama, Colombia, Ecuador, Venezuela, Peru;
L. handleyi, Colombia, Ecuador, Peru;
L. thomasi, Panama, Venezuela, Guyana, Surinam, Brazil, Ecuador, Peru, Bolivia;
L. bokermanni, southeastern Brazil.

Head and body length is about 45–65 mm, tail length is 7–10 mm, and forearm length is 30–48 mm. LaVal and Fitch (1977) reported an average weight of 16.3 grams for *L. robusta,* and Willig (1983) listed an average weight of about 8.5 grams for *L. mordax.* These bats are rusty or dark brown above and somewhat paler below.

The muzzle is elongate, the long tongue is equipped with papillae, the cheek teeth are narrow and elongate, and the short tail barely reaches to the middle of the interfemoral membrane. This genus is similar to *Glossophaga* but differs from it in the incompleteness of the zygomatic arch of the skull and in features of the incisor teeth. The nose leaf in *Lonchophylla* is high and narrow, not low and broad as in *Lionycteris.*

In Venezuela, Handley (1976) collected *L. robusta* and *L. thomasi* mostly in moist areas within forests; *L. thomasi* was found to roost in hollow trees. Bats of this genus also roost in caves. The diet includes insects, fruit, pollen, and nectar (Gardner 1977). Pregnant female *L. concava* have been recorded in March and August in Costa Rica (Wilson 1979). LaVal and Fitch (1977) found pregnant female *L. robusta* in Costa Rica in February, May, August, and October and a lactating female in January.

CHIROPTERA; PHYLLOSTOMIDAE; **Genus LIONYCTERIS**
Thomas, 1913

The single species, *L. spurrelli,* has been reported from eastern Panama, Colombia, Venezuela, Guyana, Surinam, French Guiana, northern Brazil, and Amazonian Peru (Jones and

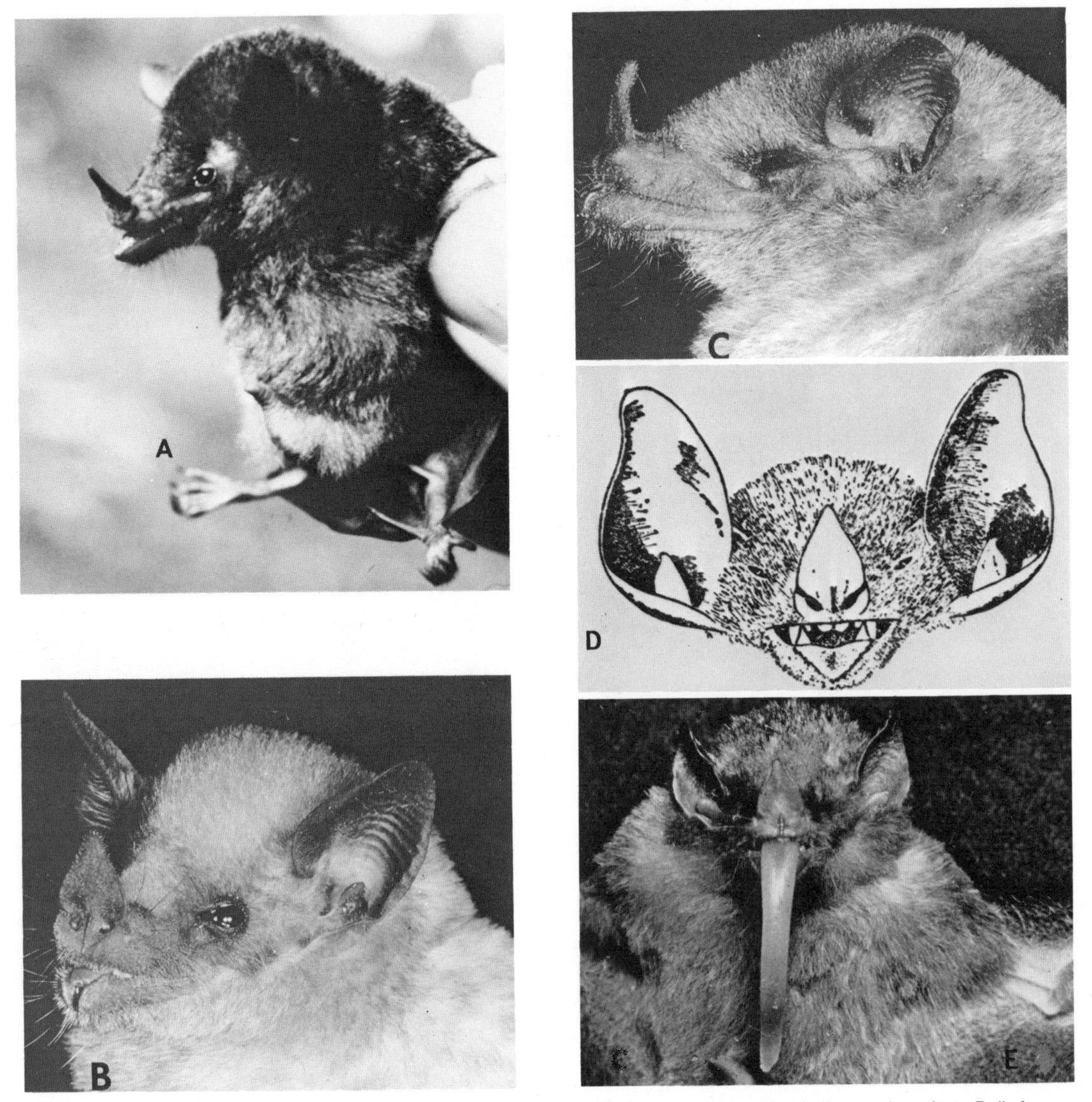

Lonchophylla robusta: A. Photo by Richard K. LaVal; B & C. Photos by David Pye; D. *Lonchophylla* sp., photo from *Bull. Amer. Mus. Nat. Hist.*, "Mammals of Costa Rica," G. G. Goodwin; E. Tongue extended *(L. mordax)*, photo by Stefan Vogel.

Carter 1976; Webster and McGillivray 1984; Williams and Genoways 1980*a*).

Head and body length is about 50 mm, tail length is about 10 mm, and forearm length is 34–36 mm. The color is reddish brown to dull brown. The muzzle is elongate, the extensible tongue is supplied with bristlelike papillae, and the cheek teeth are narrow but lack the horizontal lengthening of the cheek teeth of other bats of the subfamily Glossophaginae. The interfemoral membrane is well developed, and the tail extends about half its length.

In Venezuela, Handley (1976) collected most specimens in moist, forested areas and found this bat to roost in caves and crevices. In Peru, Tuttle (1970) caught two individuals around the edges of native villages and one among blooming cashew trees. Tuttle also collected a pregnant female with one embryo on 5 August. Graham (1987) caught two pregnant females in Peru in August.

CHIROPTERA; PHYLLOSTOMIDAE; **Genus ANOURA**
Gray, 1838

Geoffroy's Long-nosed Bats

There are four species (Anderson, Koopman, and Creighton 1982; Handley 1984; Jones and Carter 1976; Lemke and Tamsitt 1979; Nagorsen and Tamsitt 1981):

A. geoffroyi, Mexico to southeastern Brazil and northwestern Argentina, Trinidad;
A. latidens, Venezuela to central Peru;
A. caudifer, northern South America south to Bolivia and Brazil;
A. cultrata, Costa Rica, Panama, Colombia, Venezuela, Peru, Bolivia.

Lionycteris spurrelli, photo by Merlin D. Tuttle.

Head and body length is about 50–90 mm, the tail is absent or only 4–7 mm long, and forearm length is 34–48 mm. LaVal and Fitch (1977) reported the average weight of a series of specimens of *A. geoffroyi* to be 15.2 grams; *A. cultrata* weighs 14–23 grams (Tamsitt and Nagorsen 1982). Seven specimens of *A. caudifer* had a weight range of 4–8 grams. Coloration in *A. geoffroyi* is dull brown above, usually silvery gray over the shoulders and the sides of the neck, and grayish brown below. *A. caudifer* has dark brown upper parts and paler brown underparts. *A. cultrata* is shiny blackish, gray, or rich orange brown, and may be paler ventrally; the pelage is short and crisp.

Members of this genus resemble *Glossophaga* in that the muzzle is elongate, the long tongue is supplied with papillae, and the cheek teeth are narrow and elongate. *A. geoffroyi* lacks a tail and has a rudimentary calcar, whereas *A. caudifer* has a rudimentary tail and a well-developed calcar.

In Venezuela, Handley (1976) found *Anoura* to occur predominantly in moist, forested areas, often at high altitudes. Roosting was in caves and rocks. *A. cultrata* is found in forests from about 220 to 2,600 meters, roosting exclusively in caves or tunnels (Tamsitt and Nagorsen 1982). According to Gardner (1977), the diet of this genus includes fruit, pollen, nectar, and insects, and *A. geoffroyi* is considered a highly insectivorous glossophagine. Sazima (1976) found that all stomachs of *A. geoffroyi* and *A. caudifer* collected in Brazil contained nectar, pollen, and insects, including beetles and moths that were too large to have been absorbed accidentally with nectar.

In Costa Rica, Lemke and Tamsitt (1979) found that *A. geoffroyi* and *A. cultrata* roosted singly or in groups of up to 20. In Peru, Tuttle (1970) reported a colony of about 75 *A. geoffroyi* of both sexes. Data summarized by Wilson (1979) suggest that *A. geoffroyi* forms sexually segregated colonies at certain times of the year in the same caves on Trinidad. In one, there were 20 males and 25 females in June, 29 males and 1 female in October, and 32 males and 56 females in November. In this area *A. geoffroyi* apparently breeds late in the rainy season, with pregnant females having been found in November. Pregnancies in this species have also been reported in July in Nicaragua, in March in Costa Rica, and in June in Peru. Lactating female *A. geoffroyi* have been taken in Mexico in July, November, and December (Carter and Jones 1978; Wilson 1979). Pregnant or lactating female *A. caudifer* have been collected in January, February, May, June, and November (Wilson 1979). Pregnant or lactating female *A. cultrata* have been taken in Costa Rica, Colombia, and Peru in July and August; although this species usually bears a single young, one female gave birth to twin males (Tamsitt and Nagorsen 1982). One specimen of *A. geoffroyi* was still living after 10 years in captivity (Jones 1982).

CHIROPTERA; PHYLLOSTOMIDAE; **Genus SCLERONYCTERIS**
Thomas, 1912

The single species, *S. ega*, apparently is known only from three specimens: the type in the British Museum of Natural History, collected at Ega, Amazonas, Brazil; a specimen in the Museu Nacional, Rio de Janeiro, Brazil; and one in the United States National Museum of Natural History, taken at Tamatama, Rio Orinoco, T. F. Amazonas, southern Venezuela.

Measurements of the type, an adult female, are as follow: head and body length, 57 mm; tail length, 6 mm; and fore-

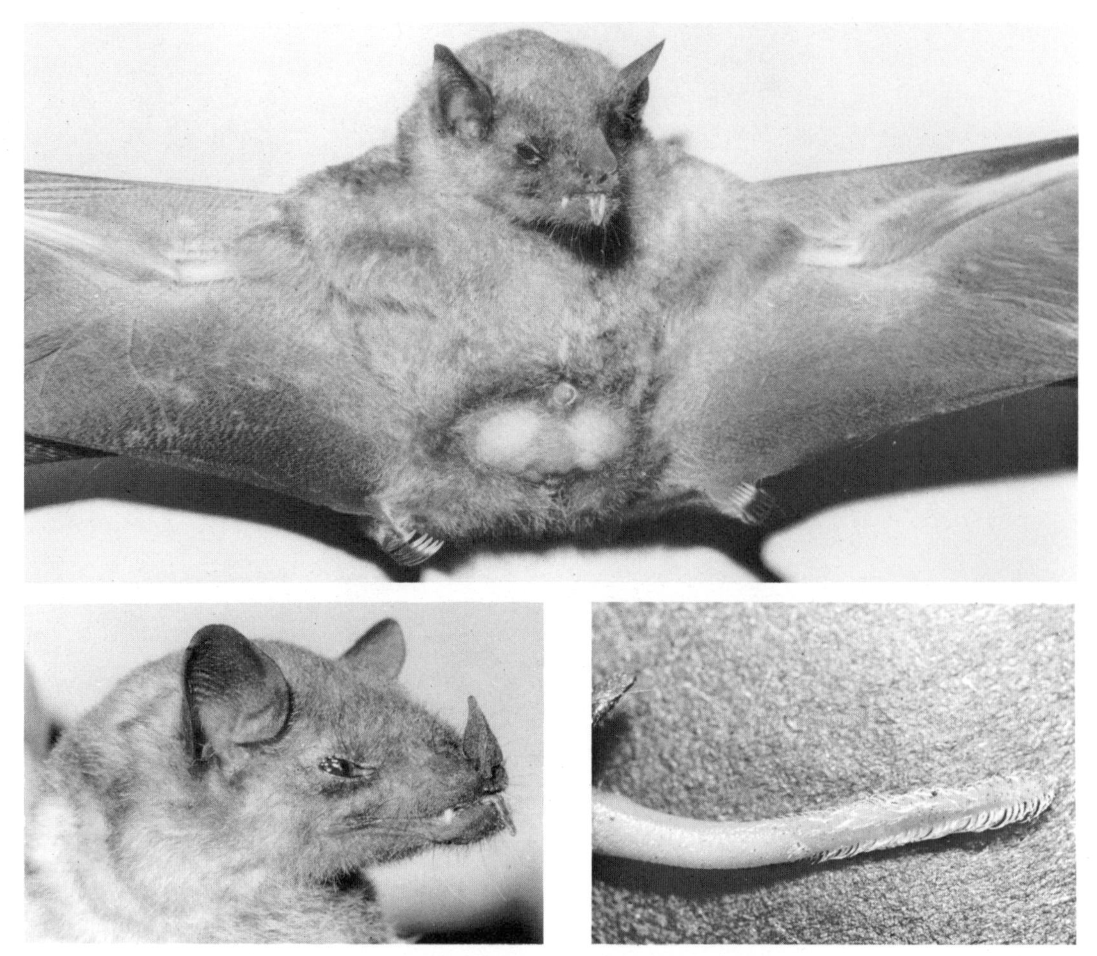

Geoffroy's long-nosed bat (*Anoura* sp.), photos by Merlin D. Tuttle. Tongue photo by P. Morris.

arm length, 35 mm. The coloration of the upper parts is dark brown; the underparts are light brown. This bat is similar to *Choeronycteris* and *Choeroniscus*, but the molar and premolar teeth are more normal in structure. The chin is unusually prominent, projecting both forward and downward. Lower incisors are not present.

Scleronycteris ega, photo by Merlin D. Tuttle, Bat Conservation International.

The specimen from Venezuela was netted in a yard near a stream in an evergreen forest at 135 meters (Handley 1976). According to Gardner (1977), the diet probably consists of fruit, pollen, nectar, and insects.

CHIROPTERA; PHYLLOSTOMIDAE; **Genus LICHONYCTERIS**
Thomas, 1895

The single species, *L. obscura*, occurs from Guatemala and Belize to Surinam, the lower Amazonian region of Brazil, and central Bolivia (Anderson, Koopman, and Creighton 1982; Gardner 1976; Hill 1985*b*; Jones and Carter 1976).

Head and body length is about 50–55 mm, tail length is about 8–10 mm, and forearm length is about 33 mm. Coloration is brownish with a yellowish suffusion, or dark brown. The muzzle is elongate, the long tongue is supplied with papillae, and the cheek teeth are elongate and slender. The interfemoral membrane is well developed. This genus differs from *Leptonycteris* in not having lower incisor teeth and in features of the upper incisors.

Specimens have been collected in tropical evergreen forest (Handley 1976; Tuttle 1970). The features of the muzzle, tongue, and teeth suggest that this genus feeds on nectar and fruit. This bat is assumed to pollinate some of the night-

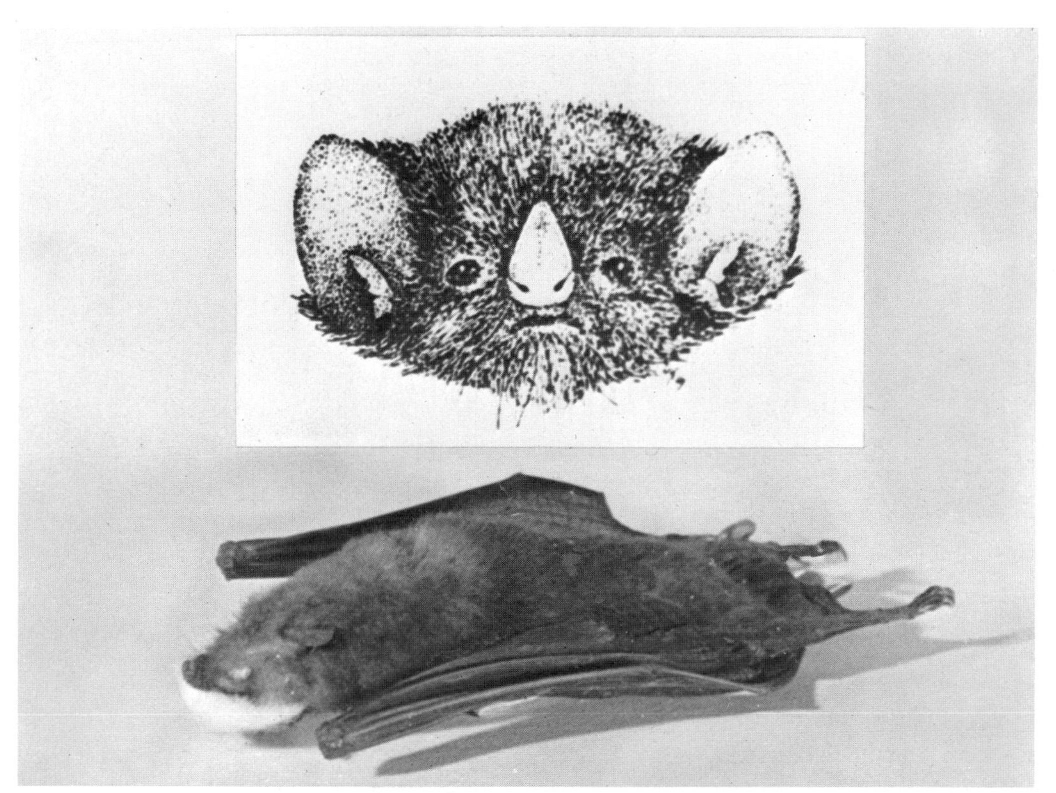

Lichonycteris sp., photo by Howard E. Uible of specimen in U.S. National Museum of Natural History. Inset: face *(L. obscura)*, photo from *Bull. Amer. Mus. Nat. Hist.*, "Mammals of Costa Rica," G. G. Goodwin.

blooming plants. Gardner, LaVal, and Wilson (1970) reported that a lactating female *L. obscura* was collected simultaneously with a flying juvenile male on 9 January in Costa Rica. Another female contained a 14-mm embryo on 27 March. Two pregnant females were taken in Guatemala in February (Wilson 1979).

CHIROPTERA; PHYLLOSTOMIDAE; **Genus HYLONYCTERIS**
Thomas, 1903

Underwood's Long-tongued Bat

The single species, *H. underwoodi*, occurs from Nayarit in western Mexico to Panama (Jones and Carter 1976; Ramirez-Pulido and Lopez-Forment 1979).

Head and body length is about 53 mm, tail length is about 6 mm, and forearm length is about 31–35 mm. Jones and Homan (1974) reported that six males weighed 6–7 grams and two females weighed 8 and 9 grams. The color is uniformly dark gray above and below; or dark brown above, with the crown of the head darker, and slightly paler below.

The muzzle is elongate, the long tongue is supplied with papillae, the cheek teeth are elongate and slender, and the lower incisors are lacking. *Hylonycteris* has three upper and three lower molars on each side, whereas *Lichonycteris* has two upper and two lower molars on each side. The interfemoral membrane is well developed. This bat resembles *Choeronycteris* and *Choeroniscus* but differs from those genera in features of the skull.

Phillips and Jones (1971) collected specimens in dense forest in Jalisco, Mexico. LaVal (1977) found groups of two and eight individuals roosting, respectively, under a log bridge and in a hollow log. This bat also roosts in caves and tunnels, apparently in small numbers. According to Gardner (1977), the diet consists of pollen, nectar, fruit, and occasionally insects. Data summarized by Wilson (1979) indicate that there is a bimodal reproductive pattern in Costa Rica, with pregnancies reported there from January to April and from August to November. Pregnant or lactating females have been found in Mexico and Guatemala in March, May, September, and November. A single young is produced at a time (Jones and Homan 1974).

Underwood's long-tongued bat *(Hylonycteris underwoodi)*, photo by Merlin D. Tuttle, Bat Conservation International.

Platalina genovensium, photo by Jaime E. Péfaur.

CHIROPTERA; PHYLLOSTOMIDAE; **Genus PLATALINA**
Thomas, 1928

The single species, *P. genovensium,* is now known from five localities in Peru and is represented in eight museum collections (Jiménez and Pefaur 1982).

The approximate measurements of the type specimen are: head and body length, 72 mm; tail length, 9 mm; and forearm length, 46 mm. In addition, Jiménez and Pefaur (1982) referred to two other specimens, each with a forearm length of about 48 mm, and to one that weighed 47 grams. In coloration, *Platalina* is pale brownish throughout.

The muzzle is elongate, the extensible tongue is supplied with bristlelike papillae, and the cheek teeth are narrow and elongate. The short tail barely reaches the middle of the interfemoral membrane. This bat differs from the other glossophagine bats in dental and skull features. The generic name refers to the broad and spatulate nature of the inner upper incisor teeth. The nose leaf is somewhat diamond-shaped.

According to A. L. Gardner (U.S. National Museum of Natural History, pers. comm., 1988), *Platalina* has been collected in relatively dry regions and in mines and grottoes, at elevations ranging from near sea level to about 2,300 meters. It has developed a highly derived feeding apparatus, independent of the specialized glossophagine genera. Although little is known of its feeding ecology, the hyoid/lingual morphology suggests that it is well adapted for the nectarivorous niche. Jiménez and Pefaur (1982) noted that it is thought to feed on the pollen and nectar of cacti. Graham (1987) collected two females in September, one of which was pregnant.

CHIROPTERA; PHYLLOSTOMIDAE; **Genus CHOERONISCUS**
Thomas, 1928

There are four species (Anderson, Koopman, and Creighton 1982; Jones and Carter 1976, 1979; Koopman 1978*a*, 1982*b*; Webster and McGillivray 1984; Williams and Genoways 1980*a*):

C. godmani, Sinaloa (western Mexico) to Colombia and Surinam;
C. minor, Colombia, Venezuela, Guyana, Surinam, French Guiana, Ecuador, Peru, Bolivia, Brazil;
C. intermedius, eastern Peru, northern Brazil, Guianas, Trinidad;
C. periosus, Pacific coast of Colombia.

Head and body length is about 50–55 mm, tail length is about 12 mm, and forearm length is 32–38 mm. LaVal and Fitch (1977) reported an average weight of 7.9 grams for a series of specimens of *C. godmani* from Costa Rica. Coloration is usually uniformly dark brown above and somewhat paler below, but the fur of the back may be bicolored.

This genus is characterized by its small size, uniformly dark color, exceptionally long muzzle, small triangular nose leaf, small ears, and short tail. The rostrum is much elongated but less than half the length of the skull. The long tongue is supplied with papillae at its tip, and the cheek teeth are narrow and elongate. There are no lower incisors.

In Venezuela, Handley (1976) collected specimens of *C. godmani* and *C. minor* mostly in moist parts of tropical evergreen forest. The only record of roosting habits appears to be the finding of eight individuals, two males and six females, on the underside of a fallen log over a stream. The diet probably consists of pollen, nectar, fruit, and insects (Gardner 1977). Pregnant female *C. godmani* have been found in Sinaloa in July, in Nicaragua in March, and in Costa Rica in December, January, February, and March (LaVal and Fitch 1977; Wilson 1979). A lactating female *C. godmani* was taken in Oaxaca, Mexico, in May, and a pregnant *C. intermedius* was taken in Trinidad in August.

CHIROPTERA; PHYLLOSTOMIDAE; **Genus CHOERONYCTERIS**
Tschudi, 1844

Mexican Long-nosed Bat, or Hog-nosed Bat

The single species, *C. mexicana,* ranges from the southern part of the southwestern United States to Honduras and Guatemala (Jones and Carter 1976).

Head and body length is about 55–80 mm, tail length is 10–16 mm, forearm length is about 43–47 mm, and weight is usually 10–20 grams. The coloration has been described as dark brown above and paler brown below. In some individuals the males are reddish brown and the females are sooty, both being lighter underneath.

This genus resembles *Choeroniscus,* but the rostrum is longer and more than half the length of the skull. As in some other bats of the subfamily Glossophaginae, the lower incisors are absent. The muzzle is elongate, the long tongue has papillae at its tip, forming a brush, and the cheek teeth are elongate and slender.

The hog-nosed bat has been recorded mainly from arid habitats at elevations of 600–2,400 meters. It roosts in caves, tunnels, and buildings, usually in fairly well-illuminated locations, and has been found under the exposed roots of trees in a shaded area. Indirect evidence indicates that colonies in southern Arizona migrate south into Mexico for the winter months. The diet consists of fruit, pollen, nectar, and probably insects (Gardner 1977). This bat is thought to pollinate certain plants. Captives lapped up the juices of grapes and plums but rejected the solid parts. No attempt at chewing was observed.

Colonies usually contain several dozen bats, though solitary individuals and groups of 2–12 have also been noted. The sexes usually roost together, though the females may

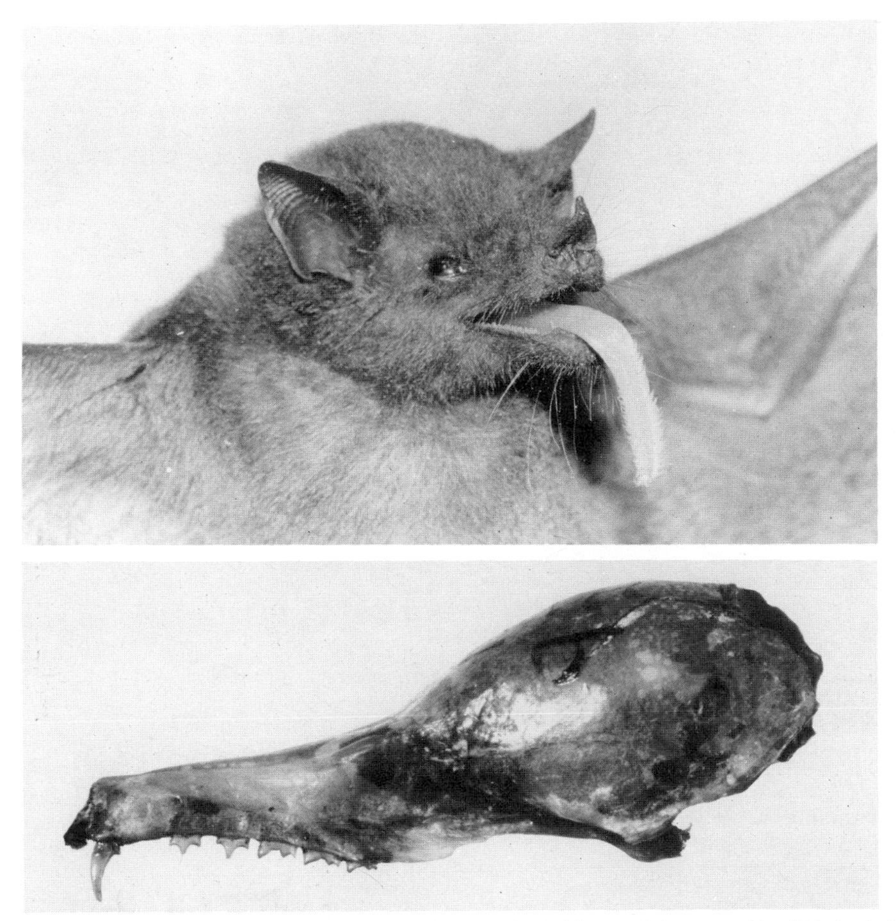

Godman's long-nosed bat (*Choeroniscus* sp.), photo by Merlin D. Tuttle. Skull of Godman's long-nosed bat *(C. minor)*, photo from British Museum (Natural History).

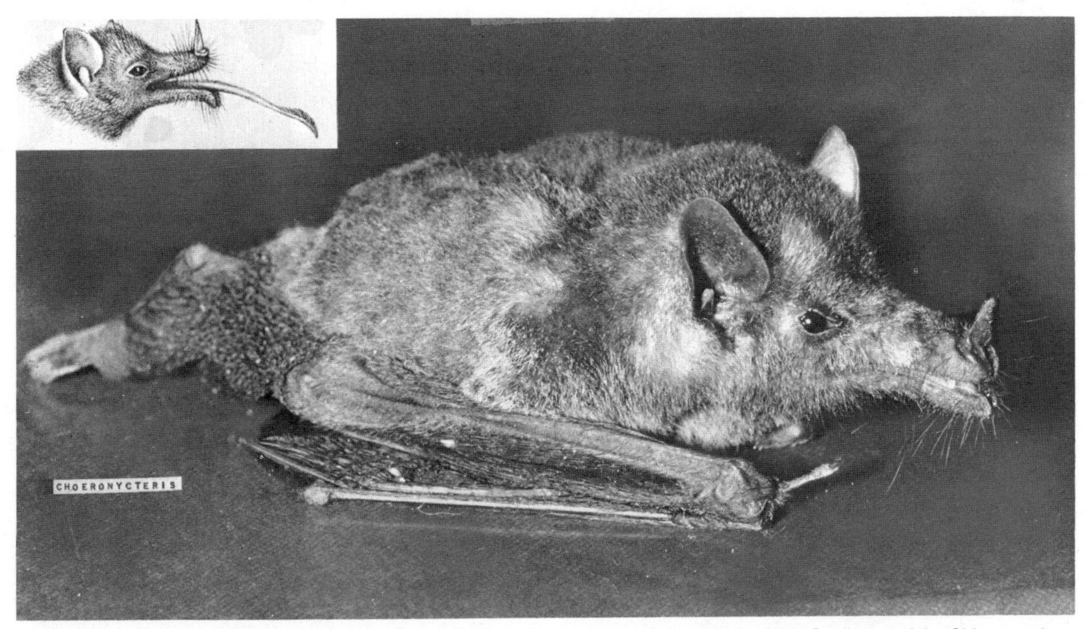

Mexican long-nosed bat *(Choeronycteris mexicana)*, photo by Lloyd G. Ingles. Inset photo from *Catalogue of the Chiroptera in the Collection of the British Museum*, G. E. Dobson.

separate from the males when the young are born and being raised (Hoffmeister 1986). In the southwestern United States and in northern and central Mexico, pregnant females have been collected in February, March, June, and September, and lactating females have been taken in June, July, and August (Wilson 1979). The litter size is 1, and the female can carry her young in flight (Barbour and Davis 1969). The young grow rapidly and probably can fly by themselves within a month of birth.

CHIROPTERA; PHYLLOSTOMIDAE; **Genus MUSONYCTERIS**
Schaldach and McLaughlin, 1960

Banana Bat, or Colima Long-nosed Bat

The single species, *M. harrisoni*, is known only from the states of Jalisco, Colima, Guerrero, and Michoacán in southwestern Mexico (Jones and Carter 1976, 1979; Webster et al. 1982). Hall (1981) and some other recent authors have treated this genus as a synonym of *Choeronycteris*. On the basis of morphological and karyological features, Webster et al. (1982) considered the two genera to be closely related but distinct.

Head and body length is 70–79 mm, tail length is 8–12 mm, and forearm length is approximately 41–43 mm. The shoulder region and the middle back are brownish light drab, the lower back is dark brown, and the underparts are colored like the shoulders.

As in other bats of the subfamily Glossophaginae, the snout is long, the tongue is extensible, and the cheek teeth are narrow and elongate. This genus is distinguished by skull and dental features, particularly by the extremely long rostrum, which is more than one-half the total length of the skull and longer than the rostrum of any other related genus except the bats of the South American genus *Platalina*, from which *Musonycteris* differs in smaller size and dental formula. *Musonycteris* seems to be most closely related to *Choeronycteris*.

Specimens of *Musonycteris* have been taken in nylon mist nets set across a small irrigation ditch in a banana grove in arid thorn forest. An extensive tropical deciduous forest was only a short distance from the thorn forest. The banana trees in the grove were blooming when the first three specimens were collected in 1958; presumably these bats were feeding on the flowers. The generic name is derived from the generic name of the banana *(Musa)* and a bat *(Nycteris)*. Gardner (1977) stated that the diet probably consists of pollen, nectar, and insects. Wilson (1979) reported that two pregnant females were taken in September in Colima.

CHIROPTERA; PHYLLOSTOMIDAE; **Genus CAROLLIA**
Gray, 1838

Short-tailed Leaf-nosed Bats

There are four species (Hall 1981; Jones and Carter 1976):

C. castanea, Honduras to Bolivia;
C. subrufa, southwestern Mexico to Nicaragua;
C. brevicauda, San Luis Potosi (eastern Mexico) to northeastern Brazil and Bolivia;
C. perspicillata, southern Mexico to Paraguay and southern Brazil, Trinidad and Tobago, Grenada in the Lesser Antilles.

Banana bat *(Musonycteris harrisoni)*, photo by John Bickman.

Short-tailed leaf-nosed bat *(Carollia perspicillata)*, photo by Ernest P. Walker. Inset photo by Harold Drysdale, Trinidad Regional Virus Laboratory, through A. M. Greenhall.

Recorded occurrences of *C. perspicillata* on Jamaica and Redonda Island are questionable.

Head and body length is 48–65 mm, tail length is 3–14 mm, forearm length is 34–45 mm, and weight is usually 10–20 grams. Coloration is generally dark brown to rusty, but 1 or 2 specimens in a series of 50 *C. perspicillata* from Central America are clear pale orange. This genus differs from *Rhinophylla* in that the lower molars are of a different form from the lower premolars and in the presence of a tail.

The species *C. subrufa* is found in relatively dry tropical deciduous forest, while the other three species occur primarily in humid tropical evergreen forest (Handley 1976; Pine 1972*b*). Reported roosting sites include caves, mines, rocks, culverts, hollow trees, logs, and buildings. In a radiotracking study in Costa Rica, Heithaus and Fleming (1978) found *C. perspicillata* to disperse nightly, with each individual going to two to six feeding areas and flying an average of 4.7 km per night. The diet seems to consist mainly of fruits such as guavas, bananas, wild figs, and plantains. Sazima (1976) observed *C. perspicillata* taking nectar from passionflowers. In this species, and probably others as well, insects also seem to be an important food (Ayala and D'Alessandro 1973; Gardner 1977).

These bats roost singly, in small groups, and in colonies of several hundred to several thousand individuals. Males and females usually live together throughout the year. In studies of both wild and captive *C. perspicillata*, Porter (1978, 1979*a*) found that colonies are divided into heterosexual groups, or harems, comprising a single adult male and from one to eight females and their infants, and that other groups are composed only of males or juveniles. Harem males appear to recruit females, vigorously defend them from other males, and exhibit territorial behavior and spatial fidelity. C. F. Williams (1986), however, collected evidence suggesting that harems are actually passive aggregations of females that form at the limited suitable roosts. The presence of harem males apparently is a result rather than a cause of female grouping. Porter (1979*b*) reported that *C. perspicillata* communicates through a rich variety of vocalizations, including warbles of greeting between male and female and screeches by which a harem male threatens other males and controls his females. Porter and McCracken (1983) added that harem males also guard the offspring of their harem females while the mothers are out foraging and apparently help to reunite separated offspring and mothers.

Data summarized by Wilson (1979) indicate that reproduction in *Carollia* generally fits a bimodal polyestrous pattern. Pregnant female *C. castanea* have been found in Central America in January–May and July–August and in South America in January–April and September–November. Pregnancies in *C. subrufa* have been recorded in Central America in December–May and July–October. Although pregnant female *C. perspicillata* have been collected in all months of the year, there are definite peak periods, occurring in February–May and June–August in Panama and somewhat earlier in other areas, depending on seasonal rainfall pattern. Pregnant female *C. brevicauda* have been found in every month from December to August in Mexico and Central America and in October in Peru. In a study in Costa Rica, LaVal and Fitch (1977) found all three resident species of *Carollia* to be seasonally polyestrous, with an apparent minimum in reproductive activity occurring late in the wet season from October to early January. The gestation period in the genus is 2.5–3 months (Porter 1978). Usually there is a single

Rhinophylla sp., photo by P. Morris.

young. Fleming (1988) determined that female *C. perspicillata* attain sexual maturity at about 1 year of age, and males between 1 and 2 years. Nearly two-thirds of births are of males, though higher male mortality evidently leads to a 1:1 sexual ratio among adults. Average life expectancy in this species is 2.6 years, though a few individuals survive for over 10 years. According to Jones (1982), a specimen of *C. perspicillata* was still living after 12 years and 5 months in captivity.

These bats, especially *C. perspicillata,* have the reputation of being pests (Pine 1972*b*). Apparently they have increased in numbers through the availability of crops as a source of food and buildings as roosting sites. Considerable damage has been reported to mangoes, coffee beans, guavas, pawpaws, almonds, and other cultivated products.

CHIROPTERA; PHYLLOSTOMIDAE; **Genus RHINOPHYLLA**
Peters, 1865

There are three species (Alberico 1987; Anderson, Koopman, and Creighton 1982; Baud 1982; Jones and Carter 1976; Webster and Jones 1984*a*; Webster and McGillivray 1984):

R. pumilio, Colombia, Venezuela, Guyana, Surinam, French Guiana, northern Brazil, eastern Ecuador and Peru, northeastern Bolivia;

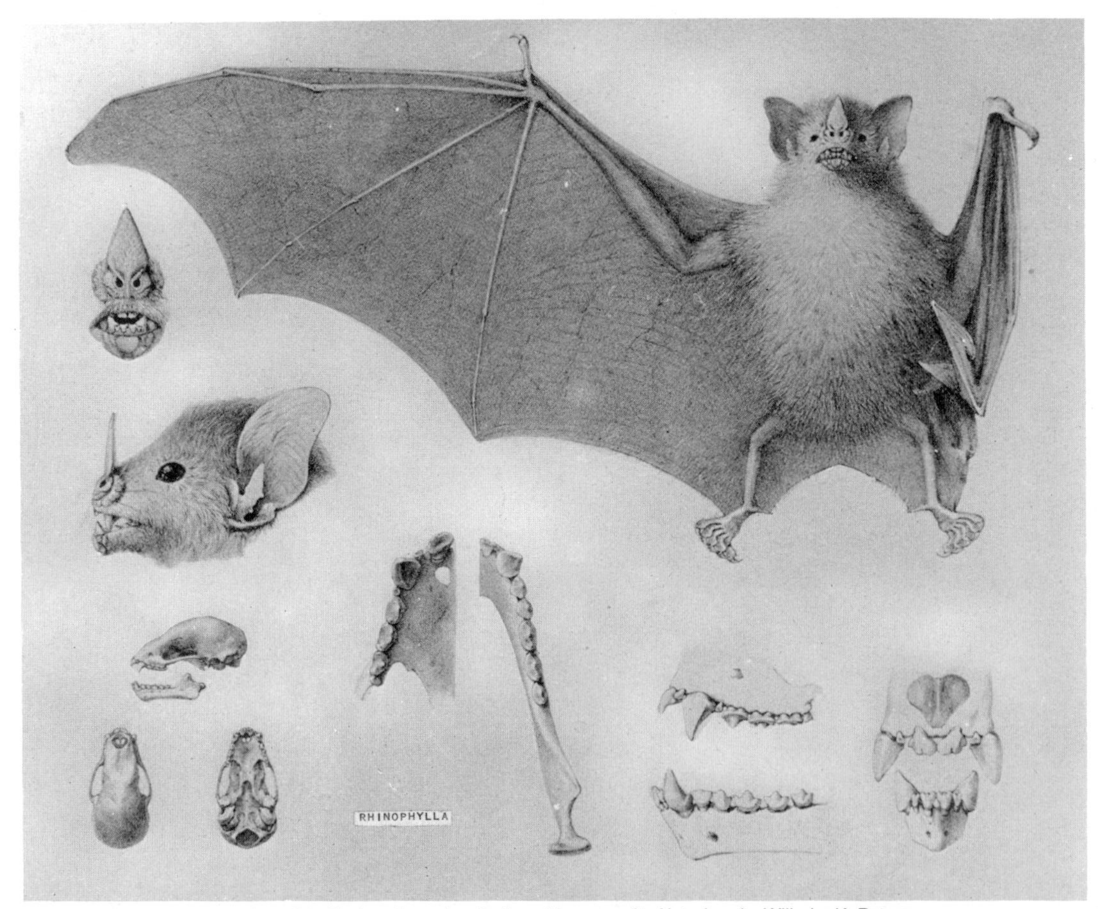

Rhinophylla pumilio, photo from *Die Fledermäuse des Berliner Museums für Naturkunde,* Wilhelm K. Peters.

R. alethina, western Colombia, northwestern Ecuador;
R. fischerae, Amazonian region of Peru and Brazil, and adjacent parts of Colombia, Ecuador, and Venezuela.

Head and body length is about 43–48 mm, there is no tail, and forearm length is about 29–37 mm. The most common coloration is grayish brown. There is a nose leaf. The members of this genus differ from *Carollia* in that the lower molars are not different in form from the lower premolars and in the absence of a tail.

In Venezuela, Handley (1976) collected *R. pumilio* mainly in moist parts of tropical evergreen forests. In Peru, Tuttle (1970) took both *R. pumilio* and *R. fischerae* most often in or near gardens where bananas and papayas were grown. Gardner (1977) stated that bats of this genus all are probably frugivores but may consume insects as well. Pregnant or lactating *R. pumilio* have been found in April, May, June, July, and December (Wilson 1979). Pregnant female *R. fischerae* have been collected in Peru in June and July (Graham 1987).

CHIROPTERA; PHYLLOSTOMIDAE; **Genus STURNIRA**
Gray, 1842

Yellow-shouldered Bats, or American Epauleted Bats

There are 13 species (Alberico 1987; Anderson, Koopman, and Creighton 1982; Davis 1980; Handley 1976; Jones and Carter 1976; Lemke et al. 1982; Marques and Oren 1987; Molinari and Soriano 1987; Soriano and Molinari 1987; Tamsitt, Cadena, and Villarraga 1986):

S. lilium, Sonora and Tamaulipas (Mexico) to northern Argentina and possibly Chile, southern Lesser Antilles, Jamaica;
S. luisi, Costa Rica to Peru;
S. thomasi, Guadeloupe Island in the Lesser Antilles;

Yellow-shouldered bat *(Sturnira lilium)*, photos by Merlin D. Tuttle, Bat Conservation International.

S. tildae, northern and central South America, Trinidad;
S. magna, Amazonian region of Colombia, Ecuador, Peru, and Bolivia;
S. mordax, Costa Rica;
S. bidens, western Venezuela, Colombia, Ecuador, Peru, northern Brazil;
S. nana, known only from the type locality at Huanhuachayo in central Peru;
S. aratathomasi, western Colombia, western Ecuador, western Venezuela;
S. ludovici, Sinaloa and Tamaulipas (Mexico) to Venezuela and Peru;
S. bogotensis, Colombia, Venezuela;
S. oporaphilum, Peru, Bolivia;
S. erythromos, presently recorded only from Colombia, eastern Peru, Bolivia, and northern and western Venezuela but probably widely distributed in northern South America.

The species *S. bidens* and *S. nana* are sometimes placed in a separate genus or subgenus, *Corvira* Thomas, 1915 (see Gardner and O'Neill 1969, 1971).

Head and body length is 51–101 mm, there is no external tail, forearm length is 34–61 mm, and adult weight is usually about 15–20 grams. According to Gardner (1976), however, the weights of four specimens of *S. magna* were 41.0–44.5 grams. Thomas and McMurray (1974) gave the weight of *S. aratathomasi* as 46.8–67.1 grams. Except in *S. bidens* and *S. nana*, conspicuous tufts of stiff yellowish or reddish hairs are present near the front of the shoulders in males. The general coloration is pinkish buff with a brown tinge, dark ochraceous brown, dark grayish brown, or dark brown; the underparts are usually paler. Some forms have two color phases: bright cinnamon brown and dull pale gray.

The ears are short, the nose leaf is normal, the interfemoral membrane is narrow and furred, and the hind limbs and feet are haired to the claws. The molar teeth are longitudinally grooved with lateral cusps. In most species there are four lower incisors, but in *S. bidens* and *S. nana* only two are present.

In Venezuela, Handley (1976) found the following habitat conditions to prevail: *S. bidens*, high cloud forest; *S. erythromos*, moist parts of mountain forests; *S. lilium*, moist parts of forests and open areas; *S. ludovici*, moist places in forested mountains and foothills; and *S. tildae*, humid lowland forests. *S. magna* is found in forests at elevations of 200–2,300 meters (Tamsitt and Hauser 1985). Roosting sites of the genus include hollow trees and buildings. The diet consists mostly of fruit.

LaVal and Fitch (1977) reported *S. ludovici* to be seasonally polyestrous in Costa Rica, with several peaks of pregnancy and lactation from April to August. Data summarized by Wilson (1979) indicate that reproduction in this species, as well as in *S. lilium*, follows a pattern of bimodal polyestry. In Colombia pregnant female *S. ludovici* have been reported from February to May and from August to December, and lactating females have been found from May to September. In Jalisco, Mexico, pregnant or lactating female *S. ludovici* have been taken in April, May, July, August, and November. Pregnant and lactating female *S. lilium* have been collected in all months of the year, but in any one region there apparently are peaks of reproductive activity. Other data listed by Wilson (1979) are: pregnant *S. tildae* in March on Trinidad and in July in Brazil; pregnant or lactating *S. mordax* in Costa Rica in February, May, and August; and pregnant *S. erythromos* in Colombia in December and in Peru in August. According to Baker, Genoways, and Patton (1978), two lactating female *S. thomasi* were collected on Guadeloupe in July.

CHIROPTERA; PHYLLOSTOMIDAE; **Genus URODERMA**
Peters, 1865

Tent-building Bats

There are two species (Anderson, Koopman, and Creighton 1982; Jones and Carter 1976, 1979; Ramirez-Pulido and Lopez-Forment 1979):

U. bilobatum, southern Mexico to Bolivia and southeastern Brazil, Trinidad;
U. magnirostrum, southern Mexico to Brazil and northern Bolivia.

Head and body length is about 54–74 mm, there is no external tail, forearm length is 39–45 mm, and adult weight is usually 13–21 grams. The coloration of the head and body is grayish brown. The ear margins are yellowish white, there are four white facial stripes, and there is a narrow white line on the middle of the lower back.

These bats may be recognized by "the naked or finely haired posterior border of the interfemoral membrane, in combination with the length of the forearm." The nose leaf consists of two parts: a horseshoe-shaped basal portion with unique rounded lobes on either side, and an erect, lancet-shaped, finely toothed portion.

In Venezuela, Handley (1976) collected *U. bilobatum* in a variety of forested and nonforested habitats and *U. magnirostrum* mostly in moist, open areas. *U. bilobatum* roosts in the leaves of trees, such as in the fronds of palms and the leaves of bananas. The bats often form their own shelter by biting across and partly through the ribs of fronds so that the terminal part hangs downward to give protection against sun, wind, and rain. The main requirement of the frond is that it be fan-shaped; these bats generally cut the fan in a semicircular or angular line so that the frond, when it collapses, becomes a rooflike shelter. The bitten spots afford good footholds, but *U. bilobatum* also hangs from above and below the cut zone. The distal part of the frond eventually dries up and drops off, and a new leaf has to be cut. Timm (1987), however, stated that since the bats bite the tissue between veins along the midrib and leave the midrib and most veins intact, they do not kill the leaves. The resulting tent is thus available for an extended period; one was observed in use for more than 60 days. Thomas Barbour, who first recorded tentmaking in *U. bilobatum*, thought that this habit developed originally in response to "the superior shade producing quality" of the leaves of introduced palms and then spread to include the native palms. The introduced palms in Panama, where Barbour made his observations, have large stiff, palmate fronds, whereas the native palms have pinnate leaves. Both species of *Uroderma* probably are mainly frugivorous, but they may also consume pollen, nectar, and insects found in flowers and fruit (Gardner 1977).

Nursing females roost in groups of 20–40 individuals and do not carry their young with them during their feeding flights. The males at this time roost singly or in smaller groups than the female groups. Data summarized by Wilson (1979) suggest that reproduction in this genus follows a pattern of bimodal polyestry. Pregnant female *U. bilobatum* have been reported from December to July in Central America and in January, July, August, September, and November in South America. Pregnant *U. magnirostrum* have been taken in El Salvador in June, in Nicaragua in March and July, in Bolivia in September, and in Brazil in June.

Tent-building bat *(Uroderma bilobatum)*, photo by Cory T. de Carvalho. Inset: *Uroderma* sp., photo by P. Morris.

CHIROPTERA; PHYLLOSTOMIDAE; **Genus VAMPYROPS**
Peters, 1865

White-lined Bats

There are eight species (Anderson, Koopman, and Creighton 1982; Barquez and Olrog 1980; Jones and Carter 1976, 1979; Koopman 1982*b*; Lemke et al. 1982; Myers and Wetzel 1983; Williams and Genoways 1980*a*; Williams, Genoways, and Groen 1983):

V. infuscus, Colombia to Bolivia and Brazil;
V. vittatus, Costa Rica to Bolivia and Venezuela;
V. dorsalis, Panama to Bolivia;
V. aurarius, mountains of southwestern Colombia, Guiana Highlands of Venezuela, central Surinam;
V. brachycephalus, Colombia, Venezuela, Guyana, Surinam, Amazonian Brazil, Ecuador, Peru;
V. helleri, southern Mexico to Bolivia and Brazil, Trinidad;
V. lineatus, Surinam to eastern Brazil and northern Argentina;
V. recifinus, eastern Brazil and possibly the Guianas.

Hall (1981) used the name *Platyrrhinus* Saussure, 1860, for this genus. Based on Handley's (1976) report of bats in Venezuela, Honacki, Kinman, and Koeppl (1982) recognized *V. umbratus* as a species distinct from *V. dorsalis* and assigned the former a range in Panama, Colombia, and Venezuela. These authorities also indicated that *V. dorsalis* occurs as far north as Costa Rica. However, Jones, Arroyo-Cabrales, and Owen (1988) did not list *V. umbratus* as a distinct species and did not include Costa Rica in the range of *V. dorsalis*. Owen (1987) listed *V. nigellus* of Peru as a species separate from *V. lineatus*, but Willig and Hollander (1987) treated the former taxon only as a subspecies.

Head and body length is 48–98 mm, there is no external tail, and forearm length is 36–64 mm. LaVal and Fitch (1977) listed the following average weights for series of specimens from Costa Rica: *V. helleri*, 15.4 grams; and *V. vittatus*, 54.7 grams. Coloration in the genus is usually dark brown or black. Pale facial and dorsal stripes are usually present. The narrow facial stripes extend from the side of the nose to the ears, and the white or gray dorsal stripe extends from between the ears to the tail membrane.

The nose leaf is well developed, and the ears are rounded and of medium size. The interfemoral membrane is narrow and fringed with hair on its posterior edge.

In Venezuela, Handley (1976) collected several species of *Vampyrops* mainly in moist, forested areas. These bats have been found under roots on canyon walls and stream banks, in moss on a tree, in clusters of leaves in the tops of trees and on the branches of trees, in caves, and in buildings. *V. lineatus* has been observed in Brazil in November and December roosting in groups of about 3–10 individuals in leaves in the tops of mango trees, and on the branches of these trees about halfway up on windy or rainy days. The females with young often roost on thick branches of the trees, 3–5 meters above the ground, apart from the males; the female *V. lineatus* without young were generally associated with males, in separate pairs. The diet consists mainly of fruit but includes some insects (Gardner 1977). In Brazil, Sazima (1976) observed *V. lineatus* taking nectar from flowers.

Willig and Hollander (1987) suggested that individual adult male *V. lineatus* maintain harems of about 6–20 females. This species evidently breeds throughout the year in southeastern Brazil; females are polyestrous and may be simultaneously pregnant and lactating. In northeastern Brazil, however, pregnancies were found only from the early dry season (July) to the end of the rainy season (February–March), and there was a bimodal distribution of breeding in this period.

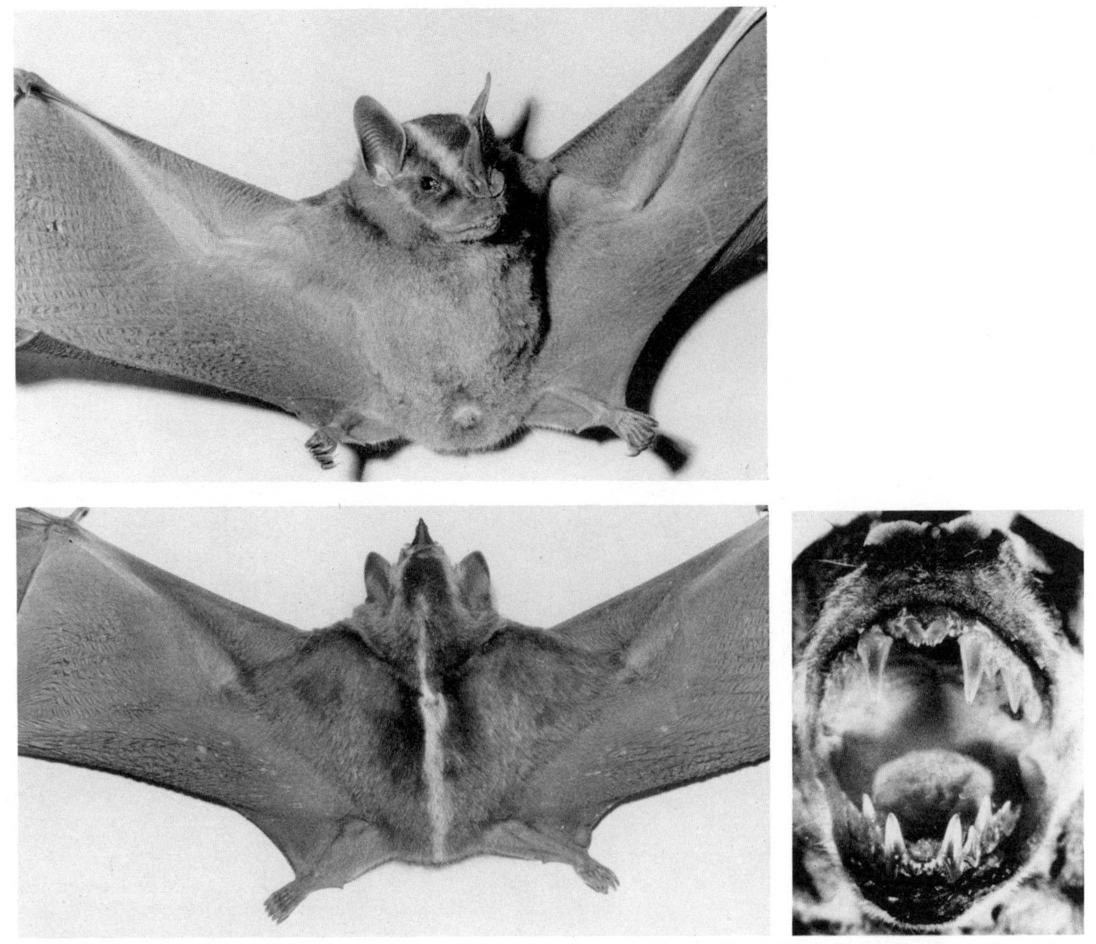

White-lined bat (*Vampyrops* sp.), photos by Merlin D. Tuttle. Inset: teeth of white-lined bat *(V. helleri)*, photo by Harold Drysdale, Trinidad Regional Virus Laboratory, through A. M. Greenhall.

Wilson (1979) listed the following additional reproductive data: *V. vittatus*, pregnant or lactating females taken in Costa Rica from March to July, pregnant females taken in Colombia in May and October; *V. dorsalis*, pregnant or lactating females found in Colombia from November to August; *V. brachycephalus*, pregnant females taken in February in Venezuela and in August in Peru, a lactating female taken in October in Venezuela; *V. helleri*, pregnant females taken in southern Mexico and Central America from March to August and in January, pregnant or lactating females taken in Colombia from January to August, pregnant females found in French Guiana in August and September, and pregnant females found in Peru in July and August; and *V. lineatus*, pregnant females found in Brazil in January, March, and December. Davis and Dixon (1976) collected pregnant female *V. helleri* and *V. brachycephalus*, each with a single large embryo, in Peru in the period 31 October–8 November. Jones (1982) reported that a specimen of *V. lineatus* was still living after 10 years and 2 months in captivity.

CHIROPTERA; PHYLLOSTOMIDAE; **Genus VAMPYRODES**
Thomas, 1900

Great Stripe-faced Bat

The single species, *V. caraccioli*, occurs from southern Mexico to northern Brazil and Bolivia and on Trinidad (Anderson, Koopman, and Creighton 1982; Carter et al. 1981; Hall 1981; Jones and Carter 1976). The name *V. major* sometimes is used for this species.

Head and body length is about 65–77 mm, there is no external tail, and forearm length is 45–57 mm. A series of 12 males from Honduras averaged 32.8 grams in weight (Valdez and LaVal 1971). Coloration is uniformly grayish brown above and below, or cinnamon brown above and grayish brown below. There are four white facial stripes and a white line extending from the top of the head down the midline of the back. This genus is similar to *Vampyrops* but has only two molar teeth on each side instead of three; these conspicuously differ from each other in form, owing to the reduction of the metaconid in the second molar to a mere trace.

In Venezuela, Handley (1976) collected this bat mainly in moist areas of tropical evergreen forest. Individuals have been found hanging from the underside of palm fronds on

Great stripe-faced bat *(Vampyrodes caraccioli)*, photo by Merlin D. Tuttle.

Trinidad. Four males were obtained on Tobago while hanging in a cluster in a shrub in virgin forest; another individual was roosting a short distance from the cluster. Several specimens have been taken in Oaxaca, Mexico, by the use of mist nets placed across small watered arroyos in dense rainforest. This bat is considered frugivorous in diet (Gardner 1977).

On forested Barro Colorado Island in Panama, Morrison (1980) found *Vampyrodes* to roost by day in dense foliage and to emerge about 45 minutes after sunset. Its initial flight to a fruiting tree covered an average of 850 meters, and it subsequently visited several more feeding areas in the course of the night. Roosts were changed almost daily, and foraging was curtailed during periods of bright moonlight, apparently as a means of avoiding predation. Individuals roosted and foraged in relatively stable groups comprising two or three adult females, their young, and probably a single adult male.

Wilson (1979) listed records of pregnant or lactating females for southern Mexico and Central America in January, June, July, and August; for Colombia from January to August and in October and November; and for Peru in July. Davis and Dixon (1976) collected pregnant females, each with a single large embryo, in Peru in the period 31 October–8 November.

CHIROPTERA; PHYLLOSTOMIDAE; **Genus VAMPYRESSA**
Thomas, 1900

Yellow-eared Bats

There are two subgenera and six species (Anderson and Webster 1983; Genoways and Williams 1979*b*; Jones and Carter 1976; Lemke et al. 1982; Myers, White, and Stallings 1983; Owen 1987; Williams and Genoways 1980*a*):

subgenus *Vampyressa* Thomas, 1900

- *V. pusilla*, southern Mexico to Peru, Paraguay, southeastern Brazil;
- *V. melissa*, eastern slopes of the Andes in southwestern Colombia and Peru;
- *V. macconnelli*, Costa Rica to Bolivia and Brazil, Trinidad;

subgenus *Vampyriscus* Thomas, 1900

- *V. nymphaea*, Nicaragua to western Colombia;
- *V. brocki*, Colombia, Guyana, Surinam;
- *V. bidens*, Colombia, Venezuela, Guyana, Surinam, northern Brazil, Ecuador, Peru, Bolivia.

A third subgenus, *Metavampyressa* Peterson, 1968, sometimes has been used for the species *V. nymphaea* and *V. brocki*, but Davis (1975) concluded that these species are sufficiently similar to *V. bidens* to warrant inclusion in *Vampyriscus*. The species *V. macconnelli* long was referred to the genus *Ectophylla* or placed in the separate genus *Mesophylla* Thomas, 1901 (Anderson, Koopman, and Creighton 1982; Hall 1981; Jones and Carter 1976, 1979). Also, however, there long were suggestions, on both karyological and morphological grounds, that this species is more closely related to *Vampyressa pusilla* than to *Ectophylla alba* (Greenbaum, Baker, and Wilson 1975; Starrett and Casebeer 1968). Owen (1987) indicated that there is uncertainty as to whether the genus *Vampyressa*, as set forth above, is a natural, monophyletic assemblage, but that, if it is, it should include *V. macconnelli;* he did not consider *Mesophylla* to warrant subgeneric rank.

Head and body length is 43 to about 65 mm, there is no external tail, and forearm length is 30–38 mm. LaVal and Fitch (1977) reported the average weight of a series of specimens of *V. pusilla* to be 8.2 grams. An individual *V. maccon-*

Vampyressa bidens, photo by Merlin D. Tuttle.

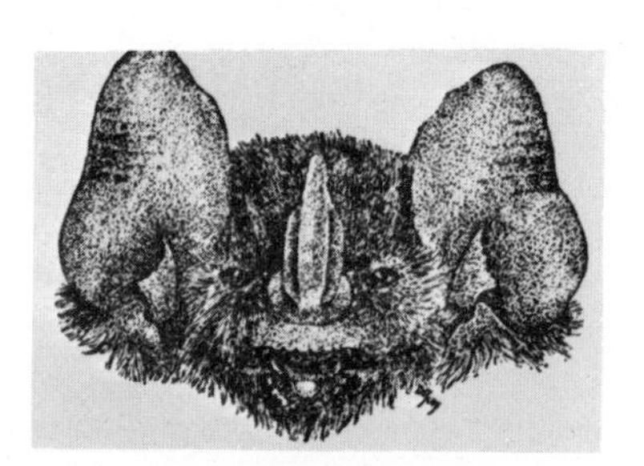

Yellow-eared bat *(Vampyressa pusilla):* Top, drawing of face from *Bull. Amer. Mus. Nat. Hist.*, "The Mammals of Costa Rica," G. G. Goodwin; Bottom, photo by Merlin D. Tuttle.

nelli weighed 7.5 grams. The coloration is smoky gray, whitish brown to pale brown, or dark brown. White facial stripes are present in some species and lacking in others. A dorsal stripe is present in *V. nymphaea* and *V. bidens*. The short, rounded ears have yellow margins. There are two upper and two lower molar teeth on each side. Each of the two middle upper incisors has two lobes at the end. In the subgenus *Vampyriscus* there usually are only two lower incisors. The occlusal surfaces of the second upper and lower molars of *V. macconnelli* lack a longitudinal ridge, such as is found in *Ectophylla* (Hall 1981).

In Venezuela, Handley (1976) collected *V. macconnelli*, *V. pusilla*, and *V. bidens* mostly near streams and in other moist parts of evergreen forest. These bats probably roost in trees and shrubs. The diet consists largely of fruit (Gardner 1977). Timm (1984) reported that *V. pusilla* constructs tents from the leaves of *Philodendron*, in much the same manner as described above in the account of *Uroderma*. Brooke (1987) described tentmaking in *V. nymphaea* and found that the number of individuals per tent varied from two to seven. These groups evidently represented harems, as they consisted of a single adult male, one to three pregnant or lactating females, and their young. Koepcke (1984) reported that *V. macconnelli* also constructed such shelters, which could be used for five to six months, and roosted therein in groups of up to eight individuals.

Wilson (1979) listed the following reproductive data: *V. pusilla*, pregnant or lactating females collected in Mexico and Central America from January to March and in July and August, and in Colombia in March, April, May, July, August, and November; *V. macconnelli*, pregnant females taken in Colombia in January and in Peru in August; *V. nymphaea*, pregnant females taken in Central America in February and April, pregnant or lactating females taken in Colombia from October to August; *V. brocki*, one lactating and two pregnant females taken in Colombia in June and July; and *V. bidens*, pregnant females taken in Peru in December. Davis and Dixon (1976) reported numerous pregnant female *V. pusilla* and *V. bidens*, each with a single large embryo, taken in Peru in the period 31 October–8 November. Genoways and Williams (1979*b*) collected a pregnant female *V. bidens* in Surinam on

Yellow-eared bat *(Vampyressa macconnelli)*, photo by Merlin D. Tuttle.

11 August. Pregnant female *V. macconnelli* also have been found in Bolivia in July (Webster and Jones 1980*a*) and on Trinidad in August (Carter et al. 1981).

CHIROPTERA; PHYLLOSTOMIDAE; **Genus CHIRODERMA**
Peters, 1860

Big-eyed Bats, or White-lined Bats

There are five species (Anderson, Koopman, and Creighton 1982; Carter et al. 1981; Jones and Baker 1979; Jones and Carter 1976; Taddei 1979):

C. doriae, Minas Gerais and Sao Paulo (eastern Brazil);
C. improvisum, Guadeloupe and Montserrat in the Lesser Antilles;
C. villosum, southern Mexico to Bolivia and Brazil, Trinidad;
C. salvini, Chihuahua (northwestern Mexico) to Bolivia;
C. trinitatum, Panama to Bolivia, Trinidad.

Head and body length is 55–87 mm, there is no external tail, and forearm length is about 37–58 mm. A female *C. villosum villosum* from Surinam weighed 23 grams (Genoways and Williams 1979*b*), and a male *C. villosum jesupi* from Barro Colorado Island weighed 13.7 grams. The upper parts are brownish, with the underparts usually paler. Facial and dorsal stripes are absent or faint in *C. villosum*. *C. salvini* and *C. trinitatum gorgasi* usually have conspicuous whitish facial and back stripes. Apparently stripes are not present in *C. doriae*. The coloration of *C. trinitatum gorgasi* is yellowish brown, brown, or grayish dorsally and paler on the abdomen; a distinct white stripe extends from the upper back to the base of the tail.

This genus may be distinguished by the absence of the nasal bones of the skull. *Chiroderma* resembles *Vampyrops* in external appearance, but the nose leaf is broader and the forearm and interfemoral membrane are more heavily furred.

Handley (1976) listed the following habitat conditions in Venezuela: *C. villosum*, humid lowlands, mostly in moist, open areas; *C. salvini*, mountains, usually in moist places within evergreen forest or forest openings; and *C. trinitatum*, usually in moist parts of evergreen forest. Watkins, Jones, and Genoways (1972) collected specimens of *C. salvini* mostly over streams or rivers in forested areas. In Peru, *C. trinitatum* was collected by Tuttle (1970) in secondary vegetation at the edge of a small garden where bananas and papayas were grown, and by Gardner (1976) in disturbed habitats such as clearings around houses and fruit groves. The diet of *Chiroderma* apparently consists largely of fruit (Gardner 1977).

Wilson (1979) listed the following reproductive data: *C. villosum*, pregnant or lactating females collected in southern Mexico and Central America in March, April, May, July, and December, pregnant females collected on Trinidad in August and September and in Colombia in January; *C. salvini*, pregnant or lactating females collected in Mexico in January and February, in Honduras in July and August, and in Colombia from January to July and in October and November; and *C. trinitatum*, pregnant or lactating females taken in Panama in February, May, and September, pregnant females taken on Trinidad in March, in Colombia in July, and in Brazil in June and July. Davis and Dixon (1976) reported

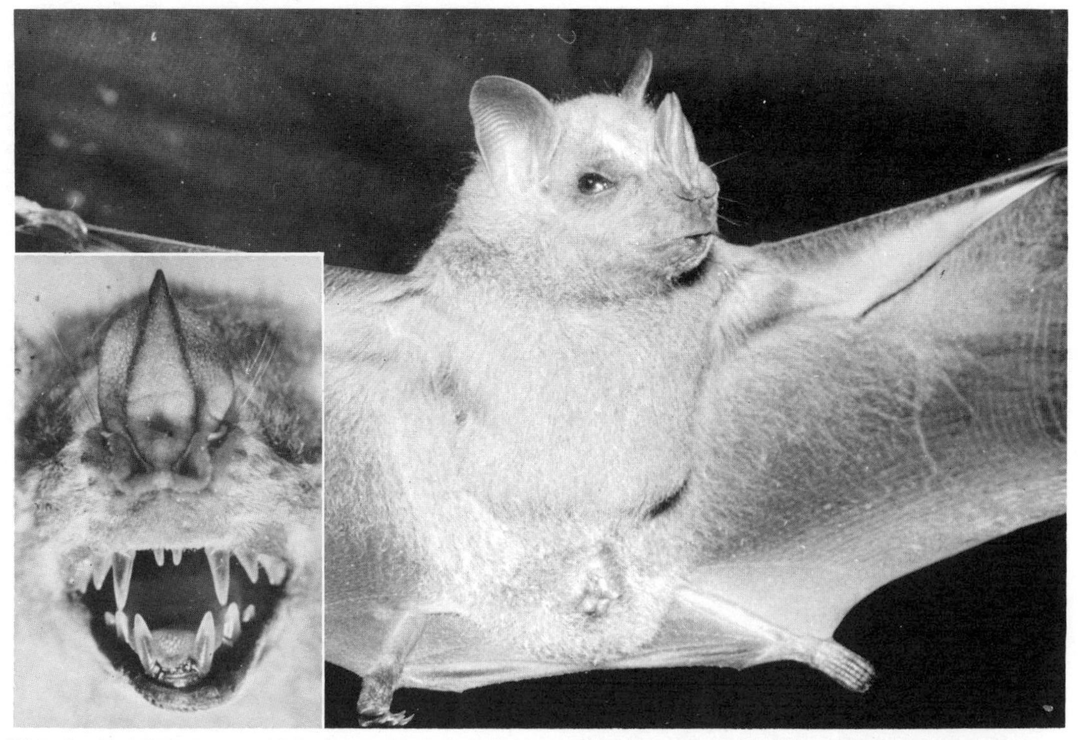

White-lined bat *(Chiroderma salvini)*, photo by Merlin D. Tuttle. Inset: *Chiroderma* sp., photo by Harold Drysdale, Trinidad Regional Virus Laboratory, through A. M. Greenhall.

pregnant female *C. villosum* and *C. trinitatum*, each with a single large embryo, in Peru in the period 31 October–8 November.

CHIROPTERA; PHYLLOSTOMIDAE; **Genus ECTOPHYLLA**
H. Allen, 1892

White Bat

The single species, *E. alba*, is found from eastern Honduras to western Panama (Timm 1982). A second species, *E. macconnelli*, recently has been transferred to the genus *Vampyressa* (see account thereof).

Head and body length is 36–47 mm, there is no external tail, and forearm length is about 25–29 mm. LaVal and Fitch (1977) reported an average weight of 5.6 grams for a series of specimens from Costa Rica.

Ectophylla resembles a small whitish *Vampyrops* in external appearance. The hair on the dorsal and ventral surfaces is lightly tipped with pale gray. The eyes are ringed with dark gray hair. The tragus, chin, ear margins, skin of the wing bones, and nose leaf are bright yellow but fade to a cream color upon death.

In Costa Rica, Gardner, LaVal, and Wilson (1970) reported catching *Ectophylla* in second-growth thickets interspersed with dense stands of wild plantain *(Heliconia)*. Timm and Mortimer (1976) found *Ectophylla* to alter the large leaves of five species of *Heliconia* for use as diurnal roosts. The bats cut the side veins extending out from the midrib, so that the two sides of the leaf fold downward to form a "tent," which the bats then get under. The bats were found roosting singly and in groups of two, four, and six. According to Gardner (1977), the diet of *Ectophylla* probably consists mainly of fruit. Pregnant females have been collected in February, March, June, July, and August in Costa Rica, and lactating females were found there in March and April. As in most phyllostomids, females apparently bear only a single young (Timm 1982).

White bat *(Ectophylla alba)*, photo by Richard S. Casebeer.

CHIROPTERA; PHYLLOSTOMIDAE; **Genus ARTIBEUS**
Leach, 1821

Neotropical Fruit Bats

There are 2 subgenera and 19 species (Anderson, Koopman, and Creighton 1982; Buden 1985; W. B. Davis 1984; Graham and Barkley 1984; Handley 1987; Jones and Carter 1976, 1979; Koepcke and Kraft 1984; Koop and Baker 1983; Koopman 1978*a*; Lazell and Koopman 1985; Myers and Wetzel 1983; Timm 1985; Webster and Jones 1984*a*; Webster and McGillivray 1984):

subgenus *Enchisthenes* Andersen, 1906

A. hartii, Jalisco and Tamaulipas (Mexico) to northwestern Peru and central Bolivia, Trinidad, a single record from southern Arizona;

subgenus *Artibeus* Leach, 1821

A. cinereus, Venezuela, Guianas, Amazonian Brazil;
A. glaucus, Colombia, Venezuela, Guyana, Ecuador, Peru, Bolivia;
A. watsoni, southern Mexico to northwestern South America;
A. gnomus, Venezuela, Guyana, Ecuador, eastern Peru, Amazonian Brazil;
A. phaeotis, Sinaloa and Veracruz (Mexico) to northeastern Peru and Guyana;
A. anderseni, Ecuador, French Guiana, Brazil, Peru, Bolivia;
A. toltecus, northern Mexico to northwestern South America;
A. aztecus, central Mexico to western Panama;
A. concolor, Colombia, Venezuela, Guianas, northern parts of Brazil and Peru;
A. hirsutus, western Mexico;
A. inopinatus, El Salvador, Honduras, Nicaragua;
A. amplus, Colombia, Venezuela;
A. jamaicensis, central Mexico to Paraguay and central Brazil, throughout Greater and Lesser Antilles, Trinidad and Tobago, Bahamas, lower Florida Keys;
A. fraterculus, Ecuador, Peru;
A. fuliginosus, Amazonian parts of Colombia, Ecuador, Peru, Venezuela, and French Guiana;
A. planirostris, Bolivia, Amazonian Brazil and adjacent regions;
A. lituratus, southern Mexico to northern Argentina, Lesser Antilles;
A. intermedius, Sinaloa and Tamaulipas (Mexico) to northern South America.

Based on an analysis of morphological characters, Owen (1987) concluded that *Artibeus*, as presently understood, is a polyphyletic assemblage. He recommended that the smaller species in the group—*A. hartii, A. cinereus, A. glaucus, A. watsoni, A. phaeotis, A. anderseni, A. toltecus, A. aztecus*, and *A. concolor*—be placed in the separate genus *Dermanura* Gervais, 1855, and that the larger species—*A. hirsutus, A. inopinatus, A. jamaicensis, A. fraterculus, A. fuliginosus, A. planirostris, A. lituratus*, and *A. intermedius*—be left in *Artibeus*. He also recognized *A. fimbriatus*, a large species in Paraguay that is related to *A. jamaicensis*. While noting that *Enchisthenes* is sometimes regarded as a distinct genus, he placed it within *Dermanura*. Neither *A. gnomus* nor *A. amplus* had been named at the time of Owen's study, though Handley's (1987) description suggests that the former is related to species that Owen put in *Dermanura*, and the latter to species that Owen left in *Artibeus*. Jones, Arroyo-Cabrales, and Owen (1988) accepted *Dermanura* as a genus and also stated that *Enchisthenes* probably deserves generic status. Humphrey and Brown (1986) questioned Lazell and Koopman's (1985) report of a resident population of *A. jamaicensis* in the Florida Keys.

In the smallest of the species listed above, *A. phaeotis*, head and body length is about 53 mm, forearm length is 35–38 mm, and weight is about 10 grams. In the largest species, *A. lituratus*, head and body length is 87–100 mm, forearm length is 64–76 mm, and weight is 44–87 grams. *Artibeus* lacks an external tail and has a narrow interfemoral membrane. The upper parts are dull brownish, grayish, or black, with a silvery tinge, and the underparts are usually paler. Four whitish facial stripes are usually present. There is no light dorsal line. The fur is short, very soft, and velvety in texture.

Artibeus differs from similar genera in its more pointed ears and in cranial and dental features. The total number of teeth, depending on the number of molars present, is 28, 30, or 32; the number of molars varies with the species and sometimes individually.

In Venezuela, Handley (1976) found these bats under the following conditions: *A. concolor*, mainly in moist, open areas; *A. jamaicensis*, mostly moist, open areas, and roosting mostly in houses; *A. fuliginosus*, moist open and forested areas; and *A. lituratus*, moist or dry areas, in both forests and openings. In Central America, W. B. Davis (1970*b*) found *A. phaeotis* and *A. toltecus* to occur in lowlands but *A. aztecus* to be restricted to high elevations. In Jalisco, Mexico, *A. phaeotis* is confined to tropical deciduous forest and thorn forest (Watkins, Jones, and Genoways 1972). *A. jamaicensis* occurs throughout the Yucatan Peninsula and roosts in caves, cenotes, and buildings (Jones, Smith, and Genoways 1973). Several species of *Artibeus* have been reported to modify palm fronds for roosting by biting around the midrib so that the leaflets fold downward to make a "tent" (Choe and Timm 1985; Foster and Timm 1976; Timm 1987). In a study in Jalisco, Mexico, Morrison (1978) found that four female *A. jamaicensis* flew an average of 8 km to forage, a distance 13 times as great as that averaged by females on Barro Colorado Island, Panama. *Artibeus* is primarily frugivorous but also consumes pollen, nectar, flower parts, and insects (Gardner 1977).

Fruits known to be taken include figs, mangoes, avocados, bananas, espave nuts, and the pulpy layer surrounding the seeds of *Acrocomia* palms. The smaller fruits are carried to feeding sites during the night, but toward morning these bats carry fruit to their regular roosts. Food passes through the digestive tract so rapidly (15–20 minutes) that there is probably little or no bacterial action. Perhaps there is some chemical or enzyme action. The newly passed fecal material often has the odor of the fruit on which the bat fed. Nuts, seeds, and fruit cores accumulate beneath roosting areas; *Artibeus* thus aids in the dissemination of seeds of tropical fruits. Control measures are sometimes necessary to protect cultivated fruit from these bats.

Some species, such as *A. lituratus*, are thought to be more solitary than others. However, Husson (1978) reported a cluster of 16 *A. lituratus* on the underside of a leaf of a coconut palm. Klingener, Genoways, and Baker (1978) observed a cluster of 6 *A. jamaicensis* roosting in a shallow embrasure. Females of the latter species have been reported to segregate into discrete groups prior to giving birth and until the young are weaned (Fenton and Kunz 1977). Morrison (1979), however, suggested a harem-type social organi-

Mexican fruit bat *(Artibeus jamaicensis)*. A. Bat hanging where it has been eating. The light-colored, irregular masses are pieces of food that it has chewed, extracted the liquid from, and rejected. The rounded portions are feces that have passed through the bat in 15–20 minutes. B. In flight, seen from below, showing greatly reduced membrane between hind legs. C. Shows rubberlike nose leaf and tiny tubercles in lower lip. Photos by Ernest P. Walker.

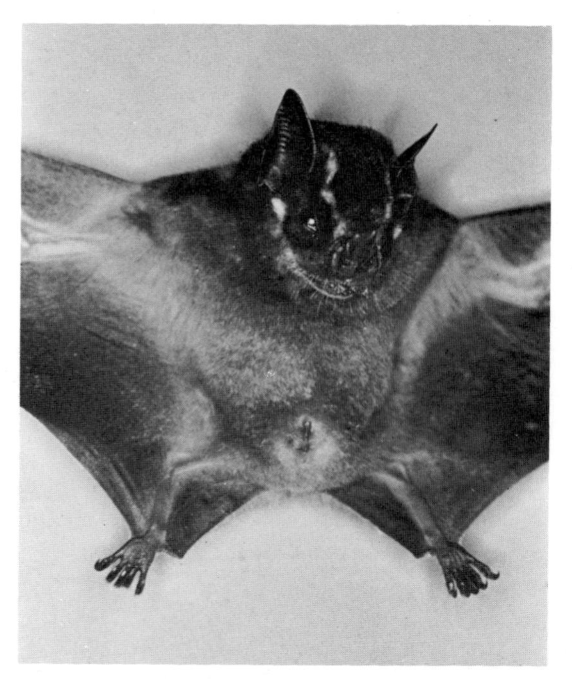

Neotropical fruit bat *(Artibeus harti)*, photo by Merlin D. Tuttle.

zation for *A. jamaicensis;* foliage roosts of this species were found to contain only 1–3 males or juveniles, but roosts in hollow trees always had a single adult male along with 4–11 females and their young. Kunz, August, and Burnett (1983) also reported evidence of harems in a cave colony of *A. jamaicensis.* Most of the bats in this group were in clusters comprising 2–14 pregnant or lactating females, their young, and a single adult male. Older and heavier males had the largest harems, and there also were groups of bachelor males and nonreproductive females.

The extensive reproductive data compiled by Wilson (1979) can be summarized briefly as follows: *A. cinereus,* bimodal polyestry suggested in Colombia, Panama, and Nicaragua, but pregnant or lactating females reported for nearly every month of the year; *A. phaeotis,* bimodal polyestry, with pregnant or lactating females reported from January to August; *A. toltecus,* apparently an extended breeding season with two births per year, pregnant or lactating females reported from January to October; *A. aztecus,* pregnancies reported during the summer in Mexico; *A. hirsutus,* no restricted breeding season, pregnancies reported in Mexico in February and from April to September; *A. inopinatus,* approximately one-month-old individuals taken in Honduras in August; *A. concolor,* a pregnant female taken in Colombia in February; *A. lituratus,* in Mexico one breeding season from about February to August, in southern Central America probably a pattern of bimodal polyestry with a quiescent period from about September to December, on Trinidad pregnancies reported from February to July and lactation from April to October, in Colombia year-round breeding with pregnancy peaks in December and May; and *A. hartii,* possibly reproductively active all year in Costa Rica, pregnancies reported in Colombia in April and May. *A. jamaicensis* has a unique seasonally polyestrous cycle in Panama. A birth peak occurs in March–April, followed by a postpartum estrus and a second birth peak in July–August. Then there is another postpartum estrus and implantation, but the blastocysts are dormant from September to November. Subsequently there is normal development, and birth in March or April. In Colombia and the Yucatan Peninsula there may be continuous or acyclic breeding by this species. Pregnant or lactating female *A. jamaicensis* have been found in various areas during all months of the year.

Davis and Dixon (1976) reported pregnant female *A. planirostris, A. fuliginosus,* and *A. lituratus,* each with a single large embryo, in Peru during the period 31 October–8 November. There is usually a single young per birth in this genus, but several cases of twins have been reported for *A. jamaicensis.* Wilson and Tyson (1970) recaptured a female *A. jamaicensis* on Barro Colorado Island 7 years after it had been banded. Jones (1982) reported that a specimen of *A. jamaicensis* was still living after 10 years in captivity.

CHIROPTERA; PHYLLOSTOMIDAE; **Genus ARDOPS**
Miller, 1906

Tree Bat

The single species, *A. nichollsi,* occurs in the Lesser Antilles from St. Eustatius to St. Vincent (Jones and Carter 1976).

Head and body length is 50–73 mm, there is no external tail, and forearm length is 42–54 mm. Jones and Genoways (1973) reported weights of 151–87 grams. The skull resembles that of *Artibeus* but is broader. The upper incisors are short and thick. The color is dark brown or grayish. Apparently, this genus lacks facial stripes and dorsal lines.

Ardops evidently roosts exclusively in trees and other kinds of arborescent vegetation (Jones and Genoways 1973). The diet presumably consists of fruit (Gardner 1977). On Dominica 5 pregnant females, each with a single embryo, were collected from 27 March to 14 April, and a lactating female was taken on 19 April. A pregnant female was collected on St. Eustatius on 9 March (Jones and Genoways 1973). Of 14 adult females taken on Guadeloupe in July, 2 were lactating and 6 were pregnant (Baker, Genoways, and Patton 1978).

CHIROPTERA; PHYLLOSTOMIDAE; **Genus PHYLLOPS**
Peters, 1865

Falcate-winged Bats

There are two living species (Jones and Carter 1976):

P. falcatus, Cuba;
P. haitiensis, Hispaniola.

The above two species are closely related and may represent a single species. A third species, *P. vetus,* is known only as a fossil from Cuba.

Head and body length is about 48 mm, there is no external tail, and forearm length is about 39–45 mm. Coloration is dark brown or grayish. Externally this genus resembles *Artibeus,* but it differs from it and other related genera in cranial and dental features.

Klingener, Genoways, and Baker (1978) reported that *P. haitiensis* was collected more frequently in thickly vegetated ravines than in drier scrub thorn habitats. *P. falcatus* occasionally enters houses, and *P. haitiensis* has been obtained in a house in a town, as well as while sleeping in mango trees. The diet is thought to consist mainly of fruit. Klingener, Genoways, and Baker (1978) recorded the following reproductive data for *P. haitiensis* in southern Haiti: all 6 females taken between 4 and 9 January were pregnant; 5 of 8

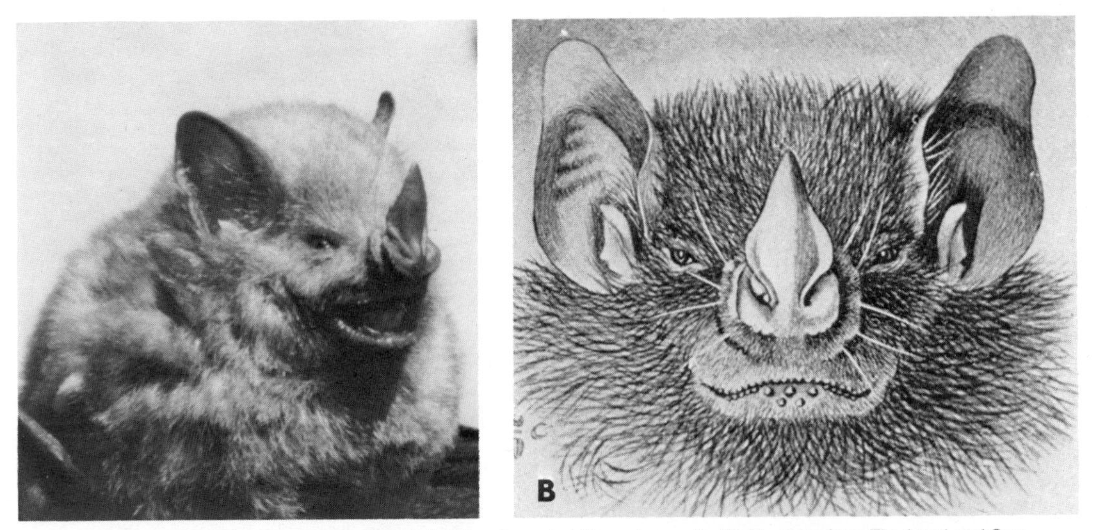

Tree bat *(Ardops nichollsi):* Left, photo by Robert J. Baker through J. Knox Jones, Jr.; Right, photo from *The Land and Sea Mammals of Middle America and the West Indies*, D. G. Elliot.

females caught on 27 May were pregnant; and of 54 females taken between 14 and 27 August, 27 were not pregnant, 8 had an enlarged uterus, 14 had embryos, and 5 had an enlarged postpartum uterus.

CHIROPTERA; PHYLLOSTOMIDAE; **Genus ARITEUS**
Gray, 1838

Jamaican Fig-eating Bat

The single species, *A. flavescens*, occurs on Jamaica (Jones and Carter 1976).

Head and body length is about 50–67 mm, there is no tail, and forearm length is about 40–44 mm. H. F. Howe (1974) reported that four specimens weighed from 9.2 to 13.1 grams. Coloration is light reddish brown above and paler below. A small white patch is present on each shoulder. Facial stripes and dorsal lines are not present. This bat resembles *Ardops* but differs from it in dental features.

H. F. Howe (1974) collected four specimens in heavily disturbed habitat and observed that *Ariteus* probably is not a cave bat. It feeds on fruits, such as the naseberry *(Achras sapota)* and the rose apple *(Eugenia jambos)*, as well as insects. It begins to fly and feed shortly after sunset.

Allen (1942) wrote that the status of this bat was not then known. However, evidently it was rare in collections and was thought likely to become scarce because of intensive agricultural development on Jamaica.

Falcate-winged bat *(Phyllops haitiensis)*, photo by Charles A. Woods.

CHIROPTERA; PHYLLOSTOMIDAE; **Genus STENODERMA**
E. Geoffroy St.-Hilaire, 1818

Red Fruit Bat

The single species, *S. rufum*, was for many years known only from a single skin and skull collected at an unknown locality. In 1916 the species was found as a fossil on Puerto Rico, and since then living specimens have been collected in Puerto Rico and on St. Thomas and St. John in the Virgin Islands.

Head and body length is about 53–73 mm, there is no external tail, and forearm length is about 46–51 mm. The upper parts are reddish brown, or nearly so, and the underparts are less reddish. Morphological comparisons indicate that this bat is most closely related to *Phyllops, Ardops*, and *Ariteus*. Like these other genera, *Stenoderma* has a white spot on both shoulders.

Jones, Genoways, and Baker (1971) collected specimens in nets set in tropical broad-leaved forest and above the forest canopy. According to Gardner (1977), the diet is composed of fruit. Three captive specimens were maintained for three weeks on a diet of mangoes, bananas, and various fruit nectars (Genoways and Baker 1972). Data summarized by Wilson (1979) suggest that *Stenoderma* is polyestrous in Puerto Rico, with pregnancies recorded in February, March, May, July, and August, and lactation in February, July, and November.

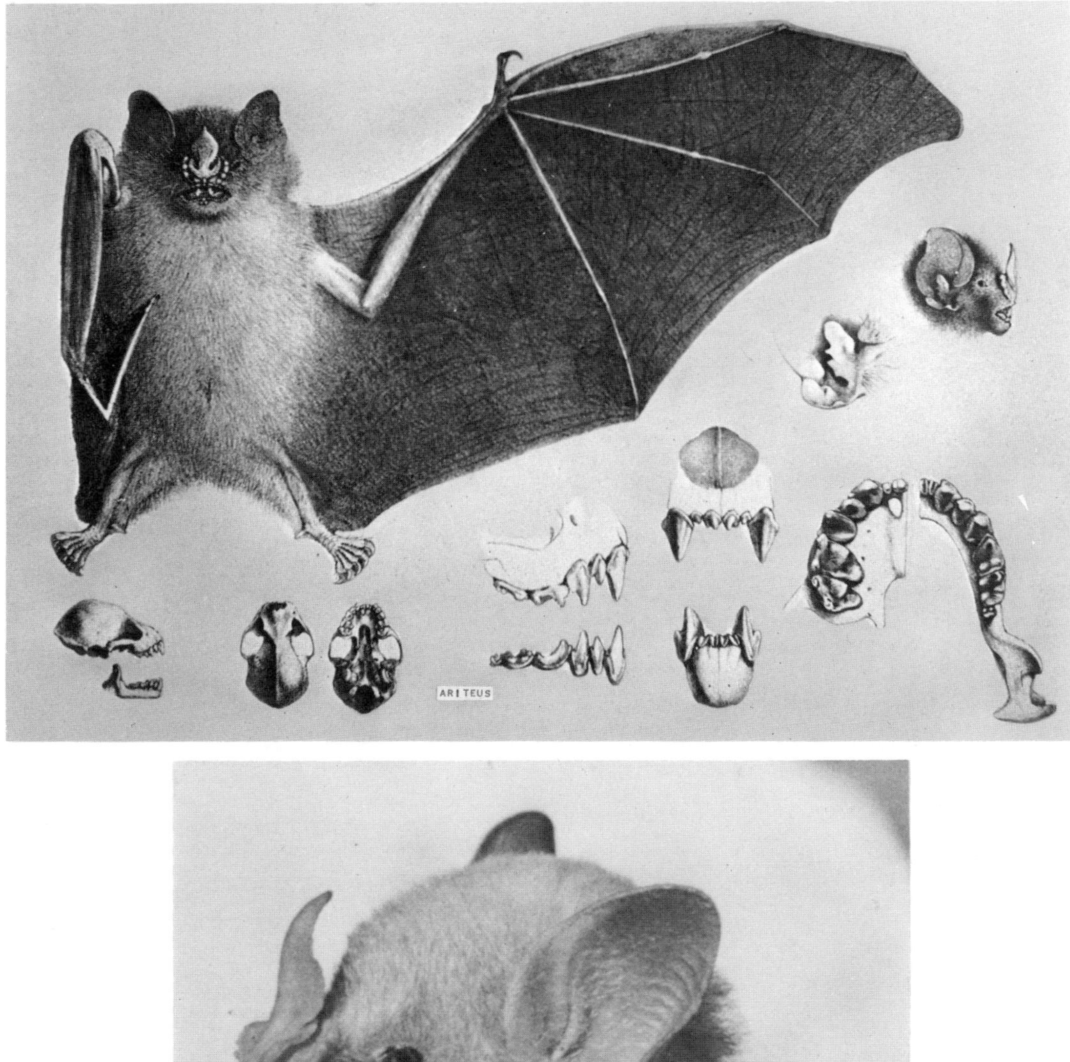

Jamaican fig-eating bat *(Ariteus flavescens):* Top, photo from *M. ber. Kö. Preuss. Akad. Wis. Berlin;* Bottom, photo by Robert J. Baker through J. Knox Jones, Jr.

Red fruit bat *(Stenoderma rufum)*, photo by J. R. Tamsitt. Inset: skull *(S. rufum)*, photo by P. F. Wright of specimen from Museum of Natural History, University of Kansas.

CHIROPTERA; PHYLLOSTOMIDAE; **Genus PYGODERMA**
Peters, 1863

Ipanema Bat

The single species, *P. bilabiatum*, has been reported from Surinam, southeastern Brazil, Bolivia, Paraguay, and adjacent Argentina; a reported occurrence in Mexico has been shown to be erroneous (Anderson, Koopman, and Creighton 1982; Jones and Carter 1976; Owen and Webster 1983). The vernacular name refers to the type locality, Ipanema, in Sao Paulo, Brazil.

In the nominate subspecies, head and body length is about 61 mm, there is no external tail, and forearm length is about 38 mm. In a newly described and larger subspecies from Bolivia, head and body length is 84 mm, forearm length is 43 mm, and weight is 27.5 grams (Owen and Webster 1983). The color is dark brown, almost black, above and grayish brown below. There is a white spot on each shoulder near the wing. *Pygoderma* may be distinguished by the unequal size of the upper incisors, the inner pair being the larger, and the greatly shortened and depressed rostrum of the skull. The palate is short and rounded. The lower lip has a central tubercle bordered by smaller protuberances.

According to Webster and Owen (1984), some specimens have been taken at night above trails and streams in mature tropical forests and secondary growth bordering forest. Others have been taken around fruit trees, and stomach contents suggest that *Pygoderma* feeds on pulpy or overripe fruit that has few fibers or seeds. Pregnant females, each with a single fetus, have been collected in March, July, and August in Paraguay and in August in Brazil.

CHIROPTERA; PHYLLOSTOMIDAE; **Genus AMETRIDA**
Gray, 1847

The single species, *A. centurio*, occurs in Venezuela, the Guianas, Brazil, Trinidad, and Bonaire Island (Jones and Carter 1976). For many years, a second species, *A. minor*, was thought to exist, but Peterson (1965*b*) showed that the small *A. minor* actually comprised the male members of *A. centurio*.

Head and body length is 35–47 mm, there is no external tail, and forearm length is 25–33 mm. Peterson (1965*b*) reported that a male weighed 7.8 grams and two females averaged 10.1 grams. The fur is sooty brown to dark brown in color, and there is a white spot on each shoulder near the wing. *Ametrida* differs from *Sphaeronycteris* in the presence of a nose leaf, a shorter facial portion of the skull, and other

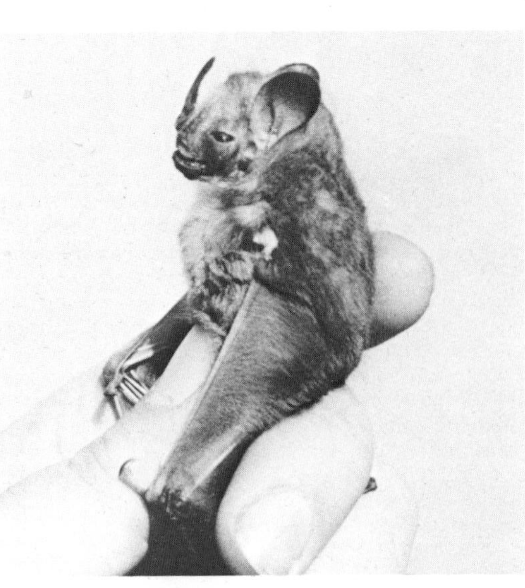

Ipanema bat *(Pygoderma bilabiatum)*, photos by R. E. Mumford.

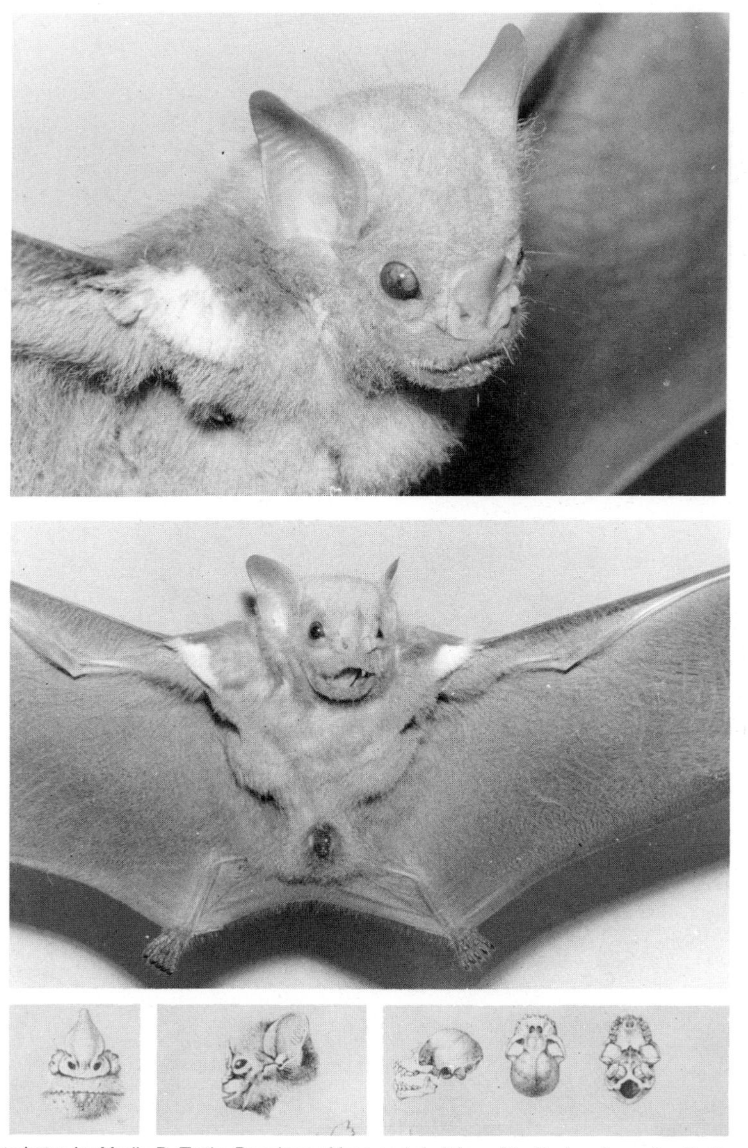

Ametrida centurio; photos by Merlin D. Tuttle. Drawings of face and skull from *Die Fledermäuse des Berliner Museums für Naturkunde*, Wilhelm K. Peters.

cranial and dental characters. The small third lower molar is present.

In Venezuela, Handley (1976) collected *Ametrida* mostly near streams and in other moist parts of evergreen forest. According to Gardner (1977), the diet is unknown but probably consists of fruit. Pregnant females, each with a single embryo, were collected on Trinidad in July and August (Carter et al. 1981).

CHIROPTERA; PHYLLOSTOMIDAE; **Genus SPHAERONYCTERIS**

Peters, 1882

The single species, *S. toxophyllum*, occurs from Colombia and Venezuela to Amazonian Peru and Bolivia (Jones and Carter 1976).

Head and body length is about 53 mm, there is no external tail, and forearm length is approximately 40 mm. The upper parts are cinnamon brown, the individual hairs in the middle of the back being whitish; the underparts are brownish white. There is a horn-shaped growth on the nose that appears to be larger on males than on females. It is probably soft flesh, like the nose-leaf structures on many other bats, but no definite statement has been found regarding this structure on live or recently killed bats of this genus.

This bat is similar to *Centurio*, but the facial portion of the skull is shorter, and the outgrowths on the face are less extreme. A third lower molar, absent in *Centurio*, is present in *Sphaeronycteris*. As in *Centurio*, there is no true nose leaf. Skull and dental features differentiate *Sphaeronycteris* from another similar genus, *Ametrida*.

In Venezuela, Handley (1976) collected numerous specimens in many kinds of habitat but mostly in moist, open areas. According to Gardner (1977), the diet is unknown but

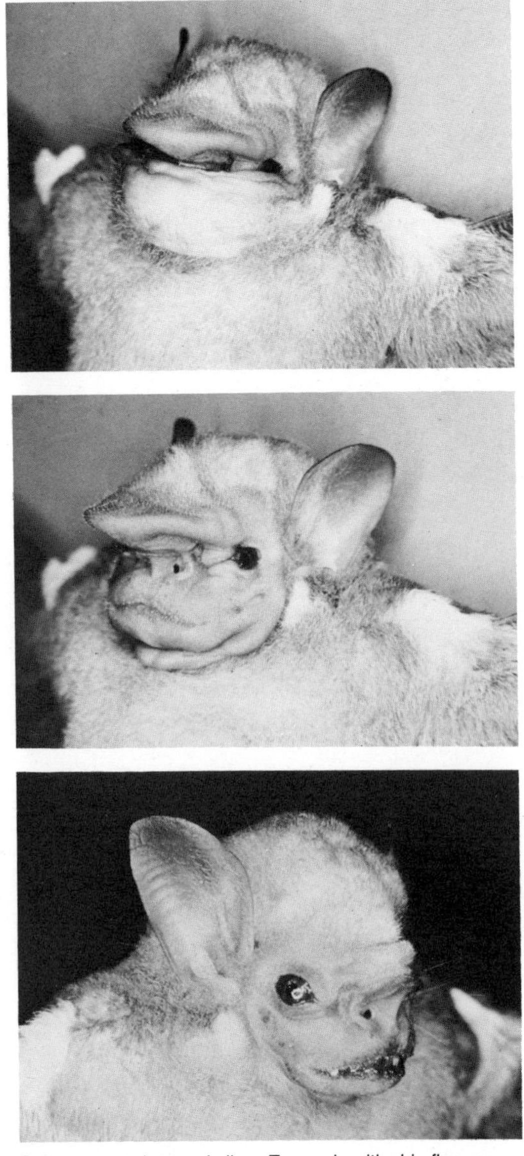

Sphaeronycteris toxophyllum: Top, male with chin flap up; Middle, male with chin flap down; Bottom, female. Photos by Merlin D. Tuttle.

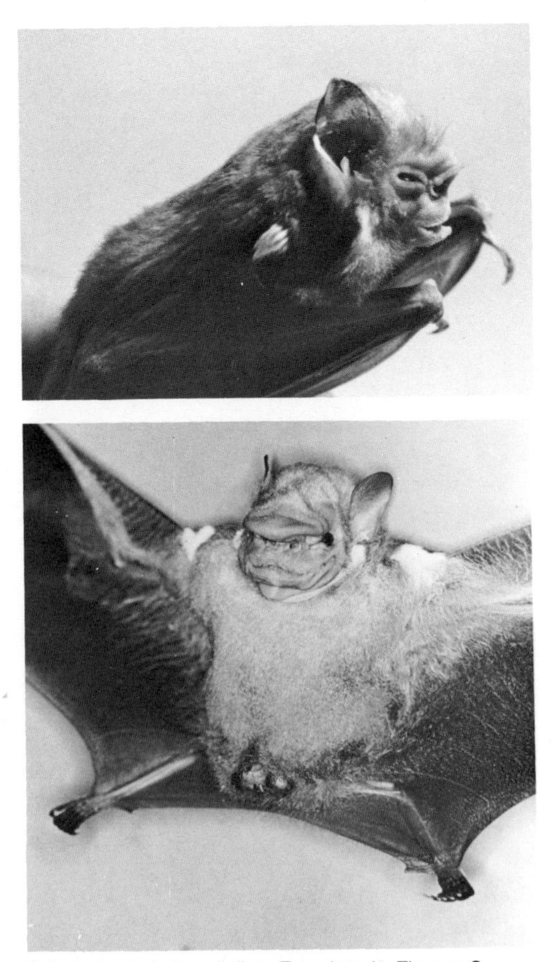

Sphaeronycteris toxophyllum: Top, photo by Thomas O. Lemke; Bottom, photo by Merlin D. Tuttle.

probably consists of fruit. Anderson and Webster (1983) collected a pregnant female in Bolivia on 3 October.

CHIROPTERA; PHYLLOSTOMIDAE; **Genus CENTURIO**
Gray, 1842

Wrinkle-faced Bat, or Lattice-winged Bat

The single species, *C. senex*, occurs from Sinaloa and Tamaulipas in northern coastal Mexico to Venezuela, as well as on Trinidad (Jones and Carter 1976, 1979).

Head and body length is 55–70 mm, there is no external tail, forearm length is approximately 41–47 mm, and adult weight is 17–28 grams. The coloration is medium brown, dark brown, or yellowish brown on the upper parts, with a white spot on each shoulder; the underparts are paler.

Externally this genus may be recognized by the grotesque facial features: the face is short, broad, naked, and covered with wrinkled outgrowths of the skin. There is no true nose leaf. Glands are present in the neck that probably secrete an odoriferous substance. The skull is characteristic in its high, rounded braincase, extremely short rostrum, and short palate and in the position of the external nares, which are directly over the roots of the upper incisors.

All the reported roosts of this genus have been in trees, such as mango, rayo (*Dracaena* sp.), and *Putranjiva*. The wrinkle-faced bat has always been found roosting singly or in twos or threes, usually under leaves, with a maximum of approximately a dozen individuals in a given tree. When roosting or when sleeping in captivity, this bat covers its face with the chin fold. This fold extends over the ears, which lie flat over the top of the head when the bat is at rest. A small projection on the top of the head acts as a sort of "doorstop," allowing the wrinkled fold of skin to be stretched taut at this point. Two areas in this facial mask, at least in some individuals, are devoid of hair; these translucent areas, or "windows," cover the bat's eyes when the mask is stretched tight, presumably allowing it to perceive light and perhaps objects even

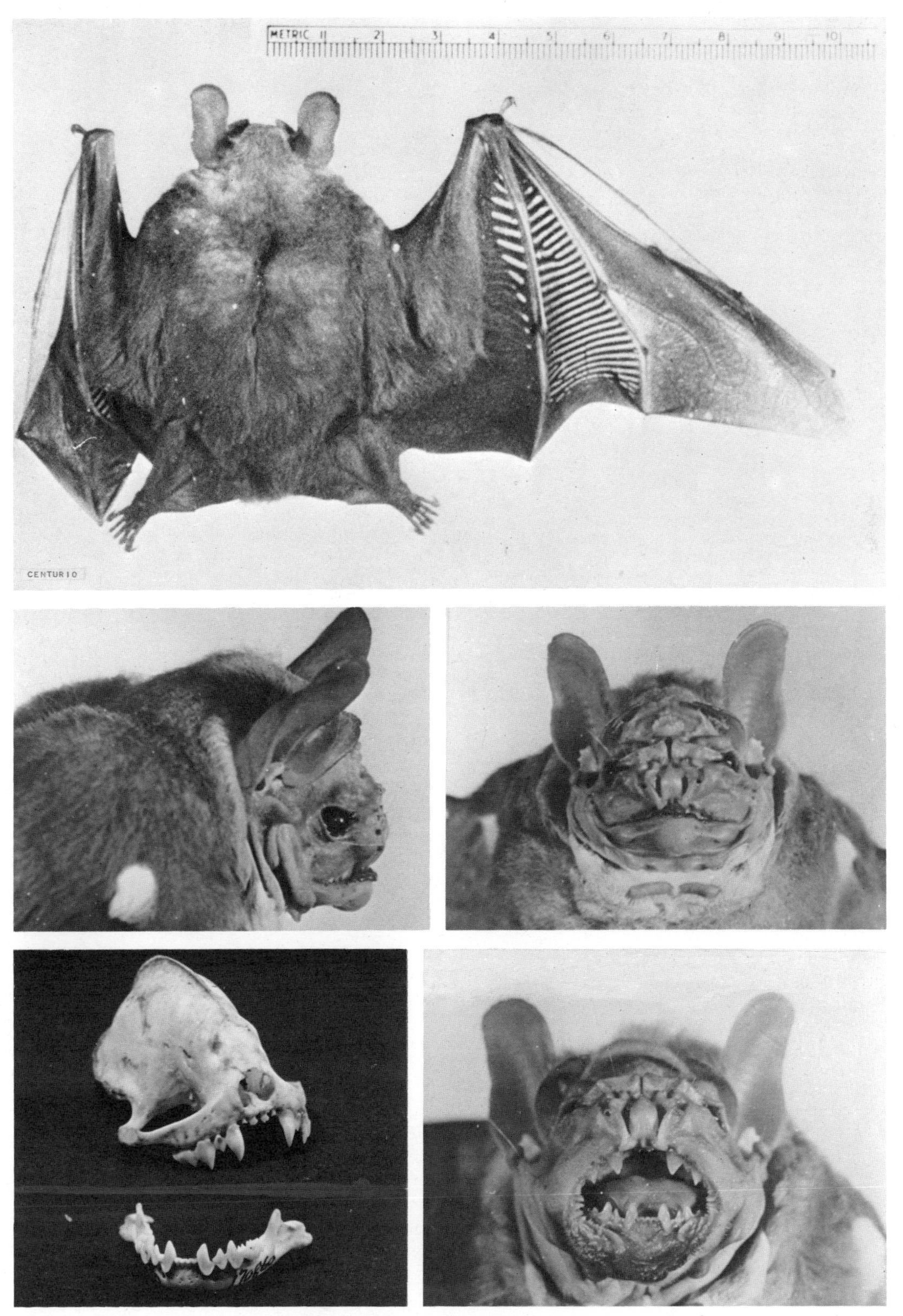

Lattice-winged bat *(Centurio senex),* photos by Harold Drysdale, Trinidad Regional Virus Laboratory, through A. M. Greenhall. Skull *(C. senex)* photo by P. F. Wright of specimen in U.S. National Museum of Natural History.

when the face is covered. When *Centurio* leaves its roost, it unmasks itself and departs in a jerky flight resembling that of a large butterfly. The facial mask then appears beneath the chin as a series of wrinkled folds of skin. The males of *Centurio* also have large lappets of loose skin on the chin that are not part of the skin fold; these lappets probably contain scent glands.

In the lowlands of western Venezuela, Handley (1976) collected specimens at a variety of moist and dry sites, in both forested and open areas. The diet is assumed to consist mainly of fruit. Captives in Trinidad preferred the soft, mushy parts of ripe bananas and pawpaws, which they appeared to suck up. The many small papillae on the skin between the lips and the gum line may act as strainers when these bats feed on soft fruit.

Pregnant females apparently shelter with males in the same trees, though they may roost by themselves. Pregnant females have been recorded in Mexico and Central America from February to August, in Colombia in April, and on Trinidad in January; lactating females have been found in Mexico in March, April, July, and August (Carter and Jones 1978; Lemke et al. 1982; Wilson 1979).

CHIROPTERA; PHYLLOSTOMIDAE; **Genus BRACHYPHYLLA**
Gray, 1834

There are two species (Swanepoel and Genoways 1978):

B. cavernarum, Puerto Rico, Virgin Islands, Lesser Antilles south to St. Vincent and Barbados;
B. nana, Cuba, Isle of Pines, Grand Cayman, Hispaniola, Middle Caicos, and as a Pleistocene or sub-Recent fossil on Jamaica.

Buden (1977) considered *B. nana* a subspecies of *B. cavernarum*, but Swanepoel and Genoways (1983*a*, 1983*b*) continued to treat the two as distinct species, noting that *B. nana* is consistently smaller than *B. cavernarum*.

Head and body length is 65–118 mm, the tail is vestigial and concealed in the base of the interfemoral membrane, and forearm length is 51–69 mm. According to Nellis and Ehle (1977), weight is about 45 grams in *B. cavernarum*. The upper parts are ivory yellow, and the hairs are tipped with sepia, except for patches on the neck, shoulders, and sides, which are paler. The underparts are usually brown.

The muzzle is somewhat conical in shape, and the lower lip has a V-shaped groove margined by tubercles. The nose leaf is vestigial, and the ears are small and separate. The interfemoral membrane is well developed. The molar teeth are broad and well ridged. The call of *B. cavernarum* is a strident, rasping squeak.

These bats are primarily cave dwellers, but they also have been found roosting in buildings and in a well. The roosts may be in either dark or well-illuminated locations, but emergence is relatively late in the evening. The diet is opportunistic, including many kinds of fruit as well as flowers, pollen, nectar, and insects (Gardner 1977; Nellis and Ehle 1977; Silva Taboada and Pine 1969; Swanepoel and Genoways 1978).

Nellis and Ehle (1977) reported a colony of about 5,000 *B. cavernarum* on St. Croix Island. In Cuba colonies of 2,000–3,000 *B. nana* are common, and five colonies have been conservatively estimated to contain 10,000 bats each (Silva Taboada and Pine 1969). About 2,000 *B. cavernarum* were killed by gassing in the ruins of an old sugar factory on St. Croix in

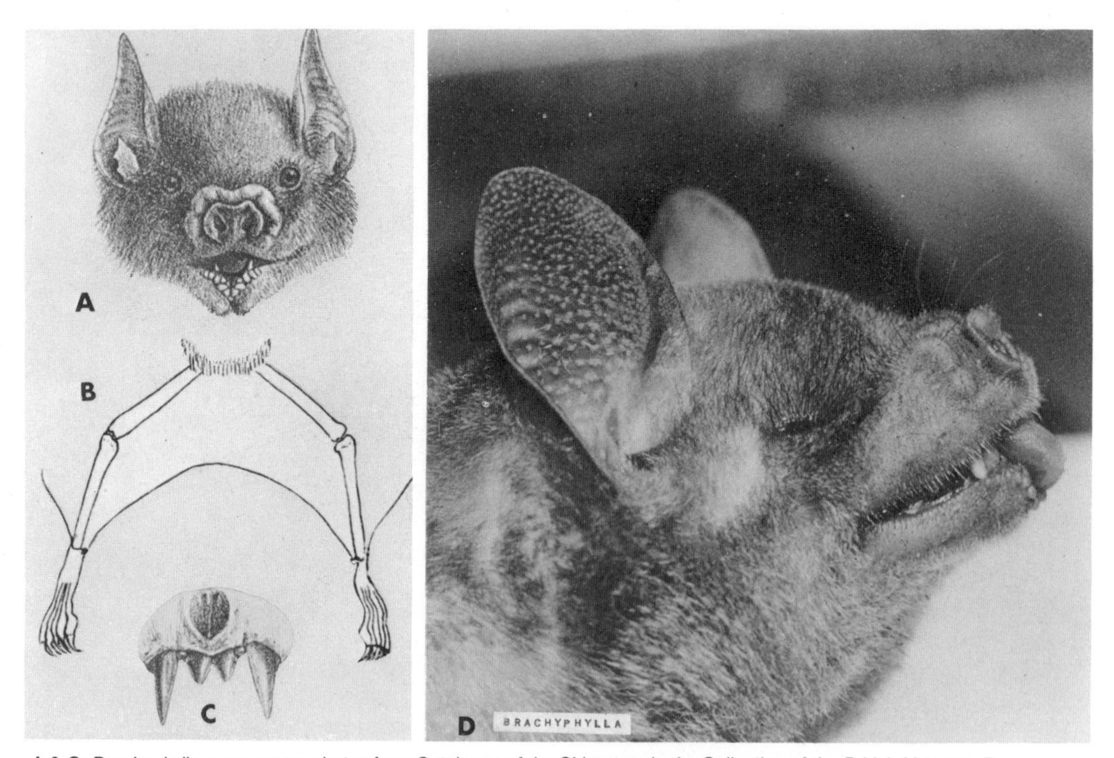

A & C. *Brachyphylla cavernarum*, photos from *Catalogue of the Chiroptera in the Collection of the British Museum*. B. *Brachyphylla* sp., photo from *Bats*, G. M. Allen. D. *B. nana*, photo by Albert Schwartz.

Brachyphylla cavernarum, photo by Robert J. Baker through J. Knox Jones, Jr.

1956 in the mistaken belief that they were *Artibeus jamaicensis,* which did considerable damage to fruit in the area. Sexing of 330 *Brachyphylla* from this colony revealed 276 females and 63 males, all adults or subadults. Adult *B. cavernarum* generally are aggressive, hitting and biting one another, especially when feeding. On St. Croix, births during the study of Nellis and Ehle (1977) occurred in a three-week period in late May or early June; colonies observed at this time comprised females with 1 young each and relatively few males and barren females. The young first flew at about two months. Other reproductive data include the taking of pregnant females in Puerto Rico in February, on St. Croix in March, and on Caicos Island in March; and lactating females in Puerto Rico in April and July, on St. Croix in April, on Montserrat in July, and on Guadeloupe in July (Baker, Genoways, and Patton 1978; Jones and Baker 1979; Wilson 1979). Klingener, Genoways, and Baker (1978) reported a lactating female *B. nana* collected in southern Haiti in August.

CHIROPTERA; PHYLLOSTOMIDAE; **Genus EROPHYLLA**
Miller, 1906

Brown Flower Bats

There are two species (Hall 1981; Jones and Carter 1976):

E. bombifrons, Hispaniola, Puerto Rico;
E. sezekorni, Bahamas, Cuba, Jamaica, Cayman Islands.

Buden (1976) considered *E. bombifrons* a subspecies of *E. sezekorni,* and this procedure was followed, with question, by Baker, August, and Steuter (1978).

Head and body length is about 65–75 mm, tail length is about 12–17 mm, and forearm length is approximately 42–55 mm. In *E. sezekorni* the upper parts are pale yellowish brown or buffy and the underparts are paler; in *E. bombifrons* the upper parts are dark brown and the underparts are slightly paler.

This genus and *Phyllonycteris* externally resemble the bats related to *Glossophaga* in that the skull is long and narrow and the tongue is long, protrusible, and armed with bristlelike papillae. In *Erophylla* the nose leaf is notched or forked at the tip, the ears are separate and about as high as they are broad, the tail projects beyond the narrow inter-

Brown flower bat *(Erophylla sezekorni),* photo by Merlin D. Tuttle, Bat Conservation International.

Brown flower bat *(Erophylla bombifrons)*, photo by Robert J. Baker through J. Knox Jones, Jr.

femoral membrane, and a short but distinct calcar is present.

Erophylla has been reported to roost in caves, both in dark interior portions and on exposed surfaces where much light penetrates (Baker, August, and Steuter 1978). The diet consists of fruit, pollen, nectar, and insects (Gardner 1977). In Cuba, colonies of *E. sezekorni* may range in size from a few hundred to several thousand (Silva Taboada and Pine 1969). A colony of several hundred *E. sezekorni* has been reported in a cave in the Bahamas. In this colony the bats hung, singly or in clusters, from the walls and ceiling. When observed in July, this colony consisted of adult males, adult females, and young. A colony of many hundreds of *E. sezekorni* in a Cuban cave in February consisted of adult males and females. Some of the females were pregnant with small embryos. A colony possibly totaling 40 *E. bombifrons* was found in a cave on Puerto Rico in July. Apparently the adults of both sexes were present, along with young from two-thirds grown to nearly full-grown. Data summarized by Wilson (1979) suggest a restricted breeding season for both *E. bombifrons* and *E. sezekorni*, with pregnant females found from February to June and lactating females from May to July. There normally appears to be a single young, but one case of twins has been reported (J. R. Tamsitt, Royal Ontario Museum, pers. comm.).

CHIROPTERA; PHYLLOSTOMIDAE; **Genus PHYLLONYCTERIS**
Gundlach, 1861

There are four species (Hall 1981; Jones and Carter 1976):

P. major, known only from cave deposits in Puerto Rico;
P. obtusa, Hispaniola;
P. poeyi, Cuba;
P. aphylla, Jamaica.

Jones and Carter (1976) considered *P. obtusa* a subspecies of *P. poeyi*. Corbet and Hill (1986) agreed, and they also placed *P. aphylla* in the subgenus *Reithronycteris* Miller, 1898.

In species known from the living state, head and body length is about 64–83 mm, tail length is about 7–12 mm, and forearm length is about 43–50 mm. H. F. Howe (1974) reported that three female *P. aphylla* weighed 14.0–14.8 grams. For *P. obtusa*, Klingener, Genoways, and Baker (1978) reported that two females and a male weighed 20.0–21.1 grams. *P. poeyi* is grayish white. The pelage of this species has a silky texture that produces silvery reflections under certain light. A specimen in alcohol of *P. aphylla* had light yellowish brown upper parts and underparts.

Both *Phyllonycteris* and *Erophylla* resemble the bats of the subfamily Glossophaginae in that the skull is long and narrow and the tongue is long, protrusible, and armed with bristlelike papillae. In *Phyllonycteris* the nose leaf is rudimentary, the ears are moderately large and separate, the tail projects beyond the narrow interfemoral membrane, and a calcar is not present. A groove on the ventral surface of the braincase in *P. aphylla* seems to be unique among mammals.

These bats roost in caves. The diet of *P. poeyi* consists of fruit, pollen, nectar, and insects (Gardner 1977). Captive *P. aphylla* thrived on a diet of bananas, mangoes, papayas, and various kinds of canned fruit nectars (Henson and Novick 1966). *P. poeyi* is gregarious, roosting by the hundreds and apparently sometimes by the thousands. Most of its young seem to be born in June. *P. aphylla* is also colonial; a female taken in January was pregnant with one embryo (Goodwin 1970). Klingener, Genoways, and Baker (1978) reported that three female *P. obtusa* taken in southern Haiti on 17 December were pregnant with one embryo each. Both *P. aphylla* and *P. obtusa* long were known only from skeletal remains and one specimen in alcohol, but recently they were rediscovered in the living state. The only major known colony of *P. aphylla* contains a few hundred individuals and is con-

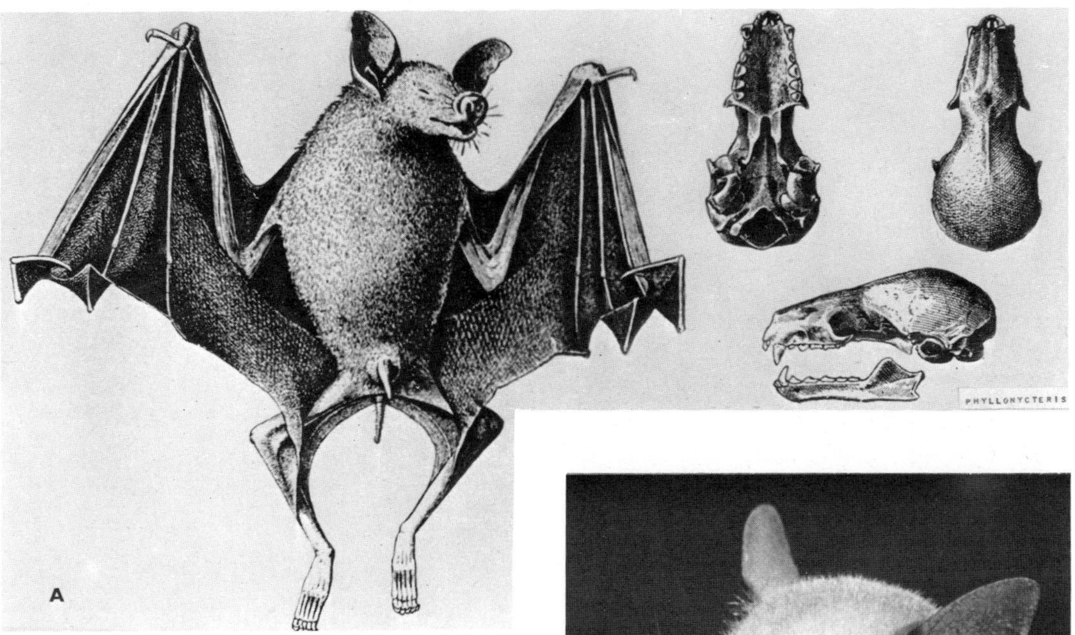

Phyllonycteris poeyi: A. Photo from *Proc. U.S. Natl. Mus.*; B. Photo by Robert J. Baker through J. Knox Jones, Jr.

sidered vulnerable to human disturbance (McFarlane 1986). *P. major* appears to be extinct.

CHIROPTERA; PHYLLOSTOMIDAE; **Genus DESMODUS**
Wied-Neuwied, 1826

Common Vampire Bat

The single species now known to be extant, *D. rotundus,* is found from northern Mexico to central Chile, Argentina, and Uruguay, as well as on the islands of Margarita and Trinidad off northern Venezuela (Koopman 1988). A closely related and larger species, *D. stocki,* occurred across much of the southern United States and Mexico during the late Pleistocene and evidently survived on San Miguel Island, off southern California, until less than 3,000 years ago (Ray, Linares, and Morgan 1988). The skeletal remains of a giant species, *D. draculae,* recently were discovered in a cave in northern Venezuela, and late Pleistocene fossils of comparable size are known from West Virginia and the Yucatan. As the remains from Venezuela are unmineralized and were found on the surface in apparent association with living species, there is a "faint possibility" that *D. draculae* still exists (Ray, Linares, and Morgan 1988).

Head and body length of *D. rotundus* is 70–90 mm, there is no tail, forearm length is 50–63 mm, and adult weight is 15–50 grams. The upper parts are dark grayish brown and the underparts are paler, sometimes with a faint buffy wash.

Desmodus is distinguished from the other true vampire bats by its pointed ears, longer thumb with a distinct basal pad, naked interfemoral membrane, and dental features. This genus has only 20 teeth, the largest being the 2 chisel-like upper incisors and the 2 upper canines.

This adaptable species inhabits both arid and humid regions of the tropics and subtropics. In Venezuela, Handley (1976) collected 57 percent of his specimens in forests of all types and 43 percent in yards, pastures, and other open habitats. Vampire bats usually reside in caves but also occupy hollow trees, old wells, mine shafts, and abandoned buildings. These retreats usually have a strong odor of ammonia from the pools of digested blood that collect in the crevices. If disturbed, the bats quickly run into more protected crevices. When moving about, they somewhat resemble large spiders. They commonly forage in an area within 5–8 km of the diurnal roost, and in some areas this distance may extend to 15–20 km (Greenhall et al. 1983). At a study site in Costa Rica, Turner (1975) estimated that a population of 100–150 vampires occupied 1,300 ha. and utilized 1,200 head of livestock as prey.

The feeding habits of the vampire have been so exaggerated and confused with Old World legends that the animal is of particular interest. Shortly after dark it leaves its roost with a silent, low flight, usually only one meter above the ground. Its usual food is the blood of horses, burros, cattle, and, occasionally, human beings. Data summarized by Gardner (1977) indicate that in some areas *Desmodus* also preys extensively on domestic turkeys and chickens. It has been conservatively estimated that each individual bat consumes 20 ml of blood per day. Greenhall (1972) listed four phases in the feeding cycle of *Desmodus,* the whole process lasting as long as two hours: (1) selection of a suitable site on a prey animal; (2) a

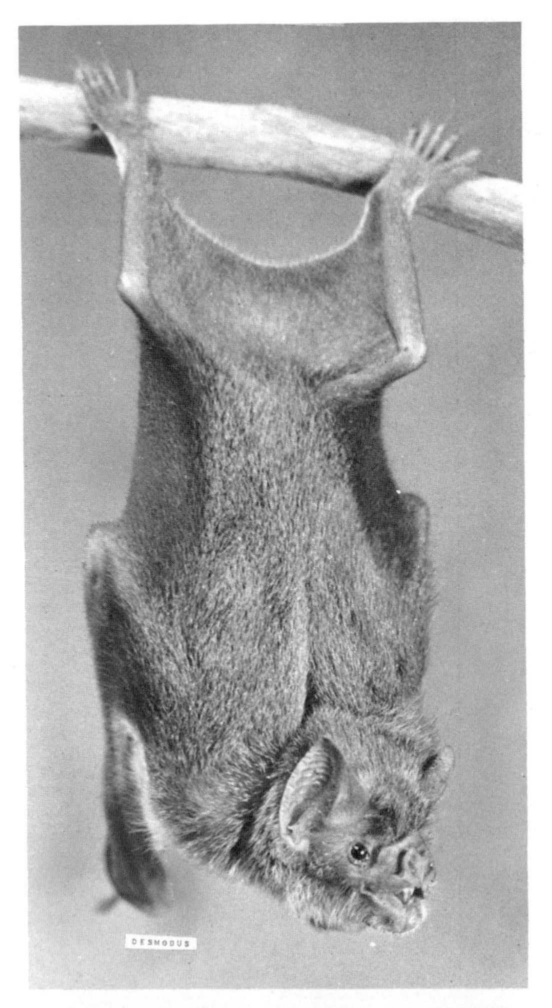

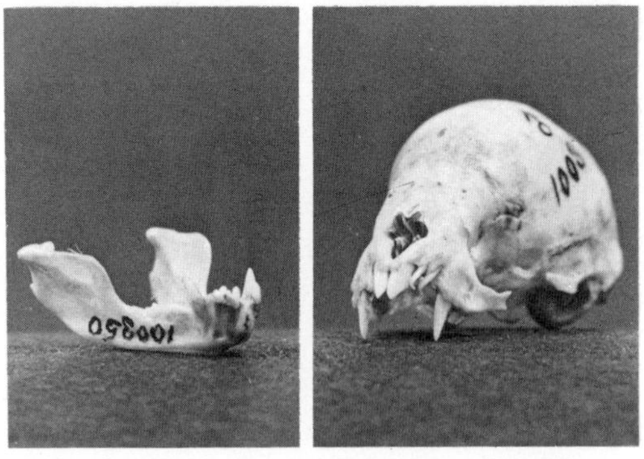

Vampire bat *(Desmodus rotundus)*, photo by Ernest P. Walker. The skull photo, by P. F. Wright of a specimen in U.S. National Museum of Natural History, shows large, sharp, middle incisors used for making incisions for taking blood. This photo, of a very young animal, is of unusual interest in that it also shows unshed milk teeth next to the permanent middle incisors.

preparation phase, during which the bat licks the site with its tongue; (3) the shearing or shaving phase, in which the hair or feathers are cut; (4) biting, typically involving the neat removal of a circular piece of skin with the upper and lower incisors. After feeding, *Desmodus* retires to its daytime retreat to digest its meal (for additional details see account of the family Phyllostomidae).

The common vampire bat may roost alone, in small groups, or in colonies of over 2,000 individuals. Most colonies contain 20–100 bats; if more than 50 are present, there are stable, identifiable social units consisting of 8–20 adult females and their young (Schmidt et al. 1978; Wilkinson 1985*a*, 1985*b*, 1988). Females within a unit regurgitate blood for their young and for each other. A single adult male resides in the vicinity of each female group and attempts to maintain his position and thus the ability to mate with the females. If the roost is within a hollow tree, this male is positioned near the top. Other males roost in groups nearby and attempt to displace the top male. They often are successful, and thus paternity varies within a social unit. Fighting between males is vicious. Sailler and Schmidt (1978) described six different vocalizations in captive *Desmodus*, mostly associated with aggressive interaction during feeding.

Wilson (1979) described the reproductive pattern of *Desmodus* as continuous polyestry, there being records from many areas suggesting year-round breeding. Some females have a postpartum estrus and produce more than one litter per year. According to Schmidt (1988), there are peak periods of birth in some areas, such as, for example, April–May and October–November in Trinidad. There normally is a single young; a case of twins has been reported, but one of the neonates was not viable. Recorded gestation periods of captive females are 205, 213, and 214 days. The young are well developed at birth, and their eyes are open. While they first feed on regurgitated blood in their second month of life and can forage for blood themselves at 4 months, they are not completely weaned before 9–10 months. Sexual maturity seems to come at about the same time. Lopez-Forment (1980) reported that wild individuals have been banded and recaptured at the same roosts nearly 9 years later and that a captive female lived for at least 19.5 years.

The vampire bat long has been considered a threat both to people and to their domestic animals in Latin America. The main problem at present is the transmission of rabies from *Desmodus* to the cattle on which it feeds. It is estimated that over 100,000 cattle die in this manner each year and that annual economic losses amount to over U.S. $40 million (Acha and Alba 1988).

CHIROPTERA; PHYLLOSTOMIDAE; **Genus DIAEMUS** *Miller, 1906*

White-winged Vampire Bat

The single species, *D. youngi*, has been recorded at scattered localities from Tamaulipas in northeastern Mexico to northern Argentina and southeastern Brazil, as well as on Margarita Island and Trinidad (Koopman 1988). *Diaemus* was included within *Desmodus* by Handley (1976), Koopman (1978*a*), and Honacki, Kinman, and Koeppl (1982), but later Koopman (1988) concluded that the two are generically distinct.

Head and body length is about 85 mm, there is no external tail, and forearm length is approximately 50–56 mm. Adults weigh approximately 30–45 grams. The pelage is usually glossy clay color, light brown, or dark cinnamon brown. The

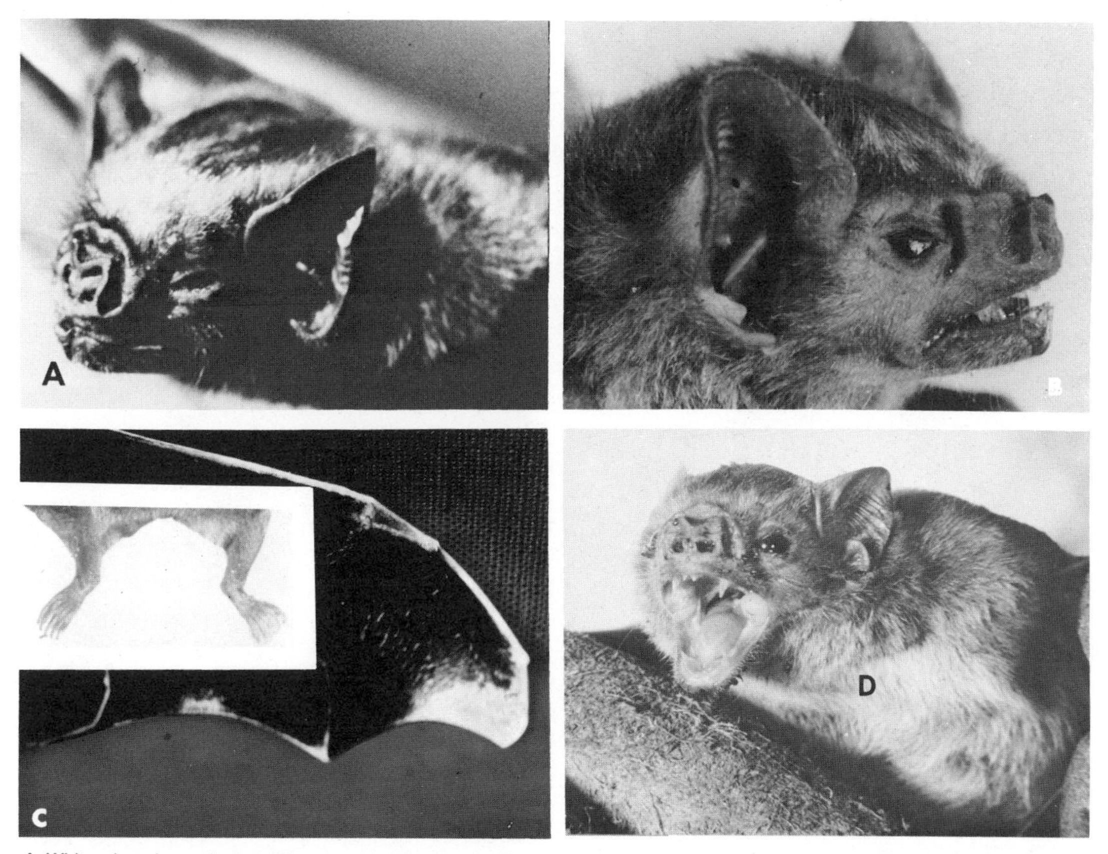

A. White-winged vampire bat *(Diaemus youngi)*, photo by Richard K. LaVal. Inset and B. Photos by Harold Drysdale, Trinidad Regional Virus Laboratory, through A. M. Greenhall. C. Photo from Pan-American Sanitary Bureau. D. Photo by David Pye.

edges of the wings are white, and the membrane between the second and third fingers is largely white.

"The peculiar short thumb with single pad under its metacarpal and the slightly recurved lower incisors with their different system of cusps are the principal characters which distinguish this genus from *Desmodus*" (Miller 1907). The thumb in *Diaemus* is about one-eighth as long as the third finger; in *Desmodus* it is about one-fifth as long. The thumb of *Diaemus* is about the same length as the thumb of *Diphylla*, but it differs from that of the hairy-legged vampire bat in the presence of the large pad at its base. *Diaemus* is the only bat known to have 22 permanent teeth. Old individuals occasionally lack the second upper molars and so have only 20 teeth.

In the lowlands of Venezuela, Handley (1976) collected specimens mainly in moist, open areas. One individual was found in a cave. *Diaemus*, as far as is known, feeds only on fresh blood, apparently preferring the blood of birds and goats. Captives would not feed on defibrinated cattle blood (even when it was mixed with chicken blood) but did feed heavily on chicken blood. A live guinea pig was attacked ravenously by a captive.

Colonies of up to 30 individuals of both sexes have been found in hollow trees on Trinidad, and a single individual was obtained in a well-illuminated cave there. In August, 2 lactating females were taken on Trinidad, and in October an immature male, 4 pregnant females, and 1 lactating female were collected (Wilson 1979).

CHIROPTERA; PHYLLOSTOMIDAE; **Genus DIPHYLLA** *Spix, 1823*

Hairy-legged Vampire Bat

The single species, *D. ecaudata*, occurs from southern Texas and eastern Mexico to northern Bolivia and south-central Brazil (Koopman 1988). The only record for Texas (and for any vampire bat anywhere within the United States in modern time) is of a single individual found in 1967 (Greenhall, Schmidt, and Joermann 1984).

Head and body length is 65–93 mm, there is no tail, and forearm length is 50–56 mm. Two females from Brazil weighed 25.5 and 26.7 grams, and two males weighed 24.0 and 24.4 grams (Russell E. Mumford, Purdue University, pers. comm.), but Greenhall, Schmidt, and Joermann (1984) stated that weight ranges up to 43 grams. The coloration is dark brown or reddish brown above and somewhat paler below.

This genus has 26 teeth, in contrast with 22 in *Diaemus* and 20 in *Desmodus*. It also differs from the other true vampire bats in its fan-shaped, seven-lobed outer lower incisor, unique among bats. This tooth resembles the lower incisor in gliding lemurs (Dermoptera). *Diphylla* looks somewhat like *Desmodus* externally but is usually smaller and has shorter and rounder ears, a shorter thumb without the basal pad, a calcar, and longer, softer fur. Another mark of recognition in

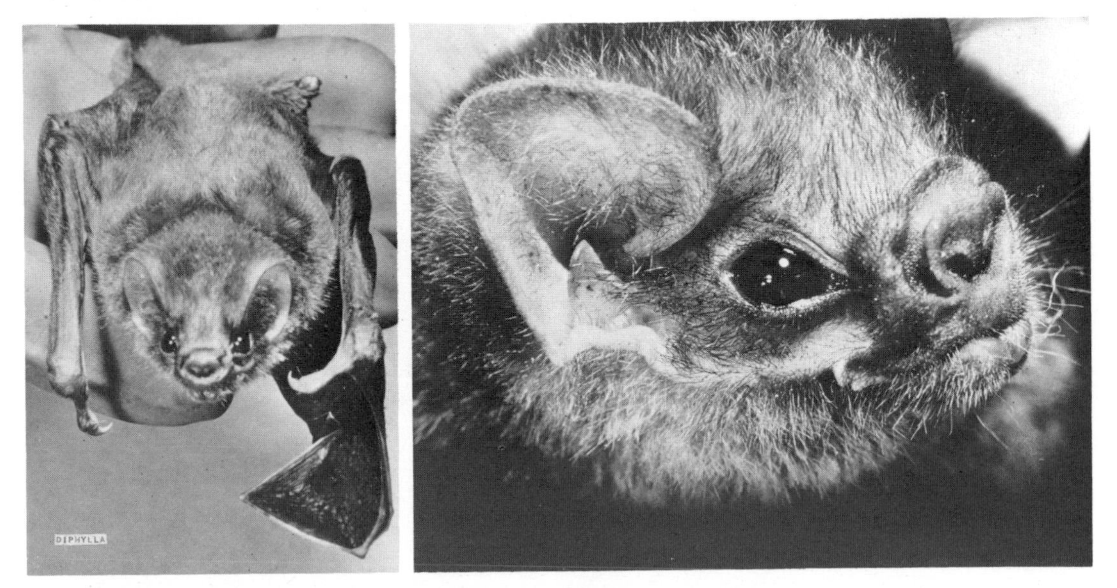

Hairy-legged vampire bat *(Diphylla ecaudata)*, photos by Ernest P. Walker.

Diphylla is the well-haired interfemoral membrane.

In northern Venezuela, Handley (1976) collected specimens in both moist and dry, forested and open areas. Individuals roosted in caves and houses. The hairy-legged vampire is shy and agile, like *Desmodus*, but usually flies to an adjacent perch when disturbed in its roost, instead of scrambling into a crevice. This bat has also been found in a mine tunnel and in hollow trees. It appears to prey mainly on birds, such as chickens. Attacks on horses, burros, and cattle also have been reported, but Greenhall, Schmidt, and Joermann (1984) stated that reliable records indicate that avian blood is the only source of nourishment. *Diphylla* bites the legs and cloacal region of chickens; secondary infections may result in the death of the bird.

This genus is not as gregarious as *Desmodus* and does not cluster in its roosts. Pools of digested blood thus do not form, and *Diphylla* may escape notice. Although as many as 35 have been found in a cave, the usual number is 12 or fewer, and often there are only 1–3. Data summarized by Wilson (1979) indicate that there may be two litters per year. Pregnant or lactating females have been reported in Mexico and Central America in March, May, July, August, October, and November.

CHIROPTERA; **Family MYSTACINIDAE; Genus MYSTACINA**
Gray, 1843

New Zealand Short-tailed Bats

The single known genus, *Mystacina*, contains two species (Flannery 1987*b*; Hill and Daniel 1985):

M. tuberculata, probably once found throughout New Zealand;
M. robusta, probably once found throughout New Zealand.

The systematic position of the Mystacinidae long has been questionable. In recent years this family usually has been placed in the superfamily Vespertilionoidea, together with the Vespertilionidae, Molossidae, and related families. However, on the basis of biochemical analysis of serum proteins, Pierson et al. (1986) showed the Mystacinidae to belong in the superfamily Phyllostomoidea and thus to have affinity to the New World families Mormoopidae, Noctilionidae, and Phyllostomidae.

Until Hill and Daniel's (1985) revision, *M. robusta* usually was treated as a subspecies of *M. tuberculata*. Daniel (1979), however, had suggested that the two might warrant elevation to specific status because of the marked difference in size between them. Adult *M. tuberculata* have a forearm length of about 40–45 mm and weigh about 12–15 grams, whereas adult *M. robusta* have a forearm length of 45–49 mm and weigh 25–35 grams. The pelage is said to be thicker than in any other species of bats. The upper parts are grayish brown to brown and the underparts are paler.

Structurally, these bats are unique: the claws are needle sharp, and the wings are remarkably transformed. The thumb has a large claw with a small talon projecting from it, and the claws of the feet also have talons. The membrane is thick and leathery along the sides of the body, forearm, and lower leg. The wings can be rolled up beneath this leathery membrane when the bat is not flying. The base of the tail membrane is also thick and wrinkled. The obliquely truncate muzzle has a rudimentary nostril pad and a scattering of stiff bristles with spoon-shaped tips. The nostrils are oblong and vertical. The ears are separate, and the tragus is long and pointed. The short, broad feet, the grooved covering of the sole of the foot, and the short, thick legs suggest climbing habits. As in the sac-winged bats (Emballonuridae), the tail perforates the tail membrane and appears on the upper surface.

The teeth are of the normal cuspidate insectivorous type. The upper premolars are well developed. The dental formula is: (i 1/1, c 1/1, pm 2/2, m 3/3) × 2 = 28.

These bats are remarkably agile, running freely on the ground and even running or climbing rapidly up sloping, smooth surfaces. This agility results from the constriction of the wing membrane; peculiar folding processes along the forearm allow the wing to be used as a normal forelimb in walking and climbing. The basal talons on the claws of the

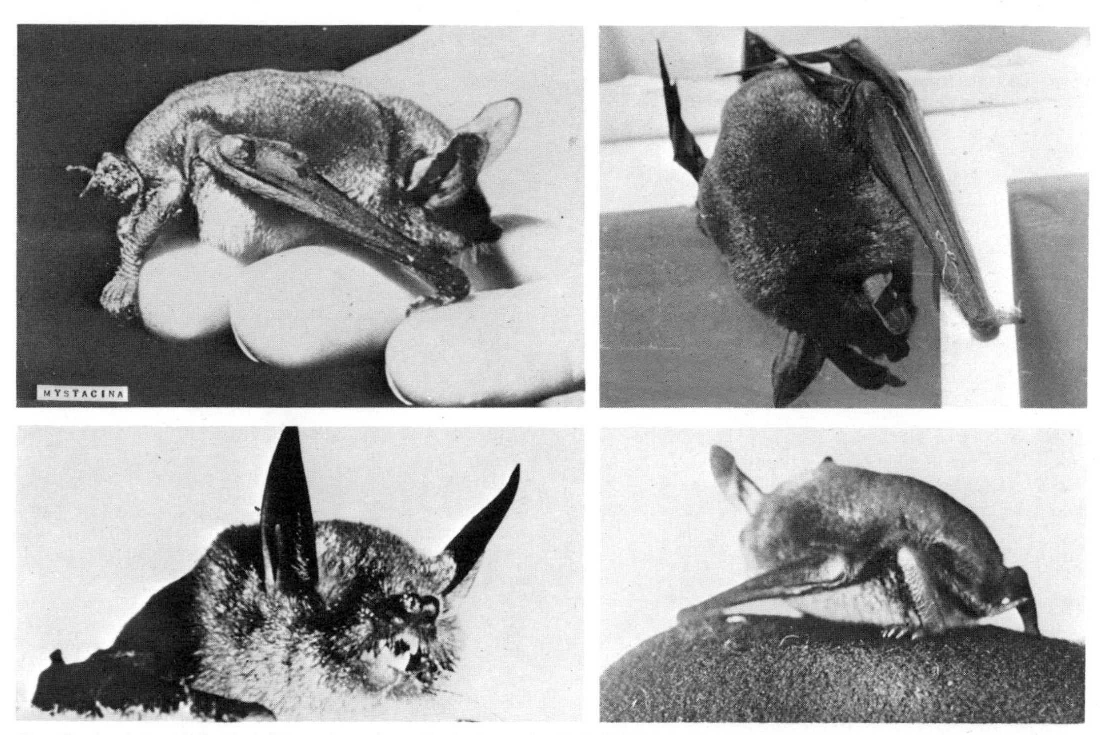

New Zealand short-tailed bat *(Mystacina tuberculata)*, photos by H. P. Collins.

thumbs and toes may be further adaptations for terrestrial, arboreal, and burrowing behavior (Daniel 1979).

Mystacina is found in forests, and roosts in hollow trees, caves, crevices, and burrows. Groups of bats apparently use their teeth to burrow roosting cavities and tunnels through the wood of trees (Daniel 1979). Emergence is relatively late in the evening. Available evidence indicates that *Mystacina* does not hibernate for a prolonged period, as does the other New Zealand bat, *Chalinolobus*, but arouses spontaneously to feed when winter weather becomes relatively mild (Daniel 1979). The diet of *Mystacina* is now known to be surprisingly broad, consisting of flying and resting arthropods, fruit, nectar, and pollen (Daniel 1976, 1979).

The first colony of *Mystacina* to be studied in detail comprised about 500 individuals of *M. tuberculata* roosting in a giant hollow kauri tree (Daniel 1979). Other observations suggest that some groups are considerably smaller. The investigated colony included about 100–150 newborn young on 16 December 1973. Apparently, *M. tuberculata* is mon-

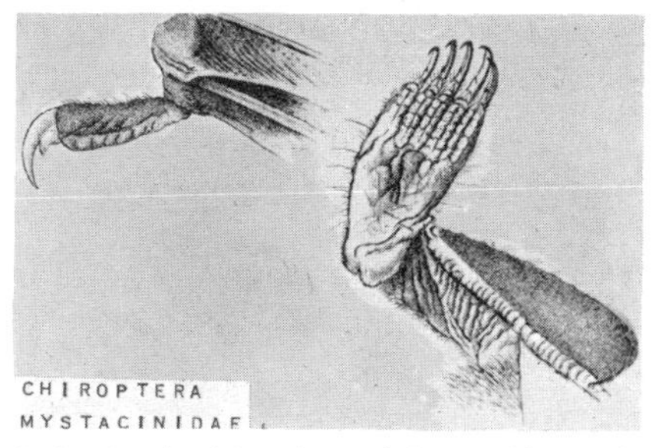

Thumb showing denticle at base of claw, and foot and part of inner surface of membrane of the New Zealand short-tailed bat *(Mystacina tuberculata)*, photo from *Proc. Zool. Soc. London.*

New Zealand short-tailed bat *(Mystacina tuberculata)*, photo by Entomology Division, D.S.I.R. (New Zealand).

estrous, with mating occurring in autumn and there being some kind of delay of development, fertilization, or implantation. The single young is born in summer (December–January). Limited observations of *M. robusta* suggest that it may be polyestrous, with births occurring from spring to autumn.

Although both species are thought once to have occurred throughout much of New Zealand, there are relatively few precise records (Daniel and Williams 1984; Flannery 1987*b*; Hill and Daniel 1985). *M. tuberculata* still is known to be present on North Island, South Island, Little Barrier Island off the north coast of North Island, and several small islands off of Stewart Island. It has been reduced, however, to perhaps 10 sporadic populations that together probably contain only a few thousand individuals and are probably in danger of extermination. From actual collection and observation of living specimens since European settlement, *M. robusta* is known only from one locality near the northern tip of South Island and from Big South Cape Island and Solomon Island, which are off Stewart Island. However, subfossil remains, which may represent populations from well within historical time, show that *M. robusta* was present at least as far north as central North Island. In any case, modern observations of this species have been restricted to Big South Cape and Solomon Islands, and it has not been seen there since 1965. It already may be extinct. Both species are thought to have declined through destruction of forest habitat, predation by introduced rats and other exotic mammals, accidental poisoning, and human disturbance of roosts. The IUCN regards *M. tuberculata* as vulnerable and *M. robusta* as extinct.

CHIROPTERA; **Family NATALIDAE; Genus NATALUS** *Gray, 1838*

Funnel-eared Bats

The single living genus, *Natalus*, contains two subgenera and five species (Cabrera 1957; Hall 1981; Ottenwalder and Genoways 1982; Williams, Genoways, and Groen 1983):

subgenus *Natalus* Gray, 1838

N. stramineus, northern Mexico to Brazil, Greater and Lesser Antilles;
N. tumidirostris, Colombia, Venezuela, Surinam, Curacao, Trinidad;
N. tumidifrons, Bahama Islands;
N. micropus, Cuba and nearby Isle of Pines, Jamaica, Hispaniola, Old Providence Island off east coast of Nicaragua;

subgenus *Nyctiellus* Gervais, 1855

N. lepidus, Bahamas, Cuba and nearby Isle of Pines.

Corbet and Hill (1986) included *N. tumidifrons* in *N. micropus*, putting the latter in a separate subgenus, *Chilonatalus* Miller, 1898.

Head and body length is 35–55 mm, tail length is approximately 50–60 mm, and forearm length is 27–41 mm. Adults usually weigh 4–10 grams. The fur is soft and long. The coloration is gray, buff, yellowish, reddish, or deep chestnut. *N. stramineus mexicanus* has two color phases—a pale phase, in which the upper parts are buffy to pinkish cinnamon, and a dark phase, in which the upper parts are rich yellowish or reddish brown; the underparts are paler in both phases.

These are slim-bodied bats, with long and slender wings, legs, and tail. The ears are large, separate, and funnel-shaped, the surface of the external ear being studded with glandular papillae as in the Old World vespertilionid genus *Kerivoula*. The tragus is short, variously thickened, and more or less triangular in shape. The small eyes are not prominent. The top of the head is considerably elevated above the concave forehead. Adult males have a large structure on the face or muzzle, the "natalid organ," which is composed of cells that show resemblance to sensory cells, but they may also have a glandular function. As far as is known, this structure is confined to the Natalidae; it is not always noticeable externally. The muzzle is elongate and lacks a nose leaf. The nostrils are oval, set close together, and open near the margin of the lip.

Funnel-eared bat *(Natalus stramineus)*, photo by Harold Drysdale, Trinidad Regional Virus Laboratory, through A. M. Greenhall.

Funnel-eared bat *(Natalus stramineus)*, photo by Merlin D. Tuttle, Bat Conservation International.

The lower lip is broad, reflected outward in front, and often has transverse grooves. The thumb is short, bound to the wing by a membrane at its base, and bears a well-developed claw. The very slender tail is completely enclosed within the large tail membrane. The membrane is pointed at its free edge at the tip of the tail. The normal-sized feet bear long, slightly recurved claws.

The upper incisors are separated from each other and from the canines. The first and second premolars are well developed in both jaws. The dental formula is: (i 2/3, c 1/1, pm 3/3, m 3/3) × 2 = 38.

In Venezuela, Handley (1976) collected *N. stramineus* primarily in lowland forest. In Jalisco, Mexico, Watkins, Jones, and Genoways (1972) found this species from sea level to about 2,500 meters. In general, bats of this genus roost in the darkest recesses of caves and mine tunnels, often with other kinds of bats. *N. tumidirostris* has been found in a hollow rubber tree. Funnel-eared bats were discovered in a dormant condition in a cave in oak forest near Ciudad Victoria, Tamaulipas, Mexico, by William J. Schaldach, Jr. They were hanging singly. The temperature on this date in January was about 12° C outside the cave and had been fairly low for about three days prior to this date. The fluttering, almost mothlike flight of funnel-eared bats is distinctive. They feed on insects.

These bats may be found in large groups, but sometimes fewer than a dozen individuals are present. Kerridge and Baker (1978) reported finding as many as several hundred *N. micropus* hanging in loose clusters from ledges in a deep cave on Jamaica. The sexes of *N. stramineus* appear to segregate when the young are born. In *N. lepidus* from Cuba and some smaller adjacent islands, there seems to be only partial segregation of the sexes, at least in July. Baker, Genoways, and Patton (1978) reported a pregnant female *N. stramineus* on Guadeloupe in August. Pregnant female *N. stramineus* also have been found in January, April, May, and June. This species breeds during the dry season in El Salvador and southern Mexico. There is one young per female.

The geological range of this family is Pleistocene to Recent in South America, Jamaica, and Cuba and Recent over the remainder of the present range.

CHIROPTERA; **Family FURIPTERIDAE**

Smoky Bats, or Thumbless Bats

This family of two genera and two species inhabits southern Central America and tropical South America. There has been controversy regarding the systematic position of this family, though generally the Furipteridae are considered to be closely related to the Natalidae, Thyropteridae, and Myzopodidae and to belong in the superfamily Vespertilionoidea (Ibáñez 1985; Koopman 1984*b*). Fossils referable to this family have not been found.

Smoky bats are small: head and body length is 33–58 mm, tail length is 24–36 mm, and forearm length is 30–40 mm. They are delicate and in general appearance resemble the funnel-eared bats (Natalidae) and the disk-winged bats (Thyropteridae). The thumb is present but is so small that it appears to be absent; it is included within the wing membrane to the base of the small, functionless claw. The wings are relatively long. The crown of the head is greatly elevated above the face line, and the snout is cut off or truncated, with the end in the form of a disk or pad. The ears are separate and funnel-shaped, as in the Natalidae and Thyropteridae. Their broad bases cover the eyes. The tragus is small, with a broad base, and somewhat triangular in shape. The nostrils are oval or triangular, are set close together, and open downward. There is no nose leaf. The tail is relatively long, but it does not extend to the free edge of the tail membrane and does not perforate it or appear on its upper surface. The legs are long and the feet are short, the latter ending in long, recurved claws. The fur tends to be coarse. Females have one pair of abdominal mammae (Ibáñez 1985). The dental formula is: (i 2/3, c 1/1, pm 2/3, m 3/3) × 2 = 36.

CHIROPTERA; FURIPTERIDAE; **Genus FURIPTERUS**
Bonaparte, 1837

The single species, *F. horrens*, occurs from Costa Rica to Peru and Brazil and on Trinidad (Cabrera 1957; Hall 1981; Tuttle 1970).

Head and body length is about 33–40 mm, tail length is 24–36 mm, forearm length is 30–40 mm, and weight is about 3 grams. Uieda, Sazima, and Storti Filho (1980) found females to be significantly larger than males. The coloration is brownish gray, dark gray, or slaty blue above and somewhat paler below. The fur on the head is long and thick and covers all the head as far as the snout, almost concealing the mouth. Both surfaces of the tail membrane are haired, and the tail is short.

This genus and *Amorphochilus* are distinguished by the reduced thumb, which is included in the wing membrane to the base of the small, functionless claw. Nasal appendages are lacking; the muzzle and lips do not have conspicuous warty outgrowths; and the tail terminates short of the posterior

Smoky bat *(Furipterus horrens):* Top, photo from *Proc. Zool. Soc. London;* Bottom, photo by Richard K. LaVal.

border of the interfemoral membrane, not perforating it or appearing on its upper surface. The snout is piglike in form and turned upward slightly. The tragus resembles a barbed arrowhead in form.

In Panama several specimens were recently collected in a cave; they were hanging in a large, domed, well-lit chamber in company with several *Mimon bennettii* (Charles O. Handley, Jr., U.S. National Museum of Natural History, pers. comm.). In Venezuela, Handley (1976) collected 6 specimens near streams or in other moist areas, 2 of them in evergreen forest and 4 in yards. In Costa Rica, LaVal (1977) found a colony of over 59 individuals, all males, on 12 May. The bats were hanging in clusters from the top of a hollow log in primary forest. Some were also observed to forage at heights of 1–5 meters above the forest floor; flight was slow and mothlike. Whenever roosting bats were approached they were alert and took flight immediately.

Uieda, Sazima, and Storti Filho (1980) observed two colonies in caves in northeastern Brazil during January and February. The bats left their caves to forage when darkness was complete, and examination of digestive tracts showed the main prey to be Lepidoptera. One colony contained about 250 individuals divided into groups of 4–30 roosting in holes in the walls. In the other cave there were 150 bats roosting separately from each other. The latter colony was composed of adult males, adult females, and young. Although the females hung in the usual head-down position, the young positioned themselves head-up on the mother's body, perhaps in response to the unusual abdominal location of the mammae in the Furipteridae.

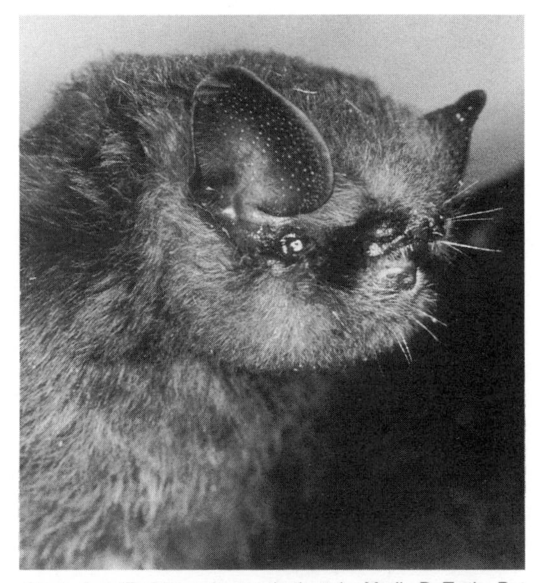

Smoky bat *(Furipterus horrens)*, photo by Merlin D. Tuttle, Bat Conservation International.

CHIROPTERA; FURIPTERIDAE; **Genus AMORPHOCHILUS**
Peters, 1877

The single species, *A. schnablii*, is known from Puna Island off the coast of Ecuador, several other points along the coast of Ecuador and Peru, an inland area in northern Peru, and northern Chile (Ibáñez 1985; Koopman 1978*a*). The type specimen is in the Berlin Museum. These bats are also represented in the collections of the Field Museum of Natural History, the American Museum of Natural History, the U.S. National Museum of Natural History, the Museum of Comparative Zoology at Harvard University, and the Lima Museum in Peru.

Head and body length is 38–58 mm, tail length is approximately 30 mm, and forearm length is 34–38 mm. The color is dark brown or slaty gray. The reduced thumb is included within the wing membrane to the base of the small, functionless claw. Nasal appendages are lacking, and the tail terminates short of the posterior border of the interfemoral membrane, not perforating it or appearing on its upper surface. This genus differs from *Furipterus* in the height of the brain-

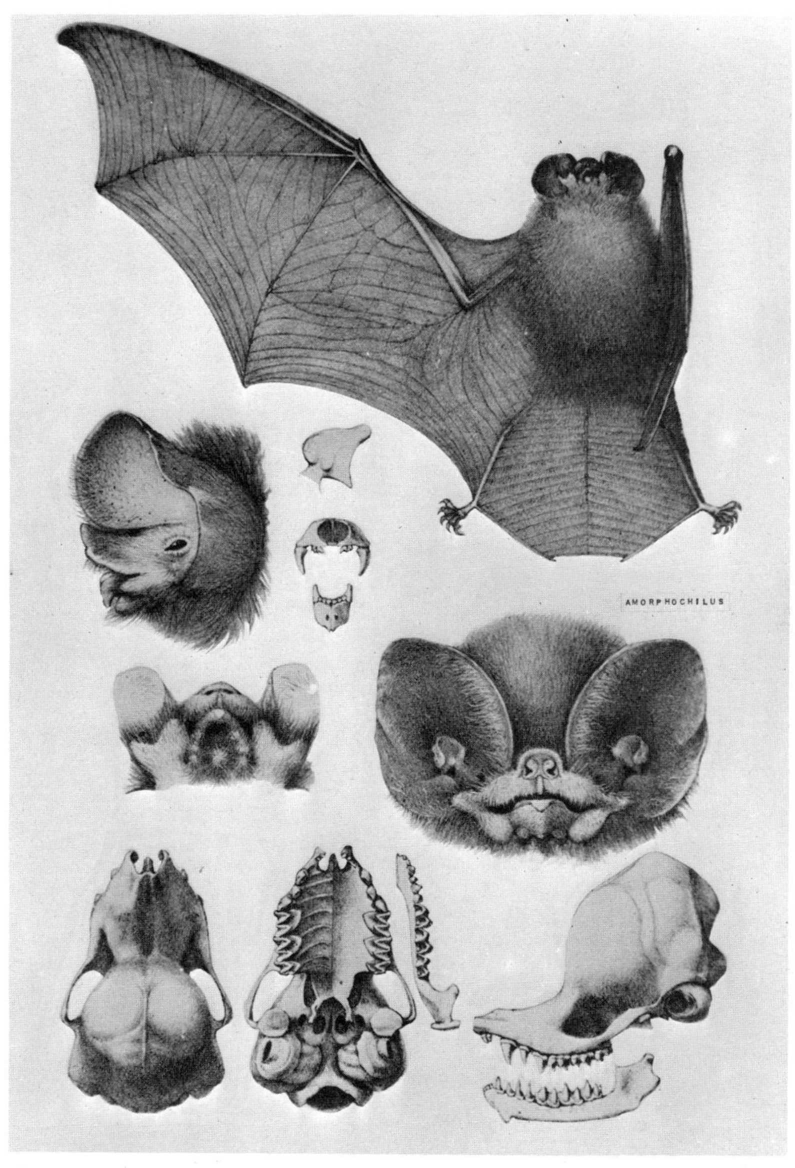

Smoky bat *(Amorphochilus schnablii)*, photo from *Die Fledermäuse des Berliner Museums für Naturkunde*, Wilhelm K. Peters.

case and in the presence of conspicuous wartlike outgrowths on the muzzle and lips.

Amorphochilus seems to occur mainly in cultivated valleys and arid regions within its range. It has been collected in an unused sugar mill, dark wine storehouses, and an irrigation tunnel; two individuals have been taken on separate occasions at 0300 hours. Ibáñez (1985) reported the discovery of a colony in a culvert at a road crossing within an area of tropical desert bush at the mouth of the Javita River, Ecuador. Analysis of the stomach contents of 5 specimens showed Lepidoptera to be the only prey. When the colony was found, on 19 November 1981, it contained 300 individuals and included both sexes, but no juveniles. None of the females were lactating, but 8 of the 10 examined contained a single fetus. On the Pacific slope of Peru, Graham (1987) collected juveniles in April and pregnant females in June, August, and October.

CHIROPTERA; **Family THYROPTERIDAE; Genus THYROPTERA**
Spix, 1823

Disk-winged Bats, or New World Sucker-footed Bats

The single known genus, *Thyroptera*, contains two species (Anderson, Koopman, and Creighton 1982; Cabrera 1957; Hall 1981; Jones, Arroyo-Cabrales, and Owen 1988; Wilson 1976):

T. discifera, southern Nicaragua to the Guianas and Peru;
T. tricolor, southern Mexico to Bolivia and southern Brazil, Trinidad.

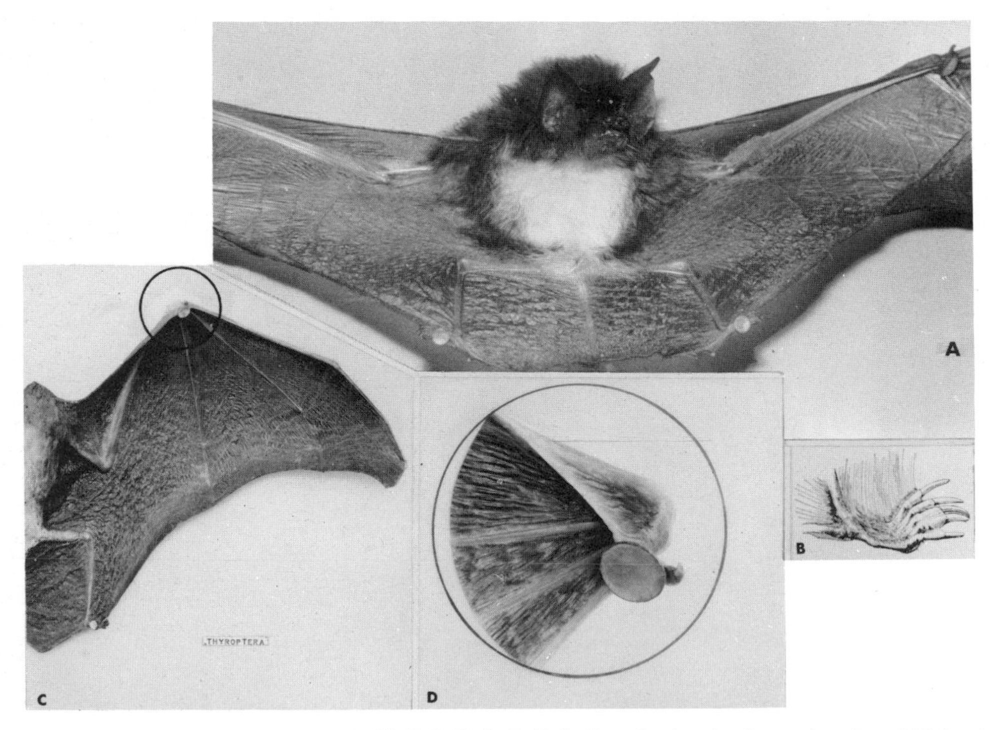

A. Disk-winged bat *(Thyroptera tricolor)*, photo by Merlin D. Tuttle. B. *T. discifera*, drawing showing syndactylism of third and fourth digits of foot, from *Proc. Biol. Soc. Washington.* C. Wing and foot showing position of thumb and disk, photo from Field Museum of Natural History. D. Enlargement of thumb disk.

Head and body length is 34–52 mm, tail length is 25–33 mm, and forearm length is 27–38 mm. Findley and Wilson (1974) reported an average weight of 4.2 grams for *T. tricolor*. This species is reddish brown or somewhat darker above, and white below, the lateral extent of the white area being variable. *T. discifera* is reddish brown above and brown below. The ears in *T. tricolor* are blackish, whereas in *T. discifera* they are yellowish and larger.

These bats have circular suction disks or cups with short stalks on the soles of the feet and at the base of the well-developed claw of the thumb. In their Old World counterpart, *Myzopoda*, of Madagascar, the suction disks are sessile and the claw on the thumb is vestigial. These concave disks are larger on the thumbs than on the feet. American sucker-footed bats are small and delicately formed. The crown of the head is considerably elevated above the concave forehead. The ears are separate and funnel-shaped, and a tragus is present. The muzzle is elongate and slender. There is a small wartlike projection above the nostrils, but there is no nose leaf. The nostrils are circular and set rather far apart. The tip of the tail extends slightly beyond the free, or rear, edge of the broad tail membrane. As in the smoky bats (Furipteridae), the third and fourth toes are joined.

The upper incisors on each side are widely spaced from each other and from the canines. The premolars are well developed in both jaws, and the molars exhibit a **W** pattern of cusps and ridges. The dental formula is: (i 2/3, c 1/1, pm 3/3, m 3/3) × 2 = 38.

In Venezuela, Handley (1976) collected most specimens in moist parts of evergreen forest. Unlike most other bats, the members of this family usually hang head upward. Disk-winged bats cling to smooth surfaces by means of the suction disks, the suction of a single disk being capable of supporting the entire weight of the bat. The usual daytime retreat is in a rolled frond or a curled leaf of some plant, such as the heliconia tree or the banana. As many as eight individuals have been found inside a curled leaf, but usually only one or two disk-winged bats rest in a given shelter. When several utilize a trumpet-shaped leaf as a roost, they space themselves evenly one above the other. Two *Thyroptera* have been found with *Rhynchonycteris*, the proboscis bat, in a curled heliconia leaf. The diet consists of insects.

In a study in Costa Rica, Findley and Wilson (1974) found the home range of *T. tricolor* to average about 3,028 sq meters, and density to average 3.3 colonies, or 19.8 bats, per hectare. Group size ranged from 1 to 9 and averaged 6. Each roosting aggregation was found to have social cohesion and would rejoin. In mid-August 1963 it was estimated that 60–80 percent of the adult females were pregnant. A lactating female *T. tricolor* carrying a young male was observed in May on Barro Colorado Island, Panama. The young, which weighed approximately 46 percent as much as its mother, remained attached to her when she flew around a room. LaVal and Fitch (1977) reported a juvenile in Costa Rica in June. In eastern Peru, Graham (1987) collected juveniles in March and October, pregnant females in July, and lactating females in December.

No fossils referable to this family have been found.

CHIROPTERA; **Family MYZOPODIDAE; Genus MYZOPODA**
Milne-Edwards and Grandidier, 1878

Old World Sucker-footed Bat

The single known genus and species, *Myzopoda aurita*, now is restricted to Madagascar. Known locality records are Majunga on the west coast and Maroantsera, Mananara, Ma-

Disk-winged bat *(Thyroptera tricolor),* photo by Merlin D. Tuttle, Bat Conservation International.

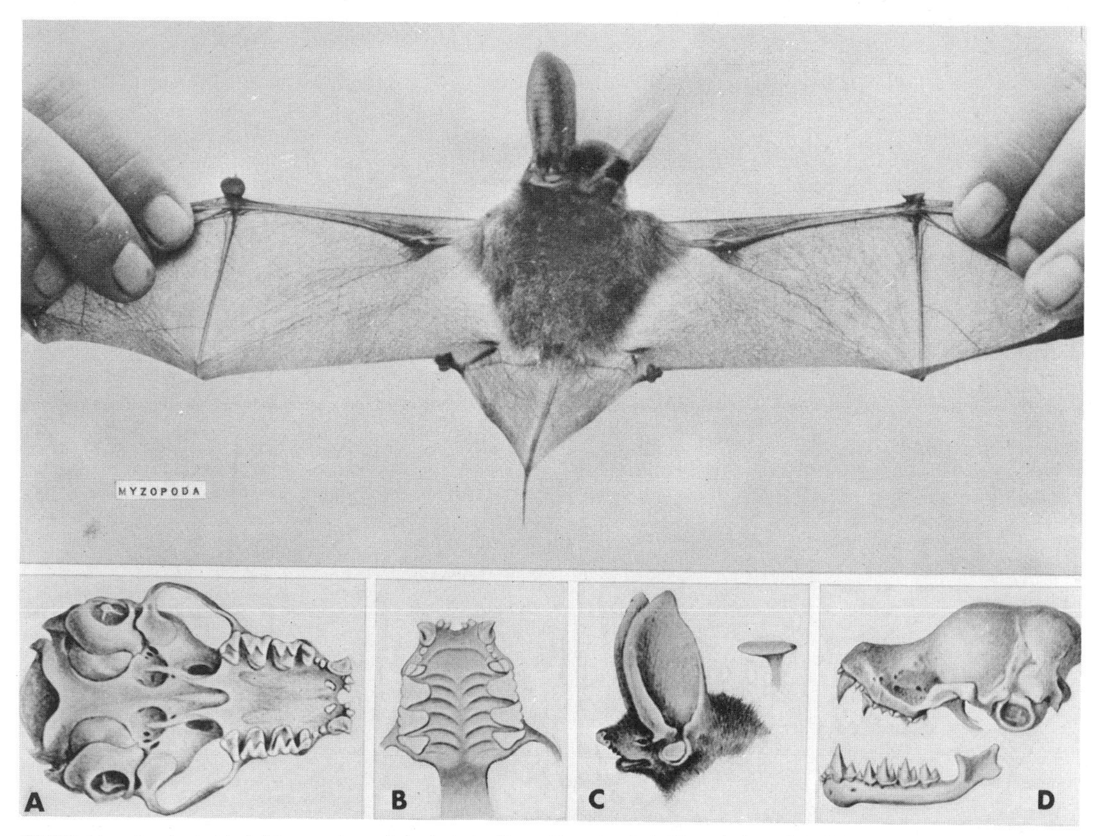

Old World sucker-footed bat *(Myzopoda aurita),* photo by Harry Hoogstraal. A. Ventral view of skull; B. Palate; C. Head and thumb disk; D. Lateral view of skull, photos from *Proc. Zool. Soc. London.*

hambo, Tamatave, Mananjary, and the northern vicinity of Fort Dauphin on the east coast (Schliemann and Maas 1978). *Myzopoda* is the only genus of bats now restricted to Madagascar, though in the Pleistocene it also occurred in East Africa (Koopman 1984*c*). Although both *Myzopoda* and *Thyroptera* possess suction disks, these organs are completely different in the two genera with regard to histological and anatomical details and probably evolved independently (Schliemann and Maas 1978).

Approximate measurements are as follow: head and body length, 57 mm; tail length, 48 mm; and forearm length, 46–50 mm. This bat may be recognized by the sessile adhesive pads or disks on the wrists and ankles, whereas the suction disks in *Thyroptera* are provided with a short stem or stalk. In addition, *Myzopoda* has a large ear with a tragus and a unique mushroom-shaped process. The lips are wide, and the upper lip extends beyond the lower. The thumb has a vestigial claw, and the tail extends beyond the tail membrane.

The skull is short, broad, and rounded. The teeth are of the normal cuspidate insectivorous type. The dental formula is: (i 2/3, c 1/1, pm 3/3, m 3/3) × 2 = 38.

The suction disks do not appear to be as efficient as those of *Thyroptera*, but the bats probably use the pads to hold on to the smooth hard stems and leaves of palms and other smooth surfaces.

CHIROPTERA; **Family VESPERTILIONIDAE**

Vespertilionid Bats

This family of 42 genera and 355 species has a worldwide distribution in the temperate and tropical regions. These bats range in altitude to the limits of tree growth and from tropical forests to arid areas. *Myotis*, a member of this family, has the widest distribution of any genus of bats. Koopman (1984*b*) listed the following five subfamilies: Kerivoulinae, with the genus *Kerivoula;* Vespertilioninae, with most genera; Murininae, with the genera *Murina* and *Harpiocephalus;* Miniopterinae, with the genus *Miniopterus;* and Tomopeatinae, with the genus *Tomopeas.* Both Gopalakrishna and Karim (1980) and Mein and Tupinier (1977) recommended that the Murininae be elevated to familial rank.

Head and body length is 32–105 mm, tail length is 25–75 mm, and forearm length is 22–75 mm. The well-developed tail extends to the rear edge of the tail membrane or slightly beyond. Adults weigh from about 4 to 50 grams. The coloration is generally blackish, gray, or various shades of brown, but a number of species are red, yellow, or orange. An ornate color pattern is present in *Scotomanes ornatus*, and the spotted bat, *Euderma*, has a characteristically arranged pattern of white spots. A nose leaf is lacking, except in the genera *Nyctophilus* and *Pharotis.* Tubular nostrils are present in two genera, *Murina* and *Harpiocephalus.* In some members of this family, large glands are present underneath the skin on the snout, producing large bulges or folds of the skin covering them. Fleshy lobes are associated with the mouth in *Chalinolobus* and *Glauconycteris.* The eyes are minute. The ears are as much as 40 mm in length, usually separate, and have a tragus and an anterior basal lobe (except in *Tomopeas*). Small suction disks or pads are present on the soles and/or wrists in *Glischropus, Eudiscopus*, and *Tylonycteris*, as well as in some species of *Pipistrellus* and *Hesperoptenus.* Wing glands are present in *Myotis vivesi* and *Cistugo.* Most females have two mammae, except for those of the genus *Lasiurus*, which have four.

The incisors are small and separated medially. The molars have a well-developed W-shaped pattern of cusps and ridges, a few genera showing a tendency toward the reduction of the cusps. The dental formula varies from (i 1/2, c 1/1, pm 1/2, m 3/3) × 2 = 28 to (i 2/3, c 1/1, pm 3/3, m 3/3) × 2 = 38.

Most vespertilionids are cave dwellers, but they also shelter in mine shafts, tunnels, old wells, rock crevices, buildings, tree hollows, the foliage of trees and bushes, hollow joints of bamboo, large tropical flowers, tall grass, abandoned bird nests, storm sewers, and culverts, as well as under rocks and loose bark on tree trunks. The members of this family generally pitch head upward when alighting, hanging by their thumbs and toes, then quickly shuffle around until they are hanging head downward by their toes, their wings folded alongside the body. Often they hang on a vertical surface rather than suspend themselves freely. Some members of this family are solitary, others roost in pairs or in small groups, and still others generally shelter in colonies. The colonial forms usually return to the same roosting site year after year. Some species remain in colonies throughout the year, whereas others congregate only in winter. A number of forms migrate between summer and winter quarters. Some of the temperate region forms with tree-dwelling habits migrate to warmer climates for the winter, whereas those having cave-dwelling habits hibernate, though they often change hibernating places during the winter. Homing ability has been demonstrated in several forms. Vespertilionids send out ultrasonic sounds through their mouths.

Nearly all members of this family feed mainly on insects. *Myotis vivesi* is definitely known to feed on fish, and two other species of *Myotis* are suspected to do so. *Antrozous pallidus* in captivity has captured and eaten lizards. Vespertilionids generally capture insects in flight, often using the well-developed tail membrane as a pouch in which to manipulate the larger kinds of prey. A fairly definite feeding territory is often established. The feeding flights generally alternate with periods of rest, when the bat hangs up to digest its catch.

Females of many colonial species segregate into maternity colonies to bear and raise the young. Males do not exhibit an active interest in the young. In those forms from temperate climates that hibernate, breeding occurs from August through October and often again in the spring, the two peri-

Spotted bat *(Euderma maculatum)*, with 24-hour-old baby, photo by David A. Easterla.

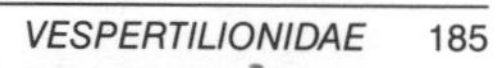

Myotis lucifugus catching a moth. Inset shows bat with the moth caught in its interfemoral membrane. Photos by Frederick A. Webster and D. A. Cahlander.

Long-eared bats of eastern North America *(Plecotus rafinesquii)* hanging in their normal position for resting: A. With ears partially uncoiled from the tight coil when sleeping; B. With ears fully extended when awake. Photos by Ernest P. Walker.

ods producing only a single litter, since the sperm from the former period are stored in the reproductive tract of the female over the winter and ovulation and fertilization occur only in the spring. *Lasiurus*, a genus that occurs in temperate regions and is thought to be migratory, breeds mainly in late summer and early fall, the births in northern latitudes taking place from late May to early July. In those forms that occur in tropical regions, breeding is immediately followed by fertilization and the development of the embryos. The gestation period in most forms is 40–70 days, but in some species it is 100 or more days. The number of young is 1–4. The life span in the wild may be no more than 4–8 years for most individuals, but there are records of banded individuals more than 21 years old, and captives are known to have lived more than 20 years.

The geological range of this family is middle Eocene to Recent in Europe, late Oligocene to Recent in North America, middle Miocene to Recent in Africa and Asia, Pleistocene to Recent in the West Indies, South America, and Australasia, and Recent over the remainder of the present range (Koopman 1984*c*).

CHIROPTERA; VESPERTILIONIDAE; **Genus MYOTIS**
Kaup, 1829

Little Brown Bats

There are 3 subgenera and 97 species (Anderson, Koopman, and Creighton 1982; Anderson and Webster 1983; Baud 1979; Blood and McFarlane 1988; Bogan 1978; Cabrera 1957; Carter et al. 1981; Chasen 1940; Corbet 1978; DeBlase 1980; Dolan and Carter 1979; Ellerman and Morrison-Scott 1966; Findley 1972; Gaisler 1983; Gauckler and Kraus 1970; Genoways and Williams 1979*a*; Hall 1981; Hayman and Hill, *in* Meester and Setzer 1977; Hill 1962, 1972*b*, 1974*b*, 1983*b*; Hill and Beckon 1978; Hill and Francis 1984; Hill, Harrison, and Jones 1988; Hill and Morris 1971; Hill and Thonglongya 1972; Hill and Topal 1973; Hoffmann, Jones, and Campbell 1987; Horacek and Hanák 1984; Koopman 1982*a*, 1984*a*; Laurie and Hill 1954; LaVal 1973*a*; Lekagul and McNeely 1977; McCarthy and Bitar 1983; Myers and Wetzel 1983; Neuhauser and DeBlase 1974; Qumsiyeh 1980, 1983; Reduker, Yates, and Greenbaum 1983; Ride 1970; Roberts 1977; Rzebik-Kowalska, Woloszyn, and Nadachowski 1978; Taylor 1934; Tupinier 1977; Van Zyll de Jong 1984; Yoshiyuki 1984):

subgenus *Myotis* Kaup, 1829

M. bechsteini, Europe, Iran;
M. myotis, central and southern Europe, Asia Minor, Lebanon, Palestine;
M. blythi, southern Europe, northern Africa, southwestern Asia, parts of central and eastern Asia;
M. chinensis, southern China, northern Thailand;
M. sicarius, Sikkim, India;
M. formosus, eastern Afghanistan, northern India to Korea, southern Japan, Taiwan, Philippines;
M. welwitschii, Zaire and Ethiopia to South Africa;
M. nattereri, Europe, Morocco, Algeria, Caucasus, Kurdistan (northern Iraq), Palestine, Jordan, Iran, Turkmen S.S.R.;

M. schaubi, Armenia, western Iran;
M. bombinus, southeastern Siberia, Manchuria, Korea, Japan;
M. pequinius, eastern China;
M. altarium, south-central China, northern Thailand;
M. emarginatus, southern Europe to Pakistan, Morocco;
M. tricolor, Zaire and Ethiopia to South Africa;
M. morrisi, Nigeria, Ethiopia;
M. goudoti, Madagascar, Comoro Islands;
M. thysanodes, western North America from southern British Columbia to southern Mexico;
M. evotis, southwestern Canada, western conterminous United States, Baja California;
M. milleri, Sierra San Pedro Martir in northern Baja California;
M. auriculus, southwestern United States to central Mexico, Guatemala;
M. keenii, southeastern Alaska, western British Columbia and Washington, Saskatchewan and Newfoundland to northwestern Florida;

subgenus *Selysius* Bonaparte, 1841

M. mystacinus, entire Palaearctic region from Ireland and Morocco to Japan and the Himalayas;
M. davidii, eastern China;
M. brandti, Europe;
M. abei, Sakhalin Island;
M. frater, Tadzhik S.S.R. to Manchuria and southeastern China, Japan;
M. insularum, supposedly Samoa;
M. muricola, Afghanistan and northern India to southern China and Malay Peninsula, islands from Sumatra to the Moluccas and possibly New Guinea;
M. ater, Siberut Island off western Sumatra, Borneo, Culion Islands (Philippines), Sulawesi, Molucca Islands;
M. australis, New South Wales, possibly northern Western Australia;
M. ikonnikovi, eastern Siberia, Mongolia, Manchuria, Korea, Sakhalin, Hokkaido;
M. hosonoi, Honshu;
M. yesoensis, Hokkaido;
M. ozensis, Honshu;
M. browni, Mindanao;
M. patriciae, Mindanao;
M. herrei, Luzon;
M. primula, northeastern India;
M. siligorensis, northern India to southern China and Malay Peninsula;
M. annectans, Bangladesh to Thailand;
M. rosseti, Thailand, Cambodia;
M. ridleyi, Malay Peninsula, probably Sumatra, Borneo;
M. ciliolabrum, southwestern Canada to western Oklahoma and central Mexico;
M. californicus, western North America from extreme southeastern Alaska to Guatemala;
M. leibii, southeastern Canada to eastern Oklahoma and Georgia;
M. scotti, highlands of Ethiopia;
M. sodalis (Indiana bat), eastern United States;
M. nigricans, east and west coasts of central Mexico to northern Argentina, Trinidad;

Little brown bat *(Myotis)*: A. *M. nattereri*, photo by Frank W. Lane; B. *M. evotis*, photo by Ernest P. Walker; C. *M. lucifugus*, photo by Ernest P. Walker.

M. carteri, western Mexico;
M. findleyi, Tres Marias Islands off Nayarit (western Mexico);
M. elegans, southern Mexico to Costa Rica;
M. dominicensis, Dominica in the Lesser Antilles;

subgenus *Leuconoe* Boie, 1830

M. horsfieldi, India, South Andaman Island, Thailand, Malay Peninsula, Hainan, Sumatra, Java, Borneo, Zamboanga Island (Philippines), Sulawesi;
M. adversus, Malay Peninsula, islands from Sumatra to the New Hebrides, Australia;
M. hasseltii, Sri Lanka, Burma, Thailand, Cambodia, Malay Peninsula, Rhio Archipelago, Sumatra, Java, Borneo;
M. bartelsi, Java;
M. hermani, Sumatra;
M. oreias, Singapore;
M. weberi, southern Sulawesi;
M. jeannei, Mindanao;
M. macropus, Kashmir;
M. bocagei, Liberia to Malawi, South Yemen;
M. montivagus, India, Burma, eastern China, Malay Peninsula, Borneo;
M. riparius, western Honduras to Uruguay, Trinidad;
M. fortidens, western and southern Mexico, Guatemala;
M. lucifugus, Alaska, Canada, conterminous United States, northern Mexico (a single record from Iceland probably represents accidental introduction by human agency [Koopman and Gudmundsson 1966]);
M. velifer, southwestern United States to Honduras;
M. grisescens (gray bat), eastern Kansas and Oklahoma to western Virginia and northwestern Florida;
M. chiloensis, Chile;
M. aelleni, northern Patagonia (Argentina);
M. nesopolus, northern Venezuela and nearby islands;
M. ruber, Brazil, Paraguay, northern Argentina;
M. peninsularis, southern Baja California;
M. cobanensis, known only from Coban in Guatemala;
M. daubentoni, Ireland to Japan;
M. nathalinae, France, Spain;
M. capaccinii, southern Europe, Palestine to Uzbek S.S.R. and southeastern China, northwestern Africa;
M. macrodactylus, Maritime Provinces in southeastern Siberia, Japan;
M. taiwanensis, Taiwan;
M. yumanensis, western North America from British Columbia to central Mexico;
M. austroriparius, southeastern United States;
M. albescens, southern Mexico to Uruguay;
M. keaysi, eastern and southern Mexico to Venezuela and southern Bolivia, Trinidad;
M. atacamensis, coastal desert of southern Peru and northern Chile;
M. simus, Amazonian region of Colombia, Ecuador, Peru, Bolivia, Paraguay, and Brazil;
M. martiniquensis, Martinique and Barbados in Lesser Antilles;
M. levis, southern Brazil, Bolivia, Paraguay, Uruguay, Argentina;
M. oxyotus, Costa Rica to Bolivia;
M. volans, western North America from southeastern Alaska to central Mexico;
M. petax, Altai region of Siberia;
M. planiceps, northeastern Mexico;
M. pruinosus, northern Honshu;
M. macrotarsus, Borneo, Philippines;
M. stalkeri, Kei Islands west of New Guinea;
M. vivesi (fishing bat), coastal parts of western Mexico and Baja California;
M. ricketti, eastern China;
M. dasycneme, eastern France to western Siberia, Manchuria.

The fishing bat, *Myotis vivesi*, sometimes is placed in a separate genus or subgenus, *Pizonyx* Miller, 1906. The name *Anamygdon solomonis*, which applies to a bat from the Solomon Islands, was considered by Phillips and Birney (1968) to be a synonym of *Myotis moluccarum*, which now is regarded as a subspecies of *M. adversus*. Lekagul and McNeely (1977) considered *M. muricola* to be a subspecies of *M. mystacinus*. Hall (1981) used the name *M. subulatus* in place of *M. leibii*. *M. australis* was considered probably conspecific with *M. ater* by Hill (1983*b*).

Head and body length is about 35–100 mm, tail length is 28–65 mm, and forearm length is 28–70 mm. *M. lucifugus*, a small North American species, usually weighs 5–14 grams. *M. siligorensis*, the smallest Old World species, weighs 2.3–2.6 grams (Lekagul and McNeely 1977). *M. myotis*, the largest species, weighs 18–45 grams. The upper parts are generally some shade of brown, and the underparts are paler. Some forms have light and dark color phases. *M. formosus* is remarkable for its orange coloration. *M. vivesi* is dark buff or pale tan above and whitish below. It differs from most *Myotis* in its greatly elongated feet and large, laterally compressed toes and hind claws. *M. macrotarsus* and *M. stalkeri* of the East Indies are also big-footed and may be closely related to *M. vivesi* (Findley 1972). The tragus of *Myotis* is erect and tapering.

This genus is the most widely distributed group of bats. It is absent only from arctic, subarctic, and antarctic regions and many oceanic islands. It probably has the widest natural distribution of any genus of terrestrial mammals except *Homo*. *M. chiloensis* is the southernmost occurring of any bat, specimens having been reported from Navarino and Wollaston islands, just south of Tierra del Fuego (see Pine, Miller, and Schamberger 1979).

There is considerable variation in habitat. In western North America, for example, *M. evotis* occupies coniferous forest, while *M. auriculus* is found in arid woodland and desert scrub (Barbour and Davis 1969). In Venezuela, Handley (1976) collected *M. albescens* mostly in moist areas, both forested and open, while *M. nesopolus* was taken mainly in dry places. A number of species, including *M. grisescens*, *M. austroriparius*, and *M. yumanensis*, seem closely associated

The face of a mouse-eared bat *(Myotis myotis)*, considerably enlarged. Note the sharp teeth, which are well adapted for piercing and chewing insects. Photo by Walter Wissenbach.

with water, roosting near and foraging over streams, ponds, or reservoirs (Barbour and Davis 1969; Tuttle 1976*b*). *M. vivesi* shelters under piles of stones along seashores or in sea caves and fissures and forages over the open sea. Patten and Findley (1970) observed a group of about 400 *M. vivesi* around a boat at least 7 km from the shore. *M. hasseltii* also is reported to forage over the ocean, and sometimes to roost in adjacent mangroves. Caves are a more common roosting site for *Myotis*, but hollow trees, rock crevices, and structures erected by people also are often used by some species. *M. formosus* has been found hanging in trees, bushes, and tall jungle grass.

All species roost by day and forage at night. The feeding flights usually alternate with periods of rest, during which the bats hang to digest their catch. The flight is described as slow and straight in *M. myotis* and erratic and faster in some other species. *M. lucifugus* attains flight speeds of up to 35 km/hr and averages about 20 km/hr (Godin 1977). Roosting sites sometimes change by season, and it is common in North American species for maternity colonies to break up in the fall and shift to hibernacula. In *M. lucifugus* there may be migration of up to 275 km between the summer and winter roosts (Godin 1977). This species, like many others found in temperate regions, is known to hibernate during the winter. Probably the greatest range of body temperature in any vertebrate occurs in *M. lucifugus*; it has been cooled to 6.5° C without apparent harm and has also been found at 54° C (Barbour and Davis 1969). Colonies of *M. grisescens* travel up to 6.6 km from roost to foraging areas; during the summer a colony occupies a definite home range and may move among six or more roosting caves. Large-scale movement by this species to hibernation sites begins in September, and this migration covers 17–437 km (Tuttle 1976*a*, 1976*b*). *M. sodalis* concentrates in caves during the fall, preliminary to winter hibernation, but disperses to other habitats for the summer. A nursery colony was found in Indiana under the loose bark of a tree, and the 50 bats present foraged over about 0.82 km of riparian habitat (Humphrey, Richter, and Cope 1977). In favorable New England habitat, *M. lucifugus* has a population density of about 10/sq km (Barbour and Davis 1969). The diet of *Myotis* usually consists predominantly of insects. *M. vivesi*, however, catches fish and small aquatic crustaceans. The exact way in which this is accomplished is unknown, but the long feet and enlarged hind claws presumably aid in the capture. Several other species, including *M. hasseltii* and *M. macrotarsus*, may also catch fish. Recently, Robson (1984) showed that fish are eaten by *M. adversus* of Australia, though its diet consists mainly of insects captured above the surface of the water.

Most species are gregarious to some extent. *M. keenii* often roosts alone but forms small nursery colonies of 1–30 bats, and its hibernating groups may consist of up to 350 individuals (Barbour and Davis 1969; Fitch and Shump 1979). In Thailand, *M. horsfieldi* roosts in caves in groups of 2–3 to over 100, and *M. hasseltii* is found in groups of up to 25 individuals of both sexes (Lekagul and McNeely 1977). Usually the females of *Myotis* form maternity colonies to bear young, with few or no males present until after the young are born. Nursery colonies of *M. lucifugus* reportedly numbered 15–1,100 in Alberta (Schowalter, Gunson, and Harder 1979) and 12–1,200 in Vermont (Godin 1977). Colonies of several other species are in the same size range, but those of *M. austroriparius* usually number many thousands, a range of 2,000–90,000 having been reported in Florida. These great aggregations disperse in October. Hibernating groups are often considerably larger than the summer colonies; a single cave in Vermont contained 300,000 hibernating *M. lucifugus*, probably the total population of the species in an area of 22,300 sq km (Barbour and Davis 1969). About 90 percent of all *M. grisescens* east of the Mississippi River and south of Kentucky are thought to hibernate in three caves, with the populations there being 125,000, 250,000, and 1,500,000 (Tuttle 1976*a*). In the summer the females collect in smaller groups at other caves to bear young, while the males assemble at sites nearby. After late July the males move into the maternity colonies. In a study of *M. bocagei* in Gabon, Brosset (1976) found the population divided largely into harems, each comprising a single adult male, 2–7 females, and recent young. Such groups roosted in the leaves of banana plants. Solitary males were also present in the population, and there was a rapid turnover of harem males, but breeding females tended to remain in the group for a considerable period. The young departed at 4–5 months, but young males sometimes returned at 12 months to try to take over the harem.

The usual reproductive pattern in temperate regions is: mating during the fall; storage of sperm in the uterus of the female through winter hibernation; ovulation and fertilization in the early spring; and birth in the late spring or early summer. Considerable variation in the period of birth in *M. volans* was indicated by the discovery of a month-old young in Washington in March and a pregnant female in Colorado in August (Barbour and Davis 1969). The gestation period is 50–60 days in *M. lucifugus* and up to 70 days in *M. myotis*. In a study of *M. velifer* in Kansas, Kunz (1973*a*) found most matings to take place in October, ovulation in April, births in June, flight capability at 3 weeks, and sexual maturity in the first year of life. O'Farrell and Studier (1973) determined that in a colony of *M. thysanodes* in New Mexico gestation was 50–60 days, births occurred synchronously in a 2-week period in late June and early July, the young could make limited flights at 16.5 days, and during the night a few females always remained in the nursery to guard the young. Most species of *Myotis* normally have a single young, but in *M. austroriparius* 90 percent of the births are twins (Barbour and Davis 1969).

Apparently there is more variation in reproductive pattern in tropical species. In the Malay Peninsula pregnant female *M. mystacinus* have been found in all months of the year, most frequently in April and May (Medway 1978). In Queensland *M. adversus* has been found to be polyestrous, with births recorded in March, September, and December (Dwyer 1970). Pregnant females have been recorded for *M. yumanensis* in Jalisco in August (Watkins, Jones, and Genoways 1972), for *M. elegans* in the Yucatan in February (Jones, Smith, and Genoways 1973), and for *M. albescens* in Costa Rica in January (LaVal and Fitch 1977). Myers (1977*a*) found *M. albescens* and *M. nigricans* to give birth during the spring (September–November) in Paraguay. *M. albescens* then had a postpartum estrus and probably bred one or two more times during the year. *M. nigricans* had no postpartum estrus but probably also bred once or twice more. Both species were found to reach sexual maturity at less than 1 year. Wilson (1971) determined that on Barro Colorado Island *M. nigricans* was a polyestrous, continuous breeder with a postpartum estrus but that mating ceased in October–December, so that no young were weaned during the dry season of January–March. The gestation period of *M. nigricans* was 50–60 days, weaning took place at 5–6 weeks, and sexual maturity was attained at 2.5–3 months in males and somewhat later in females. Life span in *Myotis* is generally about 6 or 7 years, though some individuals survive much longer in the wild. Keen and Hitchcock (1980) reported that two specimens of *M. lucifugus* were recaptured in southeastern Ontario 29 and 30 years, respectively, after banding.

There is concern for the future of several North American species of *Myotis*, especially because of the destruction or modification of roosting caves by people (Mohr 1972). *M.*

Fishing bat *(Myotis vivesi)*: A. Photo by Bruce Hayward; B. Photo by Lloyd G. Ingles; C. Photo by Robert T. Orr.

grisescens and *M. sodalis* are listed as endangered by the USDI. The IUCN classifies *M. grisescens* as endangered and *M. sodalis* as vulnerable. Both species concentrate in large numbers in relatively few caves and have declined through such factors as deliberate killing by people, disturbance by spelunkers, and commercialization of caves for tourism. Tuttle (1979) reported that the total number of *M. grisescens* in 22 major summer colonies had declined from 1,199,000 before 1968 to 293,600 in 1976. Subsequent protection allowed recovery of some of these colonies (Tuttle 1987), and overall the species numbers about 1.5 million. Based on surveys of winter colonies, Humphrey (1978) determined that the entire population of *M. sodalis* had fallen from 640,361 in 1960 to 459,876 in 1975. Although additional colonies since have been discovered (Thornback and Jenkins 1982), total numbers now have fallen to around 250,000 (Clawson 1987). The subspecies *M. lucifugus occultus* and *M. velifer velifer* of the southwestern United States and adjacent Mexico have declined drastically through disturbance of colonies and loss of riparian habitat along the Colorado River (D. F. Williams 1986).

The situation for some Old World species is even worse. *M. dasycneme*, for example, may now number only about 3,000 individuals in western Europe and fewer than 7,000 in its entire range. *M. myotis* has been nearly exterminated in Great Britain, the Low Countries, and Israel, and colonies have been drastically reduced elsewhere. Stebbings and Griffith (1986) listed those species, along with *M. emarginatus* and *M. blythi*, as endangered, *M. bechsteini* as rare, and *M. nattereri, M. capaccinii, M. mystacinus,* and *M. brandti* as vulnerable. Problems include loss of natural roosts as forests are cleared, disturbance of hibernating colonies in caves and mines, deliberate exclusion from nursery sites in castles and cathedrals, and pollution. However, the greatest immediate threat in western Europe is the remedial chemical treatment of wood in the buildings on which the bats have come to depend for roosting. The chemicals remain on the surface of treated timber for years, are absorbed through the skin and mouth of the bats, and eventually cause death or reproductive failure. The IUCN classifies *M. capaccinii* as vulnerable, and also *M. hermani* of Sumatra as rare.

CHIROPTERA; VESPERTILIONIDAE; **Genus CISTUGO**
Thomas, 1912

Wing-gland Bats

There are two species (Hayman and Hill, *in* Meester and Setzer 1977; Herselman and Norton 1985):

C. seabrai, Angola, Namibia, northwestern Cape Province in South Africa;
C. lesueuri, known only by eight specimens from the Cape Province in South Africa.

The authorities cited above treated *Cistugo* as a subgenus of *Myotis*.

Head and body length is 40–47 mm, tail length is 40–43 mm, and forearm length is 32–35 mm. The coloration in *C. seabrae* is drab brown to slaty; *C. lesueuri* has yellowish brown upper parts and yellowish white underparts.

Certain dental features and the presence of wing glands distinguish *Cistugo* from *Myotis*. In *C. seabrae* the glands are thicker, broader, and in a different position from those in *C. lesueuri*; in some specimens of the former species, two glands, set close together, are present in each wing. The wing glands of *C. lesueuri* are reported to be evident only in the living animal and in a moistened museum specimen, not in a dry skin.

These bats frequent groves of trees in arid regions. Both species begin flying soon after sunset with a fairly steady and direct flight, but they have been observed to descend later and flutter around orange trees and bushes. Perhaps this indicates that they regularly begin feeding on flying insects and later in the evening hunt for insects on foliage. The type specimen of *C. lesueuri* was taken from a cat. Although *C. seabrai* appears to be locally common, both species occupy very restricted ranges and were designated as indeterminate by Smithers (1986).

CHIROPTERA; VESPERTILIONIDAE; **Genus LASIONYCTERIS**
Peters, 1865

Silver-haired Bat

The single species, *L. noctivagans*, occurs in southeastern Alaska, southern Canada, the conterminous United States, Tamaulipas (northeastern Mexico), and Bermuda (Hall 1981; Yates, Schmidly, and Culbertson 1976). A specimen also was collected in the Caicos Islands, Bahamas (Buden 1985).

Head and body length is 55–65 mm, tail length is 38–50 mm, forearm length is 37–44 mm, and adult weight is 6–14 grams. The individual hairs are dark brown to black, with silvery white tips; the coloration of the tips of the hairs imparts the frosted or silvery appearance to the pelage. The underparts are somewhat paler, the silvery wash being less pronounced than on the upper parts. The ears are short and about as broad as they are long, and the tail membrane is furred above on the basal half.

The silver-haired bat is found along streams and rivers in wooded areas and in montane coniferous forests. It is mainly a tree dweller but sometimes hibernates in caves. During the spring and summer it shelters in tree hollows, under loose

Silver-haired bat *(Lasionycteris noctivagans)*, photos by Ernest P. Walker.

bark, among leaves, in birds' nests, in the cracks of sandstone ledges, in buildings, under loose boards of buildings, and, infrequently, in caves. Brack and Carter (1985) reported discovery of an individual in an apparent ground squirrel burrow, 30–45 cm below the surface of the ground. According to Banfield (1974), *Lasionycteris* is the first bat to appear in the evening and is often seen in broad daylight. At first it generally flies low over water to drink, but later it climbs to treetop level. In a study in Iowa, Kunz (1973*b*) determined that there was an initial foraging period within three to four hours after sunset and a distinct second period peaking six to eight hours after sunset. Barbour and Davis (1969) stated that with the possible exception of *Pipistrellus hesperus, Lasionycteris* is the slowest-flying bat in North America. Its diet consists of insects.

The silver-haired bat is a year-round resident in some parts of its range, but populations in the north and at high altitudes generally migrate to warmer areas for the winter. Hibernation occurs during the winter, at least in some places. Izor (1979) reported that *Lasionycteris* may be present during the winter in suitable caves anywhere in the conterminous United States except perhaps the extreme northern Midwest and Great Plains and that it will also use trees and buildings for hibernation north to approximately the limit of the 6.7° C mean daily minimum isotherm for January. Banfield (1974) stated that the migration to wintering sites begins in late August and September and the northward movement occurs in late May. During these periods some flocks may wander far out to sea, and such activity accounts for the records from Bermuda. *Lasionycteris* apparently has a well-developed homing instinct; Davis and Hardin (1967) reported that one specimen traveled at least 175 km to return to its home roost.

Some old reports indicate that large nursery colonies are formed by this bat, but Barbour and Davis (1969) could find no firm evidence of colonial activity. Parsons, Smith, and Whittam (1986) agreed that the early accounts of large reproductive groups are dubious but also reported the discovery of a maternity colony of about 12 adult females and their young in Ontario on 26 June 1979. Banfield (1974) noted that while *Lasionycteris* lives solitarily in the summer, it congregates in flocks for migration.

During the birth period in June and July, adult males appear to reside in areas apart from those occupied by adult females. Kunz (1982) wrote that mating probably occurs in autumn and is followed by a period of sperm storage in the females during the winter. Ovulation peaks in late April and early May, and gestation lasts 50–60 days. Easterla and Watkins (1970) collected 18 adult female *Lasionycteris* at a site in southwestern Iowa between 9 June and 2 July; 5 were lactating, 7 contained 2 embryos, 2 contained a single embryo, and the reproductive state of 4 was undetermined. The most common litter size in the genus appears to be 2. At three weeks of age the young are able to follow the mother in weak flight. Schowalter, Harder, and Treichel (1978) found a 12-year-old individual in Alberta.

CHIROPTERA; VESPERTILIONIDAE; **Genus EUDISCOPUS**
Conisbee, 1953

Disk-footed Bat

The single species, *E. denticulus,* is known from the type locality, Phong Saly in Laos, and from Pegu Yoma in south-central Burma (Koopman 1972). This species, originally described as *Discopus denticulus,* is known only from six specimens in the Field Museum of Natural History and two in the American Museum of Natural History.

Head and body length is approximately 40–45 mm, tail

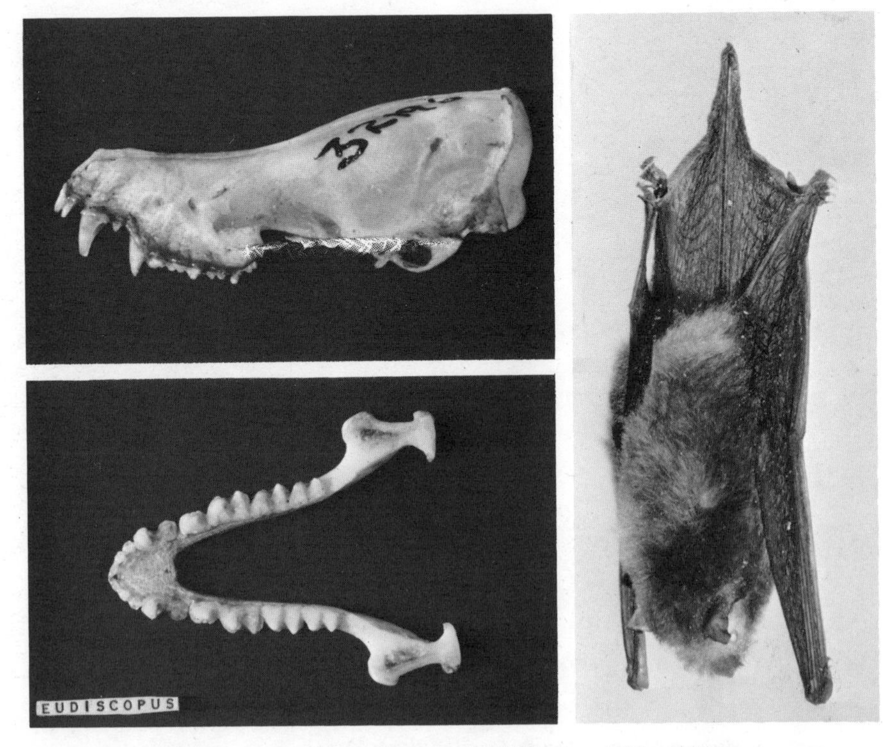

Disk-footed bat *(Eudiscopus denticulus)*, photos from Field Museum of Natural History.

length is 39–42 mm, and forearm length is 34–38 mm. The upper parts are cinnamon brown and the underparts are brighter.

Osgood (1932) described disk-footed bats as "externally similar to *Pipistrellus*, but tragus longer and more slender although not pointed at the apex; ears longer and narrowed at the tip; hind feet with highly developed disklike pads even more extreme than in *Tylonycteris* and *Glischropus*. Skull with a broad, greatly flattened braincase, somewhat as in *Tylonycteris* but with a longer, narrower rostrum; dentition with . . . three lower premolars on each side." Distinguishing features are the large adhesive disk on the foot and the three pairs of lower premolars.

The type specimen was collected at an elevation of 1,320 meters. Koopman (1972) noted that the flattened skull suggests that this bat must crawl through narrow crevices and that the foot pads indicate that it must cling to relatively smooth surfaces. The related genus *Tylonycteris* shares these two specializations and is known to roost inside hollow bamboo stems.

CHIROPTERA; VESPERTILIONIDAE; **Genus PIPISTRELLUS**
Kaup, 1829

Pipistrelles

There are 7 subgenera and 77 species (Aggundey and Schlitter 1984; Ansell and Dowsett 1988; Chasen 1940; Corbet 1978; De Vree 1972; Ellerman and Morrison-Scott 1966; Francis and Hill 1986; Gaisler 1983; Green and Rainbird 1984; Hall 1981; Hanák and Elgadi 1984; D. L. Harrison 1979; Hayman and Hill, *in* Meester and Setzer 1977; Hill 1972*b*, 1974*b*, 1976*a*, 1982*a*, 1983*b*; Hill and Francis 1984; Hill and Harrison 1987; Horacek and Hanák 1086; Ibánez and Fernandez 1985*a*; Kitchener, Caputi, and Jones 1986; Koch-Weser 1984; Kock 1981*b*; Koopman 1973, 1975, 1982*a*, 1984*a*, 1986; Koopman, Mumford, and Heisterberg 1978; Laurie and Hill 1954; Lekagul and McNeely 1977; McKean 1975; McKean, Richards, and Price 1978; Medway 1977, 1978; Qumsiyeh 1985; Qumsiyeh and Schlitter 1982; Ride 1970; Roberts 1977; Schlitter, Robbins, and Buchanan 1982; Soota and Chaturvedi 1980; Tate 1942; Taylor 1934; Thomas 1915; Thompson 1982; Varty and Hill 1988; Wang 1982; Wolton et al. 1982):

subgenus *Pipistrellus* Kaup, 1829

P. pipistrellus, Europe, Morocco, Algeria, Libya, southwestern Asia, possibly Korea;
P. nathusii, Europe, western Asia Minor;
P. permixtus, Tanzania;
P. abramus, eastern Siberia, Japan, China, Taiwan, Hainan, Viet Nam;
P. javanicus, southeastern Siberia and Japan to Thailand and Indochina, Andaman and Nicobar islands, Malaysia, Indonesia, Philippines, two specimens from uncertain localities in Australia;
P. endoi, Honshu;

A. Pipistrelle *(Pipistrellus pipistrellus)* in flight, photo by Walter Wissenbach. B. *P. subflavus* resting on hand, photo by Ernest P. Walker. C. Pipistrelle *(P. pipistrellus)*, close-up of head.

P. babu, Pakistan, Afghanistan, India, Nepal, Bhutan, Sikkim;
P. peguensis, Pegu (southern Burma);
P. paterculus, Yunnan, northern Burma;
P. tenuis, extreme southern Thailand, Malaysia, Indonesia, Philippines, Christmas Island (Indian Ocean), Papua New Guinea, Bismarck Archipelago, Solomon Islands, Louisiade and D'Entrecasteaux archipelagoes, northern Australia;
P. coromandra, Pakistan to Burma and southern China, Sri Lanka, Hainan;
P. mimus, Pakistan to Viet Nam, Sri Lanka;
P. collinus, central highlands of Papua New Guinea;
P. angulatus, Papua New Guinea, Bismarck Archipelago, Solomon Islands;
P. adamsi, northern Northern Territory, Cape York Peninsula of Queensland;
P. papuanus, Papua New Guinea and associated islands;
P. wattsi, southeast Papua New Guinea and nearby Samari Island;
P. westralis, northern coastal Australia;
P. murrayi, Christmas Island (Indian Ocean) and possibly Cocos-Keeling Islands;
P. sturdeei, Bonin Islands (south of Japan);
P. ceylonicus, Pakistan to Indochina, Sri Lanka, Borneo;
P. minahassae, northern Sulawesi;
P. ruepelli, Iraq, Algeria, Egypt, much of Africa south of the Sahara;
P. crassulus, Cameroon, southern Sudan, Zaire, northern Angola;
P. nanulus, Liberia to Zaire, Kenya, island of Bioko (Fernando Poo);
P. kuhli, southern Europe, southwestern Asia, northern and eastern Africa, Liberia, Canary Islands;
P. maderensis, Canary Islands;
P. aegyptius, Algeria, Libya, Burkina Faso, southern Egypt, Sudan, possibly Kenya;
P. rusticus, Liberia and Ethiopia south to Namibia and Transvaal;
P. aero, northwestern Kenya, possibly Ethiopia;
P. inexspectatus, Zaire, possibly Cameroon;

subgenus *Vespadelus* Troughton, 1943

P. pumilus, northern and eastern Australia;
P. vulturnus, central and southeastern Australia, Tasmania;
P. regulus, southwestern and southeastern Australia, Tasmania;
P. douglasorum, northern Western Australia, Northern Territory;
P. sagittula, southeastern Australia, Tasmania, Lord Howe Island (southwestern Pacific);

subgenus *Perimyotis* Menu, 1984

P. subflavus, eastern North America from Minnesota and Nova Scotia to Honduras and Florida;

subgenus *Hypsugo* Kolenati, 1956

P. savii, southern Europe to Mongolia and India, possibly Burma, northwestern Africa, Canary Islands, possibly Cape Verde Islands;
P. alashanicus, Mongolia, southeastern Siberia, northern China, Korea, Japan;
P. bodenheimeri, Israel, Sinai, southwestern Arabian Peninsula, possibly Socotra Island;
P. ariel, Egypt, Sudan;
P. anchietae, Zaire, Angola, Zambia;
P. austeniannus, Assam, Burma;
P. arabicus, Oman;
P. nanus, Mali to Ethiopia and south to South Africa, Madagascar;
P. musciculus, Cameroon, Gabon, Zaire;
P. helios, Kenya;
P. pulveratus, southern China, Thailand;
P. hesperus, western North America from southern Washington to central Mexico;
P. eisentrauti, Ivory Coast, Cameroon, Somalia, Kenya;
P. imbricatus, Sumatra, Java, Kangean Island, Bali, Borneo, southern Sulawesi, Luzon;
P. macrotis, Malay Peninsula, Sumatra and nearby islands;
P. curtatus, Enggano Island off western Sumatra;
P. vordermanni, Borneo, Billiton Island;
P. cadornae, northern India to Thailand;
P. lophurus, Tenasserim (southern Burma);
P. kitcheneri, Borneo;
P. joffrei, Burma;
P. stenopterus, Malay Peninsula, Rhio Archipelago, Sumatra, Borneo, Luzon;
P. anthonyi, northern Burma;

subgenus *Falsistrellus* Troughton, 1943

P. affinis, India, Nepal, Sri Lanka, Yunnan-Burma border area;
P. mordax, now thought to be restricted to Java;
P. petersi, northern Borneo, northern Sulawesi;
P. tasmaniensis, southeastern Queensland, eastern New South Wales, southern Victoria, Tasmania;
P. mackenziei, southwestern Western Australia;

subgenus *Neoromica* Roberts, 1926

P. melckorum, Zambia, South Africa, possibly Angola;
P. brunneus, Nigeria, Cameroon, Zaire;
P. capensis, throughout Africa south of the Sahara, Madagascar;
P. somalicus, Burkina Faso, Nigeria, Togo, Cameroon, Central African Republic, southern Sudan and Ethiopia to South Africa;
P. zuluensis, Zambia, Malawi, Zimbabwe, Botswana, Namibia, Transvaal;
P. guineensis, Senegal to Sudan and northeastern Zaire;
P. flavescens, Angola, Burundi;
P. rendalli, Gambia to northern Ethiopia, and south to Botswana;
P. tenuipinnis, Guinea to Uganda and Angola;

subgenus *Arielulus* Hill and Harrison, 1987

P. circumdatus, Yunnan (southern China), northern Burma, Malay Peninsula, Java, possibly India;
P. societatis, Malay Peninsula;
P. cuprosus, Sabah (northern Borneo).

This genus is sometimes considered no more than subgenerically distinct from *Eptesicus*, *Vespertilio*, and *Nyctalus*. *Pipistrellus* usually differs from *Eptesicus* in having a pair of rudimentary upper premolars, in addition to the normal pair, but this character is not constant. While acknowledging such difficulties, most recent authors have maintained *Pipistrellus* as a separate genus for purposes of convenience (see Corbet 1978; Ellerman and Morrison-Scott 1966; and Hayman and Hill, *in* Meester and Setzer 1977). A new revision by Hill and Harrison (1987), based mainly on bacular morphology and cranial characters, has clarified the situation to some extent and has resulted in the transfer of certain species, placed above in the subgenera *Vespadelus* and *Neoromica,* from *Eptesicus* to *Pipistrellus.* An assessment of karyological data by McBee, Schlitter, and Robbins (1987) tends to support this transfer with respect to *Neoromica.* Although *Perimyotis, Hypsugo,* and *Falsistrellus* were considered to be full genera by Menu (1984), Horacek and Hanák (1986), and Kitchener, Caputi, and Jones (1986) (supported by Adams et al. 1987*b*), respectively, they were treated as subgenera by Hill and Harrison. Still other subgenera have been used for *Pipistrellus,* but one, *Ia,* has been raised to generic level (Topal 1970*a*), and the single species of another, *P.* (*Megapipistrellus* Bianchi, 1916) *annectans,* is now considered to belong to *Myotis* (Topal 1970*b*). The following species were recently transferred from *Pipistrellus* to other genera: *P. rosseti* and *P. ridleyi* to *Myotis* (Hill and Topal 1973), and *P. brachyoterus* to *Philetor* (Hill 1971*b*). Heller and Volleth (1984) considered *P. societas* a synonym of *P. circumdatus* and thought that the latter should be transferred to the genus *Eptesicus,* but Hill and Francis (1984) and Hill and Harrison (1987) disagreed. The name *P. deserti,* which was synonymized under *P. aegyptius* by Qumsiyeh (1985), was used by most authorities cited in this account. Ansell and Dowsett (1988) recommended use of the name *P. africanus* in place of *P. nanus.* Based on allozyme electrophoresis, Adams et al. (1987*a*) indicated that *P. pumilus* actually is divisible into three full species and that *P. vulturnus* and *P. sagittula* are divisible into two species each.

Head and body length is 35–62 mm, tail length is 25–50 mm, forearm length is 27–50 mm, and adult weight is 3–20 grams. The ear is usually shorter and broader than in *Myotis,* and the tragus is not as sharply pointed. Pipistrelles are usually dark brown or blackish, but some are gray, chocolate brown, reddish brown, or pale brown.

There is considerable variation in habitat. In Japan, for example, *P. javanicus* forages in open areas and roosts both summer and winter in houses, while *P. endoi* is found only in mountain deciduous forests and never roosts in houses (Wallin 1969). Several European species seem to be associated with woodlands, roosting either in hollow trees or in buildings (Kowalski 1955; Ognev 1962). In Botswana, Smithers (1971) reported *P. capensis* to have a wide habitat tolerance, occurring in open dry scrub, rich riverine woodland, and *Acacia* woodland and scrub. In the northeastern United States, *P. subflavus* is often found during the summer in open woods near water; it roosts in rock crevices or, less frequently, in caves and buildings (Godin 1977). In the southern United States it commonly inhabits clusters of Spanish moss (Barbour and Davis 1969). *P. nanus* of Africa is commonly called the "banana bat" because of its habit of sheltering in the rolled leaves of bananas and plantains (Rosevear 1965).

Over most of their range, pipistrelles are among the first bats to appear in the evening. Some members of this genus occasionally fly about in bright sunlight. Their early appearance and jerky, erratic flight are characteristic. According to Barbour and Davis (1969), the fluttery flight of *P. hesperus* is the slowest and weakest of any bat in North America. Some species make spring and fall migrations between summer and winter ranges, and many apparently hibernate. In colder areas *P. subflavus* utilizes caves, mines, or deep crevices for hibernation. It enters a profound state of torpor around mid-October, from which it generally cannot be aroused until spring (Banfield 1974; Barbour and Davis 1969; Godin 1977; Schwartz and Schwartz 1959). In Louisiana, however, this species frequently moves about from one roost to another during the winter (Lowery 1974). In Britain *P. pipistrellus* goes into hibernation in late November or December, commonly spending the winter in crevices between the beams of country churches (Racey 1973). *P. mimus* of southeastern Asia also hibernates during the winter, even when temperatures are relatively warm (Lekagul and McNeely 1977). The diet of *Pipistrellus* consists of small insects caught in flight.

Pipistrellus is not as gregarious as some species of *Myotis. P. subflavus,* for example, roosts either alone or in small groups (Godin 1977). *P. hesperus* forms small maternity colonies of up to 12 individuals, including both females and young (Barbour and Davis 1969). In *P. pipistrellus,* however, the nursery colonies may contain hundreds of females (Kowalski 1955). These colonies are divided into groups of 1–10 bats that are associated with a single territorial male. This male defends a particular roosting site, allowing females to enter but driving away other males (Gerell and Lundberg 1985).

In India, *P. ceylonicus* generally occurs in small colonies in old buildings, numbering from a couple dozen to a couple hundred bats (Madhavan 1971). There is a sharply defined breeding season in this species, with mating in the first week of June, ovulation and fertilization in the second week of July, and a gestation period of 50–55 days. There are usually 2 young per birth, lactation lasts 25–30 days, and sexual maturity is attained by both sexes before 1 year. *P. mimus* breeds throughout the year in India, with the males and females remaining together at all times. Like most other species of *Pipistrellus* that have been studied, it normally produces 2 young per birth (Gopalakrishna, Thakur, and Madhavan 1975). In a study in Natal, South Africa, LaVal and LaVal (1977) found that *P. nanus* roosted within banana plants in groups usually comprising 2–3 individuals, sometimes of both sexes. Also observed, however, was one apparent maternity colony of 150 individuals roosting in the thatched roof of a hut. This species was found to be a monestrous, seasonal breeder producing 1–2 young in November or December. In the Northern Territory of Australia, Maddock and McLeod (1974) determined that *P. pumilus* was polyestrous, with periods of birth at least in August–October and March. Of 6 females with young that were examined, 5 had a single young and 1 had twins.

P. subflavus has two mating periods per year: one in the fall, probably involving only individuals over 1 year old and followed by storage of the sperm in the uterus of the female during hibernation, and a second in the spring in which the yearlings participate and which is followed by ovulation. Gestation, as measured from implantation to parturition, lasts at least 44 days. The young are born from late May to July; litter size is usually 2, rarely 1 or 3, but as many as 4 embryos have been reported; flying begins at about 3 weeks (Banfield 1974; Barbour and Davis 1969; Fujita and Kunz 1984; Schwartz and Schwartz 1959). Maximum known longevity in this species is 14.8 years (Walley and Jarvis 1971).

Stebbings and Griffith (1986) listed *P. pipistrellus, P. nathusii, P. kuhli,* and *P. savii* as vulnerable. These species are declining in Europe because of loss of remaining natural roosts in trees and the use of toxic chemicals to treat wood in the buildings on which the bats now have come to depend (see end of account of *Myotis*).

CHIROPTERA; VESPERTILIONIDAE; Genus SCOTOZOUS
Dobson, 1875

Dormer's Bat

The single species, *S. dormeri,* is found in a small area of northwestern India and adjacent Pakistan (Roberts 1977). This species usually has been placed in the genus *Pipistrellus,* but *Scotozous* was regarded as a distinct genus by Corbet and Hill (1986) and Hill and Harrison (1987).

According to Roberts (1977), head and body length in one specimen was 52 mm, tail length is about 35–38 mm, and forearm length is 34–36 mm. The dorsal fur is hoary with a scattering of silver-tipped hairs, and the ventral fur is very pale whitish gray. All species of *Pipistrellus* on the Indian subcontinent have brownish gray belly fur. *Scotozous* also differs from *Pipistrellus* in having only one pair of incisors in the upper jaw; if a second outer pair is present, the teeth are reduced in size and almost vestigial. Specimens have been collected near a grove of mango trees and roosting under roof tiles.

CHIROPTERA; VESPERTILIONIDAE; **Genus NYCTALUS**
Bowdich, 1825

Noctule Bats

There are apparently six species (Corbet 1978; Ellerman and Morrison-Scott 1966; Hanák and Elgadi 1984; Hanák and Gaisler 1983; Harrison and Jennings 1980; Hayman and Hill, *in* Meester and Setzer 1977; Ibáñez 1988; Jones 1983; Maeda 1983; Medway 1978; Mitchell 1980; Neuhauser and DeBlase 1974; Palmeirim 1982; Qumsiyeh and Schlitter 1982):

N. leisleri, Europe, Iran, eastern Afghanistan, northern India, western Himalayas, North Africa, Madeira;
N. azoreum, Azores;
N. montanus, eastern Afghanistan, northern India, Nepal;
N. noctula, Europe through most of temperate Asia to Japan and Burma, Oman, Viet Nam, Taiwan, Algeria, possibly Mozambique and Singapore;
N. lasiopterus, western Europe to the Urals and Iran, Morocco, Libya;
N. aviator, Japan, Korea, vicinity of mouth of Yangtze Kiang in China.

The species *Pipistrellus joffrei* and *P. stenopterus* have sometimes been referred to *Nyctalus* and would extend the range of the latter genus considerably farther to the southeast. Corbet (1978) listed both *N. azoreum* of the Azores and *N. verrucosus* of Madeira as subspecies of *N. leisleri,* but *N. azoreum* was considered a separate species by Hanák and Gaisler (1983).

Head and body length is 50–100 mm, tail length is 35–65 mm, and forearm length is 40–70 mm. *N. noctula,* the most widely distributed species, weighs 15–40 grams. The color is golden brown or yellowish brown to dark brown above and usually pale brown below.

These bats are generally associated with forests but may also forage in open areas and reside in or near human settlements. Roosting sites include hollow trees, buildings, and caves (Kowalski 1955; Wallin 1969). Noctule bats fly early, with quick turns. They resemble *Pipistrellus* in appearance and in the habit of beginning to fly before or just after sunset. There may be two main feeding flights of one to two hours duration each, one in the early evening and the other ending shortly before sunrise. Moore (1975) reported that *N. leisleri azoreum* flies extensively during daylight in the spring and was often observed from 0900 to 1600 hours. The voice of *N. noctula* has been described as "bursts of piercing, staccato cries" and as "a very loud, shrill, prolonged trilling sound . . . similar to the prolonged song of a bird."

N. leisleri and *N. noctula* are migratory. Three individuals of the latter species banded during hibernation in a cave in Germany were recovered the following summer north and east of the banding site. The most distant recovery was about 750 km from the cave. Medway (1978) wrote that *N. noctula* has made journeys as long as 2,347 km.

These bats are fond of beetles; in fact, there seem to be local movements of the males of *N. noctula* in the early summer in England in search of cockchafers. Noctule bats also eat winged ants, moths, and other insects. An unusual instance is the capture of house mice *(Mus musculus)* by an *N. noctula,* which then ate them.

Breeding usually takes place in September and again in the spring, but only a single litter is produced each year. The pregnant females generally assemble in groups of as many as 400 individuals in trees and buildings; the 1 or 2, rarely 3, young are born in May and June. The gestation period in *N. noctula* is 50–70 days. Adult dimensions are attained when the young are 6–7 weeks old. Kleiman and Racey (1969) found that some captive females mated during their first autumn, when only 3 months old, while males mated during their second autumn.

Stebbings and Griffith (1986) listed *N. leisleri, N. noctula,*

Noctule bat *(Nyctalus noctula),* photo by Erwin Kulzer.

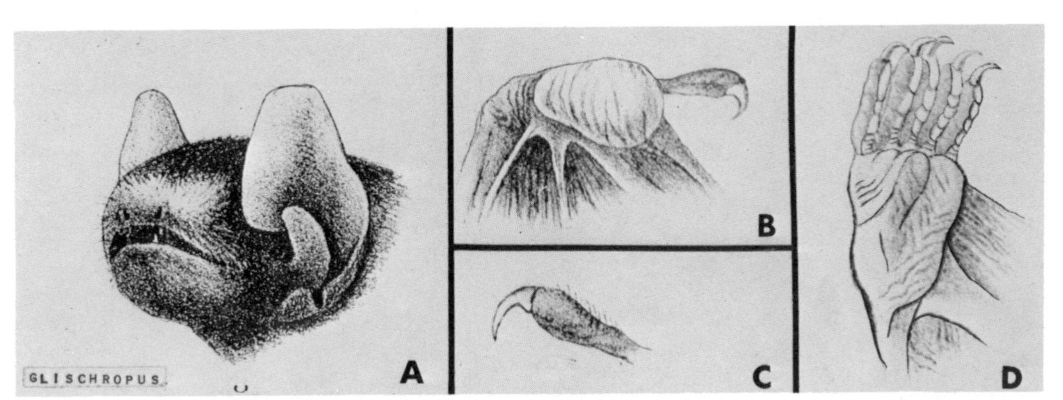

Thick-thumbed bat *(Glischropus tylopus):* A. Photo from *Catalogue of the Chiroptera in the Collection of the British Museum,* G. E. Dobson; B & C. Thumb; D. Foot, photos from *Proc. Zool. Soc. London.*

and *N. lasiopterus* as rare or vulnerable. These species are declining in Europe as natural habitat, roosting trees, and large insect prey are eliminated.

CHIROPTERA; VESPERTILIONIDAE; **Genus GLISCHROPUS**
Dobson, 1875

Thick-thumbed Bats

There are two species (Chasen 1940; Lekagul and McNeely 1977):

G. tylopus, Burma, Thailand, Malay Peninsula, Sumatra, Borneo, Philippines;
G. javanus, Java.

This genus is sometimes considered a subgenus of *Pipistrellus.* The species *Pipistrellus tasmaniensis* and *Myotis rosseti* were formerly placed in *Glischropus.*

Head and body length is about 40 mm, tail length is 32–40 mm, and forearm length is 28–35 mm. Payne, Francis, and Phillipps (1985) reported weights of 3.5–4.5 grams. The coloration is dark reddish brown to blackish above and paler below. This genus resembles *Pipistrellus,* differing in that the pads on the thumb and foot are more developed, probably as a grasping modification. The longer and pointed tragus distinguishes *Glischropus* from *Tylonycteris,* another genus of bats with pads on the hand and foot.

G. tylopus is frequently associated with bamboo, roosting in dead or damaged stalks; it also roosts in rock crevices and in new banana leaves. It tends to fly rather high (Lekagul and McNeely 1977). The type specimen of *G. javanus* was taken from the hollow top of a broken-off and partially dead bamboo stem in cultivated country, not far from the mountain forest. The diet consists of insects.

CHIROPTERA; VESPERTILIONIDAE; **Genus EPTESICUS**
Rafinesque, 1820

Big Brown Bats, House Bats, or Serotines

There are 2 subgenera and 17 species (Chasen 1940; Corbet 1978; Davis 1966*a*; Ellerman and Morrison-Scott 1966; Hall 1981; Hayman and Hill, *in* Meester and Setzer 1977; Hill and

Thick-thumbed bat *(Glischropus tylopus),* photos by Klaus-Gerhard Heller.

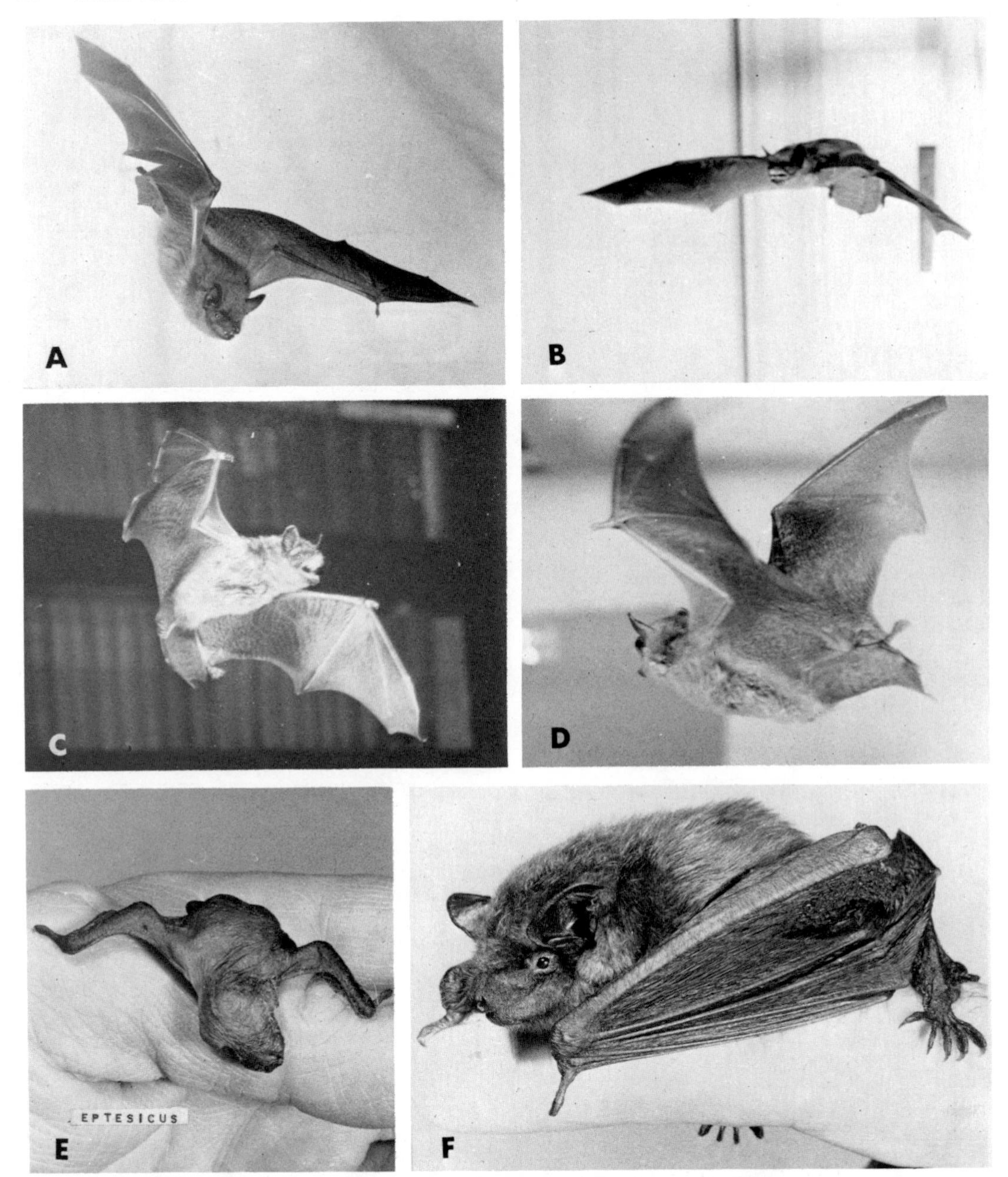

Big brown bat *(Eptesicus fuscus):* A. The upper arm and forearm are seen clearly from both above and below, and the legs are seen to be operating in unison with the wings; B. With tail and legs down to check momentum in coming in for a landing; C. Approaching for a landing using wings as well as tail membrane, but the mouth is still open, giving its ultrasonic sound, which is used in aerial navigation; D. Shows legs working in unison with wings in straightaway flight; E. Baby one day old; F. Adult in which the five equal-length toes with sharp, curved claws, the extremely long forearm, and the free thumb are clearly shown. Photos by Ernest P. Walker.

Harrison 1987; Hollander and Jones 1988; Ibáñez and Valverde 1985; Koopman 1975, 1978*a*; Lekagul and McNeely 1977; Rautenbach and Espie 1982; Schlitter and Aggundey 1986; Strelkov 1986; Williams 1978*b*):

subgenus *Eptesicus* Rafinesque, 1820

E. bobrinskoi, Kazakh S.S.R., northwestern Iran;
E. nilssonii, central Europe to Japan and Tibet;
E. gobiensis, Mongolia, Sinkiang, and adjacent parts of the Soviet Union and northern China;
E. nasutus, Arabian Peninsula to Pakistan;
E. fuscus, southern Canada to extreme northern South America, West Indies;
E. guadeloupensis, known only from Guadeloupe in the Lesser Antilles;

E. lynni, Jamaica;
E. innoxius, Pacific coast of Ecuador and Peru;
E. brasiliensis, southern Mexico to northeastern Argentina and Uruguay;
E. furinalis, Tamaulipas (eastern Mexico) to northern Argentina;
E. diminutus, central Brazil to northern Argentina and Uruguay;
E. bottae, central and southwestern Asia, northeastern Egypt;
E. serotinus, western Europe to Korea and Thailand, northern Africa, most islands of the Mediterranean, Senegal, Nigeria, Equatorial Guinea;
E. hottentotus, Kenya, Zambia, Malawi, Zimbabwe, Mozambique, South Africa, Namibia;
E. demissus, peninsular Thailand;
E. pachyotis, Assam, northern Burma, Thailand;

subgenus *Rhinopterus* Miller, 1906

E. floweri, Mali, Sudan.

Rhinopterus sometimes has been treated as a distinct genus. A number of African and Australian species recently were transferred by Hill and Harrison (1987) from *Eptesicus* to *Pipistrellus* (see account of latter). *E. loveni* of Kenya was shown by Schlitter and Aggundey (1986) to be a synonym of *Myotis tricolor*.

Head and body length is 35–75 mm, tail length is about 34–60 mm, and forearm length is 28–55 mm. Weights in the species *E. fuscus* range from 14 to 30 grams and in *E. serotinus* and *E. nilssoni* from 8 to 18 grams. The usual color is dark brown to black above and paler below. A buffy wash is often present. Some African species have white or translucent membranes. In the subgenus *Rhinopterus* the upper parts are pale fawn, the underparts are buffy, and the arms, legs, and tail are covered with small, horny, raised areas, giving a scabby appearance.

In Venezuela, Handley (1976) collected several species of *Eptesicus* in a variety of habitats, but mainly in moist, wooded areas; most roosting bats were found in holes in trees or logs. For North America, Banfield (1974) wrote that *E. fuscus* was originally a forest dweller, using hollow trees for roosting during warmer months and hibernating in caves in the winter. Many individuals still follow this way of life, but this species has become more closely associated with people than any other American bat (Barbour and Davis 1974). Colonies of *E. fuscus* are often found in attics and church belfries and behind shutters and loose boards of buildings. Hibernating sites include houses, tunnels, and storm sewers. Big brown bats usually emerge about sunset, with slow, ponderous, fluttering flight, and generally feed near the ground at lower levels than those of bats with a rapid, erratic flight. *E. fuscus* reportedly does not make substantial migrations or move very far from its place of birth. Banfield (1974) reported the average distance traveled probably to be under 32 km. The record natural movement for the species is 288 km (Mills, Barrett, and Farrell 1975). Hibernation in *E. fuscus* is not very profound and may be relatively brief, lasting only from December to April in Canada (Banfield 1974). The diet of *Eptesicus* consists of insects. Rydell (1986) reported that female *E. nilssonii* defended small feeding territories, even against members of their own nursery colony, by means of chases and shrill, clearly audible calls.

The females of *E. fuscus* form maternity colonies to rear young, and at this time the males roost alone or in small groups. Later during the summer both sexes are found roosting together. In hibernating colonies apparently there are usually more males than females (Goehring 1972). According to Barbour and Davis (1974), summer colonies in Kentucky have ranged in size from 12 adults to about 300, the most common number being 50–100. In a study in Ohio, Mills, Barrett, and Farrell (1975) found nursery colonies to contain 8–700 bats, with an average of 154. In northern temperate regions the young of *Eptesicus* are usually born from April through July, following mating in the fall and storage of sperm in the uterus over the winter. The number of young is usually 1 or 2. The species *E. fuscus* usually has a single young in the Rocky Mountains and westward, and twins in that part of its range east of the Rockies. In Nicaragua, Jones, Smith, and Turner (1971) found a pregnant female *E. furinalis* with 2 embryos on 22 April and a lactating female on 5 July. *Eptesicus* has a potentially long life span, there being numerous records of wild *E. fuscus* recaptured after being banded over 10 years before. Individuals that had survived at least 19 years were reported by Hitchcock (1965) and by Schowalter, Harder, and Treichel (1978).

According to Barbour and Davis (1974), *E. fuscus* is generally beneficial to people because of its insectivorous diet. Banfield (1974), writing on the mammals of Canada, stated that this aspect of the species is overbalanced by its nuisance value in occupying buildings and its menace to health as a carrier of rabies. This last point should be put in perspective, however, by noting that since 1925 there have been only 3 documented human deaths in Canada resulting from contact with a rabid bat (Rosatte 1988*a*). For the United States, Tuttle and Kern (1981) stated that only 9 cases of human rabies from bats had been reported in more than 30 years.

CHIROPTERA; VESPERTILIONIDAE; **Genus IA**
Thomas, 1902

Great Evening Bat

The single species, *Ia io*, is known from central and southern China, Assam, northern Thailand, and northern Viet Nam (Lekagul and McNeely 1977). Ellerman and Morrison-Scott (1966) and various other authors treated *Ia* as a subgenus of *Pipistrellus*, but Topal (1970*a*) showed that *Ia* should be considered a separate genus, closely related to *Eptesicus*.

According to Lekagul and McNeely (1977), head and body length is 89–104 mm, tail length is 61–83 mm, and forearm length is 71.5–80 mm. The upper parts are uniform sooty brown, and the underparts are dark grayish brown. The face is relatively hairless, but the insides of the ears are densely haired near the tip. The tip of the tail extends slightly beyond the interfemoral membrane. According to Topal (1970*a*), *Ia* has the same dental formula as *Pipistrellus*, while *Eptesicus* differs from both in having only a single upper premolar. Nonetheless, in other critical characters, especially the structure of the baculum, *Ia* resembles *Eptesicus* and differs from *Pipistrellus*.

Data summarized by Topal (1970*a*) indicate that specimens of *Ia* have been taken in caves at altitudes of 400–1,700 meters. Some specimens were observed in the process of returning to caves at only about 1730 hours, thus indicating that their initial period of foraging is relatively early. Banding studies suggest the possibility that *Ia* undertakes lengthy migrations.

Frosted bat, or particolored bat *(Vespertilio murinus)*, photo by Liselotte Dorfmüller.

CHIROPTERA; VESPERTILIONIDAE; **Genus VESPERTILIO**
Linnaeus, 1758

Frosted Bats, or Particolored Bats

There are three species (Corbet 1978):

V. murinus, Europe to southeastern Siberia and Afghanistan;
V. superans, northeastern China, Manchuria, Ussuri region of southeastern Siberia, Korea, Japan;
V. orientalis, eastern China, Japan, Taiwan.

Head and body length is 55–75 mm, tail length is 35–50 mm, and forearm length is 40–60 mm. *V. murinus* weighs about 14 grams. The coloration is reddish brown to blackish brown above and usually dark brown or gray below. These bats may have a "frosted" appearance from white-tipped hairs.

The ear in this genus is shorter and broader than in *Eptesicus,* and the facial part of the skull is flattened, with a deep pit on each side that is lacking in *Eptesicus.*

According to Wallin (1969), *V. orientalis* occurs mainly within broad-leaved deciduous forests in mountainous areas. This species has been found roosting in bushes during the day, in a more exposed manner than *V. murinus.* It apparently does not spend the winter in caves, but has been found in buildings at this time. *V. murinus* often inhabits large forest areas but sometimes occurs on steppes, and in some regions it is often found in towns and cities. When sleeping, *V. murinus* usually lies in the narrowest crevices. This may be a species that prefers the vicinity of rocks for roosting sites; it has been suggested that the present extensive range of *V. murinus* is the result of adapting to the utilization of human structures as shelters.

V. murinus appears late in the evening and flies at a considerable height, usually more than 20 meters above the ground. This species, in flight in the fall and winter, "utters a strong, shrill, grinding or whistling . . . cry. This is sometimes combined with a continuous buzzing sound (which latter is possibly produced with the wings)" (Ryberg 1947). This is believed to be a calling and mating cry.

Several thousand *V. murinus* often summer together in such localities as the protecting walls of old fortresses. Male colonies of this species, numbering about 15 bats in each of two cases, have been observed in May in Germany. This species begins to disperse in the fall and is thought to be migratory in the sense that it travels back and forth between summer and winter retreats, though the movements may be to or from suitable hibernation sites. The one or two young are born in May and June after a gestation period of 40–50 days.

CHIROPTERA; VESPERTILIONIDAE; **Genus LAEPHOTIS**
Thomas, 1901

Four species are currently recognized (Ansell and Dowsett 1988; Fenton 1975; Herselman and Norton 1985; Hill 1974*a*; Peterson 1971*b*, 1973; Rautenbach, Fenton, and Braack 1985; Setzer 1971):

- *L. wintoni*, Ethiopia, Kenya, Cape Province of South Africa;
- *L. namibensis*, Namibia;
- *L. angolensis*, southern Zaire, northeastern Angola, Zimbabwe;
- *L. botswanae*, southern Zaire, Zambia, Malawi, northwestern Zimbabwe, Botswana, eastern Transvaal.

The listing of *L. wintoni* in South Africa is based on a single specimen. Both Corbet and Hill (1986) and Honacki, Kinman, and Koeppl (1982) indicated that this specimen actually represents *L. namibensis*. However, Herselman and Norton (1985) indicated that the measurements of the specimen are by far closest to *L. wintoni*. It may be that the genus *Laephotis* actually comprises only a single highly variable species.

Head and body length is 45–59 mm, tail length is 35–47 mm, and forearm length is about 32–38 mm. A male specimen of *L. botswanae* weighed 6 grams (Smithers 1971). Two female *L. angolensis* averaged 7.7 grams (Fenton 1975). The upper parts are tawny olive or coppery brown and the underparts are pale brown, dark brown, or gray.

The bats of this genus are quite similar to those of *Histiotus* of South America, differing mainly in skull features. The teeth are similar to those of *Histiotus*. The ears of *Laephotis* are separate and 15–19 mm long, relatively shorter than in *Histiotus*.

A female has been found "under the bark of a dead tree limb with another (presumably a male) which flew away and was not collected." Thirteen specimens from Zaire, 12 of them females, were found under the bark of trees.

CHIROPTERA; VESPERTILIONIDAE; **Genus HISTIOTUS**
Gervais, 1855

Big-eared Brown Bats

Four species are recognized (Cabrera 1957; Koopman 1982*b*):

- *H. montanus*, Colombia through western and southern South America to Tierra del Fuego;
- *H. macrotus*, Peru, Bolivia, Argentina, Chile;
- *H. alienus*, southeastern Brazil, Uruguay;
- *H. velatus*, Brazil, Paraguay.

An unnamed species of this genus occurs in Venezuela (Handley 1976; Linares 1973). Corbet and Hill (1986) listed *H. laephotis* of Bolivia as a species distinct from *H. montanus*.

Head and body length is 54–70 mm, tail length is 45–55 mm, and forearm length is 42–52 mm. The upper parts are light brown, grayish brown, or dark brown and the underparts are grayish brown, whitish gray, or dark brown.

This genus closely resembles *Laephotis* of Africa. *Histiotus* resembles *Eptesicus* in dental and skull characters but differs from that genus in its much larger ears, which are at least as long as the head. In *H. macrotus* the ears are connected by a low band of skin.

These comparatively rare bats apparently occur over a wide area in a variety of habitats, sometimes being found in forests or at high elevations in the mountains. They may roost in buildings, and Baker (1974) discovered a specimen of *H. montanus* in Ecuador in a hole in a cliff at an altitude of

Laephotis wintoni, photo by John Visser.

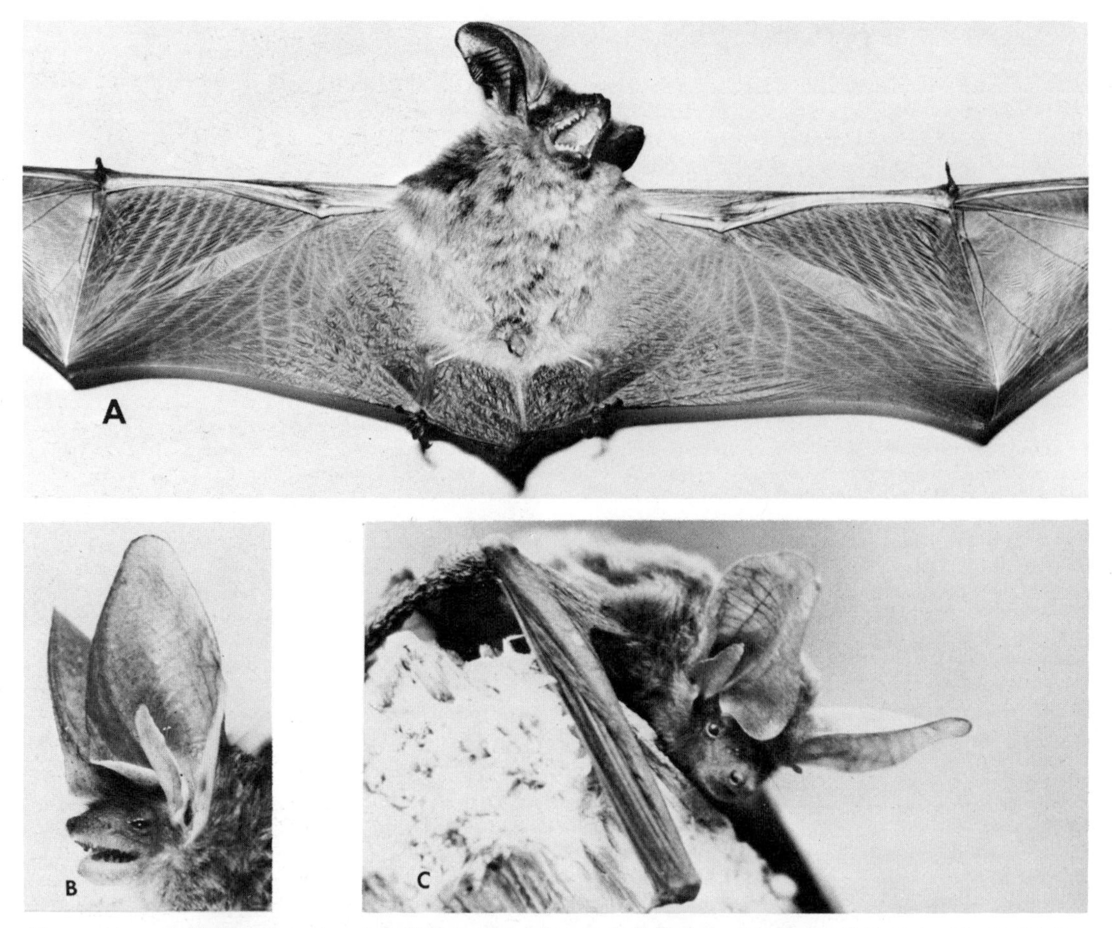

Big-eared brown bat *(Histiotus montanus)*: A. Photo by Alfredo Langguth; B. Photo by Abel Fornes; C. Photo by John P. O'Neill.

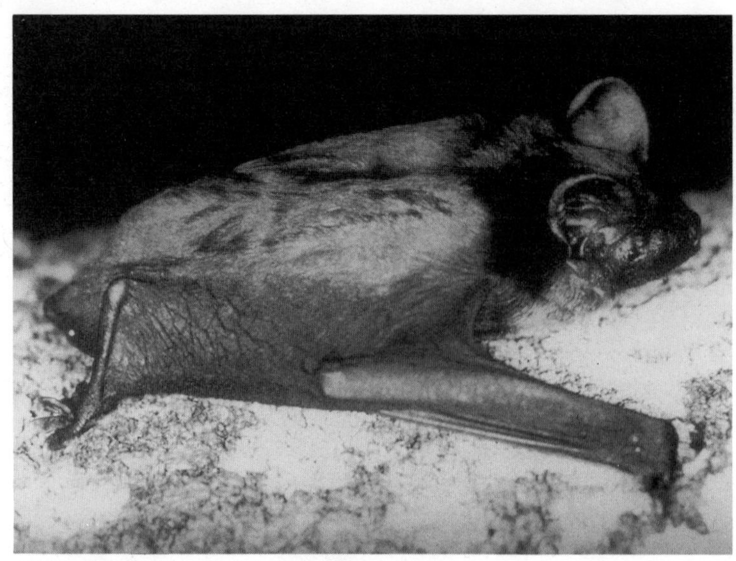

Philetor brachypterus, photo by Klaus-Gerhard Heller.

4,117 meters. In southeastern Brazil, Mumford and Knudson (1978) found *H. velatus* to roost in clusters of 6–12 individuals in the attics of buildings. In July the clusters comprised both males and nongravid females. On 31 October 6 females were taken, each with a suckling young. Peracchi (1968) also found a colony of this species in October; it comprised adults and young of various ages. In Colombia in July, Arata and Vaughan (1970) collected a pregnant female *H. montanus*, 7 lactating females, and 4 immatures. In Malleco Province, Chile, in December, Greer (1965) caught 6 immature *H. montanus* and 11 adult females, 4 of which were lactating.

CHIROPTERA; VESPERTILIONIDAE; **Genus PHILETOR**
Thomas, 1902

The single species, *P. brachypterus*, now is known from Nepal, the Malay Peninsula, Sumatra, Borneo, Mindanao, New Guinea, New Britain Island in the Bismarck Archipelago, and possibly Java and Bangka (Hill 1974*b;* Hill and Francis 1984; Kock 1981*a;* Koopman 1983; Medway 1977). This species formerly was known as *P. rohui*, but Hill (1971*b*) showed that the species then called *Pipistrellus brachypterus* actually is referable to *Philetor*, that it is conspecific with *P. rohui*, and that its name has priority. The species formerly called *Eptesicus verecundus* also now is considered part of *Philetor brachypterus* (Hill 1966, 1971*b*). Still another species that may be referable to *Philetor* is *Pipistrellus anthonyi* of northeastern Burma (Koopman 1983).

Head and body length is about 52–64 mm, tail length is 30–38 mm, forearm length is 30–38 mm, and weight is 8–13 grams. Coloration is reddish brown to dark brown. The wings are described as being relatively short. The muzzle is broad, and the skull is short and rounded, with a large, rounded braincase.

On Borneo, Lim, Shin, and Muul (1972) collected most specimens of *Philetor* from tree holes about 1.5–4.5 meters above the ground. The stomachs of three specimens contained insects such as Coleoptera and Hymenoptera. A pregnant female, with a single embryo, was taken in the period from late May to mid-June.

CHIROPTERA; VESPERTILIONIDAE; **Genus TYLONYCTERIS**
Peters, 1872

Club-footed Bats, or Bamboo Bats

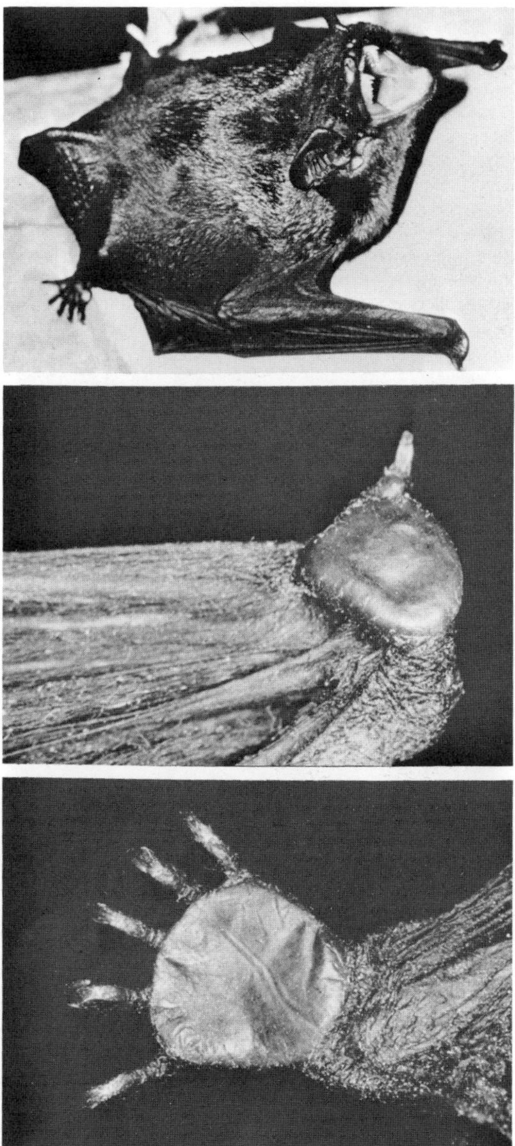

Club-footed bat: Top (*Tylonycteris* sp.), photo by H.-G. Heller; Middle and bottom *(T. robustula)*, photos by David Pye.

There are two species (Heaney and Alcala 1986; Lekagul and McNeely 1977; Medway 1973):

T. pachypus, India to southern China and the Malay Peninsula, Sumatra, Java, Borneo, Philippines;
T. robustula, southern China, Indochina, Thailand, Malaysia, Indonesia, Philippines.

Head and body length is 35–50 mm, tail length is 24–33 mm, and forearm length is 22–33 mm. Medway (1972) listed weights of 3.5–5.8 grams for *T. pachypus* and 7.1–11.2 grams for *T. robustula*. However, Heaney and Alcala (1986) noted that *T. pachypus* weighs only about 2 grams and is a competitor with *Craseonycteris thonglongyai* for the title of world's smallest bat. Coloration is reddish brown or dark brown above and paler below.

The bats of this genus may be recognized by the greatly flattened skull and the presence of cushions or pads on the thumb and foot. The ears are about as long as the head, and the tragus is short and bluntly rounded at the tip. *Mimetillus*, from Africa, has been referred to as the African counterpart of *Tylonycteris*, as it also has a flattened and broadened skull. The special adaptations of *Tylonycteris* may be another case of similar or parallel development that fits the animal for a certain mode of life. These bats are remarkably adapted for gaining access to and roosting in the hollow joints of bamboo stems. The small size and flattened skull facilitate their entrance through cracks in the stem, and the suction pads enable them to hang up in the joint.

The following natural history information is taken from Medway (1972, 1978), Medway and Marshall (1972), and Lekagul and McNeely (1977). There appears to be considerable overlap in the ecological niches of the two species. Both characteristically roost in the internodal spaces of standing culms of the large bamboo, *Gigantochloa scortechinii,* access to which is provided by a narrow vertical slit originating as the pupation chamber and emergence hole of the chrysomelid beetle, *Lasiochila goryi.* The larger of the two bats, *T. robustula,* tends to use slightly larger roosting sites than those sometimes entered by *T. pachypus* and may be found in other kinds of bamboo and between rocks. The two species have been seen together feeding on swarms of termites. These bats are gregarious, sometimes roosting in groups of 40 or more individuals. Most groups apparently have a harem-type arrangement, though there are also lone individuals and groups composed of only a few males. One complete roosting group collected had a single adult male, 12 adult females, and 24 infants. In a study in the Malay Peninsula, however, average composition was 1.7 males and 3.2 females for *T. pachypus* and 1.4 males and 2.1 females for *T. robustula.* Marking investigations have shown that there are frequent changes in group composition and roosting sites. There seems to be a restricted annual breeding season, with births generally occurring over a 1-month period that may take place from February to May, but there is one record of a pregnant female *T. robustula* in August. The gestation period in both species is about 12–13 weeks, and there are usually twin births. The young are carried by the mother for the first few days and then are left at the roost until weaning and independence at about 6 weeks. Both sexes attain sexual maturity in their first year of life.

CHIROPTERA; VESPERTILIONIDAE; **Genus MIMETILLUS**
Thomas, 1904

Moloney's Flat-headed Bat, or Narrow-winged Bat

The single species, *M. moloneyi,* is found from Sierra Leone to Ethiopia and south to Angola and Zambia (Hayman and Hill, *in* Meester and Setzer 1977; Largen, Kock, and Yalden 1974).

Head and body length is 48–58 mm, tail length is 26–38 mm, and forearm length is 27–31 mm. The wingspan is only about 175 mm. Kingdon (1974*a*) gave the weight as 6.0–11.5 grams. The pelage is short and rather scanty. Coloration is dark brown to blackish, or somewhat brighter with a chestnut tinge.

This bat differs from all others in its greatly reduced wings, relative to the size of head and body. It has a remarkably

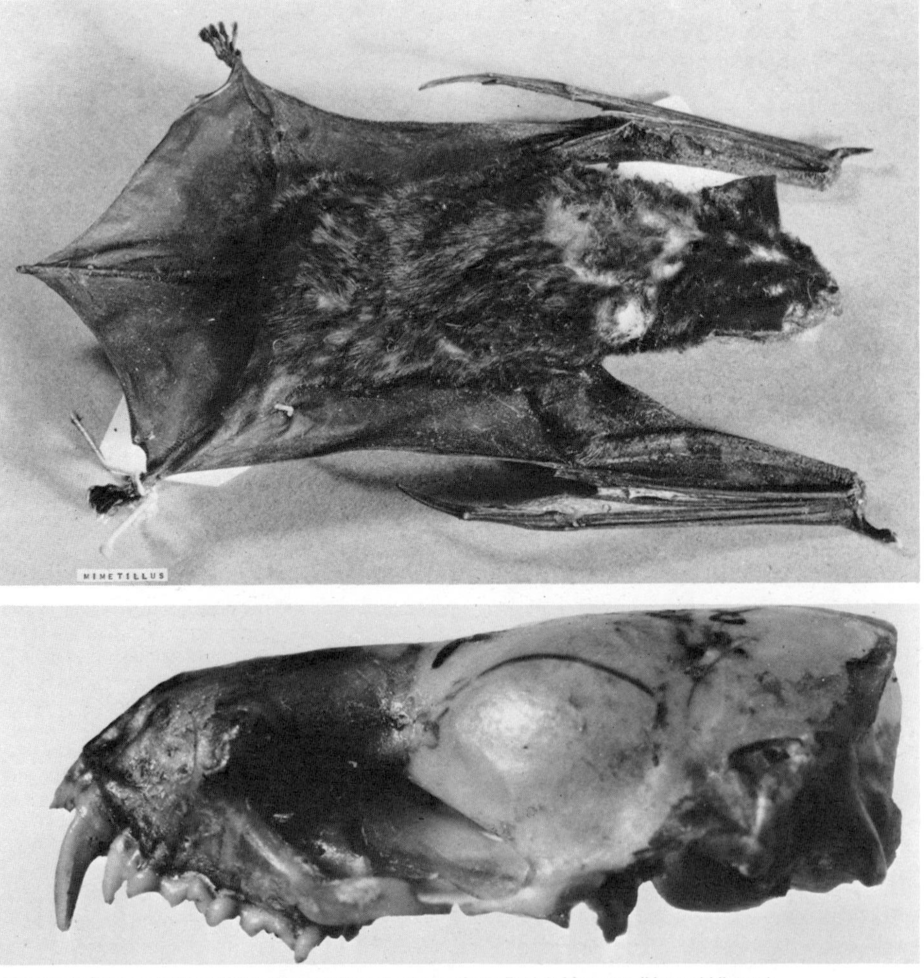

Moloney's flat-headed bat *(Mimetillus moloneyi),* photos from British Museum (Natural History).

flattened and broadened skull. The legs are short and stout. The short, triangular ears have rounded tips.

Kingdon (1974*a*) provided the following information. *Mimetillus* is found across tropical Africa; it is not restricted to forests and occurs in wooded country to an altitude of 2,300 meters. It roosts in cracks beneath the bark of dead trees. It forages at dusk, often while there is still light, with very rapid flight. This bat apparently needs to drop some distance before it can achieve flight, and it must return to its roost to rest every 10–15 minutes because it must beat its wings so fast to fly. The diet consists of small flying insects. *Mimetillus* occurs in colonies of 9–12 individuals. Births take place biannually at the end of dry spells, in February–March and August.

CHIROPTERA; VESPERTILIONIDAE; **Genus HESPEROPTENUS**
Peters, 1869

There are two subgenera and five species (Hill 1976*b*, 1983*b*; Hill and Francis 1984; Mitchell 1980):

subgenus *Hesperoptenus* Peters, 1869

H. doriae, known only by the holotype from Sarawak in Borneo and a single specimen from Selangor in the Malay Peninsula;

subgenus *Milithronycteris* Hill, 1976

H. tickelli, India, Nepal, Sri Lanka, Andaman Islands, southeastern Burma, Thailand, possibly China;
H. tomesi, Malay Peninsula, Borneo;
H. gaskelli, central Sulawesi;
H. blanfordi, Thailand, southeastern Burma, Malay Peninsula, Borneo.

Head and body length is 40–75 mm, tail length is 24–53 mm, and forearm length is 25 to about 60 mm. Payne, Francis, and Phillipps (1985) reported weights of 6.1–6.4 grams for *H. blanfordi* and 30–32 grams for *H. tomesi*. *H. tickelli* is grayish yellow to golden brown, *H. blanfordi* is reddish brown, *H. tomesi* is dark brown, and *H. doriae* is light brown. The black wings of *H. tickelli* are usually marked with white.

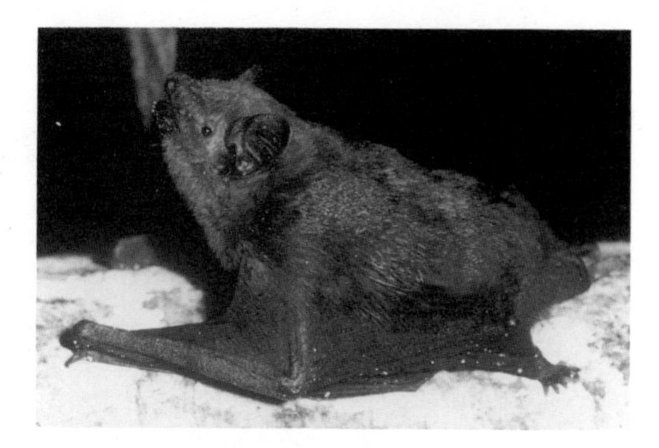

Hesperoptenus blanfordi, photo by Klaus-Gerhard Heller.

Externally these bats are like *Eptesicus*, differing from that and similar genera in cranial and dental features. *H. tickelli* has friction pads on the thumbs.

Bats of the genus *Hesperoptenus* roost singly or in small groups of both sexes, often in the foliage of trees, and begin to fly early in the evening, soon after *Pipistrellus*. *H. tickelli* usually establishes a feeding territory, which it patrols night after night in search of insects. Except when food is abundant, this species will chase away other individuals of the same species when they transgress the limits of the territory. Sometimes a male and a female share the circumscribed area. Most young of *H. tickelli* are born in May or June, but one pregnant female was noted in December.

CHIROPTERA; VESPERTILIONIDAE; **Genus CHALINOLOBUS**
Peters, 1866

Lobe-lipped Bats, Groove-lipped Bats, or Wattled Bats

There are six species (Koopman 1971, 1984*a*; Ride 1970; Ryan 1966; Van Deusen and Koopman 1971):

C. gouldii, Australia, Tasmania, Norfolk Island, New Caledonia;

Hesperoptenus doriae, photo by Klaus-Gerhard Heller.

Wattled bat *(Chalinolobus nigrogriseus)*, photo by B. Thompson / National Photographic Index of Australian Wildlife.

C. morio (chocolate bat), southern Australia, Tasmania;
C. picatus, southwestern Queensland, northwestern New South Wales;
C. nigrogriseus, Papua New Guinea, northeastern Western Australia to Queensland and northeastern New South Wales;
C. dwyeri, central New South Wales and adjacent part of Queensland;
C. tuberculatus, New Zealand.

The genus *Glauconycteris* of Africa is sometimes considered a subgenus of *Chalinolobus*. The population of *C. gouldii* on New Caledonia in the South Pacific formerly was considered a separate species, *C. neocaledonicus*. This population may have become extinct in recent years through habitat loss (Tidemann 1986).

Head and body length is 43–67 mm, tail length is 32–60 mm, and forearm length is 30–50 mm. These bats are chocolate brown, dark brown, or black, sometimes with a reddish tinge. The hairs may be tipped with white. *C. picatus* is called the pied wattled bat because it is black above and has a whitish fringe of hairs below. This genus is characterized by wattlelike, fleshy, outwardly projecting lobes at the corners of the mouth. Glandular swellings are present on the short, broad muzzle.

These bats roost in the foliage of trees, hollow trees, caves, and mine tunnels. Like most other insectivorous bats, they become active at dusk and often visit water holes to drink and hunt insects. Through radiotracking, Lunney, Barker, and Priddel (1985) determined that *C. morio* has a large home range, at least 5 km long.

Both *C. gouldii* and *C. morio* have been reported to roost in groups of 30–50 individuals. Kitchener (1975), however, stated that while *C. gouldii* is known to have established small maternity colonies, it is thought to be solitary for most of the year. In a study in Western Australia, Kitchener determined that female *C. gouldii* are monestrous. In the southern part of the state, mating occurs at the beginning of winter (about July), sperm is stored by the females over the winter, ovulation and fertilization start at the end of winter (about September), gestation lasts about three months, and births occur from late November to mid-January; in the central and northern parts of the state, births begin around October. The normal litter size is 2 young. According to Kitchener and Coster (1981), *C. morio* also is monestrous, but there normally is a single young, and births take place from mid-September to mid-November. Dwyer (1966) reported that mating in *C. dwyeri* occurs in late summer or autumn, births take place in early December, and average litter size is 1.8. Menzies (1971) studied a maternity colony of *C. nigrogriseus* in the roof of a house in Port Moresby, Papua New Guinea. Several hundred bats were present in November 1969, roosting in tight clusters of 10–40 individuals. All 35 adults collected were lactating females, and 76 juveniles were also taken.

C. tuberculatus of New Zealand is very rare and may be in danger of extinction. For several decades, only two small colonies were known, each containing 5–10 individuals, but Daniel and Williams (1983) reported discovery of a group of 200–300 on North Island. Observations indicate that these bats leave their cave at night, fly up to 2 km, and feed mainly on mosquitoes, moths, and midges caught in the air. They apparently hibernate during the winter. Their breeding season evidently extends from October to February (spring and summer), with most females giving birth to a single young in January or February.

CHIROPTERA; VESPERTILIONIDAE; **Genus GLAUCONYCTERIS**
Dobson, 1875

Butterfly Bats, or Silvered Bats

There are nine species (Ansell and Dowsett 1988; Baeten, Van Cakenberghe, and De Vree 1984; Hayman and Hill, *in*

Butterfly bat *(Glauconycteris superba sheila)*, photo by George S. Cansdale.

Meester and Setzer 1977; Koopman 1971; Meester et al. 1986; Peterson 1982; Peterson and Smith 1973):

G. variegata, Ghana to Ethiopia, and south to South Africa;
G. gleni, known only from Cameroon and Uganda;
G. argentata, Cameroon, Zaire, Rwanda, northeastern Angola, Malawi, Kenya, Tanzania;
G. superba, Ghana, northeastern Zaire;
G. poensis, Sierra Leone to Zaire, Bioko (Fernando Poo);
G. alboguttatus, Ituri Forest of Zaire;
G. beatrix, Cameroon and Gabon to Uganda;
G. egeria, known only by the type specimen from Cameroon;
G. kenyacola, known only by the type specimen from the east coast of Kenya.

Honacki, Kinman, and Koeppl (1982), Koopman (1971), and Ryan (1966) considered *Glauconycteris* only a subgenus of *Chalinolobus*, but Corbet and Hill (1986), Hayman and Hill (*in* Meester and Setzer 1977), Peterson (1982), and Peterson and Smith (1973) retained it as a distinct genus. According to Corbet and Hill (1986), *G. variegata* and *G. poensis* occur as far west as Senegal, and *G. superba* and *G. beatrix* occur as far west as Ivory Coast.

Head and body length is 35–68 mm, tail length is 35–55 mm, and forearm length is 35–50 mm. Peterson and Smith (1973) listed the following weight ranges: *G. argentata*, 6–12 grams; *G. gleni*, 8.5–15 grams; and *G. variegata*, 6–14 grams. The ground color is usually cream, dusky brown, grayish, or black; the individual hairs may be tricolored—black, white, and gray; and a variable pattern of white spots and flank stripes is usually present. The wings are usually pigmented and veined, the membrane venation being darkly outlined.

Members of this genus are characterized externally by a fleshy lobe at the base of the mouth that is connected by a ridge with the lobe of the ear. This genus is similar to *Chalinolobus* from Australia and New Zealand, differing in its dental formula, the extreme shortening of the muzzle and toothrow, and the distinctive external patterns. The pronounced membrane venation may have protective value by simulating a withered or skeletonized leaf.

Habitats include rainforest, open woodland, and savannah. These attractive bats roost in palm fronds, banana and plantain leaves, bushes, holes in trees, and native huts. Often only 2–3 *Glauconycteris* shelter together in vegetation, but as many as 9 individuals have been found hanging in a row from a palm frond, and at least 30 have been seen on a palm at a height of 4 meters from the ground. The latter group remained in place despite being approached to within 1.5 meters and photographed.

Glauconycteris often flies 17 or more meters above the ground. The flight is reported as butterflylike or mothlike. Often it begins to feed when it is still bright daylight, particularly on overcast days. The diet consists of small insects.

Kingdon (1974*a*) provided the following reproductive information: sexually active male *G. variegata* caught in Uganda in July and October; two lactating female *G. gleni* taken in January; and four of eight female *G. argentata* taken in March in Kenya carried young. Litter size in the genus is one or two.

CHIROPTERA; VESPERTILIONIDAE; **Genus NYCTICEIUS**
Rafinesque, 1819

Evening Bats

There are two species (Hall 1981):

N. humeralis, extreme southern Ontario, eastern United States, northeastern Mexico;
N. cubanus, Cuba.

Evening bat *(Nycticeius humeralis)*, photo by Richard K. LaVal.

Until recently, *Nycticeius* was thought to comprise the subgenus *Nycticeius*, with the above species plus *N. schlieffeni* of Africa, and also the subgenera *Scoteanax* and *Scotorepens* of Australia and New Guinea. However, Hill and Harrison (1987) placed *N. schlieffeni* in a distinct genus, *Nycticeinops*, and Kitchener and Caputi (1985) elevated *Scoteanax* and *Scotorepens* to generic rank. Still another former subgenus, *Scoteinus* Dobson, 1875, with the type species *Nycticeius emarginatus*, is actually a synonym of *Scotomanes* (Sinha and Chakraborty 1971).

Head and body length is 45–65 mm, tail length is 36–41 mm, and forearm length is 34–38 mm. The weight of *N. humeralis* is 6–12 grams. The upper parts are brown and the underparts are usually paler. The ears and tragus are more rounded in this genus than in *Myotis*. *Nycticeius* can be distinguished from *Scoteanax* and *Scotorepens* by its interdental palate, which is clearly longer than broad.

N. humeralis frequents cultivated and natural clearings in hardwood forests of southeastern North America (Banfield 1974). This species roosts in the cavities and hollows of trees, under loose bark, and in buildings; it is seldom found in caves. Its flight has been described as slow and steady; it often flies quite high in the evening and much lower later in the night. In the fall, usually by mid-October, the more northerly populations migrate southward. One individual was recovered 547 km south of the place of banding (Watkins 1972).

The diet consists of insects, probably small, soft-bodied forms. *N. humeralis* does well in captivity on a fine mixture of eggs, cheese, and mealworms, and it readily accepts small pieces of vegetables, fruits, and unsalted meats.

The sexes of *N. humeralis* generally segregate during the period when the young are born, and females form nursery colonies ranging in size from fewer than 6 to about 950 individuals (Watkins 1972). Mating probably takes place in the late summer and early fall, when the sexes are together. Evidently the sperm is stored in the reproductive tract of the female over the winter, and ovulation and fertilization occur in the spring. In a study of a group of *N. humeralis* in Florida, Bain (1978) found that males were present from October to March, and the colony apparently had a harem-type organization. Five solitary males competed for a cluster of 20 females at the roost, and the males excluded one another. Mating was observed on 2 December, births took place in May, and the young males left the area after 6 weeks; the young females, however, remained in the colony. In general, births in this species occur from May to mid-June, sometimes as late as early July. There usually are twin births, but from 1 to 4 embryos have been found. Average life span in wild *N. humeralis* seems to be about 2 years, with a few individuals known to have survived for over 5 years (Watkins 1972).

CHIROPTERA; VESPERTILIONIDAE; **Genus NYCTICEINOPS**
Hill and Harrison, 1987

Schlieffen's Twilight Bat

The single species, *N. schlieffeni*, is found in Sinai, the southwestern Arabian Peninsula, and open country from Mauritania and Egypt to northern Namibia and the Transvaal (Hayman and Hill, *in* Meester and Setzer 1977; Smithers 1983). This species long was placed in the genus and subgenus *Nycticeius*, together with the species *N. humeralis* and *N. cubanus* of North America. A number of authorities, such as Kitchener and Caputi (1985), questioned this rather unusual arrangement, and finally Hill and Harrison (1987) erected the separate genus *Nycticeinops* for *N. schlieffeni*. Moreover, Hill and Harrison did not consider *Nycticeinops* and *Nycticeius* to be closely related, and placed each in a different tribe of the subfamily Vespertilioninae.

Head and body length is 40–56 mm, tail length is 26–37 mm, forearm length is 29–35 mm, and weight is 6–9 grams (Kingdon 1974*a*). The upper parts are brown to pale brown, the underparts being paler brown to grayish white. The cranium is similar to that of *Nycticeius*, but the rostrum is shorter and more narrowed anteriorly, the maxillary toothrows are much more convergent, and the anterior palatal emargination extends farther posteriorly. The baculum is distinctive, with an expanded base and long fluted shaft, while that of *Nycticeius* is slipperlike and elevated proximally and distally (Hill and Harrison 1987).

According to Smithers (1983), *Nycticeinops* tends to be

found in relatively dry savannah, woodland, and open country and is absent from the high forest zone. It roosts by day in hollow trees, crevices in branches, buildings, and cellars. It emerges before dusk and pursues insects with an erratic flight. Although usually solitary, numbers may congregate to forage.

Van der Merwe and Rautenbach (1986) reported *Nycticeinops* to be monestrous and to breed during the spring in the Transvaal. Pregnant females were taken there from September to November, and lactating females in November. Litter size evidently ranged from one to four young, with three being the most common number.

CHIROPTERA; VESPERTILIONIDAE; **Genus SCOTEANAX**
Troughton, 1943

Greater Broad-nosed Bat

The single species, *S. rueppelli,* occurs in coastal eastern Australia from northeastern Queensland to southern New South Wales. *Scoteanax* long was considered by most authorities to be only a subgenus of *Nycticeius* but was given full generic rank by Kitchener and Caputi (1985). This position was supported by Baverstock et al. (1987) and Hill and Harrison (1987).

Head and body length is 63–73 mm, tail length is 44–58 mm, forearm length is 51–56 mm (Kitchener and Caputi 1985), and weight is 25–35 grams (Richards, *in* Strahan 1983). The upper parts are hazel to cinnamon brown and the underparts are tawny olive. From *Nycticeius, Scoteanax* differs in its generally greater size, more robust skull with much more pronounced occipital helmet, and considerably reduced third upper molar tooth.

Scoteanax favors tree-lined creeks and the junction of woodland and cleared paddocks but also forages in rainforests. It usually roosts in tree hollows but also has been found in the roof spaces of old buildings. It emerges just after sundown and usually flies slowly and directly at a height of 3–6 meters. The diet includes beetles and other large, slow-flying insects, as well as small vertebrates. A single young is produced in January (Richards, *in* Strahan 1983).

CHIROPTERA; VESPERTILIONIDAE; **Genus SCOTOREPENS**
Troughton, 1943

Lesser Broad-nosed Bats

Baverstock et al. (1987) and Kitchener and Caputi (1985) recognized four species:

S. orion, coastal southeastern Australia;
S. balstoni, southeastern New Guinea, Australia except extreme north and south;
S. sanborni, northern Western Australia, northern Queensland;
S. greyi, throughout mainland Australia except southwestern Western Australia, Victoria, and northern Queensland.

Scotorepens long was considered by most authorities to be only a subgenus of *Nycticeius* but was given full generic rank by Kitchener and Caputi (1985). Hill and Harrison (1987) agreed and placed the two genera in separate tribes of the subfamily Vespertilioninae. Koopman (1978*b,* 1984*a*) considered *orion* and *sanborni* to be part of *S. balstoni* and also recognized *S. influatus* of inland Queensland as a species separate from *S. balstoni.*

The information for the remainder of this account was taken from Hall (*in* Strahan 1983), Kitchener and Caputi (1985), and Richards (*in* Strahan 1983). Head and body length is 37–65 mm, tail length is 25–42 mm, forearm length is 27–40 mm, and weight is 6–18 grams. The upper parts are brown, gray-brown, or tawny olive, and the underparts are paler brown or buff. *Scoteanax* resembles *Nycticeius* in having a broad, square muzzle and only a single upper incisor on each side but differs in having a considerably reduced third upper molar.

Lesser broad-nosed bats occur in a variety of habitats. They often are found in areas that are generally dry but where there is access to watercourses, lakes, and ponds. Small colonies roost by day in hollow trees and buildings and emerge at dusk to pursue small insects that congregate over water. Mating evidently occurs just before winter (about May in Australia), at least in temperate areas, and the young

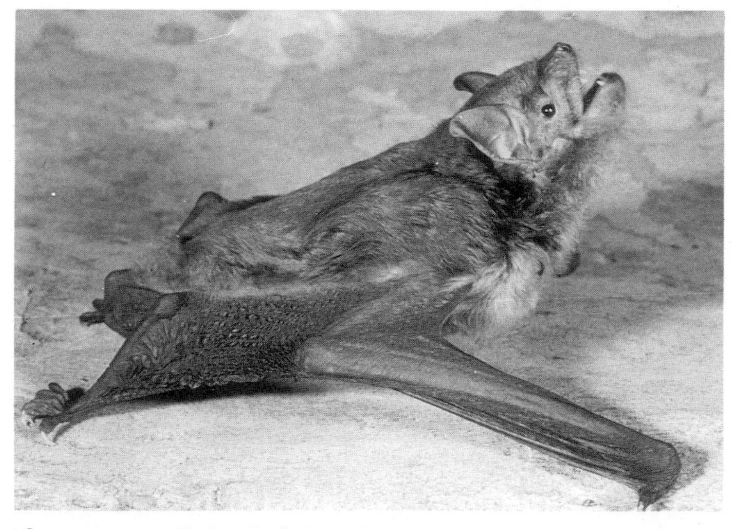

Scoteanax rueppelli, photo by G. A. and M. M. Hoye / National Photographic Index of Australian Wildlife.

Lesser broad-nosed bat *(Scotorepens greyi)*, photo by Stanley Breeden.

are born around November. In *N. balstoni* pregnancy lasts a total of 7 months and litters contain 1–2 young. They cling to their mother for 10 days and subsequently are left at the roost until they are ready to fly on their own, at about 5 weeks.

CHIROPTERA; VESPERTILIONIDAE; **Genus SCOTOECUS**
Thomas, 1901

There are three species (Hayman and Hill, *in* Meester and Setzer 1977; Hill 1974*c*; Kingdon 1974*a*; Koopman 1986; McLellan 1986; Robbins 1980):

S. pallidus, Pakistan, India;
S. albofuscus, Senegal to Malawi;
S. hirundo, Senegal to Ethiopia, and south to Angola and Zambia.

Scotoecus sometimes is treated as a subgenus of *Nycticeius*. *S. hindei*, found from Nigeria to Somalia and southward, sometimes is considered a species separate from *S. hirundo*.

Head and body length is 46–68 mm, tail length is 28–40 mm, and forearm length is 28–38 mm. The pelage is fine and silky. The upper parts are brownish, reddish, or dark gray. In the African *S. hirundo* the belly is pale and the wings are dark, while in *S. albofuscus* the belly is dark and the wings are pale. *S. pallidus* of Asia has brownish white underparts. *Scotoecus* is distinguished from *Nycticeius* by its broader rostrum and the broad, flat anterior surface of its canines.

The African species occur mostly in open woodland, and *S. pallidus* occupies semidesert (Hill 1974*c*). According to Roberts (1977), the latter species has been collected in roof crevices, old tombs, and pump houses. It flies slowly and close to the ground. It is gregarious, with both sexes occurring together in mid-March but separately during the summer, when the young are born. Kingdon (1974*a*) wrote that in western Uganda *S. hirundo* appears to have a well-defined breeding season; pregnant females, each with two embryos, have been taken in March, and lactating females have been taken in May.

CHIROPTERA; VESPERTILIONIDAE; **Genus RHOGEESSA**
H. Allen, 1866

Rhogeëssa Bats, or Little Yellow Bats

There are two subgenera and seven species (R. J. Baker 1984; Carter et al. 1981; Handley 1976; Jones, Arroyo-Cabrales, and Owen 1988; LaVal 1973*b*):

subgenus *Baeodon* Miller, 1906

R. alleni, mountains of western Mexico from Zacatecas and Jalisco to central Oaxaca;

subgenus *Rhogeessa* H. Allen, 1866

R. gracilis, mountains of western Mexico;
R. parvula, western Mexico from central Sonora to Isthmus of Tehuantepec;

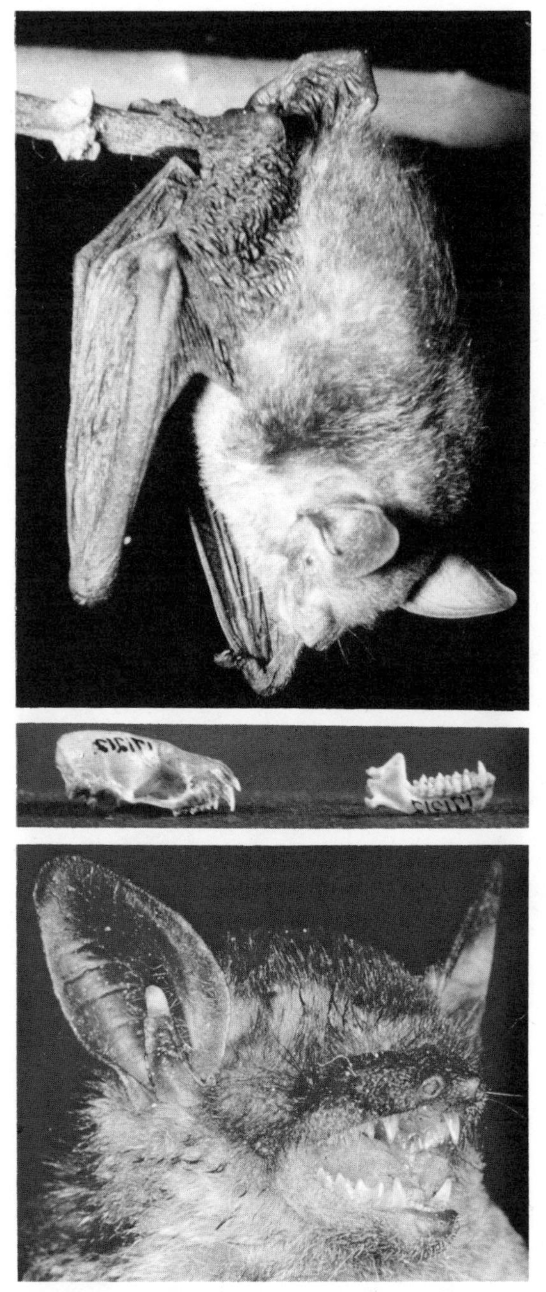

Rhogeëssa bat *(Rhogeessa parvula):* A. Photo by Bruce J. Hayward; skull photo by P. F. Wright of specimen in U.S. National Museum of Natural History; B. *R. tumida*, photo by David Pye.

R. mira, known only from the type locality in Michoacán (central Mexico);
R. tumida, Tamaulipas (northeastern Mexico) to Bolivia and southern Brazil, Trinidad;
R. genowaysi, Chiapas (southern Mexico);
R. minutilla, northeastern Colombia, northern Venezuela.

Baeodon was considered a distinct genus by Corbet and Hill (1986) but not by Honacki, Kinman, and Koeppl (1982) or Jones, Arroyo-Cabrales, and Owen (1988). Smith and Genoways (1974) thought that *R. minutilla* eventually would prove to be a geographic race of *R. parvula*. Based on chromosomal data, there have been suggestions that *R. tumida* actually comprises several biological species that are difficult, if not impossible, to distinguish by examination of skins and skulls (Baker, Bickham, and Arnold 1985; Honeycutt, Baker, and Genoways 1980). This view was confirmed with the description by R. J. Baker (1984) of *R. genowaysi*, which can be separated from *R. tumida* solely by chromosomal and genic means. This discovery serves as a warning that there may be numerous cryptic species of mammals that cannot be distinguished by classical systematic methods.

Head and body length is 37–50 mm, tail length is 28–48 mm, and forearm length is 25–35 mm. Adults usually weigh 3–10 grams. Coloration is yellowish brown or light brown above and slightly paler below. The genus is distinguished by features of the skull and dentition.

Little yellow bats probably occur in a variety of habitats. In Mexico, *R. gracilis* has a known altitudinal range of 600–2,000 meters (J. K. Jones 1977). In Venezuela, Handley (1976) found *R. tumida* mainly in humid lowlands, both in open and forested sites, and *R. minutilla* largely in arid thorn forest. These bats roost mostly in hollow trees but have also been found under palm fronds, in thatched roofs, and under boards. They have been observed in San Luis Potosi in slow, fluttery flight resembling that of *Pipistrellus*, usually from 1 to 4 meters above the ground. Definite hunting areas are generally established. When hunting over clearings, they usually fly in circles from 15 to 30 meters in diameter, but when over trails and roads, they fly back and forth for distances of about 30 meters. A given area may be shared by 2–5 bats.

LaVal (1973*b*) listed the following reproductive data: *R. gracilis*, pregnant females taken 15 May, two subadults taken 27 July; *R. parvula*, pregnant females collected from late February to early June, lactating females taken from late April to early July, and flying young taken from June to September; and *R. tumida*, pregnancy and lactation extending from mid-February to mid-July in Mexico, Central America, and Venezuela. Apparently, the usual litter size for the genus is two.

CHIROPTERA; VESPERTILIONIDAE; **Genus SCOTOMANES**
Dobson, 1875

Harlequin Bats

There are two species (Lekagul and McNeely 1977; Sinha and Chakraborty 1971):

S. ornatus, eastern India to southeastern China;
S. emarginatus, known only by the type specimen from an unknown locality in India.

A specimen from Indochina given the subgeneric designation *Parascotomanes* Bourret, 1942, and the specific name *beaulieui*, which sometimes has been referred to *Scotomanes*, was shown by Topal (1970*a*) to belong to the species *Ia io*. *S. emarginatus* was formerly placed in the genus *Nycticeius* and subgenus *Scoteinus* but was shown by Sinha and Chakraborty (1971) to belong to *Scotomanes*.

Head and body length is about 72–78 mm, tail length is 50–62 mm, and forearm length is 50–60 mm. *Scotomanes ornatus* can be identified by its unusual and attractive color pattern. In this species the upper parts are russet brown and

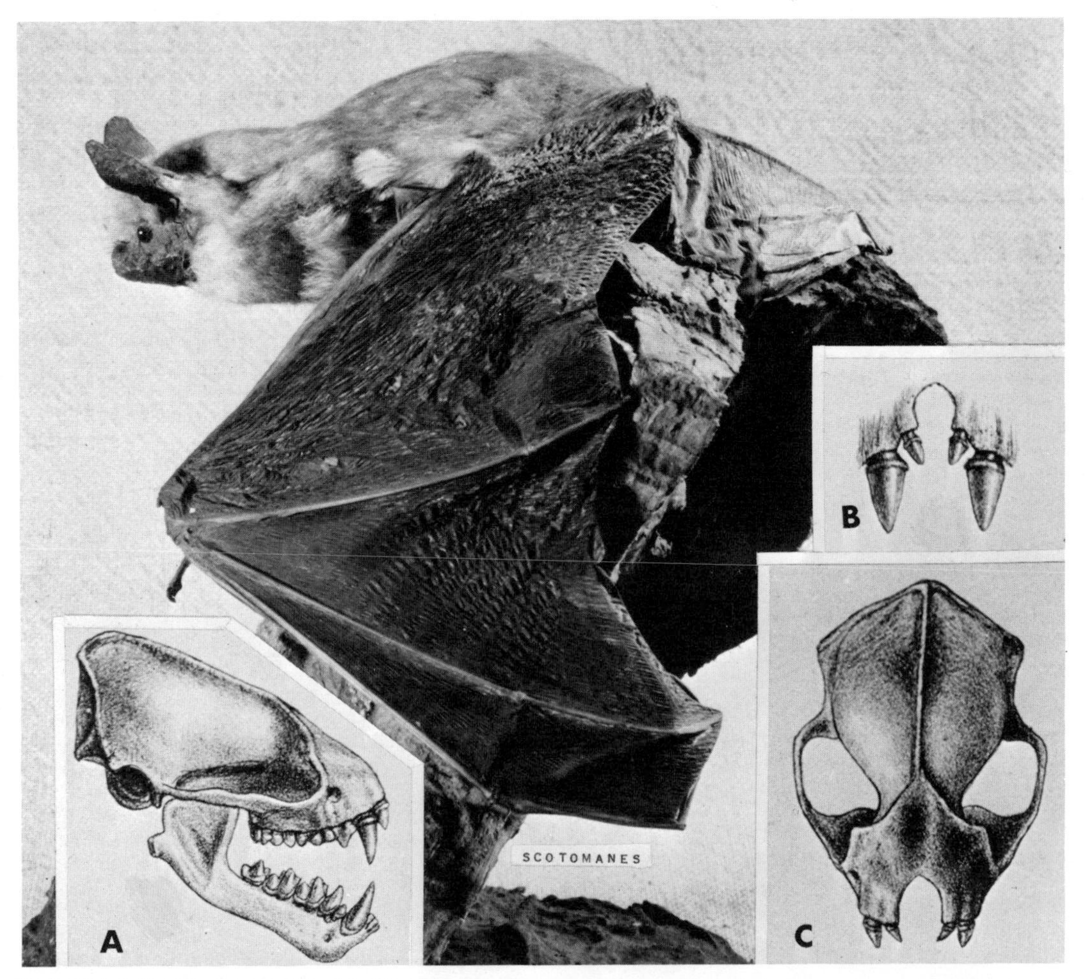

Harlequin bat *(Scotomanes ornatus)*, photo from Royal Scottish Museum through Tom Scott. A. Lateral view of skull; B. Upper incisors; C. Dorsal view of skull, photos from *Catalogue of the Chiroptera in the Collection of the British Museum*, G. E. Dobson.

the underparts are dark brown along the midline and whitish along the sides. There is a white stripe down the middle of the back, a patch of white hairs on the crown of the head, and two white spots just above the wing, behind each shoulder.

One specimen was taken in a cave, but the usual roosting site of *S. ornatus* is in the foliage and on the branches of trees, often from two to four meters above the ground. Because of their peculiar markings, a cluster of these bats hanging in a tree could be mistaken for fruit. One specimen was taken from a folded plantain leaf. Another individual flew into an inhabited cabin in the early evening, apparently in pursuit of insects.

CHIROPTERA; VESPERTILIONIDAE; **Genus SCOTOPHILUS**
Leach, 1821

House Bats, or Yellow Bats

There are 11 species (Corbet 1978; De Vree 1973; Ellerman and Morrison-Scott 1966; Hayman and Hill, *in* Meester and Setzer 1977; Hill 1980*b*; Koopman 1986; Laurie and Hill 1954; Lekagul and McNeely 1977; McLellan 1986; Meester et al. 1986; Robbins 1978, 1980, 1984; Robbins, De Vree, and Van Cakenberghe 1985; Schlitter et al. 1986):

S. nigrita, Senegal, Ghana, Benin, Nigeria, Sudan, Zaire, Kenya, Zimbabwe, Malawi, Mozambique;
S. nux, Sierra Leone to Kenya;
S. nucella, high forest zone of Ghana and Uganda, and probably similar habitat in other parts of West, Central, and East Africa;
S. dinganii, Africa south of the Sahara;
S. robustus, Madagascar;
S. leucogaster, drier areas south of the Sahara from Mauritania and Senegal to Ethiopia and northern Kenya, and from southern Angola and Namibia to Zambia, possibly South Yemen;
S. viridis, savannah zones from Senegal to Central African Republic and from southern Sudan to eastern South Africa, Madagascar;
S. borbonicus, Madagascar, Reunion in the Indian Ocean;
S. kuhlii, Pakistan to Indochina and Malay Peninsula, Sri Lanka, Taiwan, Hainan, Sumatra, Java, Bali, Borneo, Philippines, many small islands of East Indies;

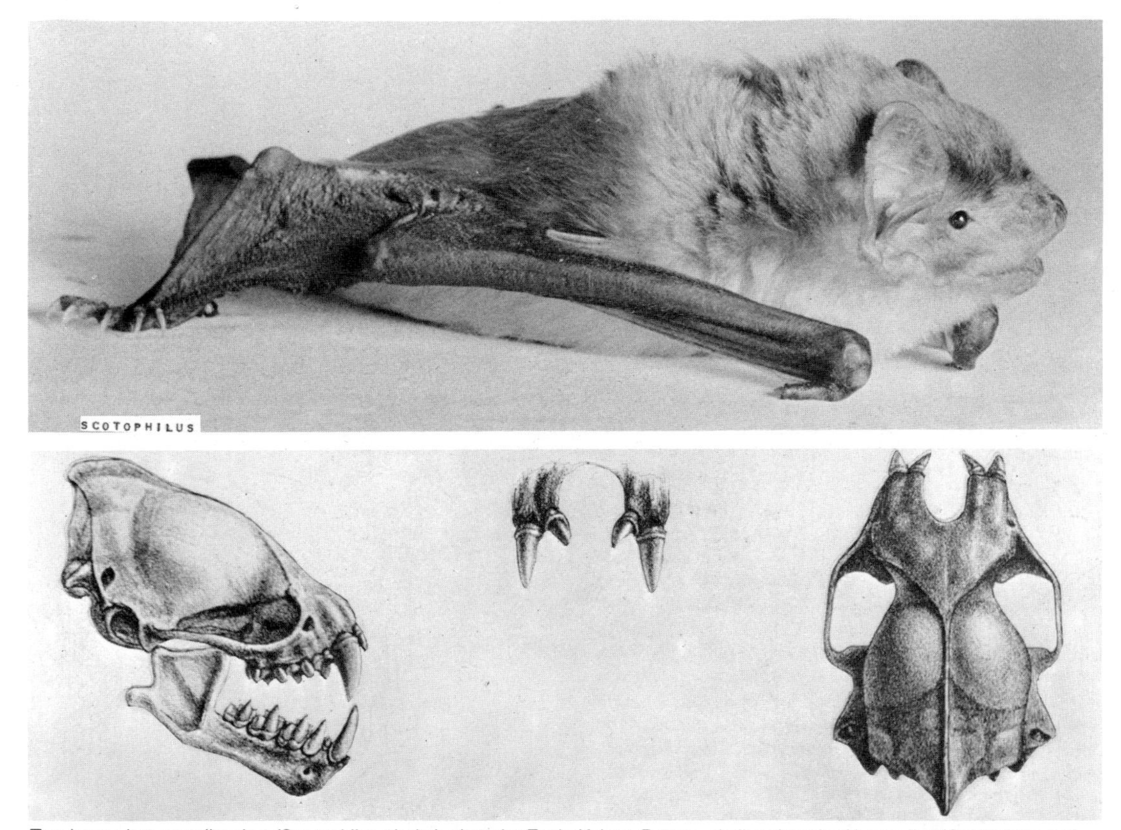

Top, house bat, or yellow bat *(Scotophilus nigrita)*, photo by Erwin Kulzer. Bottom, skull and teeth of house bat (*Scotophilus* sp.), from *Catalogue of the Chiroptera in the Collection of the British Museum*, G. E. Dobson.

S. heathi, Pakistan to southern China and Indochina, Sri Lanka, Hainan;
S. celebensis, Sulawesi.

Kitchener and Caputi (1985) indicated that *S. nigrita* does not belong in *Scotophilus*. Robbins, De Vree, and Van Cakenberghe (1985) did recognize *S. nigrita* as a very distinctive species of *Scotophilus*, though they noted that the name *S. gigas* often is applied to this species. There is much additional controversy regarding the systematics of this genus, especially as to whether various African mainland forms are conspecific with *S. borbonicus*. The originally described subspecies of the latter, *S. b. borbonicus* of Reunion, may have been extinct for over a century (Hill 1980*b*).

Head and body length is 60–117 mm, tail length is 40–65 mm, and forearm length is 42–89 mm. Lekagul and McNeely (1977) gave the weight of *S. kuhlii* as 15–22 grams. Fenton (1975) gave the average weight of *S. nigrita* as about 19 grams. The upper parts are often yellowish brown, and the underparts buffy, yellow, or white, but the coloration is quite variable in a given species. In Java, and perhaps throughout its range in Indonesia, *S. kuhlii* has two color phases, one a brilliant rufous, the other a dull olive brown.

These are rather heavy bodied, strongly built bats with powerful jaws and teeth. *Scotophilus* is distinguished by the structure of the molar teeth and the dental formula.

The members of this genus are common house-roosting bats over most of their range, usually sheltering in attics, often in those that are roofed with corrugated iron and having extremely high temperatures. Hollow trees, often palms, are also used as roosting sites, and a colony was found among the leaf stalks of a fan palm. In southern Africa they sometimes frequent abandoned woodpecker and barbet nests. These bats appear about dusk, in fairly steady and strong flight, and generally feed from 3 to 12 meters above the ground. The diet consists of beetles, termites, moths, and other insects.

In some areas the roosting colonies are fairly small, generally fewer than 20 individuals, while in other areas they may number in the hundreds. Female *S. kuhlii* in India are reported to assemble in maternity colonies comprising several hundred individuals. This species breeds in March. The gestation period is about 105–15 days, and the young number 1–2.

Another monestrous Indian species, *S. heathi*, mates in January or early February; sperm is then stored 60–80 days, ovulation takes place in early March, gestation lasts about 116 days, parturition is in July, two young are usually produced, and lactation lasts 2–3 weeks (Krishna and Dominic 1981). In Zimbabwe, pregnant or lactating female *S. nigrita* have been taken from October to December, and the normal number of embryos is two (Smithers 1971). Births of this species, often twins, have been recorded in July in Sudan and in January–March in Tanzania and Uganda, there possibly being a second period of births in August–September. Pregnant female *S. leucogaster*, each with two embryos, have been taken in February–March (Kingdon 1974*a*). Pregnant female *S. dinganii*, also with two embryos each, were collected in Zambia in August and September, and a lactating female of that species was taken there in December (Ansell 1986).

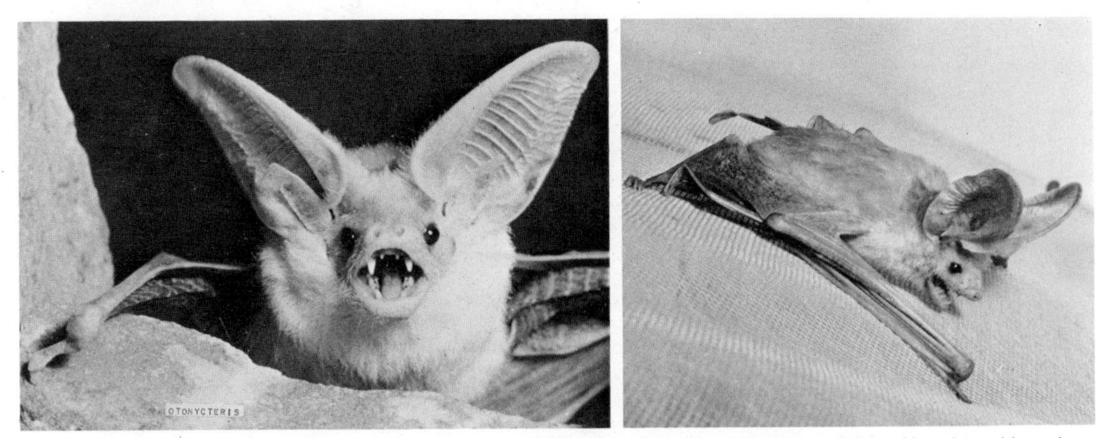

Hemprich's long-eared bat *(Otonycteris hemprichi)*, photos by Erwin Kulzer. The skin of the wings and feet of bats is so thin and tender that cages for newly captured bats should be lined with coarse mesh cloth, as in these pictures.

CHIROPTERA; VESPERTILIONIDAE; **Genus OTONYCTERIS**
Peters, 1859

Desert Long-eared Bat

The single species, *O. hemprichi*, occurs in the desert zone from Morocco and northern Niger through Egypt and the northern Arabian Peninsula to Kazakh S.S.R. and Pakistan (Aulagnier and Mein 1985; Corbet 1978; Fairon 1980; Roberts 1977; Shaimardanov 1982).

Head and body length is about 73–81 mm, tail length is about 47–70 mm, and forearm length is 57–67 mm. Gaisler, Madkour, and Pelikan (1972) gave the weight of two adult males as 18 and 20 grams. The coloration above is pale sandy to dark brown; the underparts are usually whitish.

The large ears, about 40 mm in length, are directed nearly horizontally and are connected across the forehead by a low band of skin. Five female specimens revealed two pairs of pectoral mammae, a unique condition in mammals. It is not known whether both pairs are functional. The skull and teeth resemble those of *Eptesicus*.

Harrison (1964) stated that this bat is capable of inhabiting extremely barren and arid regions. In the Negev Desert a pair was found roosting in a rocky crevice on a hill. This bat has also been found in buildings and reportedly has a slow, floppy flight. Based on an analysis of its body mass, low aspect ratio, and low relative wing loading, Norberg and Fenton (1988) predicted that *Otonycteris* will be found to be carnivorous in diet. Three pregnant females, each with two embryos, were found in a deserted hut in Jordan on 2 May (Atallah 1977). Another female with two fetuses was taken in Soviet Central Asia on 12 June (Roberts 1977).

CHIROPTERA; VESPERTILIONIDAE; **Genus LASIURUS**
Gray, 1831

Hairy-tailed Bats

There are 13 species (Baker et al. 1988; Cabrera 1957; Dinerstein 1985; Genoways and Baker 1988; Hall 1981; Koopman and Gudmundsson 1966; Krzanowski 1977; Webster, Jones, and Baker 1980; Wilkins 1987):

L. intermedius, Atlantic and Gulf coasts of United States, western and eastern coasts of Mexico, southern Mexico to Honduras, Cuba and nearby Isle of Pines;
L. egregius, eastern Panama to southern Brazil;
L. ega (yellow bat), southern Texas and eastern Mexico to northeastern Argentina;
L. xanthinus, southwestern United States, central and western Mexico, Baja California;
L. cinereus (hoary bat), north-central and southern Canada, conterminous United States, Mexico, Guatemala, Colombia and Venezuela to Chile and Argentina, Bermuda, Hawaii, Iceland, Orkney Island north of Scotland;
L. brachyotis, Galapagos Islands;
L. castaneus, Costa Rica, eastern Panama;
L. seminolus, Pennsylvania, southeastern United States, possibly northeastern Mexico, occasionally Bermuda;
L. pfeifferi, Cuba;
L. degelidus, Jamaica;
L. minor, Bahamas, Hispaniola, Puerto Rico;
L. borealis (red bat), Alberta and Nova Scotia to Chihuahua (northern Mexico) and Florida;
L. blossevillii, southern British Columbia through Utah and Mexico to Chile and Argentina.

Hall (1981) used the name *Nycteris* Borkhausen, 1797, for this genus. Hall and most other American authors have not continued to recognize *Dasypterus* Peters, 1871, as a separate genus for the yellow bats, *L. intermedius*, *L. egregius*, and *L. ega*. Some authorities, including Buden (1985), Corbet and Hill (1986), and Honacki, Kinman, and Koeppl (1982), have included *L. pfeifferi*, *L. degelidus*, and *L. minor* within *L. borealis*, but Bickham (1987) disagreed on the basis of electrophoretic data. Chromosomal data led Baker et al. (1988) to recognize *L. degelidus* as a full species and to suggest that it, *pfeifferi*, and *minor* might actually be subspecies of *L. seminolus*. Corbet and Hill (1986) also listed *L. insularis* of Cuba as a species distinct from *L. intermedius*.

Head and body length is 50–90 mm, tail length is 40–75 mm, and forearm length is 37–58 mm. Adults weigh from 6 to 30 grams. The red bats are brick red to rusty red, usually washed with white; the hoary bats receive their common name from the silver frosting on the yellowish brown to mahogany brown hairs of the upper parts; and the yellow bats are whitish buff, yellowish, or orange, usually with a blackish wash. Male red bats tend to be more brightly colored than the females. The tail membrane is well furred above.

Red bats *(Lasiurus borealis)*, mother with nursing baby, photo by Ernest P. Walker.

These bats generally occur in wooded areas and roost in foliage, or occasionally in tree holes and buildings. The distribution of *L. seminolus* nearly coincides with that of Spanish moss, within clumps of which it roosts for most of the year. Hairy-tailed bats usually appear early in the evening. Feeding flights are usually 6–15 meters above the ground. Most insects are captured in flight, but *L. borealis* will alight on vegetation to pick off insects.

Populations of *L. seminolus* may shift southward in the fall, and this species is known to become torpid in cold weather (Barbour and Davis 1969; Lowery 1974). Both *L. borealis* and *L. cinereus* are highly migratory, sometimes moving south many hundreds of kilometers during the fall and occasionally landing on ships at sea or on oceanic islands, especially Bermuda. Migratory movements of these species are not well documented, but it appears likely that from September to November populations in Canada and the northern United States move south to the central and southern United States. Winter populations of *L. borealis* build up in Missouri and Kentucky but may not increase substantially farther to the south. During the winter limited hibernation may take place, but the bats awake and forage on warmer days. Northward migration begins as early as March or April and reaches Canada by late May (Banfield 1974; Barbour and Davis 1969, 1974; Lowery 1974; Schwartz and Schwartz 1959). The population density of *L. borealis* was estimated at 1 per 0.4 ha. in part of Iowa and 1 per ha. in Indiana (Mumford 1973).

Available evidence suggests that these bats are generally solitary but that the females of some species form small nursery colonies and that flocks of up to several hundred individuals form for migration. During the summer the sexes of *L. cinereus* apparently segregate, with males concentrated in the western United States and females and young in the east. Barbour and Davis (1969) gave the following information on *L. intermedius*. At least 50 females with their young were once found in less than 0.4 ha. of live oaks; during the summer female aggregations are formed, but males are rare therein; and during the winter males apparently congregate. In populations of *Lasiurus* in temperate regions, mating takes place in the late summer or fall, sperm is stored over the

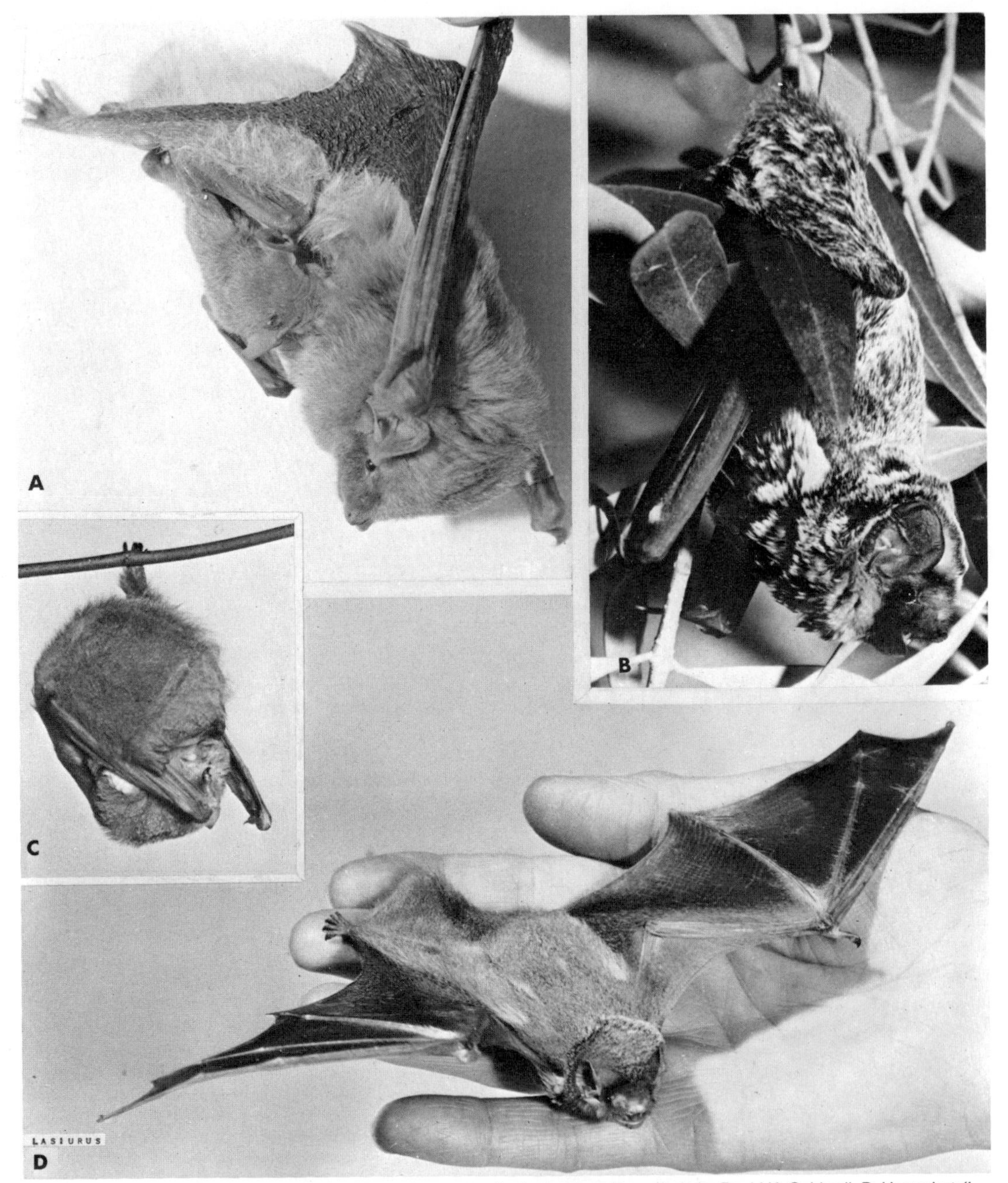

A. Hairy-tailed bat *(Lasiurus intermedius)*, with nursing young clinging to its mother, photo by David K. Caldwell. B. Hoary bat *(L. cinereus)*, photo by Bruce Hayward. Red bat *(L. borealis):* C. In sleeping position; D. Resting on hand, photos by Ernest P. Walker.

winter in the uterus, ovulation and fertilization occur in the spring, and births take place from late May to early July. Barbour and Davis (1969) gave the estimated gestation period of *L. borealis* as 80–90 days. Greer (1965) collected female *L. borealis*, each with 2 young, in Chile in December and January.

Lasiurus is the only genus of bats in which there are commonly more than two young per birth. *L. cinereus, L. intermedius,* and *L. seminolus* are known to produce litters of up to four young (Banfield 1974; Barbour and Davis 1969; Lowery 1974), while there are two records of female *L. borealis* being found with five clinging young (Hamilton and Stalling 1972). The normal litter size in all species apparently is two or three, but single young have been reported for *L. ega, L. intermedius,* and *L. cinereus* (Birney et al. 1974; Bogan 1972). Old observations suggested that females initially carried their litters during foraging, but R. Davis (1970) found no evidence for the transportation of young by *Lasiurus* or any other North American bat, except in cases of disturbance.

The Hawaiian hoary bat *(L. cinereus semotus)* is desig-

Barbastelle *(Barbastella barbastellus)*, photo by Erwin Kulzer.

nated as endangered by the USDI and as indeterminate by the IUCN. Comparatively little is known about this subspecies, but it is thought to have declined because of loss of its forest habitat. A few thousand individuals may still survive. The species *L. brachyotis*, one of the few mammals native to the Galapagos Islands, was observed regularly in the late nineteenth century (Allen 1942) but was listed as extinct by Goodwin and Goodwin (1973).

CHIROPTERA; VESPERTILIONIDAE; **Genus BARBASTELLA**
Gray, 1821

Barbastelles

There are two species (Corbet 1978; Ibáñez and Fernandez 1985*a*):

B. barbastellus, Europe and the larger Mediterranean islands, Morocco, Canary Islands;
B. leucomelas, from the Caucasus, through Iran and Central Asia, to Japan and Indochina.

Corbet (1978) also mentioned old records from Sinai, Senegal, and Eritrea in Ethiopia, each of which is dubious or, if valid, might apply to either species. Qumsiyeh (1985), however, considered *leucomelas* to be only a subspecies of *B. barbastellus* and accepted the presence of this subspecies in Sinai and southern Israel.

Head and body length is 43–60 mm, tail length is 40–55 mm, and forearm length is 35–45 mm. Adults usually weigh 6–10 grams. The pelage is soft and long. In *B. barbastellus* the coloration is blackish, the tips of the hairs being whitish or yellowish to impart a "silvered" or "frosted" appearance, and the underparts are somewhat paler. *B. leucomelas* is uniformly dark brown above and below, the tips of the hairs being gray to buffy on the upper parts and whitish on the underparts.

The fur extends onto the tail membrane and the wings. The nostrils of *Barbastella* open upward and outward behind a median pad. The ears, conspicuously notched on their external margins, are short, broad, and united by a low band across the forehead. This genus resembles *Plecotus* and *Euderma* in most skull features and in dental characters but lacks the extreme auditory specializations of those genera.

The flight is described as relatively slow and flapping. Barbastelles often fly near the ground but sometimes feed at fairly high altitudes, particularly during good weather. *B. barbastellus* is reported to appear early, not later than sunset, and *B. leucomelas* is said to emerge rather late in the evening. There is some indication that these bats wander considerably, and they sometimes emerge from hibernation and fly around during the winter. They hibernate in caves, usually in drier regions, from late September to early April. They have been found in caves with *Myotis, Pipistrellus,* and *Plecotus.* The common roosts in summer are trees, generally on and under bark, and buildings. From one to about half a dozen individuals are usually found in a given roost, but during the breeding season barbastelles apparently congregate in fairly large numbers. Breeding habits are not known. One captive barbastelle picked houseflies off a ceiling and also fed on mealworms, and another individual readily accepted small pieces of meat.

Stebbings and Griffith (1986) designated *B. barbastellus* as vulnerable worldwide and as possibly endangered in western Europe. One of the rarest bats in Europe, it appears to be declining through habitat disturbance, pollution, and loss of hollow trees.

CHIROPTERA; VESPERTILIONIDAE; **Genus PLECOTUS**
E. Geoffroy St.-Hilaire, 1818

Lump-nosed Bats, Long-eared Bats, or Lappet-eared Bats

There are two subgenera and six species (Corbet 1978; Hall 1981; Ibáñez and Fernandez 1985*b*):

subgenus *Plecotus* E. Geoffroy St.-Hilaire, 1818

P. auritus, Europe to Japan and the Himalayas;
P. teneriffae, Canary Islands;
P. austriacus, southern Europe and northern Africa to Mongolia and western China, Cape Verde Islands, Senegal;

subgenus *Corynorhinus* H. Allen, 1865

P. mexicanus, northern and central Mexico, northern Yucatan;
P. rafinesquii, Indiana, southeastern United States;
P. townsendii, southwestern Canada, western conterminous United States, Ozark and central Appalachian regions, Mexico.

Corynorhinus formerly was treated as a separate genus; both it and *Idionycteris* came to be regarded as subgenera of *Plecotus,* but *Idionycteris* now again is usually considered distinct. Corbet (1978) listed *P. teneriffae* as part of *P. austriacus,* but Ibáñez and Fernandez (1985*b*) regarded it as a distinct species with closer affinity to *P. auritus.*

Head and body length is 45–70 mm, tail length is 35–55 mm, forearm length is 35–52 mm, and adult weight is 5–20 grams. The upper parts are brown, the underparts are paler, and the ears and membranes are brownish. These bats may be distinguished by their large ears, up to 40 mm in length, and by the glandular masses on the muzzle, which in some forms rise to above the muzzle as flaplike lumps.

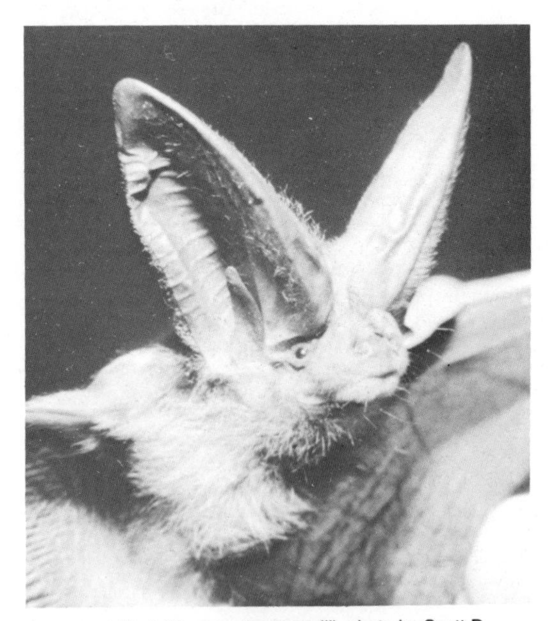

Long-eared bat *(Plecotus townsendii)*, photo by Scott D. Keefer.

Barbour and Davis (1969) listed the following habitat conditions: *P. rafinesquii,* forest; and *P. townsendii* desert scrub, pinon-juniper or pine forest. Lump-nosed bats roost in caves, tunnels, buildings, and trees. In Europe *P. auritus* is usually found in buildings and trees in the summer and in caves in the winter, whereas *P. austriacus* usually occurs in buildings during the summer and in buildings or cellars in the winter (Horacek 1975). *P. townsendii* generally shelters in caves throughout the year, except in parts of the west, where it inhabits buildings (Barbour and Davis 1969). *P. rafinesquii* most frequently roosts in partially lighted, mostly unoccupied buildings but is also found in caves, trees, and other natural shelters (C. Jones 1977). These bats do not begin flying until after dark. Their flight is slow, and they are able to hover like a butterfly at a point that interests them. The ears are thrown forward when they search for insects in foliage. These bats also hunt insects in flight. The prey is often picked off the foliage of trees and bushes and the walls of buildings while the bat is hovering, hummingbirdlike. All investigated species have been found to hibernate for at least part of the winter. While there may be changes between summer and winter roosting sites, apparently there are not extensive migrations. Humphrey and Kunz (1976) found *P. townsendii* in western Oklahoma and Kansas to be relatively sedentary, with over 80 percent of 88 bats being recovered at the same site as banded. Overall population density in this region varied from 1 per 40 ha. to 1 per 53 ha.

These bats are gregarious. Horacek (1975) found the usual number of individuals in summer colonies to be 5–10 in *P. auritus* and 10–20 in *P. austriacus.* C. Jones (1977) stated that *P. rafinesquii* roosts singly or in clusters of 2–100, with females greatly outnumbering males in most colonies. Barbour and Davis (1969) wrote that maternity colonies of *P. townsendii* contain up to 1,000 females; this species is known to roost in clusters. Based on a study in the karst region of western Oklahoma and Kansas, Humphrey and Kunz (1976) provided the following social information on *P. townsendii:* during the summer males roost alone or in groups of up to 6 bats, nursery colonies contain 17–40 adult females, and winter hibernacula have 1–62 individuals.

P. auritus breeds in the fall and spring. In *P. townsendii* most mating occurs in the winter roost, and the sperm then stored in the reproductive tract of the female remains motile for 76 days. The gestation period is 56–100 days, apparently being dependent upon body temperature, and the young are born from late May through July. In *P. rafinesquii* the females give birth to a single young in late May or early June, and the young is capable of nonagile flight at 15–18 days (C. Jones 1977). The longevity record for the genus appears to be held by a female *P. rafinesquii* that had a life span of at least 10 years and 1 month (Paradiso and Greenhall 1967).

Horacek (1975) suggested that *P. austriacus* may have spread through Europe in historical time, because of its ability to utilize human structures for roosting. Nonetheless, there is concern about the future of this and several other species of *Plecotus.* Piechocki (1966) stated that numbers of *P. austriacus* and *P. auritus* in central Germany had declined considerably in recent years because of environmental disruption. Stebbings and Griffith (1986) designated both species as vulnerable throughout their ranges. A new problem is the use of toxic chemicals to treat the wood in the buildings on which the bats have come to depend (see end of account of *Myotis*).

Humphrey and Kunz (1976) observed that the total population of *P. townsendii* on the southern plains of the United States is at most 14,000, that the species is barely persisting because of precise natural requirements, and that avoidance

A & D. *Plecotus auritus,* photos by Walter Wissenbach. B & C. Eastern lump-nosed bat *(P. rafinesquii),* photos by Ernest P. Walker.

of human disturbance is essential for its continued existence. The entire species is classified as indeterminate by the IUCN. The subspecies *P. townsendii ingens* of the Ozark region and *P. t. virginianus* of the central Appalachians are listed as endangered by the USDI. The former subspecies may number only about 700 individuals, the latter about 11,000. These bats have declined because of direct killing by people and through abandonment of roosting caves when disturbed by explorers and vandals.

CHIROPTERA; VESPERTILIONIDAE; **Genus IDIONYCTERIS**
Anthony, 1923

Allen's Big-eared Bat

The single species, *I. phyllotis*, is found in mountainous regions from the southwestern United States to central Mexico (Czaplewski 1983). Although Hall (1981) considered *Idionycteris* a subgenus of *Plecotus*, there now has been general acceptance of the suggestion by Williams, Druecker, and Black (1970), on the basis of karyotype, that the two are generically distinct (Corbet and Hill 1986; Honacki, Kinman, and Koeppl 1982; Jones et al. 1986). The information for the remainder of this account was taken from Czaplewski (1983).

Head and body length is about 50–60 mm, tail length is 44–55 mm, forearm length is 42–49 mm, and weight is 8–16 grams. The dorsal pelage is long, soft, and basally blackish with tips a contrasting yellowish gray. The ventral hairs are black basally with pale buffy tips. The ears, which are 34–43 mm long, are as large as those of *Plecotus*, but *Idionycteris* is distinguished by a pair of fleshy lappets projecting over the forehead from the anterior bases of the ears. *Plecotus* also differs in having keeled calcars and nostrils that are not elongated posteriorly.

Allen's big-eared bat dwells primarily in forested mountainous areas, occasionally in desert scrub, and usually in the vicinity of rocks or cliffs. Maternity colonies have been found roosting in tunnels and piles of boulders. Open-air flight is fast, direct, and often accompanied by loud "peeps." Foraging flight is slow, highly maneuverable, and characterized by long-constant frequency echolocation. The diet consists largely of moths.

Seasonal movements and winter location and activity are unknown. During the summer the sexes segregate, with females gathering into maternity colonies of about 25–100 individuals and males possibly remaining solitary. Pregnant females, each with a single embryo, have been collected in June in New Mexico, Arizona, and Durango. Lactating females have been reported from June until early August, and flying young as early as 31 July.

CHIROPTERA; VESPERTILIONIDAE; **Genus EUDERMA**
H. Allen, 1892

Spotted Bat, or Pinto Bat

The single species, *E. maculatum*, is found in southern British Columbia, the western conterminous United States, and northern Mexico (T. L. Best 1988; Hall 1981; Watkins 1977).

Head and body length is 60–77 mm, tail length is 47–51 mm, forearm length is 44–55 mm, and weight averages about 15 grams; the ears, larger than those of any other American bat, measure 34–50 mm from notch to tip (T. L. Best 1988; Watkins 1977). The color is dark reddish brown to black, with a characteristic white spot on each shoulder and a white spot at the base of the tail. The hairs of the underparts are tipped with white, and the ears and membranes are grayish. The color pattern of *Euderma* is unique.

The spotted bat is known from 57 meters below sea level to the high transition zone of Yosemite National Park in California. It has been reported from a wide variety of habitats but has been collected most often in dry, rough desert country (Watkins 1977). Several have been taken in and on houses; others have been obtained in caves or cavelike structures and around water. Easterla (1973) observed 13 banded *Euderma* after release to ignore trees and fly to cliffs, where many entered crevices or other retreats under loose rocks or boulders. He suggested that distribution is determined mainly by availability of suitable cliff habitat. Poché and Bailie (1974) netted 4 male *Euderma* over scattered pools in Utah. When one was released, it was seen to drop to the ground, where it seized and ate a grasshopper. Other observations indicate that the spotted bat feeds mainly on moths and other insects caught in flight. Indeed, recent studies (Leonard and Fenton 1983; Woodsworth, Bell, and Fenton 1981) show that the spotted bat is not a ground feeder or gleaner but a fast-flying, high-level forager. One individual was seen to fly around a 50-ha. area for about an hour, always remaining at or above treetop height, 10–30 meters above the ground; it made six attempts to catch insects during a 44-minute period.

Although four individuals once were reported hibernating on the walls of a cave above a large pool of water in Utah, the recent studies confirm that *Euderma* is usually solitary and roosts alone in steep cliff faces. It also defends an exclusive foraging territory. Apparently there is a single young. Births have been observed in western Texas in early June; lactating females have been taken in Texas and New Mexico in June and early July and in Utah in August (Barbour and Davis 1969; Easterla 1973, 1976).

The spotted bat was unknown until 1890; the second specimen was found 13 years later; and the first collections in Canada, Oregon, and Colorado have been reported only in the last decade (Finley and Creasey 1982; McMahon, Oakley, and Cross 1981; Woodsworth, Bell, and Fenton 1981). The species sometimes has been called America's rarest mammal. There have been an increasing number of observations recently, and Easterla (1973) netted 54 individuals in Big Bend National Park, Texas. Nonetheless, a range-wide survey involving the monitoring of echolocation calls (Fenton, Tennant, and Wyszecki 1987) confirmed that the species is generally rare and is perhaps numerous only in a few restricted localities.

CHIROPTERA; VESPERTILIONIDAE; **Genus MINIOPTERUS**
Bonaparte, 1837

Long-winged Bats, or Bent-winged Bats

There are 11 species (Aggundey and Schlitter 1984; Baeten, Van Cakenberghe, and De Vree 1984; Corbet 1978; Ellerman and Morrison-Scott 1966; Hayman and Hill, *in* Meester and Setzer 1977; Hill 1971*c*, 1974*b*, 1982*a*, 1983*b*; Hill and Beckon 1978; Koopman 1982*a*, 1984*a*; Laurie and Hill 1954; Lekagul and McNeely 1977; McKean 1972; Medway 1977; Qumsiyeh and Schlitter 1982; Ride 1970; Taylor 1934):

M. minor, Kenya, Tanzania, lower Congo River, Sao Tome Island, Madagascar, Comoro Islands;
M. inflatus, Liberia, Cameroon, Gabon, southern Zaire, western Kenya, Rwanda, Ethiopia;

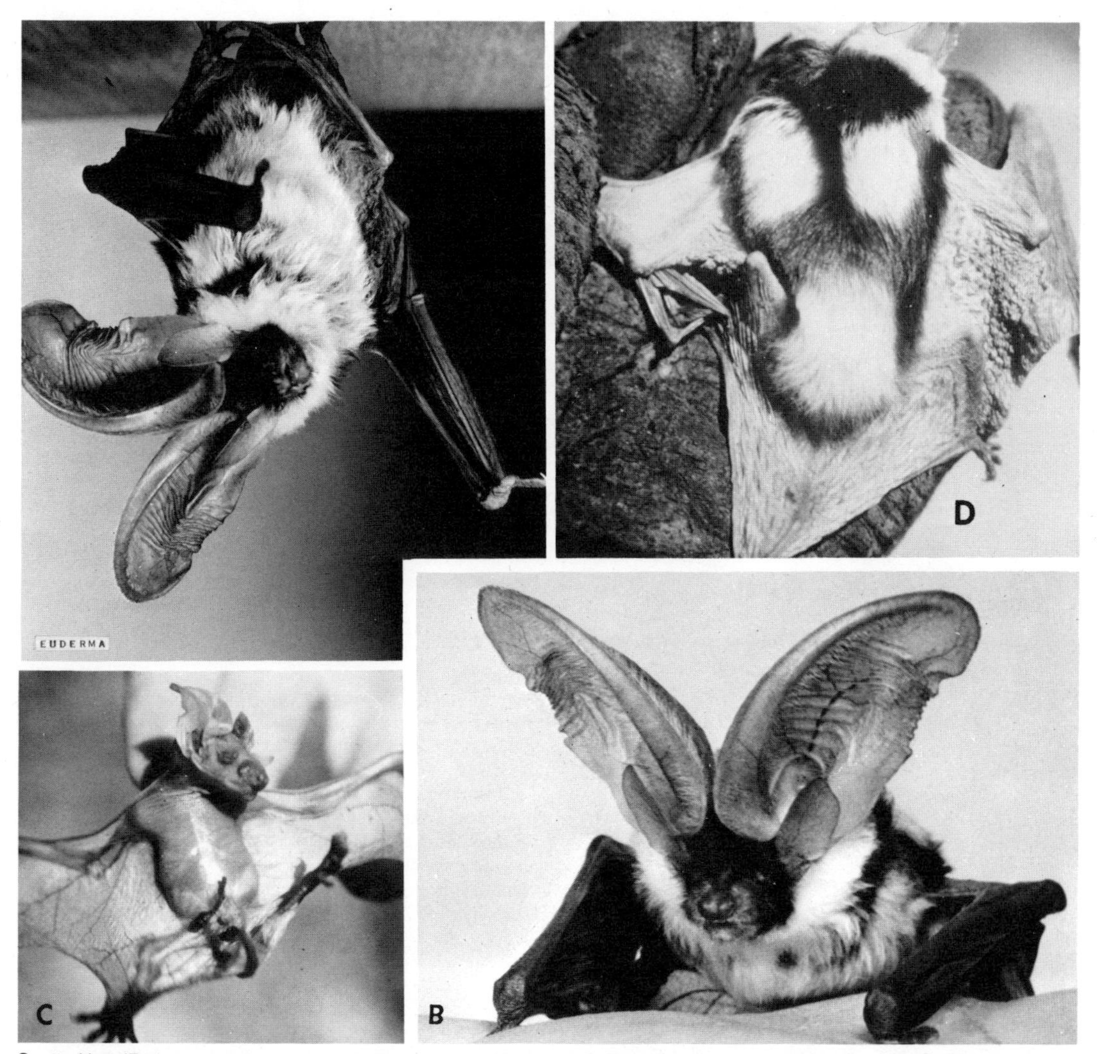

Spotted bat *(Euderma maculatum):* A & B. Photos by Sam Learned; C. Baby, 24 hours old, photo by David A. Easterla; D. Photo by David A. Easterla.

M. fraterculus, South Africa, southern Malawi, possibly Kenya and Zambia;
M. schreibersi, southern Europe to Japan and the Solomon Islands, Philippines, northern Africa, most of Africa south of the Sahara, northern and eastern Australia;
M. magnater, Yunnan (southern China), Hong Kong, Burma, Thailand, Malay Peninsula, Sumatra, Java, New Guinea, Bismarck Archipelago;
M. australis, Philippines, islands from Java and Borneo to New Caledonia and the Loyalty Islands, eastern Australia;
M. pusillus, southern India, Thailand, Nicobar Islands, Sumatra, Java, possibly Benguet Island (Philippines), Sulawesi, Moluccas, New Guinea, Solomons, New Hebrides, New Caledonia, Loyalty Islands;
M. medius, Malay Peninsula, Java, Borneo, Sulawesi, New Guinea;
M. fuscus, Ryukyu Islands, Okinawa;
M. tristis, Philippines, Sulawesi, New Guinea, Solomon Islands, New Hebrides;
M. robustior, Loyalty Islands (South Pacific).

There is considerable disagreement regarding the classification of this genus. Maeda (1982) described and/or distinguished 11 species in addition to those listed above, but Hill (1983*b*) accepted none of them. *M. blepotis* (Malay Peninsula, Borneo, Java, Amboina and Kei islands), *M. eschscholtzi* (Philippines, Australia), and *M. haradai* (Thailand) were considered by Maeda to form a related group, but they were regarded as components of *M. schreibersi* by Hill. *M. macrodens* (Burma and southern China to Java and Borneo), *M. oceanensis* (eastern Australia), and *M. fuliginosus* (Afghanistan and India to Japan and northern Viet Nam) were considered by Maeda to form another group that also included *M. magnater,* but while Hill did regard *macrodens* as a subspecies of *M. magnater,* he considered *oceanensis* and *fuliginosus* to be components of *M. schreibersi. M. solomonensis* (Solomon Islands) and *M. paululus* (Philippines, Java, Rennell Island in the Solomons) were considered by Maeda to be species related to *M. australis,* but Hill considered them to be only subspecies thereof. *M. macrocneme* (New Guinea to New Caledonia), considered by Maeda to be a species related to *M. fuscus,* was designated a subspecies of

Long-winged bat, or long-fingered bat (*Miniopterus* sp.), photo by Howard E. Uible. Inset: face *(M. minor)*, photo by David Pye.

M. pusillus by Hill. *M. melanesiensis* (Solomon Islands, New Hebrides) and *M. bismarckensis* (Bismarck Archipelago) were described by Maeda as species related to *M. tristis*, but Hill synonymized *melanesiensis* under *M. tristis* and suggested that *bismarckensis* is part of *M. magnater.* Koopman (1984*a*) supported recognition of *M. paululus* and *M. macrodens* as full species and indicated that the range of both includes Timor. Several reports of *Miniopterus* in South Australia, cited by Maeda, were not accepted by Hill or Koopman. Peterson (1981) described a new species, *M. propitristis*, from Sulawesi, New Guinea, and other islands extending eastward to the New Hebrides, but Hill (1983*b*) considered it a subspecies of *M. tristis.* Peterson (cited by Wilson 1985) accepted *M. oceanensis* as a valid taxon.

Head and body length is 40–78 mm, tail length is 40–67 mm, forearm length is 37–55 mm, and adult weight is usually 6–20 grams. The coloration is reddish, reddish brown, dark brown, grayish brown, or grayish.

The second bone of the longest finger is about three times as long as the first bone. When the bat hangs by its hind feet, this lengthened terminal part of the third finger folds back upon the wing. The tail is completely enclosed within the interfemoral membrane and is proportionately longer than in other bats of the same size.

The members of this genus usually roost in caves but have also been found in rock clefts, culverts, eaves and roofs of buildings, and crevices of trees. They are often associated with *Notopteris* and species of *Myotis* in their daytime retreats. They appear early in the evening, with a rapid, jerky flight. They feed on small beetles and other insects, usually at heights of 10–20 meters. *Miniopterus* hibernates in the cooler parts of its range.

In a study of *M. schreibersi* in South Africa, Van der Merwe (1975) found that seasonal migrations occurred. From late winter to late spring there was a movement of pregnant females from wintering caves in the southern Transvaal to maternity caves in the north. In late summer the females and weaned young moved back to the south. Studies of this species in India showed that the population of a given area tended to be centered in one large cave but that individuals spent part of their time in secondary roosts within a 70-km radius (Lekagul and McNeely 1977).

These bats may be highly gregarious. Van der Merwe (1978) reported that in one maternity cave of *M. schreibersi* in the Transvaal the juveniles alone numbered 110,000 (each female gives birth to a single young from early November to early December). His studies, and others in Asia (Lekagul and McNeely 1977), showed that young are not carried by the mother but are deposited in a large communal nursery, in clusters separate from those of the adults. Lekagul and McNeely stated that the typical roosting group of *M. australis* consists of only one male and six females and that such groups are transitory, breaking up when the bats go out to feed. Dwyer (1968), however, reported a nursery colony of 4,000 *M. australis*, including 1,800 young, associated with a much larger group of *M. schreibersi* in New South Wales. Males were found to disappear from the colony by December, when the young were born. In Europe mating occurs in August and September, fertilization occurs shortly thereafter, embryonic development is retarded through hibernation, and the young are not born until spring. Females are ready for breeding again after the young are weaned. In a study in eastern Australia, Richardson (1977) found both *M. schreibersi* and *M. australis* to be monestrous. In *M. schreibersi* mating took place in the fall (late May–early June), with fertilization and development to the blastocyst stage following immediately; implantation was then delayed until August, and births occurred in December. In *M. australis* mating was in August, implantation occurred by mid-September with little or no delay, and births took place in December.

Brosset and Saint Girons (1980) found *M. inflatus* to be monestrous and strictly a seasonal breeder in Gabon; there was no delayed implantation, females gave birth to a single young in October, and pregnancy and lactation together lasted six months. The longevity record in the genus is nine years.

Stebbings and Griffith (1986) regarded *M. schreibersi* as endangered in western Europe and possibly throughout the

Tube-nosed insectivorus bat *(Murina leucogaster)*, photo by Modoki Masuda, Nature Productions, Japan.

world. Several colonies that formerly contained thousands of individuals have almost or entirely disappeared. This species is very sensitive and may be eliminated if its roosts in caves and mines are disturbed by human workers or tourists.

CHIROPTERA; VESPERTILIONIDAE; **Genus MURINA**
Gray, 1842

Tube-nosed Insectivorous Bats

There are 2 subgenera and 14 species (Chasen 1940; Corbet 1978; Ellerman and Morrison-Scott 1966; Hill 1962, 1963*b*, 1972*b*; Hill and Francis 1984; Laurie and Hill 1954; Maeda 1980; Medway 1977, 1978; Sly 1975; Taylor 1934; Yoshiyuki 1983):

subgenus *Murina* Gray, 1842

M. aurata, mountains of Sichuan and Yunnan (south-central China), Nepal, Sikkim, northern India, Burma;
M. ussuriensis, Ussuri region of southeastern Siberia, Manchuria, Korea, Sakhalin, Kuril Islands;
M. silvatica, Japan;
M. leucogaster, southern Siberia to central China and Japan;
M. tenebrosa, known only by a single specimen from Tsushima Island (Japan), possibly also on Yakushima Island;
M. tubinaris, Kashmir, northeastern India, Burma, Laos, northern Viet Nam;
M. suilla, Malay Peninsula, Sumatra and nearby Nias Island, Java, Borneo;

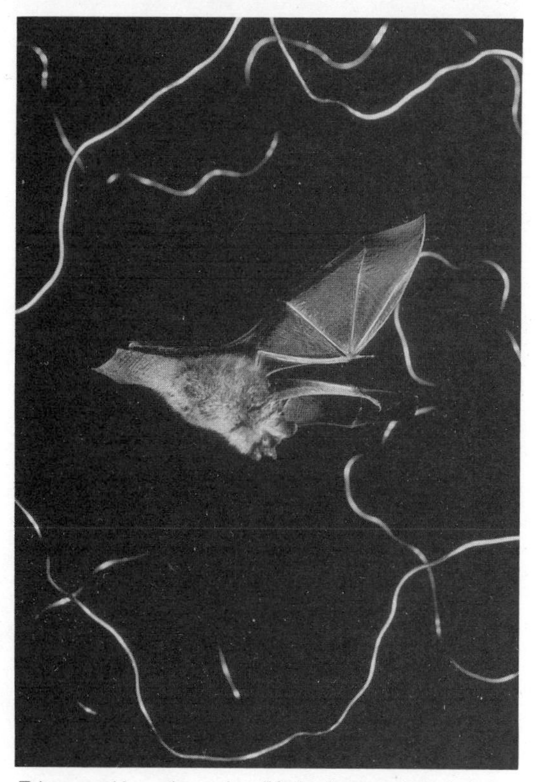

Tube-nosed insectivorus bat *(Murina leucogaster)*, photo by Modoki Masuda, Nature Productions, Japan.

M. florium, Sumbawa, Flores, Buru, Ceram, Goram, possibly Amboina and New Guinea, Queensland;
M. huttoni, northern India to southeastern China, Malay Peninsula;
M. cyclotis, eastern India to Indochina and Malay Peninsula, Sri Lanka, Hainan, Borneo, Mindanao;
M. puta, possibly Taiwan;
M. aenea, Malay Peninsula;
M. rozendaali, Borneo;

subgenus *Harpiola* Thomas, 1915

M. grisea, northern India.

Although Hill and Francis (1984) indicated that *M. suilla* includes *M. balstoni* of Java and *M. canescens* of Nias Island, Corbet and Hill (1986) continued to list each as a species.

Head and body length is 33–60 mm, tail length is 30–42 mm, and forearm length is 28–45 mm. Medway (1978) listed the following weights: *M. suilla*, 3–4 grams; *M. cyclotis*, 9–10 grams; *M. huttoni*, 6.4 grams; and *M. aenea*, 7.5 grams. The fur is usually thick and woolly. The coloration is often brownish or grayish, but in some forms it is reddish, and in *M. aurata* it is golden yellow above. The nostrils, which are located at the ends of tubes, are somewhat like those of *Harpiocephalus* and *Paranyctimene* of the family Pteropodidae.

Murina is often found in hilly areas. These bats seem to be low-flying, as they have been noted skimming over the surface of crops and grass in their feeding flights. Members of this genus have been found roosting in the dead dry leaves of cardamon plants and in caves. Several usually roost together. Medway (1978) wrote that pregnant *M. cyclotis*, carrying two fetuses each, were taken in February and May in the Malay Peninsula.

CHIROPTERA; VESPERTILIONIDAE; **Genus HARPIOCEPHALUS**
Gray, 1842

Hairy-winged Bat

There apparently are two species (Ellerman and Morrison-Scott 1966; Hill and Francis 1984):

H. harpia, eastern and southern India, Indochina, Taiwan, Sumatra, Java, Borneo, Amboina Island in the Moluccas;
H. mordax, northern Burma, Borneo.

Head and body length is 55–75 mm, tail length is 37–55 mm, and forearm length is 40–54 mm. Coloration above is chestnut, orange-red, or rusty, usually mixed with gray; the underparts are grayish buffy. As in *Murina*, the fur is thick and woolly. The legs, tail membrane, and occasionally part of the wings are covered with hair.

Four genera of bats possess tubular nostrils—*Nyctimene* and *Paranyctimene* in the family Pteropodidae and *Murina* and *Harpiocephalus* in the family Vespertilionidae. *Harpiocephalus* is distinguished from *Murina* mainly by dentition. The cheek teeth of *Harpiocephalus* are massive, powerful, and supplied with blunt cusps.

Little information has been recorded on the habits and life history of these bats. However, they have been observed to frequent hilly country, and they probably roost in vegetation. Beetle remains were found in the stomach of one individual, perhaps an indication that they feed on chitinous insects, as their dentition suggests.

CHIROPTERA; VESPERTILIONIDAE; **Genus KERIVOULA**
Gray, 1842

Painted Bats, or Woolly Bats

There are 2 subgenera and 21 species (Bergmans and Van Bree 1986; Chasen 1940; Ellerman and Morrison-Scott 1966; Hayman and Hill, *in* Meester and Setzer 1977; Hill

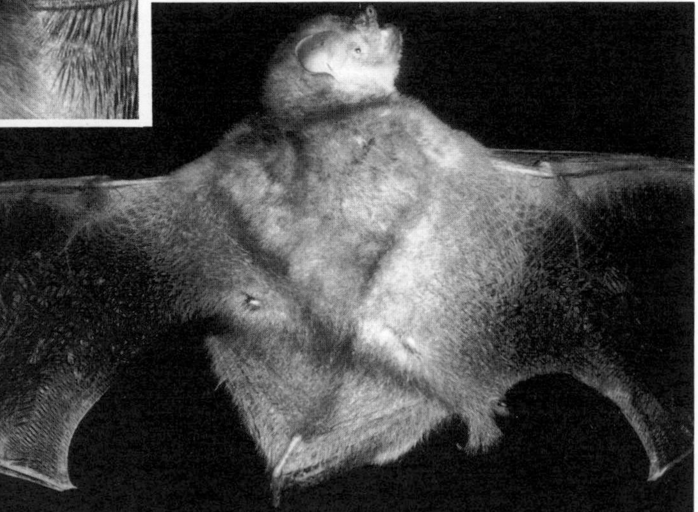

Hairy-winged bat *(Harpiocephalus harpia)*, photos by Paul D. Heideman.

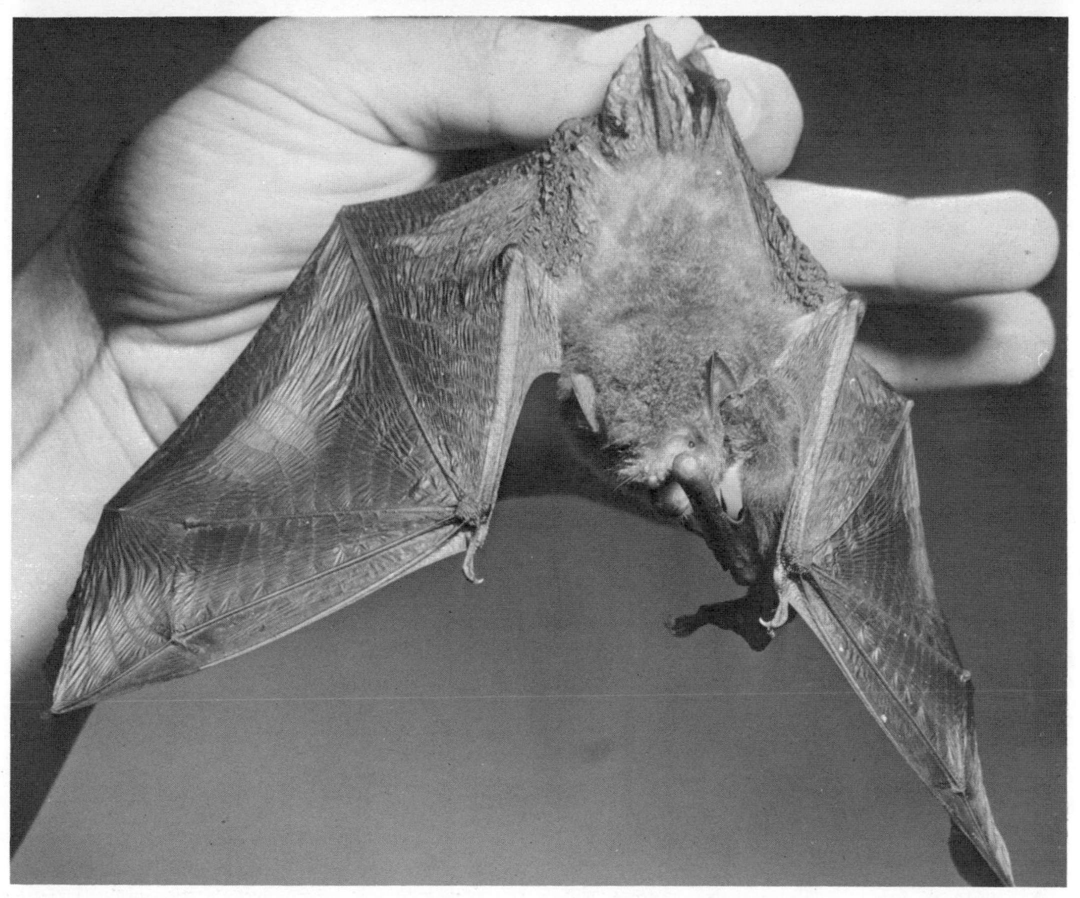

Woolly bats, or painted bats *(Kerivoula papillosa).* A small baby is between the wing and head of the mother in the lower picture. This is not the normal position in which baby bats are carried; it probably resulted from the handling of the bats. Photos by Lim Boo Liat.

1965, 1977*b*; Hill and Francis 1984; Laurie and Hill 1954; Lekagul and McNeely 1977; Lunney and Barker 1986; Medway 1977, 1978; Ride 1970; Taylor 1934):

subgenus *Kerivoula* Gray, 1842

K. argentata, southern Kenya to Namibia and eastern South Africa;
K. lanosa, Liberia to Ethiopia, and south to South Africa;
K. smithii, Nigeria, Cameroon, eastern Zaire, Kenya;
K. cuprosa, southern Cameroon, Kenya, possibly Zaire;
K. phalaena, Liberia, Ghana, Cameroon, middle Congo River;
K. africana, Tanzania;
K. whiteheadi, Malay Peninsula, Borneo, Philippines;
K. picta, southern and eastern India to southern China and Malay Peninsula, Sri Lanka, Hainan, Sumatra, Java, Bali, possibly Borneo, Moluccas;
K. muscina, Papua New Guinea;
K. agnella, Sudest (Tagula) and St. Aignan's (Misima) islands southeast of New Guinea;
K. minuta, Malay Peninsula, Borneo;
K. intermedia, Malay Peninsula, Borneo;
K. pellucida, Malay Peninsula, Sumatra, Jolo (southern Philippines);
K. hardwickei, southern and eastern India to southern China and Malay Peninsula, Sri Lanka, Sumatra, Java, Kangean Islands, Bali, Borneo, Sulawesi, Philippines;
K. myrella, Admiralty Islands northeast of New Guinea;
K. papillosa, eastern India, Indochina, Malay Peninsula, Sumatra, Java, Borneo, Sumbawa, Sumba, Flores;

subgenus *Phoniscus* Miller, 1905

K. jagorii, Java, Bali, Borneo, Philippines;
K. rapax, Sulawesi;
K. papuensis, Papua New Guinea, eastern Queensland, southeastern New South Wales;
K. atrox, Malay Peninsula, Sumatra, Borneo;
K. aerosa, reputedly from east coast of South Africa but possibly from Sulawesi.

Phoniscus was considered a distinct genus by Hill (1965) but not by Koopman (1979, 1982*a*).

Head and body length is 31–57 mm, tail length is 32 to about 55 mm, and forearm length is 27–45 mm. Medway (1978) listed weights of 9–10 grams for *K. papillosa* and 4–6 grams for *K. hardwickei*. Hill and Francis (1984) reported that the smallest examples of *K. minuta* are little larger than *Craseonycteris*, probably the smallest of bats. The painted bat, *K. picta*, is bright orange or scarlet, with black wings and orange along the fingers. It thus resembles *Myotis formosus* in its color pattern, but the painted bat is a smaller species. *K. argentata* of Africa also has a striking appearance: this bat is bright orange rufous interspersed with black and frosted with whitish tips to the hairs. The other forms of *Kerivoula* are reddish brown, yellowish brown, olive brown, dark brown, or grayish; shining gray hairs, simulating a tuft of moss, are sometimes present in the pelage. This genus is characterized by the long, woolly, rather curly hair; the delicate form; the rather large and somewhat pointed, funnel-shaped ears; and the presence of 38 teeth.

Painted bats are forest inhabitants. Some of the African species often roost and apparently raise their young in the abandoned hanging nests of weaverbirds and sunbirds. Tree hollows and trunks, foliage, huts, and buildings are also used as roosts by *Kerivoula*. *K. picta*, from Asia, shelters among the dry leaves of vines and other plants, in plantain fronds, and in flowers. One individual of this species was noted in the foliage of a longan tree, an evergreen that retains decaying orange and black leaves throughout the year; *K. picta* could readily be mistaken for just another russet leaf in such a roost. These bats emerge late in the evening, have a rather weak, fluttering flight, and usually fly in circles close to the ground.

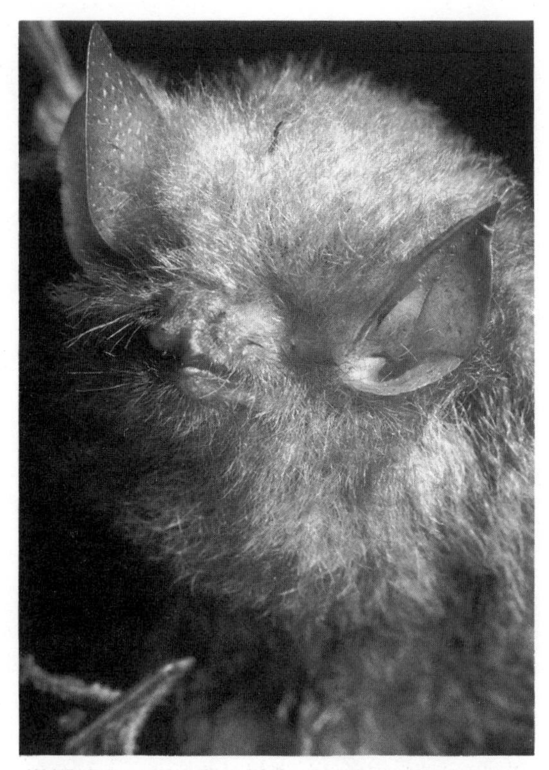

Woolly bat, or painted bat (*Kerivoula* sp.), photo by Lawrence R. Heaney.

Kerivoula roosts singly or in groups of as many as six individuals. Four individuals of *K. argentata* were clinging together so tightly under the eaves of a hut in Africa that they resembled the mud nest of a wasp. A group of seven *Kerivoula* in the Philippines were noted flying together in the daytime and performing aerial gyrations between periodic returns to a roosting site in foliage; this behavior may possibly have been associated with breeding. Young *Kerivoula* have been found in October in Zaire and Sri Lanka. Medway (1978) stated that a lactating female *K. hardwickei* with a nearly full-grown young was taken in January in the Malay Peninsula. Lim, Shin, and Muul (1972) reported the collection of two female *K. papillosa*, each with one embryo, in mid-June on Borneo.

CHIROPTERA; VESPERTILIONIDAE; **Genus ANTROZOUS**
H. Allen, 1862

Pallid Bat, or Desert Bat

The single species, *A. pallidus*, is found from southern British Columbia and Montana to central Mexico and also is known from a few localities on Cuba (Hermanson and O'Shea 1983;

Shryer and Flath 1980). *A. p. koopmani* of Cuba was listed as a full species by Hall (1981) but was shown to be a subspecies of *A. pallidus* by Martin and Schmidly (1982). *Bauerus* (see account thereof) formerly was considered a subgenus or synonym of *Antrozous.*

Head and body length is 60–85 mm, tail length is 35–57 mm, forearm length is 45–60 mm, and adult weight is usually 17–28 grams. The woolly fur is creamy, yellowish, or light brown on the upper parts and paler, sometimes almost whitish, on the underparts. Distinctive features of this genus are the large ears and the presence of a small horseshoe-shaped ridge on the squarely truncate muzzle. The nostrils are located beneath the ridge on the front of the muzzle. The ears are large and separate, though not as large, proportionately, as in *Plecotus* and *Euderma.*

The pallid bat favors rocky outcrops with desert scrub but commonly ranges up to forested areas with oak and pine. It roosts in caves, rock crevices, mines, hollow trees, and buildings; emergence is fairly late in the evening (Barbour and Davis 1969). There is evidence for both migration and hibernation in various areas. In a study in central Arizona, Vaughan and O'Shea (1976) found *A. pallidus* to arrive in March or April and to disappear in November. All daytime retreats were crevices or chambers in vertical or overhanging cliffs. During the hottest periods shelter was sought in deep crevices, where body temperatures of 30° C could be maintained passively. O'Shea and Vaughan (1977) found that *A. pallidus* had two nightly foraging periods with an intervening roosting period. This roosting interval was longer in the autumn than in warmer months. Foraging took place at a height of 0.5–2.5 meters above the ground, with a flight pattern consisting of dips, rises, and low gliding swoops. This flight style was well suited to the taking of relatively large, substrate-roving or slow-flying prey. Food items included Coleoptera, Orthoptera, Homoptera, Lepidoptera, arachnids, and a lizard. Barbour and Davis (1969) stated that *A. pallidus* takes food primarily from the ground but also forages in foliage and alights on flowers in search of insects.

These bats appear to be highly social. O'Shea and Vaughan (1977) stated that after the initial foraging period individual *A. pallidus* located one another through vocal communication and gathered in night roosting clusters, where they entered torpor. Brown (1976) listed four main kinds of adult vocalization: a directive call, used for orientating individuals to one another; squabble notes, used for spacing bats when roosting; an irritation buzz, used in agonistic intraspecific encounters; and ultrasonic orientation pulses, used to communicate exploratory activity to other individuals. According to Barbour and Davis (1969), maternity colonies of *A. pallidus* begin to form in early April and number about 12–100 bats. Males have been reported both from nursery colonies and in separate groups. Vaughan and O'Shea (1976) found that 95 percent of a studied population roosted in groups of 20 or more, the largest colony numbering 162 individuals. O'Shea and Vaughan (1977) observed the following conditions in central Arizona: during most of the summer adult males seemed to occur separately; in July and August the females and young appeared to forage together and maintain large colonies; and in mid-August a postbreeding dispersal occurred.

Barbour and Davis (1969) wrote that mating in *A. pallidus* begins in late October and probably takes place sporadically in the winter; sperm is retained in the uterus of the female through the winter; and gestation is 53–71 days. The young are born in May and June; usually there are twins, but 20

Pallid bat, or desert bat *(Antrozous pallidus)*, photo by W. G. Winkler and D. B. Adams.

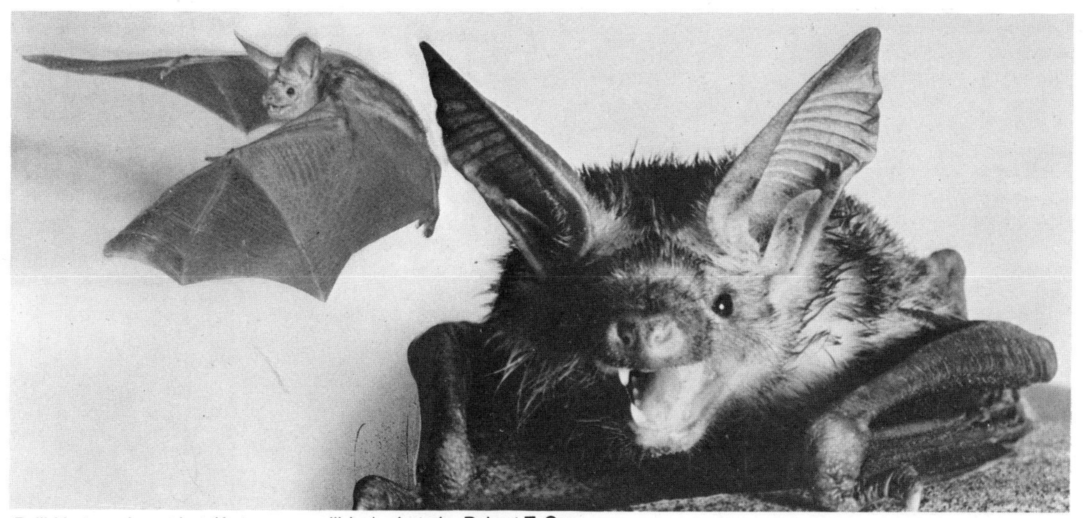

Pallid bat, or desert bat *(Antrozous pallidus)*, photo by Robert T. Orr.

percent of births are single. Manning et al. (1987) reported that of 12 pregnant females collected in Texas on 11 May, 1 had a single fetus, 8 had 2 fetuses, 1 had 3 fetuses, and 2 had 4 fetuses. The young open their eyes by about 5 days, begin to fly at 4–5 weeks, are weaned after 6–8 weeks, and may breed in their first year of life (Hermanson and O'Shea 1983). A wild *A. pallidus* is known to have lived for a minimum of 9 years and 1 month (Cockrum 1973).

CHIROPTERA; VESPERTILIONIDAE; **Genus BAUERUS**
Van Gelder, 1959

Van Gelder's Bat

The single species, *B. dubiaquercus,* is known from mainland Mexico in Jalisco, Chiapas, southern Veracruz, Jalisco, and Quintana Roo, from the Tres Marias Islands off the west coast of Nayarit, and from Belize, Honduras, and Costa Rica (Engstrom, Lee, and Wilson 1987; Juarez G., Jiménez A., and Navarro L. 1988). *Bauerus* originally was described as a subgenus of *Antrozous,* and such status was supported by Pine, Carter, and LaVal (1971). However, based on several morphological and karyological distinctions, Engstrom and Wilson (1981) and Martin and Schmidly (1982) considered the two taxa to be generically distinct. This view was accepted by Corbet and Hill (1986) and Jones, Arroyo-Cabrales, and Owen (1988), though not by Honacki, Kinman, and Koeppl (1982).

Head and body length is about 57–75 mm, tail length is 46–57 mm, forearm length is 48–57 mm, and weight is 13–20 grams (Engstrom and Wilson 1981; Juarez G., Jiménez A., and Navarro L. 1988). The pelage is soft, lax, and dark brown. *Bauerus* is distinguished from most other North American vespertilionids by its darker coloration and larger ears. It resembles *Antrozous* externally but is darker and has slightly smaller ears.

Cranially it is distinguished from *Antrozous* by its more pronounced sagittal crest, relatively small auditory bullae, and anteriorly inflected upper toothrow. Most specimens have a spiculelike third lower incisor, whereas *Antrozous* has only two lower incisors on each side (Engstrom, Lee, and Wilson 1987).

Bauerus has been found in a variety of tropical forest habitats at elevations from 370 to 1,450 meters. Most specimens have been taken on the Tres Marias Islands. Roosting sites are unknown. On the basis of morphology, it has been speculated that this bat feeds on large insects and takes food exclusively in flight. A pregnant female with one fetus was collected in Honduras in April, lactating females were taken in Chiapas in April and in Quintana Roo in June, and two postlactating females were found in Costa Rica in July (Engstrom, Lee, and Wilson 1987; Juarez G., Jiménez A., and Navarro L. 1988).

CHIROPTERA; VESPERTILIONIDAE; **Genus NYCTOPHILUS**
Leach, 1821

New Guinean and Australian Big-eared Bats

There are eight living species (Churchill, Hall, and Helman 1984; Hill and Pratt 1981; Koopman 1984*a;* Laurie and Hill 1954; Maddock, *in* Strahan 1983; Richards, *in* Strahan 1983; Ride 1970):

N. walkeri, northeastern Western Australia, northern Northern Territory, northwestern Queensland;
N. arnhemensis, northeastern Western Australia, northern Northern Territory;
N. microtis, Papua New Guinea;
N. gouldi, southwestern Western Australia, southeastern Australia, Tasmania;
N. bifax, northern Western Australia to eastern Queensland;
N. timoriensis, New Guinea, Western Australia, northwestern Northern Territory, South Australia, eastern Queensland, eastern New South Wales, Victoria, Tasmania, possibly Timor;
N. geoffroyi, throughout Australia except northern Queensland, Tasmania;
N. microdon, northeastern New Guinea.

An additional species, *N. howensis* from Lord Howe Island, east of Australia, was described by McKean (1975). Although known only from fossil remains, it may have occurred in

Van Gelder's bat *(Bauerus dubiaquercus),* photo by D. S. Rogers through Mark D. Engstrom and courtesy of Carnegie Museum.

Australian big-eared bat *(Nyctophilus timoriensis)*, photo by Stanley Breeden.

modern times, since there are reports of the former presence of a large bat on the island. Koopman (1984*a*) did not consider *N. bifax* to be a species distinct from *N. gouldi*. The genus and species *Lamingtona lophorhina* were described by McKean and Calaby (1968) on the basis of six specimens from Mount Lamington in southeastern Papua New Guinea. *Lamingtona* was considered to be related to *Nyctophilus* but to be distinguished by several morphological characters, including the absence of a band of integument connecting the ears across the forehead and a basally more slender tragus. Hill and Koopman (1981) considered *Lamingtona lophorhina* to closely resemble *Nyctophilus microtis* in these and other characters, concluded that the two are conspecific, and provisionally recognized *lophorhina* as a subspecies of *N. microtis*.

Head and body length is 38–75 mm, tail length is 30–55 mm, forearm length is 31–47 mm, and weight is 4–20 grams (Allison, *in* Strahan 1983; Churchill, Hall, and Helman 1984; Kitchener, *in* Strahan 1983; McKenzie, *in* Strahan 1983; Maddock, *in* Strahan 1983; Richards, *in* Strahan 1983). The coloration is light brown, cinnamon brown, orange brown, dark brown, or ashy gray. *Nyctophilus* and *Pharotis* are the genera in the family Vespertilionidae that combine long united ears with a small, horseshoe-shaped nose leaf. The ears, about 25 mm in length, usually are united by a low membrane. There is a fleshy disk behind the nose leaf, and the muzzle is abruptly truncate.

These bats have been observed in forested areas, scrub country, and arid regions. Daytime retreats are small caverns, crevices in rocks, tree hollows, and under the bark of trees. *N. geoffroyi* seems to have adapted well to human presence and may be found in cities roosting within the roofs of buildings (Maddock, *in* Strahan 1983). Some populations are thought to be active throughout the year, but Phillips and Inwards (1985) concluded that in southeastern Australia *N. gouldi* undergoes lengthy bouts of hibernation during the colder months from April to September. Most, if not all, of the species have a rather slow and fluttering flight. They are highly maneuverable and pursue a variety of insects in the open air but also glean from foliage and tree branches. Ride (1970) stated that *N. geoffroyi* will land on the ground to pick up beetles and other insects. He also said (pers. comm.) that this species will often come to an observer's hand to take a proffered insect, preferably one that is so held as to kick, flutter, or "buzz" and thus attract the bat's attention. The bat hovers in front of the hand and at times will even land gently on it to take the insect. Big-eared bats can also fly off a horizontal surface by leaping into the air with a sudden downward cupping of the outstretched wings, as does *Antrozous*.

Social structure seems to vary, with *N. timoriensis* reported to roost alone or in pairs and *N. geoffroyi* forming maternity colonies of 10 to over 100 individuals (Maddock, *in* Strahan 1983; Richards, *in* Strahan 1983). On the mainland these colonies are made up almost entirely of pregnant females, but in Tasmania they contain adults of both sexes. Births, commonly of 2 young, take place in the colonies in late spring or early summer. In New South Wales, McKean and Hall (1964) found pregnant females or newborn young in October and November. Studies of both wild and captive *N. gouldi* in southeastern Australia (Phillips and Inwards 1985) showed that mating takes place within hibernating colonies during the winter. Females are monestrous and store the sperm until ovulation in September or early October. The 1 or 2 young are born in late October and November; they are never carried by the mother, begin practice flights at 4–5 weeks, and are weaned after 6 weeks. Females reach sexual maturity at 7–9 months, males at 12–15 months.

CHIROPTERA; VESPERTILIONIDAE; **Genus PHAROTIS**
Thomas, 1914

The single species, *P. imogene*, is known only by a series of specimens taken about a century ago, at elevations below 1,000 meters, in southeastern New Guinea (Koopman 1982*a*).

Measurements of the type specimen, an adult female, are as follows: head and body length, 50 mm; tail length, 42 mm; and forearm length, 37.5 mm. The coloration is dark brown both above and below.

This bat is similar to the bats of the genus *Nyctophilus*, differing externally from that group in the shorter muzzle and the larger ears and nose leaf. *Pharotis* and *Nyctophilus* are the only genera of the family Vespertilionidae that combine long ears united at the base with a small, horseshoe-shaped nose leaf.

CHIROPTERA; VESPERTILIONIDAE; **Genus TOMOPEAS**
Miller, 1900

The single species, *T. ravus*, is restricted to the arid and semiarid coastal region of Peru (W. B. Davis 1970*a*; Koopman 1978*a*).

W. B. Davis (1970*a*) listed the following measurements: total length, 73–85 mm; tail length, 34–45 mm; forearm length, 31.2–34.5 mm; and weight, 2–3.5 grams. The coloration is pale brown above and dull buff to whitish cream buff below. The basal half of the fur is dull slaty gray. The face, ears, and membranes are black. The ears lack an anterior

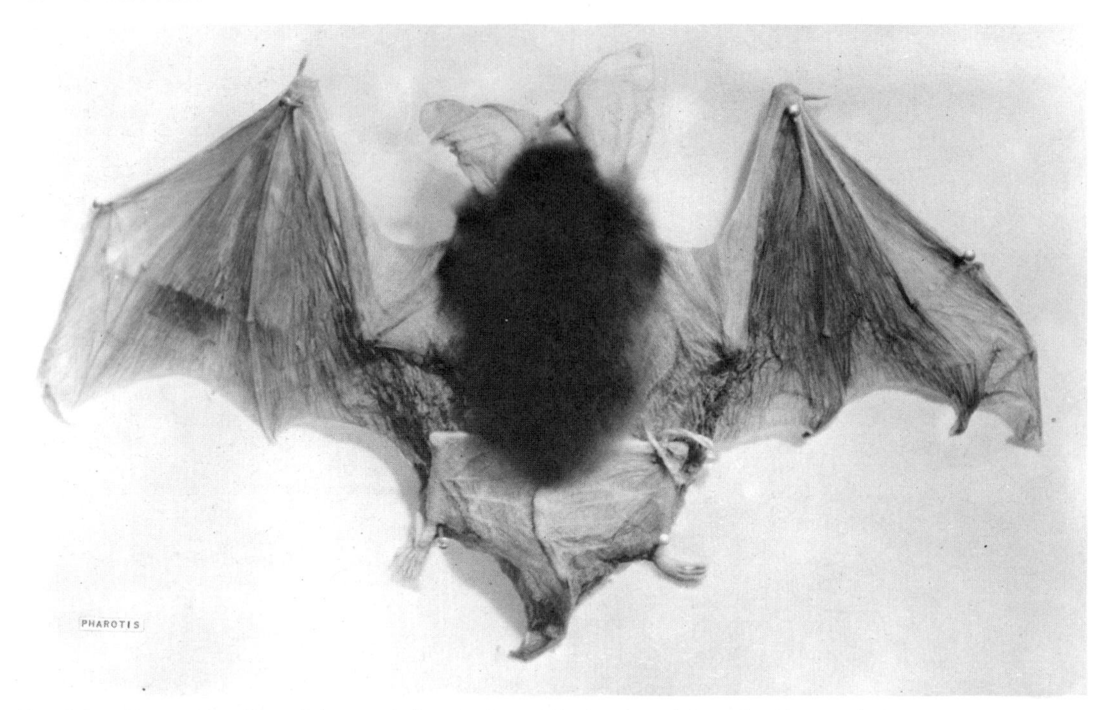

New Guinea big-eared bat *(Pharotis imogene)*, photo by Howard Hughes through Australian Museum, Sydney.

basal lobe but have a rudimentary keel, and the tragus is short and blunt.

This genus has been placed in a separate subfamily of the family Vespertilionidae because *Tomopeas* has characters that are suggestive of both the Vespertilionidae and the Molossidae. Allen (1939*a*:208) stated: "It has the delicate form and the long tail wholly included in the membrane as in the former, and the peculiar ear structure and the fusion of the last cervical with the first thoracic vertebra rather characteristic of the latter. It may thus be thought of as a very ancient and ancestral type, partly bridging the gap between the two families."

This bat has a known altitudinal range from sea level to 1,000 meters. Specimens have been taken in mist nets set among large mesquite trees and have been found roosting under the exfoliations of granite boulders and outcroppings. Juvenile females, as well as adults of both sexes, were collected in July and August (W. B. Davis 1970*a*). Based on the presence of juveniles in August but only nonpregnant females in August and November, Graham (1987) suggested that parturition was possible in the coastal dry season from May to September.

CHIROPTERA; **Family MOLOSSIDAE**

Free-tailed Bats and Mastiff Bats

This family of 16 genera and 86 species is found in the warmer parts of the world, from southern Europe and southern Asia south through Africa and Malaysia and east to the Fiji Islands, and from the central United States south through the West Indies, Mexico, and Central America to the southern half of South America. The sequence of genera presented here follows that suggested in Freeman's (1981) discussion of generic considerations for the Molossidae, and that was adopted by Koopman (1984*b*), except that the highly distinctive *Cheiromeles* is placed last. A somewhat different sequence was proposed by Legendre (1984), who recognized three subfamilies: Molossinae, with the genera *Molossops* (including *Cynomops* and *Neoplatymops* as subgenera), *Myopterus, Molossus, Eumops,* and *Promops;* Tadaridinae, with *Mormopterus* (including *Micronomus, Platymops,* and *Sauromys* as subgenera), *Nyctinomops, Rhizomops* (including only the species here called *Tadarida brasiliensis*), *Otomops,* and *Tadarida* (including *Chaerephon* and *Mops* as subgenera); and Cheiromelinae, with the single genus *Cheiromeles.*

Head and body length is 40 to approximately 130 mm, tail length is 14–80 mm, and forearm length is 27–85 mm. The tail projects far beyond the free edge of the narrow tail membrane, hence the common name "free-tailed bats" for this family. The short body hair has a velvetlike texture. In one genus *(Cheiromeles)* the hair is so short that the animal appears to be naked. Some species of *Tadarida* have an erectile crest of hairs on the top of the head. The coloration is

Mastiff bat *(Eumops perotis)*, photo by Abel Fornes.

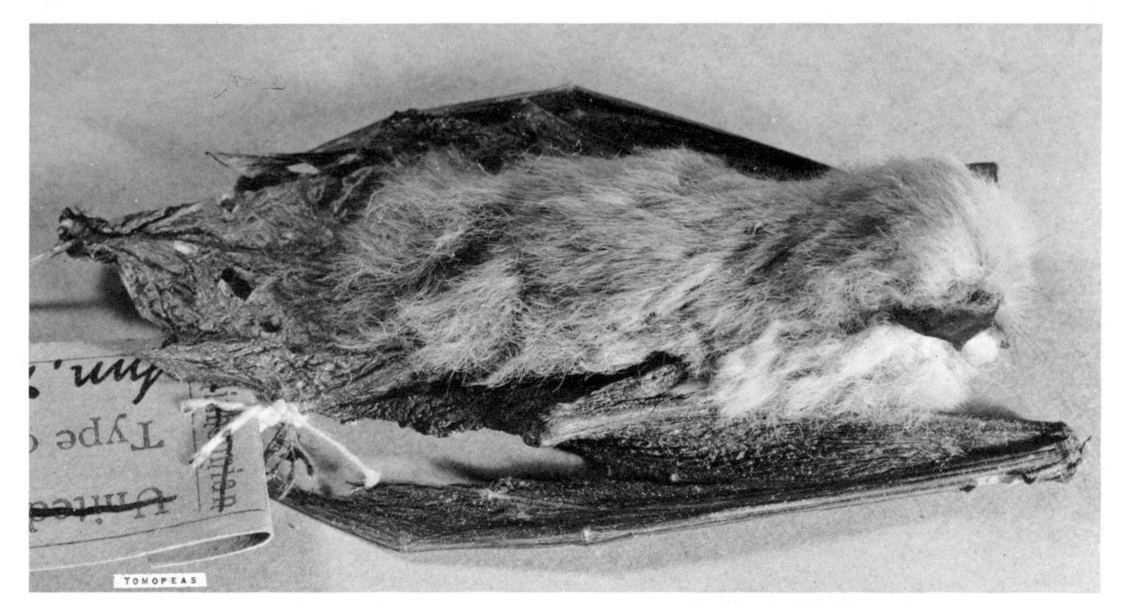

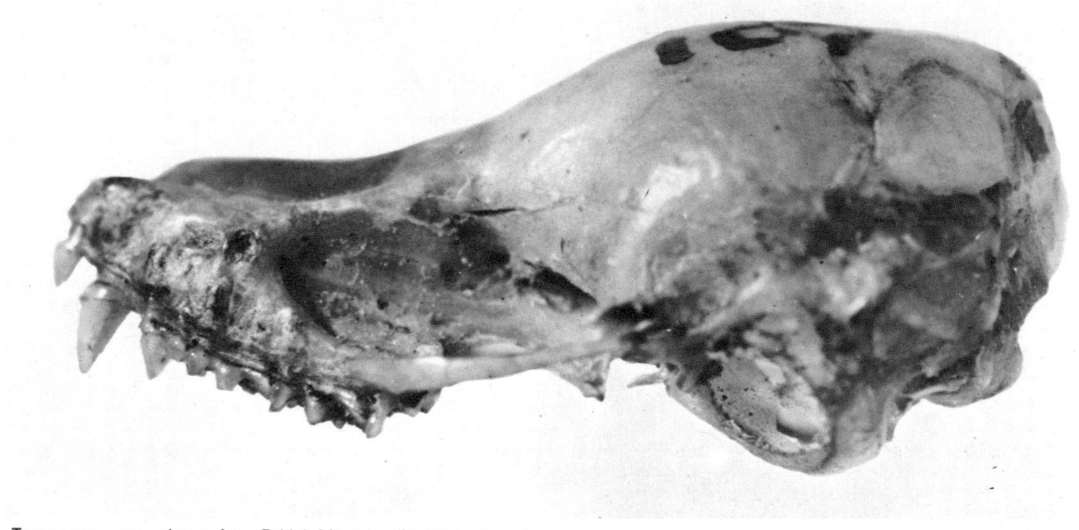

Tomopeas ravus, photos from British Museum (Natural History).

usually brown, buff, gray, or black. Some species, particularly in the genus *Molossus*, have two color phases. The head is rather thick and the muzzle is broad and obliquely truncate, usually with a scattering of short hairs having spoon-shaped tips. The eyes are small. The ears are thick and rather leathery, variable in size and form, and often united across the forehead and directed forward. A tragus is present. The nostrils usually open on a pad, the upper surface of which is often adorned with small hornlike projections. There is no nose leaf. The lips are large; in some genera the upper lip is furrowed by vertical wrinkles. Throat glands are often present. The wings are long, narrow, thick, and leathery. The legs are short and strong, and the foot is broad. Curved, spoon-shaped bristles are present on the outer toes of each foot; these are used by the bat in cleaning and grooming its fur.

The teeth are of the normal cuspidate insectivorous type. The dental formula varies from (i 1/1, c 1/1, pm 1/2, m 3/3) × 2 = 26 to (i 1/3, c 1/1, pm 2/2, m 3/3) × 2 = 32. The braincase is thick, flat, and broad.

Molossid bats roost in caves, tunnels, buildings, hollow trees, foliage (at least *Molossus*), the decayed wood of logs (reported for *Molossops*), the crevices of rock cliffs, and holes in the earth (reported for *Cheiromeles*). They also shelter under bark and rocks and often under the corrugated iron roofing of human-made structures; they prefer the high temperatures, up to 47° C and above in such places. A characteristic strong musky odor generally permeates their retreats. Some species live in groups of hundreds of thousands or even millions, others associate in smaller groups, and some forms are solitary. The colonial species generally return to their favored haunts year after year. *Nyctinomops femorosaccus* and *Eumops perotis* have been found roosting together in a crack in a large granite boulder; the two genera were segregated, the *Nyctinomops* hanging in the higher, narrower part of the crevice and the *Eumops* sheltering in the lower, wider part.

The members of this family tend to be active throughout the year. The northernmost species may be inactive for short

Mexican free-tailed bat *(Tadarida brasiliensis)* with right wing extended, showing that it is proportionately narrower than that of most bats except other members of the family Molossidae. The rate of wingbeat of these bats is faster than that of most other insectivorous bats. Photo by Ernest P. Walker.

periods during the winter, but there is no definite evidence of true hibernation. Rather, these species usually make local movements or, as in *Tadarida brasiliensis* of the New World, a seasonal migration. Compared with the flight of many insectivorous bats, their flight is swift and relatively straight. They fly with their mouths open and send out ultrasonic sounds. The diet consists of insects, often hard-shelled forms.

These bats usually have one young per litter and one litter per year. Two young are born on rare occasions, and *Molossus ater*, in Trinidad, possibly has two litters per year. Two successive pregnancies have been reported in a female *Tadarida* from Africa. Breeding generally takes place just before ovulation in late winter and spring. There may be partial or complete segregation of the sexes in some species.

During the Civil War, the Confederate Army used guano, probably produced by *Tadarida brasiliensis*, as a source of niter (sodium nitrate) for gunpowder. Guano for fertilizer was obtained in commercial quantities for 20 years from Carlsbad Caverns, New Mexico. Glover M. Allen stated that during the period of greatest activity, for about 15 years, this guano was obtained from September to March; one to three carloads were shipped daily, each weighing about 40 tons. Allen reported that the prices realized for it were said to be about $20–$80 per ton. The attempt of C. Campbell to increase the number of *Tadarida brasiliensis* by erecting wood "bat roosts" proved to be a failure. He had hoped to secure guano in commercial quantities and also thought that these bats would eradicate mosquitoes and malaria. Only small amounts of guano were obtained, and mosquitoes are seldom eaten by this species. The attempted utilization of this same species as a carrier of small incendiary bombs during World War II was also abandoned.

The geological range of this family is late Eocene to Recent in Europe, late Oligocene or early Miocene to Recent in South America, middle Miocene to Recent in Africa, Pleistocene to Recent in Asia, Australia, North America, and the East and West Indies, and Recent over the remainder of the present range (Koopman 1984*c*).

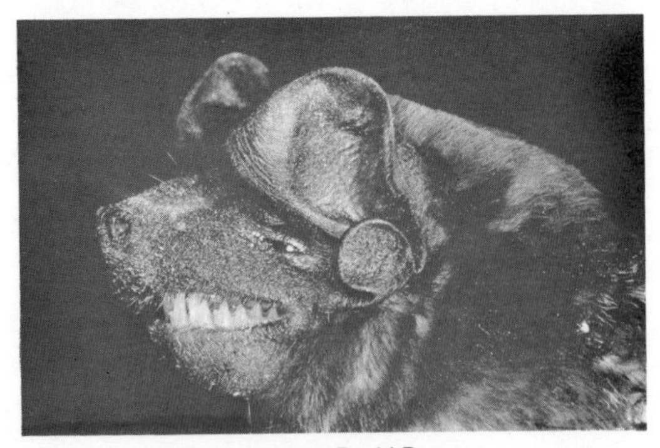

Molossus molossus, photo by David Pye.

CHIROPTERA; MOLOSSIDAE; **Genus MORMOPTERUS**
Peters, 1865

Little Goblin Bats

There are 2 subgenera and 10 species (Allison, *in* Strahan 1983; Cabrera 1957; Freeman 1981; Hall 1981; Hayman and Hill, *in* Meester and Setzer 1977; Hill 1983*a*; Koopman 1982*a*, 1982*b*, 1984*a*; Legendre 1984; Peterson 1985; Richards, *in* Strahan 1983):

subgenus *Mormopterus* Peters, 1865

M. acetabulosus, Ethiopia, Natal (South Africa), Mauritius and Reunion (Indian Ocean);
M. jugularis, Madagascar;

Little goblin bat *(Mormopterus planiceps)*, photo by G. B. Baker / National Photographic Index of Australian Wildlife.

M. doriae, Sumatra;
M. kalinowskii, coastal Peru, northern Chile;
M. phrudus, Peru;
M. minutus, Cuba;

subgenus *Micronomus* Troughton, 1943

M. planiceps, central and southern Australia;
M. loriae, Papua New Guinea, northern and eastern Australia;
M. norfolkensis, southeastern Queensland, coastal New South Wales, possibly Norfolk Island (southwestern Pacific);
M. beccarii, Amboina Island, New Guinea, Louisiade Archipelago, northern Australia.

Mormopterus sometimes has been treated as a subgenus of *Tadarida,* but there now has been almost universal acceptance that the two are generically distinct. On the other hand, there has been disagreement concerning whether *Platymops* and *Sauromys* (see accounts thereof) are subgenera of *Mormopterus.* In addition, *Micronomus* was recognized as a subgenus by Legendre (1984) but was considered only a synonym of *Mormopterus* by Freeman (1981). Koopman (1982*a,* 1984*a*) regarded *M. loriae* as a subspecies of *M. planiceps.* Peterson (1985) implied agreement that *M. loriae* and *M. planiceps* are conspecific and indicated further that the resulting species should have the name *M. petersi.* He also suggested that *M. astrolabiensis* of the Cape York Peninsula of northern Queensland is a species distinct from *M. beccarii* and that another species, with affinity to *M. jugularis,* is present at Hermannsburg in the south-central Northern Territory. Based on electrophoretic analyses, Adams et al. (1988) suggested that there are at least five species of *Mormopterus* in Australia, but they could not equate those species with the ones discussed above.

Based on the Australian and Cuban species, head and body length is 43–65 mm, tail length is 27–40 mm, forearm length is 29–41 mm, and weight is 6–19 grams; the upper parts are dark brown, grayish brown, or charcoal, and the underparts are usually paler (Allison, *in* Strahan 1983; Hall 1981; Richards, *in* Strahan 1983). *Mormopterus* is characterized by small overall size, separated and erect ears, unwrinkled lips, medium-thick jaws, a skull with a tall and posteriorly curving coronoid process, shallow basisphenoid pits, and a well-developed third upper molar tooth (Freeman 1981).

The small amount of available natural history information is restricted mainly to the Australian species (Allison, *in* Strahan 1983; Richards, *in* Strahan 1983). These bats, found in tropical forests, woodlands, open areas, and cities, roost mainly in roofs and tree hollows. They usually forage for insects above the tree canopy, water holes, and creeks but sometimes scurry after prey on the ground. Flight is swift and direct. Colonies vary in number from fewer than 10 to several hundred individuals. A single young is produced in December or January (summer). Crichton and Krutzsch (1987) reported that female *M. planiceps* stored sperm for at least 2 months prior to ovulation, which took place in late July and August, and that gestation then lasted for 3–5 months. Development of young was slow, but females reached sexual maturity before 1 year. Graham (1987) collected a lactating female *M. kalinowskii* in coastal Peru in May.

CHIROPTERA; MOLOSSIDAE; **Genus SAUROMYS** *Roberts, 1917*

South African Flat-headed Bat

The single species, *S. petrophilus,* is found in Namibia, Zimbabwe, Botswana, South Africa, the Tete district of western Mozambique, and possibly Ghana (Meester et al. 1986). Both Freeman (1981) and Legendre (1984) concluded that *Sauromys* is a subgenus of *Mormopterus.* This position was followed by Honacki, Kinman, and Koeppl (1982) and Koopman (1984*b*) but not by Corbet and Hill (1986), Meester et al. (1986), or Peterson (1985).

Head and body length is 50–60 mm, tail length is 30–40 mm, and forearm length is 35–42 mm; weight is 9–22 grams (Smithers 1983). The upper parts are brownish gray to tawny olive and the underparts are paler.

Sauromys differs from *Platymops* in the absence of wart-like granulations on the forearm, the absence of a well-developed throat sac, a well-developed anterior premolar (reduced or absent in *Platymops*), the strong development of a distinct hypocone on the first and second upper molars, and other cranial details. *Sauromys* differs from *Tadarida* in the flatness of the skull, the conformation of the anterior nasal aperture, the weakly bicuspidate upper incisors, the wide degree of separation of the ears, and other details. Freeman (1981) indicated that *Sauromys* shares the characters of *Mor-*

Sauromys petrophilus, photo by John Visser.

mopterus, as given in the account thereof, and that the bats of both genera (or subgenera) are flat-headed to some degree. However, Meester et al. (1986) suggested that flattening of the skull is more conspicuous in *Sauromys*.

Specimens have been taken by overturning slabs of rocks in hilly regions. *Sauromys*, with its flattened skull, apparently is adapted to such situations. It rests during the day under slabs of exfoliated rock or in narrow rock crevices or fissures. It generally occurs in small numbers, most records being of up to four (Smithers 1983).

CHIROPTERA; MOLOSSIDAE; **Genus PLATYMOPS**
Thomas, 1906

Flat-headed Free-tailed Bat

The single species, *P. setiger*, is found in southeastern Sudan, southern Ethiopia, and Kenya (Hayman and Hill, *in* Meester and Setzer 1977). Peterson (1985) recognized *P. macmillani* of Ethiopia as a separate species. Both Freeman (1981) and Legendre (1984) concluded that *Platymops* is a subgenus of *Mormopterus*. This position was followed by Honacki, Kinman, and Koeppl (1982) and Koopman (1984*b*) but not by Corbet and Hill (1986), Meester et al. (1986), or Peterson (1985).

Head and body length is 50–60 mm, tail length is 22–36 mm, and forearm length is 29–36 mm. A small, tufted sac is present in the throat region of both sexes. This genus resembles some species of *Tadarida* in external appearance, but in *Platymops* the skull is flattened, so the depth of the braincase is equal to only about a third of its width. The ears are separate. In some individuals the forearm is roughened with warty protuberances. Freeman (1981) indicated that *Platymops* shares the general characters of *Mormopterus*, as given in the account thereof, but its lips appear to be wrinkled and are covered with unusually thick hairs; she considered it a highly derived subgenus.

Platymops does not appear to be gregarious. Usually one to five individuals shelter under rocks and slabs, the flattened skull being an adaption for hiding in crevices. Individuals have been seen over a small permanent water hole and marsh in rapid erratic flight at heights of nine meters and less. The bats appeared when it was almost dark and were feeding on small beetles.

CHIROPTERA; MOLOSSIDAE; **Genus MOLOSSOPS**
Peters, 1865

Broad-faced Bats

There are two subgenera and five species (Alberico 1987; Cabrera 1957; Goodwin and Greenhall 1961; Hall 1981; Handley 1976; Ibáñez and Ochoa G. 1985; Koopman 1978*a*; Myers and Wetzel 1983; Williams and Genoways 1980*a*, 1980*b*):

subgenus *Molossops* Peters, 1865

M. temminckii, Venezuela and Peru to central Brazil and northern Argentina;
M. neglectus, Surinam;

subgenus *Cynomops* Thomas, 1920

M. planirostris, Panama to Paraguay and northern Brazil;
M. greenhalli, southwestern Mexico to Surinam, Trinidad;
M. abrasus, Colombia, Venezuela, Guyana, Peru, Brazil, Paraguay, northern Argentina.

Cynomops sometimes is considered a distinct genus, and *Neoplatymops* (see account thereof) sometimes is designated

Flat-headed free-tailed bat *(Platymops setiger)*, photo by Bruce J. Hayward.

a subgenus of *Molossops.* Handley (1976) recognized an additional species in Venezuela, *M. paranus;* it was not considered distinct from *M. planirostris* by Koopman (1978*a*), though Ochoa G. and Ibáñez (1985) suggested that the name may be valid. Corbet and Hill (1986) did list *M. paranus* as a species and included central Mexico within its range, but Jones, Arroyo-Cabrales, and Owen (1988) did not mention the presence of such a species in Mexico or Central America. The name *M. brachymeles* sometimes has been used in place of *M. abrasus,* but Williams and Genoways (1980*a*) considered the latter to be correct.

Head and body length is 40 to about 95 mm, tail length is 14–37 mm, and forearm length is 28–51 mm. Vizotto and Taddei (1976) listed weights of about 5–7 grams for *M. temminckii* and 6–10 grams for *M. planirostris.* The upper parts are yellowish brown, russet, chocolate brown, dark brown, or slate black and the underparts are usually yellowish white, gray, or slaty; some species have a white throat.

The bats of this genus resemble those of *Molossus* in external features, "but more conspicuous lines of fur diverging from the angle in the bend of the wing along the forearms and fourth finger are usually distinctive." According to Freeman (1981), both *Molossops* and *Myopterus* are characterized by a broad face, widely separated ears, no development of wrinkles on the lips, thickish dentaries, and no anterior palatal emargination. *Molossops* lacks basisphenoid pits, whereas *Myopterus* has very deep ones.

In Venezuela, Handley (1976) found most specimens of *M. planirostris* roosting in rotting snags in swamps. *M. temminckii* also has been found sheltering in decaying logs. *M. greenhalli* generally roosts in colonies of 50–75 individuals in the hollow branches of large trees. The males and females of

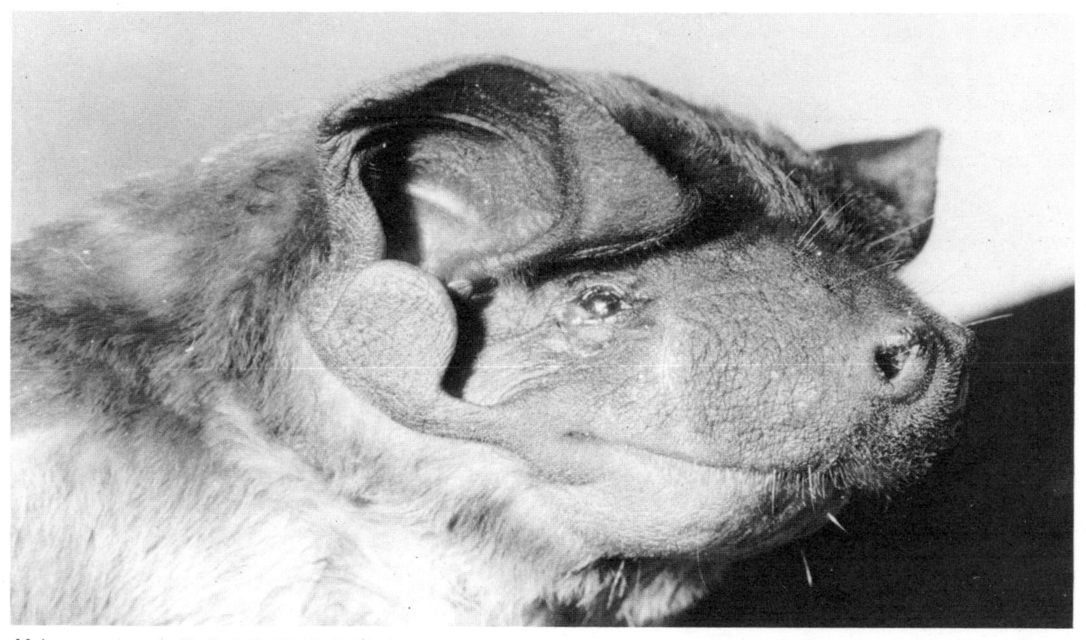

Molossops greenhalli, photo by Merlin D. Tuttle.

this species appear to remain together throughout the year.

Vizotto and Taddei (1976) reported *M. temminckii* and *M. planirostris* in Brazil to roost mainly in hollow fence posts in groups of up to 8 individuals; pregnant females were found there from September to January, and lactating females in February. Taddei, Vizotto, and Martins (1976) collected pregnant female *M. abrasus* from October to December in southeastern Brazil. Pregnant or lactating females of *M. greenhalli* have been taken in May, June, and July (Gardner, LaVal, and Wilson 1970; Valdez and LaVal 1971).

CHIROPTERA; MOLOSSIDAE; **Genus NEOPLATYMOPS**
Peterson, 1965

South American Flat-headed Bat

The single species, *N. mattogrossensis,* is known from five localities in southern Venezuela, southern Guyana, the Matto Grosso region of Brazil, and east-central Brazil, but it probably has a wide distribution in the Amazon Basin and the eastern Brazilian highlands (Willig 1985*a*). Although Peterson (1965*a*) originally described *Neoplatymops* as a full genus, it was regarded only as a subgenus of *Molossops* by both Freeman (1981) and Legendre (1984). Honacki, Kinman, and Koeppl (1982) agreed, but not Corbet and Hill (1986). Except as noted, the information for the remainder of this account was taken from Willig (1985*a*) and Willig and Jones (1985), who also considered *Neoplatymops* generically distinct, on the basis of morphological and karyological data.

Neoplatymops is one of the smallest molossids. Average measurements for males, followed by those for females, are: head and body length, 52.6 mm and 51.1 mm; tail length, 25.5 mm and 25.7 mm; forearm length, 30.1 mm and 30.2 mm; and weight, 6.1 grams and 5.4 grams. The pelage is short and sparse. The upper parts are generally pale brownish, the ears and membranes are dark brown, and the underparts are whitish to grayish. The ears are widely separated, there is a gular gland on the throat, the skull is conspicuously flattened, and there is no sagittal crest.

Neoplatymops has some resemblance to *Molossops* but has two (rather than one) well-developed upper premolar teeth, a flatter skull, and a series of small, wartlike granular structures on the dorsal surface of the forearm. These forearm granulations, which are not found in any other New World molossid, may function to provide anchorage on the porous surfaces of the rocks on which *Neoplatymops* roosts, thereby reducing the likelihood of being pulled off by a predator.

The local distribution of *Neoplatymops* is restricted to areas containing rocky outcrops, where it roosts, close to the ground, in narrow horizontal crevices beneath granitic exfoliations. In this regard, it is convergent with the Old World *Platymops* and *Sauromys*. Its morphology suggests that it is capable of maneuverable flight and the utilization of both soft and hard-bodied prey. Stomach contents of specimens show a wide variety of small insects, with beetles being most common.

Three or four individuals sometimes roost under the same granitic exfoliation. Females are monotocous and seasonally monestrous. In eastern Brazil pregnancies have been found initially in August, during the middle of the dry season. Synchronized births occur there in November and December, during the transition from the dry to the wet season, and lactation persists through the wet season from December to April.

CHIROPTERA; MOLOSSIDAE; **Genus CABRERAMOPS**
Ibáñez, 1980

Ibáñez (1980) erected this genus for the species originally described as *Molossops aequatorianus,* known only by four specimens collected in 1864 at Babahoyo in west-central Ecuador.

In two specimens, head and body length was 50 and 52 mm, tail length was 29 and 31 mm, and forearm length was 37.6 and 35.9 mm. From *Molossops, Cabreramops* differs in having the anterior bases of the ears near together on the forehead, a wrinkled upper lip (though more weakly than in *Tadarida*), and well-developed basisphenoid pits. There is no anterior palatal emargination in the skull, the sagittal crest is absent, and there are only two lower incisors and a single upper premolar on each side.

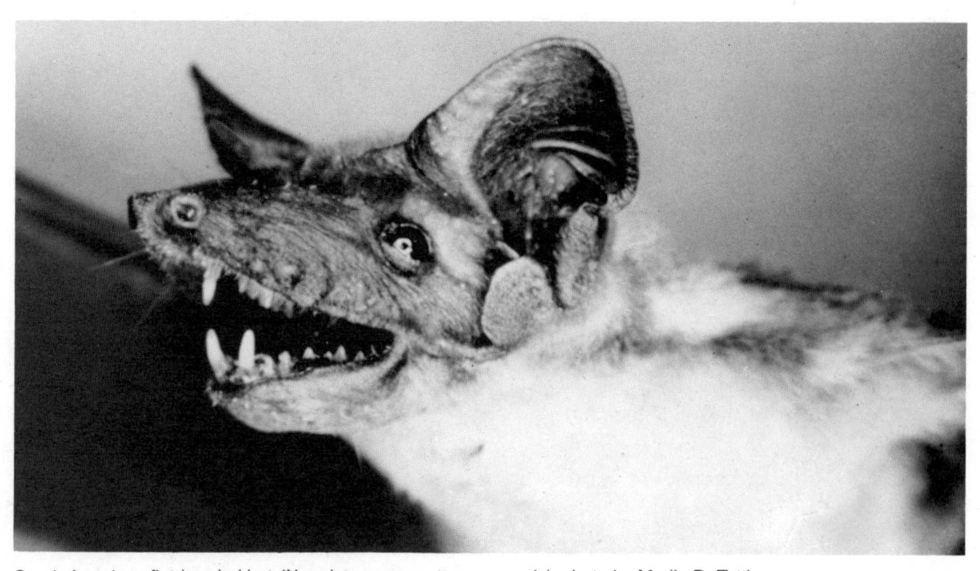

South American flat-headed bat *(Neoplatymops mattogrossensis)*, photo by Merlin D. Tuttle.

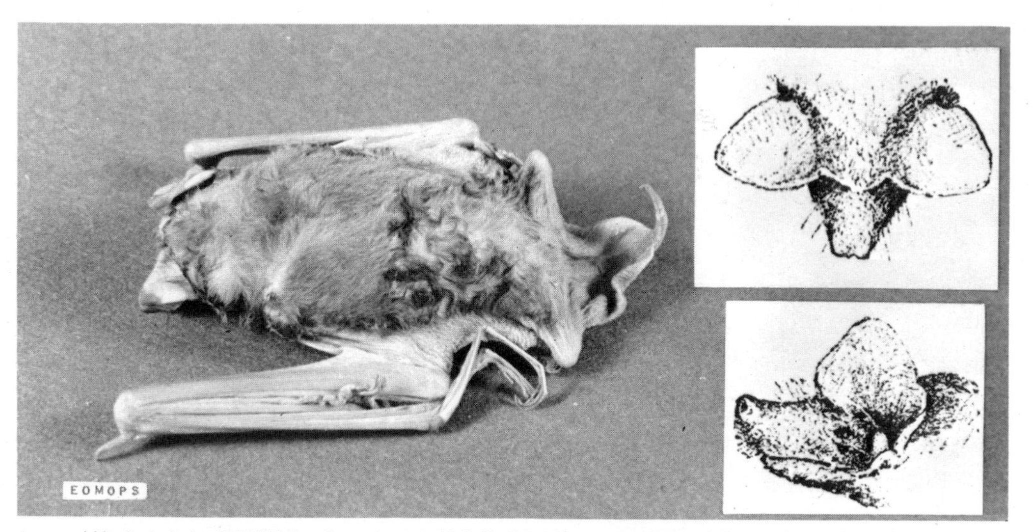

Myopterus whitleyi, photo by P. F. Wright of specimen in U.S. National Museum of Natural History. Inset photos from *Ann. & Mag. Nat. Hist.*

CHIROPTERA; MOLOSSIDAE; **Genus MYOPTERUS**
E. Geoffroy St.-Hilaire, 1818

Three species were listed by Hayman and Hill (*in* Meester and Setzer 1977):

M. whitleyi, Ghana, Nigeria, Cameroon, Zaire, Uganda;
M. albatus, Ivory Coast, northeastern Zaire;
M. daubentonii, Senegal.

The validity of the last species is questionable; it may be conspecific with *M. albatus*. The name *Eomops* Thomas, 1905, has sometimes been used in place of *Myopterus*.

Head and body length is about 56–66 mm, tail length is about 25–33 mm, and forearm length is about 33–37 mm. The upper parts are dark brown and the underparts are light reddish yellow to white. This genus resembles some of the other molossid bats in external features, but it differs from them in cranial and dental characters. Some external features of *Myopterus* are as follows: the ears are shorter than the head; the ears and the short tragus are rounded above; the muzzle projects beyond the jaws; the end of the nose is separate from the upper lip; and the nostrils open almost laterally.

According to Happold (1987), *M. whitleyi* lives only in the rainforest zone. It seems to be solitary; individuals have been found clinging to the bark of a tree, among the leaves of plantains, and in the roof of a shed. At night it flies in the forest, into gardens, and around houses. Freeman (1981) speculated that *Myopterus* is probably more maneuverable in flight than are most molossids, that *M. whitleyi* takes small, soft-bodied prey, and that *M. albatus* is able to consume large, hard-bodied prey.

CHIROPTERA; MOLOSSIDAE; **Genus TADARIDA**
Rafinesque, 1814

Free-tailed Bats

There are 2 subgenera and 8 species (Aggundey and Schlitter 1984; Ansell 1986; Ansell and Dowsett 1988; Cabrera 1957; Corbet 1978; Ellerman and Morrison-Scott 1966; Freeman 1981; Hall 1981; Harrison 1975*b*; Harrison and Bates 1984; Hayman and Hill, *in* Meester and Setzer 1977; Hill 1961*b*, 1982*a*, 1983*a*; Hill and Morris 1971; Kock 1975; Koopman 1975, 1982*a*, 1984*a*; Largen, Kock, and Yalden 1974; Laurie and Hill 1954; Legendre 1984; Lekagul and McNeely 1977; McKean and Calaby 1968; Medway 1977, 1978; Peterson 1971*c*, 1974*a*; Pienaar 1972; Rautenbach, Fenton, and Braack 1985; Ride 1970; Schlitter et al. 1986):

subgenus *Tadarida* Rafinesque, 1814

T. teniotis, Mediterranean zone of Europe and northern Africa to Japan and Taiwan;
T. lobata, Kenya, Zimbabwe;
T. fulminans, eastern Zaire, Rwanda, Kenya, Tanzania, Zambia, Malawi, Zimbabwe, Madagascar;
T. ventralis, southern Sudan, Ethiopia, eastern Zaire, Kenya, Malawi, Zambia, Transvaal;
T. aegyptiaca, Algeria, Egypt, Nigeria and Sudan to South Africa, southern Arabian Peninsula, Iran, Pakistan, India, Sri Lanka;
T. australis, Western Australia to southern Queensland and Victoria;
T. kuboriensis, Kubor Range in central Papua New Guinea;

subgenus *Rhizomops* Legendre, 1984

T. brasiliensis, southern United States to Chile and Argentina, West Indies.

Tadarida often is considered to include as subgenera what are here regarded as the separate genera *Mormopterus*, *Chaerephon*, *Mops*, and *Nyctinomops* (see accounts thereof). *Rhizomops*, described as a full genus by Legendre (1984), was not accepted as such by Corbet and Hill (1986) or Jones, Arroyo-Cabrales, and Owen (1988). Koopman (1982*a*, 1984*a*) considered *T. kuboriensis* a subspecies of *T. australis*, but Freeman (1981) and Legendre (1984) regarded the two as distinct species. The name *T. africana* often is used in place of *T. ventralis*.

In *T. australis*, one of the largest species, head and body length is 85–100 mm, tail length is 40–55 mm, forearm length is 57–63 mm, and weight is 25–40 grams (Richards, *in*

Free-tailed bat *(Tadarida aegyptiaca)*, photo by Erwin Kulzer.

Strahan 1983). In African species, head and body length is 65–100 mm, tail length is 30–59 mm, forearm length is 45–66 mm, and weight is 14–39 grams (Kingdon 1974*a*; Smithers 1983). In the New World *T. brasiliensis*, head and body length is 46–66 mm, tail length is 29–42 mm, forearm length is 36–46 mm, and weight is 10–15 grams (Hall 1981; Hoffmeister 1986; Lowery 1974). Coloration in the genus ranges from reddish brown to almost black. *Tadarida* is characterized by wrinkled lips, deep anterior palatal emargination, relatively thin jaws, and a third upper molar with an N-shaped occlusal pattern. Most species have ears that are joined by a band of skin across the top of the head. In *T. brasiliensis* and *T. aegyptiaca* the ears are separated, but not so widely as in *Mormopterus*, and are large and forward-facing as in other *Tadarida* (Freeman 1981).

Habitat varies considerably. African species live in either forest or open country and generally are reported to roost in trees and buildings (Hayman and Hill, *in* Meester and Setzer 1977; Rosevear 1965; Smithers 1971). In Venezuela, Handley (1976) found most *T. brasiliensis* roosting in houses. Barbour and Davis (1969) stated that *T. brasiliensis* roosts in buildings on the west coast and in the southeastern United States, and mainly in caves in the southwest. Colonies of the latter species may make spectacular mass exits from their caves after sunset and then fly up to 65 km to foraging areas. The diet consists mostly of small moths and beetles. Based on radar observations in the southwest, Williams, Ireland, and Williams (1973) determined the average speed of groups of *T. brasiliensis* to be 40 km/hr (7–105 km/hr) and the average maximum altitude to be 2,300 meters (600–3,100 meters). Some populations of this species also make lengthy migrations. In a banding study, Cockrum (1969) found that maternity colonies began assembling in the southwestern United States in April and disappeared by mid-October. He determined that populations in California, western Arizona, and southern Arizona made only localized movements in the spring and fall, or relatively short migrations to the south or west. Populations from central Arizona to Kansas and Texas, however, migrated deep into Mexico, sometimes traveling over 1,600 km. Less is known about southeastern populations. Lowery (1974) described a colony of 20,000 *T. brasiliensis* in New Orleans that existed for at least 35 years, moving to an unknown location in the fall and then returning in the spring.

The large summer groups of *T. brasiliensis* are basically maternity colonies, consisting mainly of females. In the southeastern United States such colonies usually number over 1,000 bats, but in the southwest there may be millions of individuals in a single cave. During the 1960s about 100 million *T. brasiliensis* occupied 13 caves in Texas, and an estimated 25 million to 50 million bats were in Eagle Creek Cave in Arizona (Barbour and Davis 1969; Cockrum 1969). Considering only sheer numbers, there was no larger con-

A. Free-tailed bat *(Tadarida aegyptiaca)* at rest on a horizontal surface with its two-day-old baby on its back. This position is normal for the young when these bats are on ledges in caves and tombs, but the babies of most bats cling beneath the mother so they do not interfere with her wings. B. Two-day-old young of an Egyptian tomb bat *(T. aegyptiaca)*. Photos by Erwin Kulzer.

centration of mammalian life known to exist. Some males are always present in the large nursery colonies, but most tend to gather in relatively small groups nearby. Cockrum (1969) found summer male colonies usually to number only 10–300 bats in the southwestern United States. It is possible that some males do not even migrate northward for the summer, as a group of 40,000 was discovered in late June in Chiapas, Mexico. In late summer, after the young are full-grown, the sexes begin to reassociate. Most other species are not known to be as gregarious as *T. brasiliensis*. For example, *T. australis* seldom is found in groups of more than 10 (Richards, *in* Strahan 1983).

In *T. brasiliensis* in North America, mating takes place in February–March, ovulation occurs around late March, gestation lasts about 77–84 days, and the single young is usually born in June or July (Barbour and Davis 1969). Since maternity colonies may consist of millions of tightly packed females and young, it once was believed that mothers nursed offspring indiscriminately during the five-week lactation period. New investigations (McCracken 1984; McCracken and Gustin 1987), however, show that each female usually locates her own pup, perhaps through scent or vocalization, though occasionally another young will "steal" her milk as she searches through the colony. In a study in New Orleans, Pagels and Jones (1974) found that the gestation period was about 11 weeks, the young were capable of maneuverable flight at 38 days, and full size was attained by 60 days. Short (1961) reported that males reached sexual maturity at about 18–22 months. LaVal (1973*c*) reported that a banded *T. brasiliensis* lived at least 8 years.

In recent years there has been increasing concern about drastic declines in some populations of *T. brasiliensis* in the southwestern United States (Geluso, Altenbach, and Wilson 1976, 1981; Gosnell 1977; Humphrey 1975; Mohr 1972). The most famous colony, that of Carlsbad Caverns National Park in New Mexico, fell from an estimated 8.7 million bats in 1936 to only 200,000 in 1973. The even larger group in Eagle Creek Cave, Arizona, was reduced to only about 600,000 individuals in 1970. The exact cause of such declines is unknown, but the most likely factor is poisoning through accumulation of organochlorine residues in the bats as a result of the spraying of their prey insects with DDT. There may have been a moderate recovery of the Carlsbad Caverns population since DDT was banned in the United States in 1972, but this insecticide is still being used in agricultural operations in Mexico, where the bats spend the winter. Fortunately, at least one of the giant summer colonies still exists, as McCracken and Gustin (1987) reported the presence of about 20 million *T. brasiliensis* at Bracken Cave in central Texas.

CHIROPTERA; MOLOSSIDAE; **Genus CHAEREPHON**
Dobson, 1874

Lesser Mastiff Bats

There are 13 species (Aggundey and Schlitter 1984; Ansell 1986; Chasen 1940; Corbet 1978; Eger and Peterson 1979; Fenton and Peterson 1972; Harrison 1975*b*; Hayman and Hill, *in* Meester and Setzer 1977; Hill 1961*b*, 1982*a*, 1983*a*; Hill and Beckon 1978; Hill and Morris 1971; Kock 1975; Koopman 1975, 1982*a*; Largen, Kock, and Yalden 1974; Lekagul and McNeely 1977; Medway 1978; Nader and Kock 1980; Peterson 1971*a*, 1972; Peterson and Harrison 1970; Ride 1970; Schlitter et al. 1986; Schlitter, Robbins, and Buchanan 1982):

C. bivittata, Ethiopia to Zambia and Mozambique;
C. ansorgei, Cameroon and Ethiopia to Angola and Transvaal;
C. bemmelini, Liberia, Cameroon, eastern Zaire, southern Sudan, Kenya, Uganda, Tanzania;

Chaerephon pumila, photo by John Visser.

C. *nigeriae*, southwestern Saudi Arabia, Ghana and Niger to Ethiopia and Botswana;
C. *major*, savannah zones from Mali and Liberia to southern Sudan and Tanzania;
C. *pumila*, southwestern Arabian Peninsula, most of Africa south of the Sahara, Madagascar, Aldabra Island;
C. *chapini*, Zaire, Uganda, Zambia, Angola, Namibia;
C. *russata*, Ghana, Cameroon, northeastern Zaire, Kenya;
C. *aloysiisabaudiae*, Ghana, Gabon, Zaire, Uganda;
C. *gallagheri*, eastern Zaire;
C. *plicata*, India to Malay Peninsula, Sri Lanka, Hainan, Sumatra, Java, Borneo, Philippines;
C. *johorensis*, Malay Peninsula, Sumatra;
C. *jobensis*, New Guinea and nearby Japen (Jobi) Island, Solomon Islands, northern Australia, New Hebrides, Fiji Islands.

Chaerephon was regarded as a full genus by Freeman (1981), Honacki, Kinman, and Koeppl (1982), Koopman (1984*b*), and Richards (*in* Strahan 1983) but only as a subgenus of *Tadarida* by Corbet and Hill (1986), Legendre (1984), and Meester et al. (1986). The species *C. bivittata, C. ansorgei*, and *C. bemmelini* sometimes are placed in *Tadarida*, but Freeman (1981) assigned them to *Chaerephon*.

In *C. plicata* of Southeast Asia, head and body length is 65–75 mm, tail length is 30–40 mm, forearm length is 40–50 mm, and weight is 17–31 grams. The pelage is dense and soft, the upper parts are dark brown, and the underparts are slightly paler. The face is covered with stiff, short, black bristles, and the ears are thick, round, broad, and joined on the front of the muzzle (Lekagul and McNeely 1977). In *C. jobensis* of Australia, head and body length is 80–90 mm, tail length is 35–45 mm, forearm length is 46–52 mm, and weight is 20–30 grams. Coloration is chocolate to gray brown above and slightly grayer below (Richards, *in* Strahan 1983). In various African species, head and body length is about 50–80 mm, tail length is 25–45 mm, forearm length is 35–53 mm, and weight is 8–26 grams. The upper parts are dark brown or black, the underparts are slightly paler, and there are sometimes white markings on the sides or belly (Happold 1987; Kingdon 1974*a*; Smithers 1983). Several species have tufts of glandular hairs, arising from the crown, behind the ears, and several have a heavy crest of long straight hairs on the back of the membrane uniting the ears. These patches of specialized hairs are often restricted to the males, as is the saclike throat gland.

Chaerephon differs from *Tadarida* in having ears that are joined by a band of skin, usually a more elevated mandibular condyle, and broader wing tips. *Chaerephon* differs from *Mops* in having less robust jaws and more constricted anterior palatal emargination. There are usually five upper cheek teeth, and the last upper molar has an N-shaped occlusal surface (Freeman 1981; Kingdon 1974*a*).

These bats inhabit open forests, savannahs, and agricultural areas, sometimes in the mountains. They roost in hollow trees, crevices, and caves. Some species have adapted well to human presence and can be found in roofs, rafters, thatch, and other suitable parts of buildings (Richards, *in* Strahan 1983; Smithers 1983). *C. pumila* is known to be a fast flier and to hunt above the forest canopy and buildings at heights of over 70 meters; it takes a wide variety of small insects (Kingdon 1974*a*).

Social structure appears to vary. Rosevear (1965), for example, reported that *C. pumila* is usually found roosting singly. Smithers (1983), however, noted that while *C. bivittata* usually does not occur in groups of more than about 6 individuals, *C. pumila* may occur in colonies of hundreds in favorable areas. Kingdon (1974*a*) observed that the larger groups are found in buildings, where there is much available roosting space. A colony of 350 *C. jobensis* was found in a building in a town (Richards, *in* Strahan 1983). *C. plicata* of southeastern Asia generally is found in caves in groups of 200,000 or more individuals (Lekagul and McNeely 1977; Medway 1978).

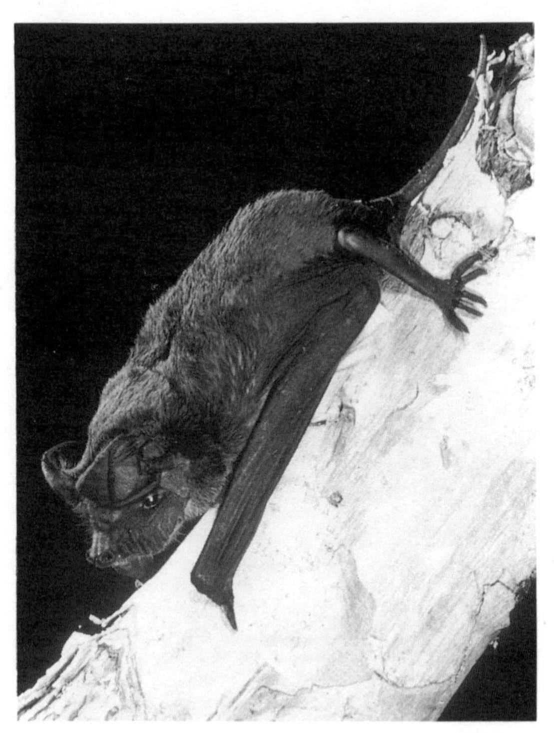

Chaerephon jobensis, photo by B. G. Thomson / National Photographic Index of Australian Wildlife.

In a detailed study of *C. pumila* in Ghana, McWilliam (1988) found harems of up to 21 adult females and their young, attended by a single adult male. Group composition was stable, with some individuals remaining together at the same site throughout the 16-month study period. Three cohorts of young were produced during the wet season from April to October. A few young females stayed in their natal groups to replace the loss of older females, but most young of both sexes dispersed during the dry season, after they had attained reproductive maturity.

Young *C. jobensis* have been found in December and January (Richards, *in* Strahan 1983), and a pregnant female *C. bivittata* was taken in March (Kingdon 1974*a*). In Uganda, *C. pumila* is a continuous breeder, the peaks coinciding with maximum rainfall in October–November and April–May (Mutere 1973*a*). Studies of the same species in the Transvaal (Van der Merwe, Rautenbach, and Van der Colf 1986) have suggested less of a correlation with rainfall. In that area there is an extended breeding season of 8 months per year, with birth peaks in November, January, and April; this period evidently is associated with higher minimum temperatures and thus greater availability of insects. Females were found to be polyestrous and able to have three pregnancies per season. The gestation period is about 60 days, and a single young is born. Weaning apparently occurs before the age of 21 days, and females become sexually mature at 5–12 months.

CHIROPTERA; MOLOSSIDAE; **Genus MOPS**
Lesson, 1842

Greater Mastiff Bats

There are 2 subgenera and 14 species (Aggundey and Schlitter 1984; Chasen 1940; Corbet 1978; El-Rayah 1981; Harrison 1975*b*; Hayman and Hill, *in* Meester and Setzer 1977; Hill 1961*b*; Hill and Morris 1971; Jones 1971*a*; Kock 1975; Koopman 1975; Koopman, Mumford, and Heisterberg 1978; Largen, Kock, and Yalden 1974; Laurie and Hill 1954; Medway 1977; Peterson 1972; Schlitter, Robbins, and Buchanan 1982):

subgenus *Xiphonycteris* Dollman, 1911

M. spurrelli, Ivory Coast, Ghana, Cameroon, Central African Republic, Equatorial Guinea, Zaire;
M. nanulus, forest zone from Sierra Leone to Ethiopia and Kenya;
M. petersoni, Ghana, Cameroon;
M. leonis, forest zone from Sierra Leone to eastern Zaire, island of Bioko (Fernando Poo);
M. brachyptera, Mozambique, possibly Kenya;
M. thersites, forest zone from Sierra Leone to Rwanda, possibly Mozambique and Zanzibar;

subgenus *Mops* Lesson, 1842

M. condylurus, most of Africa south of the Sahara, Madagascar;
M. niveiventer, Zaire, Tanzania, Angola, Zambia, northern Botswana;
M. demonstrator, Burkina Faso, Sudan, northeastern Zaire, Uganda;
M. mops, Malay Peninsula, Sumatra, Borneo;
M. sarasinorum, Sulawesi, Mindanao;
M. trevori, northeastern Zaire, Uganda;
M. congica, southern Cameroon, northeastern Zaire, western Uganda;
M. midas, southwestern Saudi Arabia, much of savannah zone of Africa south of the Sahara, Madagascar.

Mops was regarded as a full genus by Freeman (1981), Honacki, Kinman, and Koeppl (1982), and Koopman (1984*b*) but only as a subgenus of *Tadarida* by Corbet and Hill (1986), Legendre (1984), and Meester et al. (1986). *Xiphonycteris* was treated as a full genus, with only the one species *M. spurrelli*, by Hayman and Hill (*in* Meester and Setzer 1977). The above use and composition of *Xiphonycteris* as a subgenus is based primarily on Koopman (1975); neither Freeman (1981) nor Legendre (1984) recognized this subgenus, but Corbet and Hill (1986) and El-Rayah (1981) did. Another name, *Philippinopterus* Taylor, 1934, is a synonym of *Mops*, and its only species, *M. lanei* of Mindanao, was considered conspecific with *M. sarasinorum* by Koopman (1975). Freeman (1981) regarded *M. niangarae* of northeastern Zaire as a species distinct from *M. trevori*.

Mops condylurus, photo by Richard K. LaVal.

The information for the remainder of this account was compiled from Freeman (1981), Happold (1987), Kingdon (1974*a*), Medway (1978), Rosevear (1965), and Smithers (1983). Head and body length is 52–121 mm, tail length is 34–56 mm, forearm length is 29–66 mm, and weight is 7–64 grams. The upper parts are often dark brown but vary in color both within and among species from reddish to almost black. The underparts are usually paler, and in some species there are white markings.

Mops is characterized by ears that are joined over the top of the head by a band of skin, very wrinkled lips, a robust skull, medium to thick jaws, a last upper molar that is reduced to a V pattern, and usually only two lower incisors on each side. From *Tadarida* and *Chaerephon*, it is distinguished by having more developed sagittal and lambdoidal crests in the skull, a generally thicker dentary with a higher coronoid process, usually more anterior palatal emargination, and reduced dentition.

Habitat varies widely within species and includes forest, woodland, savannah, and dry brushland. These bats roost in caves, mines, culverts, hollow trees, crevices, attics, and thatched roofs. They emerge after sundown and fly high and fast in pursuit of insects. Their powerful jaws suggest a diet of hard-bodied prey, such as beetles.

Colonies range in size from fewer than 10 to several hundred individuals. Studies in Uganda have indicated the following reproductive patterns: *M. nanulus*, two breeding seasons, with fertilization in January and July, and synchronized births in March–April and September, when insect availability is greatest; *M. midas*, a well-defined breeding season, with pregnancy in January and lactation in March, and possibly a second season, as a lactating female was taken in October; *M. congica*, two breeding seasons, with pregnancies having been reported in March and September; and *M. condylurus*, two seasons, with most mating in April–May and November–December and birth peaks coinciding with maximum rainfall in July–August and February–March. Pregnant females of *M. midas* and *M. condylurus* have been taken in Botswana from December to February. The gestation period of the latter species is about two months, and a single young is born. Females are thought to be able to breed in the season following their birth.

CHIROPTERA; MOLOSSIDAE; **Genus OTOMOPS**
Thomas, 1913

Big-eared Free-tailed Bats

There are five widely scattered species (Ansell 1974; Chasen 1940; Ellerman and Morrison-Scott 1966; Hayman and Hill, *in* Meester and Setzer 1977; Hill 1983*a*; Hill and Morris 1971; Largen, Kock, and Yalden 1974; Laurie and Hill 1954):

O. martiensseni, Central African Republic, Djibouti, Ethiopia, Zaire, Kenya, Tanzania, Angola, Zimbabwe, Malawi, Natal (South Africa), Madagascar;

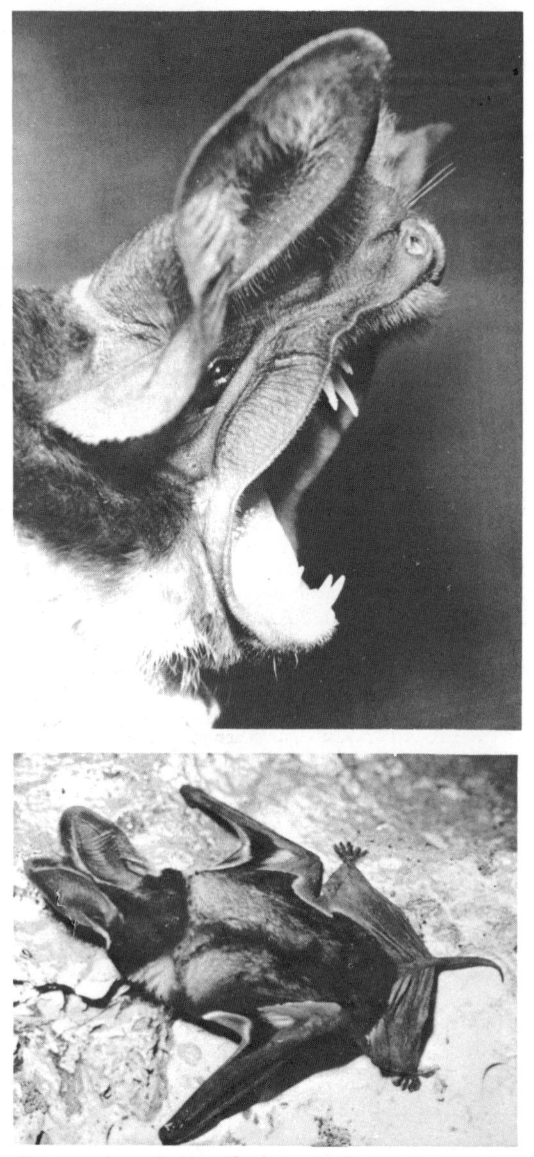

Big-eared free-tailed bat *(Otomops martiensseni)*, photos by D. W. Yalden.

O. wroughtoni, southern India;
O. formosus, Java;
O. papuensis, Papua New Guinea;
O. secundus, northeastern New Guinea.

Head and body length is 60 to about 100 mm, tail length is 30–50 mm, and forearm length is 49 to about 70 mm. Ansell (1974) gave the weights of a male and a female *O. martiensseni* as 36 grams and 27 grams, respectively. The coloration is reddish brown, pale brown, or dark brown. Most species have a grayish or whitish area on the back of the neck and upper back.

There is a series of small spines along the anterior borders of the ears. The ears, 25–40 mm in length, are united by a low membrane. A glandular sac is sometimes located in the lower throat region.

These bats roost in caves, hollow trees, and human-made structures and reportedly are usually solitary or associate in small groups. According to Kingdon (1974*a*), however, the few known colonies of *O. martiensseni* contain many hundreds of bats packed close together. There are particularly large numbers in the lava tunnels on Mt. Suswa in Kenya. Mutere (1973*b*) studied two populations of this species in Kenya. In the group at Ithunda, to the south, adult females showed evidence of pregnancy from October to January, and adult males had a peak in their sexual cycle in August. In the Suswa population, to the north, the same pattern was evident, but there were also a few isolated pregnancies in May and June. Brosset (1962) reported that females of *O. wroughtoni* collected in December in India had newborn young, while others were on the verge of delivery. Specimens taken in May had no young, nor were any females pregnant. Brosset suggested that the breeding season was near the end of autumn.

CHIROPTERA; MOLOSSIDAE; **Genus NYCTINOMOPS**
Miller, 1902

New World Free-tailed Bats

There are four species (Cabrera 1957; Hall 1981; Handley 1976; Husson 1978; Koopman 1978*a*, 1982*b*; Ochoa G. 1984; Taddei and Garutti 1981):

N. aurispinosus, east and west coasts of Mexico, southern Mexico, Colombia, Peru, Venezuela, eastern Brazil;
N. femorosaccus, southwestern United States, northern and western Mexico;
N. laticaudatus, eastern and southern Mexico to northern Argentina and southern Brazil, Cuba;
N. macrotis, Iowa, southeastern Kansas, southwestern United States, northern and central Mexico, most of South America, Cuba, Dominican Republic, Jamaica.

Nyctinomops was regarded as a full genus by Freeman (1981), Honacki, Kinman, and Koeppl (1982), Koopman (1984*b*), and Legendre (1984) but only as a subgenus of *Tadarida* by Corbet and Hill (1986) and Jones, Arroyo-Cabrales, and Owen (1988). *N. europs* of Venezuela, Surinam, Brazil, and Trinidad and *N. gracilis* of Venezuela and Brazil were listed as synonyms of *N. laticaudatus* by Freeman (1981).

Head and body length is 54–84 mm, tail length is 34–57 mm, and forearm length is 41–64 mm (Hall 1981; Hoffmeister 1986). The upper parts are brown or reddish and the

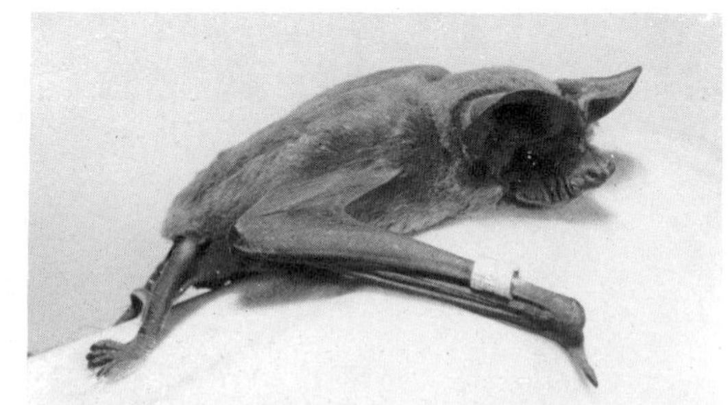

A New World free-tailed bat *(Nyctinomops laticaudatus)* wearing an aluminum band on its right forearm. The band bears a number and the name of the organization to which it should be reported if captured. Note that the very delicate skin of the forearm is being injured by the band, which emphasizes the fact that bands must be very carefully placed; otherwise the bat may not survive and banding will be futile. Photo by Ernest P. Walker.

underparts are paler. Although the species of *Nyctinomops* formerly were joined with *Tadarida brasiliensis* in the genus and subgenus *Tadarida,* Freeman (1981) indicated that *T. brasiliensis* actually has closer affinity to certain African *Tadarida.* From *T. brasiliensis, Nyctinomops* is distinguished by its short second phalanx of the fourth digit of the wing, narrower rostrum of the skull, loss of the third lower incisor tooth, well-joined (rather than separated) ears, and slightly narrower anterior palatal emargination.

In Venezuela, Handley (1976) found most *N. laticaudatus* roosting in rocks. In the Yucatan, Jones, Smith, and Genoways (1973) reported this species to frequent buildings. According to Hoffmeister (1986), *N. femorosaccus* inhabits rocky cliffs and slopes in the southern desert of Arizona but also makes use of buildings. *N. macrotis* has been taken in a variety of habitats in Arizona, including ponderosa pine, Douglas fir, and desert scrub, but apparently requires rocky cliffs, with crevices and fissures, for roosting. Both species apparently migrate from Mexico into the United States and form maternity colonies during the summer, but there are also some overwintering groups. They emerge from their roosts late in the evening and fly fast and high in pursuit of insects, mainly large moths.

Nyctinomops is not so gregarious as *Tadarida brasiliensis.* According to Schmidly (1977), colonies of *N. femorosaccus* usually number fewer than 100 individuals, and a nursery colony of *N. macrotis* contained 130. Such groups consist almost entirely of adult females and their young; the males roost separately. Barbour and Davis (1969) stated that both species give birth to one young during June or July. Lactating females have been taken in Arizona in August (Hoffmeister 1986). In the Yucatan, Jones, Smith, and Genoways (1973) found pregnant female *N. laticaudatus,* each with a single embryo, in April and lactating females in August.

CHIROPTERA; MOLOSSIDAE; **Genus EUMOPS**
Miller, 1906

Mastiff Bats, or Bonneted Bats

There are nine species (Anderson, Koopman, and Creighton 1982; Dolan and Carter 1979; Eger 1977; Graham and Barkley 1984; Hall 1981; Myers and Wetzel 1983):

E. auripendulus, southern Mexico to Paraguay and southern Brazil, Jamaica, Trinidad;
E. underwoodi, southern Arizona to Nicaragua;
E. glaucinus, southern Florida, central Mexico to Paraguay and southeastern Brazil, Cuba, Jamaica;
E. maurus, Guyana, Surinam;
E. dabbenei, Magdalena River Valley of Colombia, northern Venezuela, central Paraguay, Chaco Province of northern Argentina;
E. bonariensis, southern Mexico to central Argentina;
E. hansae, Costa Rica, Panama, Venezuela, Guyana, north-central Peru, northern Brazil;
E. perotis, southwestern United States, northern Mexico, Venezuela, Ecuador, Peru, Bolivia, Paraguay, eastern Brazil, northern Argentina;
E. trumbulli, Amazonian region of South America.

Hall (1981) considered *E. nanus,* found from southern Mexico to northern South America, to be specifically distinct from *E. bonariensis.* Koopman (1978*a*) regarded *E. trumbulli* as a subspecies of *E. perotis.*

Head and body length is 40–130 mm, tail length is 35–80 mm, and forearm length is 37–83 mm. *E. perotis* is the largest bat found in the United States (Barbour and Davis 1969). Adults of *E. underwoodi sonoriensis,* which have a head and body length of about 110 mm, weigh from 40 to about 65 grams. The upper parts are brownish, grayish, or black and the underparts are slightly paler.

The large ears are rounded or angular in outline and usually connected across the head at the base. Throat sacs are present in some of these bats. When present, they are more developed in the males than in the females. Apparently, when the males are sexually active, these sacs swell and secrete material that has an odor that may attract females.

These bats usually roost in crevices in rocks, tunnels, trees, and buildings. Like most molossid bats, *Eumops* often shelters under the corrugated iron roofing of human-made structures. The roosts, at least in the larger species, are usually six meters or more above the ground. Because of their size and long narrow wings, these bats require considerable space to launch themselves into flight. Although *E. perotis* appears to be nonmigratory, it moves to different roosting sites with the changing seasons, at least in the northern parts of its range. Some individuals of this species become inactive for short periods in winter in the southwestern United States. *E. per-*

Mastiff bat, or bonneted bat *(Eumops perotis),* photo by Lloyd G. Ingles. page 348

otis emits loud, high-pitched "peeps" while in flight. This species feeds on small insects, mainly members of the order Hymenoptera, probably catching them from near ground level to treetop height.

Groups of 10–20 bats are usual in this genus, though some individuals may roost alone and some colonies may consist of as many as 70 bats. The adult males and females do not segregate. In the United States births of *E. perotis* occur from June to August, and those of *E. underwoodi* and *E. glaucinus* take place in June and July. Usually a single young is produced by these species (Barbour and Davis 1969). In Costa Rica, Gardner, LaVal, and Wilson (1970) recorded pregnant female *E. glaucinus*, each with 1 embryo, on 19 May and 30 December and lactating females in April, May, and August. In the Yucatan, Birney et al. (1974) collected 7 adult female *E. glaucinus* on 17 May; 1 was not pregnant, 5 had 1 embryo each, and 1 had 2 embryos. Dolan and Carter (1979) collected 3 lactating female *E. underwoodi* in Nicaragua on 27 July. In Peru, Graham (1987) collected juvenile *E. auripendulus* in September and *E. glaucinus* in August.

CHIROPTERA; MOLOSSIDAE; **Genus PROMOPS**
Gervais, 1855

Domed-palate Mastiff Bats

There are two species (Cabrera 1957; Freeman 1981; Genoways and Williams 1979*b*; Hall 1981; Handley 1976; Koopman 1978*a*, 1982*b*; Marinkelle and Cadena 1972):

P. centralis, southern Mexico to Surinam and Peru, Trinidad, Paraguay;
P. nasutus, northern Peru and Surinam to northern Argentina and central Brazil, Trinidad.

P. pamana, known by a single incomplete specimen from central Brazil and sometimes designated a full species, was listed as a synonym of *P. nasutus* by Freeman (1981).

Head and body length is 60–90 mm, tail length is 45–75 mm, and forearm length is 43–63 mm. *P. centralis* is the largest species. A male and two females of this species from Trinidad weighed 14.4–17.2 grams. Coloration is drab brown to glossy black above and somewhat paler below. The short, broad skull, with its strongly domed palate, in combination with certain dental features, is characteristic. The short, rounded ears meet on the forehead, and a small muzzle pad (without processes) and throat sacs are present.

The members of this genus do not seem to be as gregarious as some of the other molossid bats. Colonies of approximately half a dozen individuals have been found roosting in hollow trees and on the underside of palm leaves. A colony of six *P. centralis*, all females, found on the underside of a palm leaf in Trinidad, contained two lactating individuals; the collection date was in April. In a study of a small group of *P. nasutus* roosting in the roof of a house in southern Brazil, Sazima and Uieda (1977) found parturition and lactation periods to coincide with the rainy season (November–December). Based on the collection of pregnant and lactating female *P. centralis* in April and October, Graham (1987) considered parturition to be possible in both the wet and dry seasons of coastal Peru. As far as is known, the diet consists of insects.

Domed-palate mastiff bat (*Promops* sp.), photos by Merlin D. Tuttle.

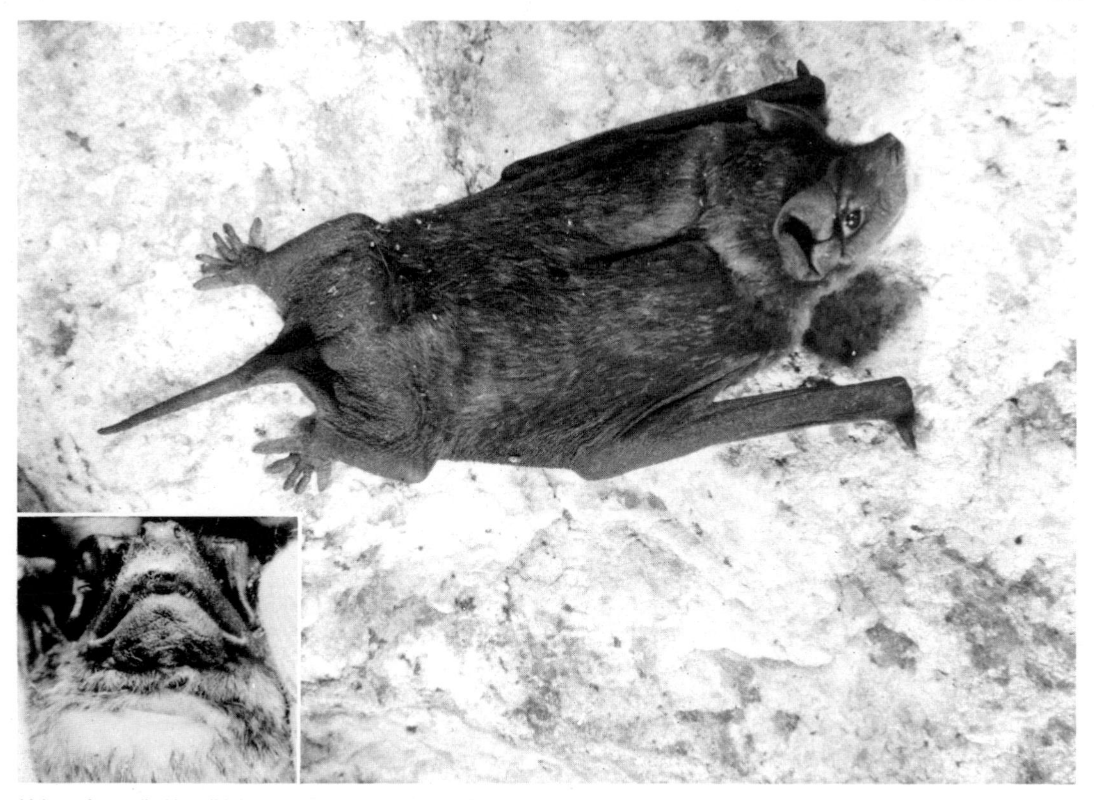

Velvety free-tailed bat *(Molossus molossus)*, photo by Bruce J. Hayward. Inset: *M. molossus*, photo by David Pye.

CHIROPTERA; MOLOSSIDAE; **Genus MOLOSSUS**
E. Geoffroy St.-Hilaire, 1805

Velvety Free-tailed Bats

There are seven species (Anderson, Koopman, and Creighton 1982; Cabrera 1957; Carter et al. 1981; Dolan and Carter 1979; Freeman 1981; Goodwin and Greenhall 1961; Hall 1981; Handley 1976; Koopman 1978*a*, 1982*b*; Marinkelle and Cadena 1972; Myers and Wetzel 1983):

M. ater, northern Mexico to northern Argentina, Trinidad
M. pretiosus, southern Mexico to Colombia and northern Guyana;
M. sinaloae, western and southern Mexico to northern Costa Rica;
M. trinitatus, southern Costa Rica to Surinam, Trinidad;
M. bondae, Honduras to Venezuela;
M. molossus, Mexico to Paraguay and Surinam, West Indies;
M. barnesi, French Guiana, Brazil.

Freeman (1981) listed *M. barnesi* as a synonym of *M. molossus*, and *M. macdougalli* of southern Mexico as a species apart from *M. pretiosus*. Jones, Arroyo-Cabrales, and Owen (1988) and Koopman (1982*b*) treated *M. trinitatus* as a subspecies of *M. sinaloae*, but Freeman's (1981) studies indicated that the two are specifically distinct.

Head and body length is 50–95 mm, tail length is 20–70 mm, forearm length is 33–60 mm, and adult weight is usually 10–30 grams. The general coloration is reddish brown, dark chestnut brown, dark brown, rusty blackish, or black. Many, perhaps all, of the species have two color phases. Some species have a long, bicolored pelage, whereas others have short, velvety, unicolored fur.

Externally, these bats resemble *Tadarida* and *Molossops*, but they differ in cranial and dental characters. The bases of the ears meet on the forehead. Throat sacs may be present.

In Venezuela, Handley (1976) collected several species of *Molossus*, mostly in moist areas but often at dry sites, within a variety of forested and open habitats; roosts included buildings, hollow trees and logs, holes in trees, and rocks. The genus is also known to roost in the fronds of palm trees and in caves. It is often found in attics with galvanized roofing, where the temperature may reach 55° C. *M. ater ater*, in Trinidad, rests in a horizontal rather than a vertical position, and individuals of other species probably do the same. Species of *Molossus* and *Noctilio albiventris* often are found roosting in the same trees and buildings. These bats begin to fly early in the evening, often before sunset. *M. ater* is a fast and erratic flier, sometimes flying high in the air and sometimes near the ground, depending on where insects are to be found. Both sexes of *M. ater* have large internal cheek pouches; when these are filled to capacity, the bat returns to the roost to chew and swallow its catch.

Molossus roosts in groups of up to hundreds of individuals. The adult males and females of *M. ater* may segregate and live apart even when roosting at the same site. This species may produce two litters per year on Trinidad. Pregnant female *M. ater* have been collected in Nayarit, Mexico, in July; in the Yucatan from April to August (Bowles 1972; Jones, Smith, and Genoways 1973); in Guatemala in March (Jones 1966); in El Salvador in November; in Nicaragua in March and July–August (Jones, Smith, and Turner 1971); and in Costa Rica in February and March (LaVal and Fitch 1977). Of 7 female *M. ater* taken in Hidalgo, Mexico on 30 July, 4 were lactating (Watkins, Jones, and Genoways 1972).

Of 10 female *M. ater* taken in Coahuila on 14 May, 2 were pregnant and 5 were lactating (Ramirez-Pulido and Lopez-Forment 1979). In a study of a colony of about 500 *M. ater* in a house in Manaus, Brazil, Marques (1986) found females to be polyestrous. Pregnancies occurred almost throughout the several years of study, but successful reproduction and lactation was limited to a period of favorable environmental conditions from July to November 1980. The gestation period was estimated to last 2–3 months.

Studies by Krutzsch and Crichton (1985) of a population of *M. molossus* on Puerto Rico (which they treated as a separate species, *M. fortis*) also indicate polyestry. Mating occurs in February or March, there is a birth season in June, and then a postpartum estrus, renewed mating in June or July, and another birth season in September. A single young is born, and lactation lasts about six weeks. Of 32 adult *M. molossus* caught in a church in Jalisco, Mexico, on 7 August, 31 were females, and 27 of these were pregnant (Watkins, Jones, and Genoways 1972). In Nicaragua *M. molossus* reportedly gives birth from early March to mid-July (Jones, Smith, and Turner 1971). In southwestern Colombia, pregnant females and immatures of this species were taken in July and August (Arata and Vaughan 1970). Of 18 female *M. molossus* taken in late July on Montserrat, 7 were pregnant and 11 were lactating (Jones and Baker 1979). Of 42 female *M. molossus* taken in late July on Guadeloupe, 18 were pregnant (Baker, Genoways, and Patton 1978).

Pregnant or lactating female *M. pretiosus* have been taken in Nicaragua in March, April, June, July, and August (Jones, Smith, and Turner 1971) and in Costa Rica in May, July, August, and October (Dolan and Carter 1979; LaVal and Fitch 1977). Pregnant or lactating female *M. sinaloae* have been taken in the Yucatan from March to May (Birney et al. 1974; Bowles 1972), in Honduras in July (Dolan and Carter 1979), and in Nicaragua in February and July (Jones, Smith, and Turner 1971). LaVal and Fitch (1977) thought that *M. sinaloae* might breed throughout the year in Costa Rica. Pregnant female *M. bondae* have been collected in Costa Rica in January, March, and August (Gardner, LaVal, and Wilson 1970; LaVal and Fitch 1977) and in Nicaragua in late July (Dolan and Carter 1979). All reproductive records for the genus refer to a single young.

CHIROPTERA; MOLOSSIDAE; **Genus CHEIROMELES**
Horsfield, 1824

Naked Bats, or Hairless Bats

There are two species (Lekagul and McNeely 1977):

C. torquatus, Malay Peninsula, Sumatra, Java, Borneo, Philippines;
C. parvidens, Sulawesi, Philippines.

Freeman (1981) regarded *Cheiromeles* as the most extreme member of the Molossidae but considered it phenetically closest to *Molossops*. Corbet and Hill (1986) listed *Cheiromeles* last among the Molossidae, and Legendre (1984) placed this genus in its own new subfamily. J. F. Koopman, cited by Freeman (1981), questioned whether *C. torquatus* and *C. parvidens* are specifically distinct.

Head and body length is 115–45 mm, tail length is 50–71 mm, and forearm length is 70–86 mm. Medway (1978) gave the weight of *C. torquatus* as 167–96 grams. These bats are

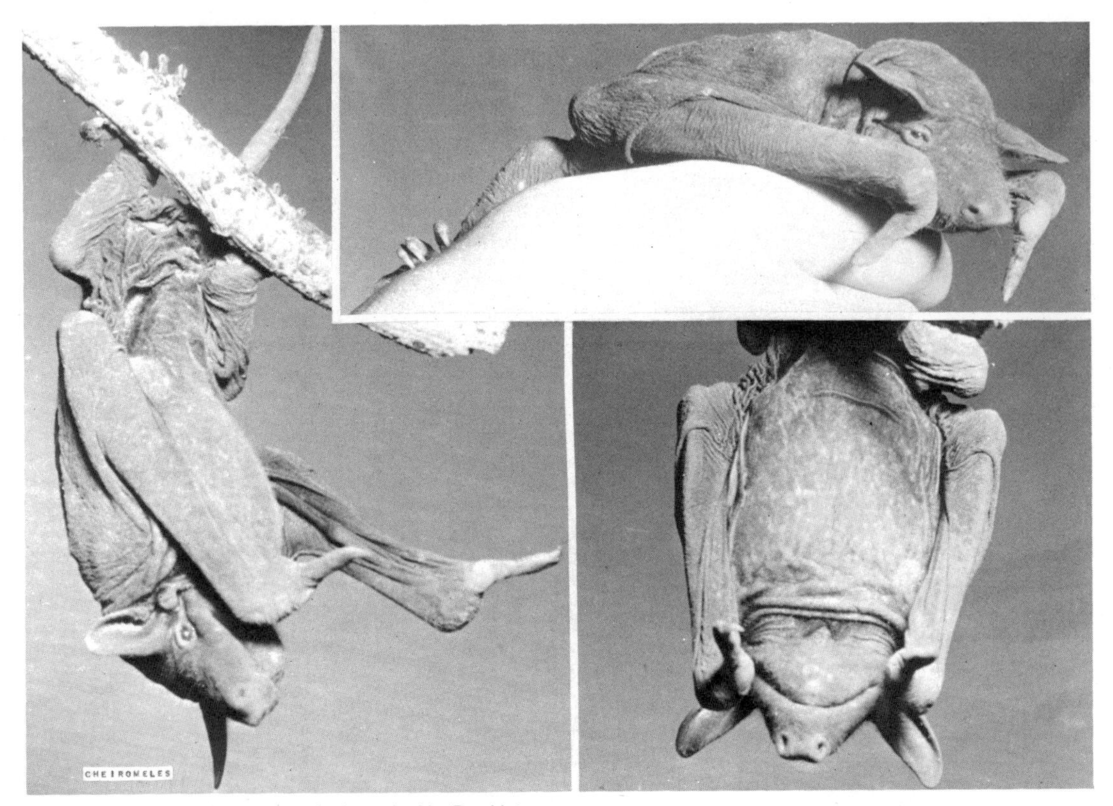

Naked bat *(Cheiromeles torquatus)*, photos by Lim Boo Liat.

nearly devoid of hair. Fine, short hairs are present on the head, tail membrane, and underparts, and a ruff of black bristles is present on the lower neck in the vicinity of a conspicuous glandular throat sac, where a strong-smelling secretion is produced. The skin is remarkably thickened and elastic. It is almost black in *C. torquatus* and dark brown in *C. parvidens*. The wings are attached to the back near the midline, and unlike in most Molossidae, the ears are separate and the lips are smooth.

This genus is unique among bats in several respects: the essentially naked appearance, the great development of the throat sacs, and the wing pouches. A pouch is present in both sexes along the sides of the body, formed by an extension of a fold of skin to the upper arm bone and to the upper leg bone. This pouch opens toward the rear and is 25–50 mm deep. The folded wings are pushed into these pouches by the hind feet; the first toe of each hind foot is opposable to the other toes and is supplied with a flattened nail instead of a claw. When the membrane portions of the wings are folded within the pouch, the bat can move about relatively freely on all four limbs. "The supposition is that, living in large, hollow trees, the bulldog bat must do a fair amount of climbing to find a suitable place to rest," and "perhaps the folding of the wings into pouches is to give elbow-room" (Burton 1955).

Naked bats roost in hollow trees, rock crevices, and holes the earth. They are not rare, as nearly a thousand have been noted in a hollow tree and a colony of about 20,000 was observed in a cave in Borneo (Freeman 1981). *Cheiromeles* has been seen at dusk flying high in the air with a rapid flight. Medway (1978) reported that the diet consists of termites and other insects that are hunted in the open air, either above the forest canopy or over clearings or paddy fields. A captive individual was maintained for several weeks on a diet of grasshoppers and moths.

There are usually two offspring (Freeman 1981). The mammae are located near the opening of the pouch, and it was formerly thought that the young were carried and nursed in this pocket. The young are probably left in the roost by the parents when they leave on their evening flights.

World Distribution of Bats

For maximum usefulness, it has been necessary to devise the simplest practicable outline of the approximate distribution of the genera in the sequence used in the text. The tabulation should be regarded as an index guide to groups of bats or to geographic regions. At the same time it gives a good overall picture of the general distribution of bats.

The major geographic distribution of the genera of Recent bats that appears in the tabulation is designed to show their natural distribution at the present time or within comparatively recent times. It should be noted that most of the animals occupy only a portion of the geographic region that appears at the head of the column. Some are limited to the tropical portion, others to temperate zones, and still others to the colder areas. Also, many restricted ranges cannot be designated either by letters to show the general area or by footnotes because of limited space on the tabulations. *It therefore should not be assumed that a mark indicating that an animal occurs within a geographic region implies that it inhabits all that area.* For more detailed outlines of the ranges of the respective genera, it is necessary to consult the generic texts.

Explanation of Geographic Column Headings

Europe and Asia constitute a single land mass, but this land mass has widely different types of zoogeographic areas created by high mountain ranges, plateaus, latitudes, and prevailing winds. The general distribution of Recent bats can be shown much more accurately by using two columns, Europe and Asia, than with a single one headed Eurasia.

Most islands are included with the major land masses nearby unless otherwise specified, though in many instances some of the bats indicated for the continental mass do not occur on the islands.

With Europe are included the British Isles and other adjacent islands, including those in the Arctic.

With Asia are included the Japanese Islands, Taiwan, Hainan, Sri Lanka, and other adjacent islands, including those in the Arctic.

With North America are included Mexico and Central America south to Panama, adjacent islands, the Aleutian chain, the islands in the Arctic region, and Greenland, but not the West Indies.

With South America are included Trinidad, the Netherlands Antilles, and other small adjacent islands, but not the Falkland and Galapagos islands unless named in footnotes.

With Africa are included only Zanzibar Island and small islands close to the continent, but not the Cape Verde or Canary islands.

The island groups treated separately are:

Southeastern Asian Islands, in which are included the Andamans, Nicobars, Mentawais, Sumatra, Java, Lesser Sundas, Borneo, Sulawesi, Moluccas, and the many other adjacent small islands;

New Guinea and small adjacent islands;

Australian region, in which are included Australia, Tasmania, and adjacent small islands;

Philippine Islands and small adjacent islands;

West Indies;

Madagascar and small adjacent islands;

other islands that have only one or a few forms of bats and are named in footnotes.

Footnotes indicate the major easily definable deviations from the distribution indicated in the tables.

Explanation of symbols used in World Distribution Table

■	The bats occur on the land or in the water area.
N	Northern portion
S	Southern portion
E	Eastern portion
W	Western portion
Ne	Northeastern portion
Se	Southeastern portion
Sw	Southwestern portion
Nw	Northwestern portion
C	Central portion

Examples: N, C for northern and central, Nc for north-central. Numerals refer to footnotes indicating clearly defined limited ranges within the general area.

WORLD DISTRIBUTION TABLE

Genera of Recent Bats	North America	West Indies	South America	Madagascar	Africa	Europe	Asia	Southeast Asia Islands	Philippine Islands	New Guinea	Australian Region	Antarctic Region	Arctic Region	Atlantic Ocean	Indian Ocean	Pacific Ocean
CHIROPTERA PTEROPODIDAE																
Eidolon				■	■		■ Sw									
Rousettus				■	■		■ S	■	■	■ 2						
Boneia								■ 3								
Myonycteris					■											
Pteropus				■			■ S	■	■	■ 2	■ N,E				■	■ 4
Acerodon								■ E	■							
Neopteryx								■ 3								
Pteralopex										■ 5						■ 6
Styloctenium								■ 3								
Dobsonia								■ E	■ C	■ 2	■ Ne					
Aproteles										■ C						
Harpyionycteris								■ 3	■ S							
Plerotes					■ Sc											
Hypsignathus					■ W,C											
Epomops					■											
Epomophorus					■											
Micropteropus					■											
Nanonycteris					■ W											
Scotonycteris					■ W											
Casinycteris					■ C											
Cynopterus							■ S	■	■							
Megaerops							■ Se	■	■ S							
Ptenochirus									■							
Dyacopterus							■ 7	■ 8	■ N							
Chironax							■ 7	■								
Thoopterus								■ 9								
Sphaerias							■ S									
Balionycteris							■ 7	■ 10								

2. And the Bismarck Archipelago and Solomon Islands. 3. Sulawesi only. 4. East to the Cook Islands. 5. Solomon Islands only. 6. Fiji Islands. 7. Malay Peninsula only. 8. Sumatra and Borneo only. 9. Sulawesi and Morotai only. 10. Borneo only.

Genera of Recent Bats	North America	West Indies	South America	Madagascar	Africa	Europe	Asia	Southeast Asia Islands	Philippine Islands	New Guinea	Australian Region	Antarctic Region	Arctic Region	Atlantic Ocean	Indian Ocean	Pacific Ocean
CHIROPTERA PTEROPODIDAE Continued																
Aethalops							■1	■								
Penthetor							■1	■2								
Haplonycteris									■							
Otopteropus									■N							
Alionycteris									■S							
Latidens							■Sc									
Nyctimene								■E	■C	■3	■Ne					■4
Paranyctimene										■						
Eonycteris							■Se	■	■							
Megaloglossus					■W,C											
Macroglossus							■Se	■	■S	■3	■N					
Syconycteris								■E		■5	■E					
Melonycteris										■3						
Notopteris																■Sw
CHIROPTERA RHINOPOMATIDAE																
Rhinopoma					■N		■S	■6								
CHIROPTERA EMBALLONURIDAE																
Taphozous				■	■		■	■	■		■				■	
Saccolaimus					■		■S	■	■	■7	■N					
Emballonura				■			■1	■	■	■3						■8
Coleura					■		■Sw								■9	
Rhynchonycteris	■S		■N													
Centronycteris	■S		■N													
Balantiopteryx	■S		■10													
Saccopteryx	■S		■N													
Cormura	■S		■N													
Peropteryx	■S		■													
Cyttarops	■S		■N													
Diclidurus	■S		■N													
CHIROPTERA CRASEONYCTERIDAE																
Craseonycteris							■11									
CHIROPTERA NYCTERIDAE																
Nycteris				■	■		■S	■								
CHIROPTERA MEGADERMATIDAE																
Megaderma							■S	■	■							
Cardioderma					■E											
Macroderma											■					
Lavia					■											
CHIROPTERA RHINOLOPHIDAE																
Rhinolophus					■	■	■	■	■	■5	■E					
CHIROPTERA HIPPOSIDERIDAE																
Hipposideros				■	■		■S	■	■	■3	■N					■12
Asellia					■N		■Sw									
Anthops										■13						
Aselliscus							■Se	■E		■3						■12
Rhinonycteris											■N					
Triaenops				■	■C,E		■Sw									
Cloeotis					■E,S											
Coelops							■Se	■	■S							
Paracoelops							■14									

1. Malay Peninsula only. 2. Borneo only. 3. And the Bismarck Archipelago and Solomon Islands. 4. Santa Cruz Islands. 5. And the Bismarck Archipelago. 6. Sumatra only. 7. And the Solomon Islands. 8. East to Samoa. 9. Seychelles Islands only. 10. Ecuador only. 11. Thailand only. 12. East to the New Hebrides. 13. Solomon Islands only. 14. Viet Nam only.

Genera of Recent Bats	North America	West Indies	South America	Madagascar	Africa	Europe	Asia	Southeast Asia Islands	Philippine Islands	New Guinea	Australian Region	Antarctic Region	Arctic Region	Atlantic Ocean	Indian Ocean	Pacific Ocean
CHIROPTERA MORMOOPIDAE																
Pteronotus	■S	■	■N													
Mormoops	■S	■	■N													
CHIROPTERA NOCTILIONIDAE																
Noctilio	■S	■	■													
CHIROPTERA PHYLLOSTOMIDAE																
Micronycteris	■S		■													
Macrotus	■S	■														
Lonchorhina	■S		■N													
Macrophyllum	■S		■													
Tonatia	■S		■													
Mimon	■S		■													
Phyllostomus	■S		■													
Phylloderma	■S		■													
Trachops	■S		■													
Chrotopterus	■S		■													
Vampyrum	■S		■													
Glossophaga	■S	■	■													
Monophyllus		■														
Leptonycteris	■S		■N													
Lonchophylla	■S		■													
Lionycteris	■1		■													
Anoura	■S		■													
Scleronycteris			■Nc													
Lichonycteris	■S		■													
Hylonycteris	■S															
Platalina			■2													
Choeroniscus	■S		■N													
Choeronycteris	■S															
Musonycteris	■3															
Carollia	■S	■4	■													
Rhinophylla			■													
Sturnira	■S	■	■													
Uroderma	■S		■													
Vampyrops	■S		■													
Vampyrodes	■S		■													
Vampyressa	■S		■													
Chiroderma	■S	■4	■													
Ectophylla	■S															
Artibeus	■S	■	■													
Ardops		■4														
Phyllops		■5														
Ariteus		■6														
Stenoderma		■7														
Pygoderma			■													
Ametrida			■N													
Sphaeronycteris			■													
Centurio	■S		■N													
Brachyphylla		■														
Erophylla		■														
Phyllonycteris		■														
Desmodus	■S		■													

1. Panama only. 2. Peru only. 3. Mexico only. 4. Lesser Antilles only. 5. Cuba and Hispaniola only. 6. Jamaica only. 7. Puerto Rico and the Virgin Islands only.

Genera of Recent Bats	North America	West Indies	South America	Madagascar	Africa	Europe	Asia	Southeast Asia Islands	Philippine Islands	New Guinea	Australian Region	Antarctic Region	Arctic Region	Atlantic Ocean	Indian Ocean	Pacific Ocean
CHIROPTERA PHYLLOSTOMIDAE Continued																
Diaemus	■S		■													
Diphylla	■S		■													
CHIROPTERA MYSTACINIDAE																
Mystacina																■1
CHIROPTERA NATALIDAE																
Natalus	■S	■	■N													
CHIROPTERA FURIPTERIDAE																
Furipterus	■S		■N													
Amorphochilus			■W													
CHIROPTERA THYROPTERIDAE																
Thyroptera	■S		■													
CHIROPTERA MYZOPODIDAE																
Myzopoda				■												
CHIROPTERA VESPERTILIONIDAE																
Myotis	■	■2	■	■3	■	■	■	■	■	■4	■					■5
Cistugo					■S											
Lasionycteris	■													■6		
Eudiscopus							■7									
Pipistrellus	■			■	■	■	■	■	■	■8	■9			■10		
Scotozous							■Sc									
Nyctalus						■	■							■11		
Glischropus							■Se	■	■							
Eptesicus	■	■	■		■	■	■									
Ia							■Se									
Vespertilio						■	■									
Laephotis					■											
Histiotus			■													
Philetor							■Se	■	■	■12						
Tylonycteris							■S	■	■							
Mimetillus					■W,C											
Hesperoptenus							■S	■								
Chalinolobus											■9					■13
Glauconycteris					■											
Nycticeius	■C,E	■14														
Nycticeinops					■		■Sw									
Scoteanax											■E					
Scotorepens										■Se	■					
Scotoecus					■		■Sc									
Rhogeessa	■S		■													
Scotomanes							■Se									
Scotophilus				■	■		■S	■	■						■15	
Otonycteris					■N		■Sw									
Lasiurus	■	■	■16											■17		■18
Barbastella					■Nw	■	■							■10		
Plecotus	■				■N	■	■							■19		
Idionycteris	■S															
Euderma	■Wc															
Miniopterus				■3	■	■S	■	■	■	■8	■					■5
Murina							■	■	■S							
Harpiocephalus							■Se	■								
Kerivoula					■		■S	■	■	■	■Ne					

1. New Zealand. 2. Lesser Antilles only. 3. And the Comoro Islands. 4. And the Solomon Islands. 5. East to the New Hebrides. 6. Bermuda and the Bahamas. 7. Burma and Laos only. 8. And the Bismarck Archipelago and Solomon Islands. 9. And Tasmania. 10. Canary Islands. 11. Azores and Madeira Islands. 12. And the Bismarck Archipelago. 13. New Caledonia and New Zealand. 14. Cuba only. 15. Reunion Island. 16. And the Galapagos Islands. 17. Bermuda, Iceland, and Orkney Island. 18. Hawaii only. 19. Canary and Cape Verde Islands.

Genera of Recent Bats	North America	West Indies	South America	Madagascar	Africa	Europe	Asia	Southeast Asia Islands	Philippine Islands	New Guinea	Australian Region	Antarctic Region	Arctic Region	Atlantic Ocean	Indian Ocean	Pacific Ocean
CHIROPTERA VESPERTILIONIDAE Continued																
Antrozous	■ W.S	■ 1														
Bauerus	■ S															
Nyctophilus										■	■ 2					
Pharotis										■ Se						
Tomopeas			■ 3													
CHIROPTERA MOLOSSIDAE																
Mormopterus		■ 1	■ W	■	■			■ 4		■	■				■ 5	
Sauromys					■ S											
Platymops					■ Ec											
Molossops	■ S		■													
Neoplatymops			■ N.C													
Cabreramops			■ 6													
Myopterus					■ W.C											
Tadarida	■ C.S	■	■	■	■	■	■			■	■					
Chaerephon				■	■		■ S	■	■	■ 7	■ N					■ 8
Mops				■	■		■	■	■ S							
Otomops				■	■		■ Sc	■ 9		■						
Nyctinomops	■ C.S	■	■													
Eumops	■ S	■	■													
Promops	■ S		■													
Molossus	■ S	■	■													
Cheiromeles							■ 10	■	■							

1. Cuba only. 2. And Tasmania. 3. Peru only. 4. Sumatra only. 5. Mauritius and Reunion. 6. Ecuador only. 7. And the Solomon Islands. 8. East to Fiji. 9. Java only. 10. Malay Peninsula only.

Appendix

GEOLOGICAL TIME

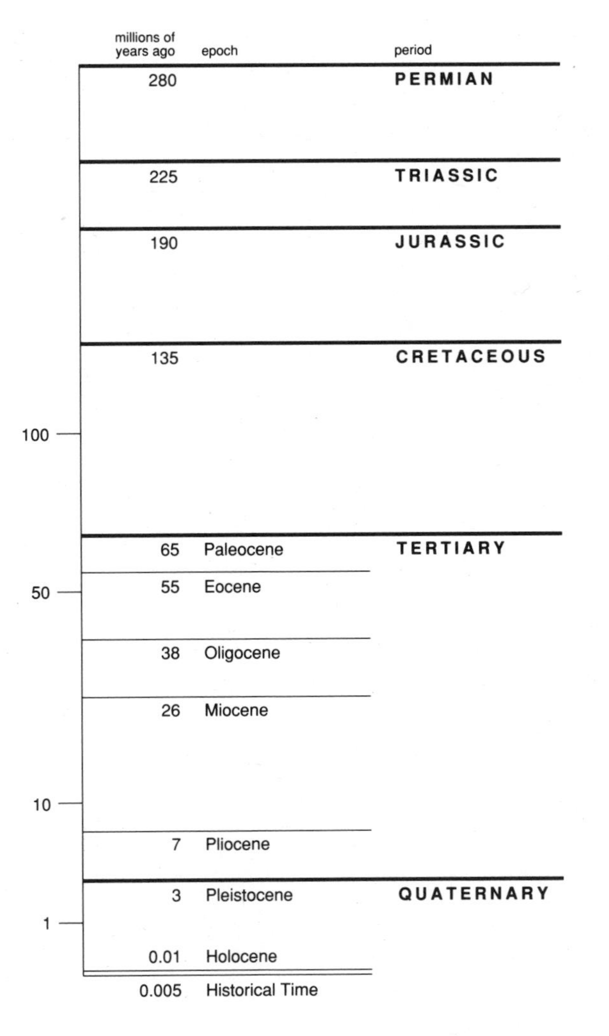

LENGTH

scales for comparison of metric and U.S. units of measurement

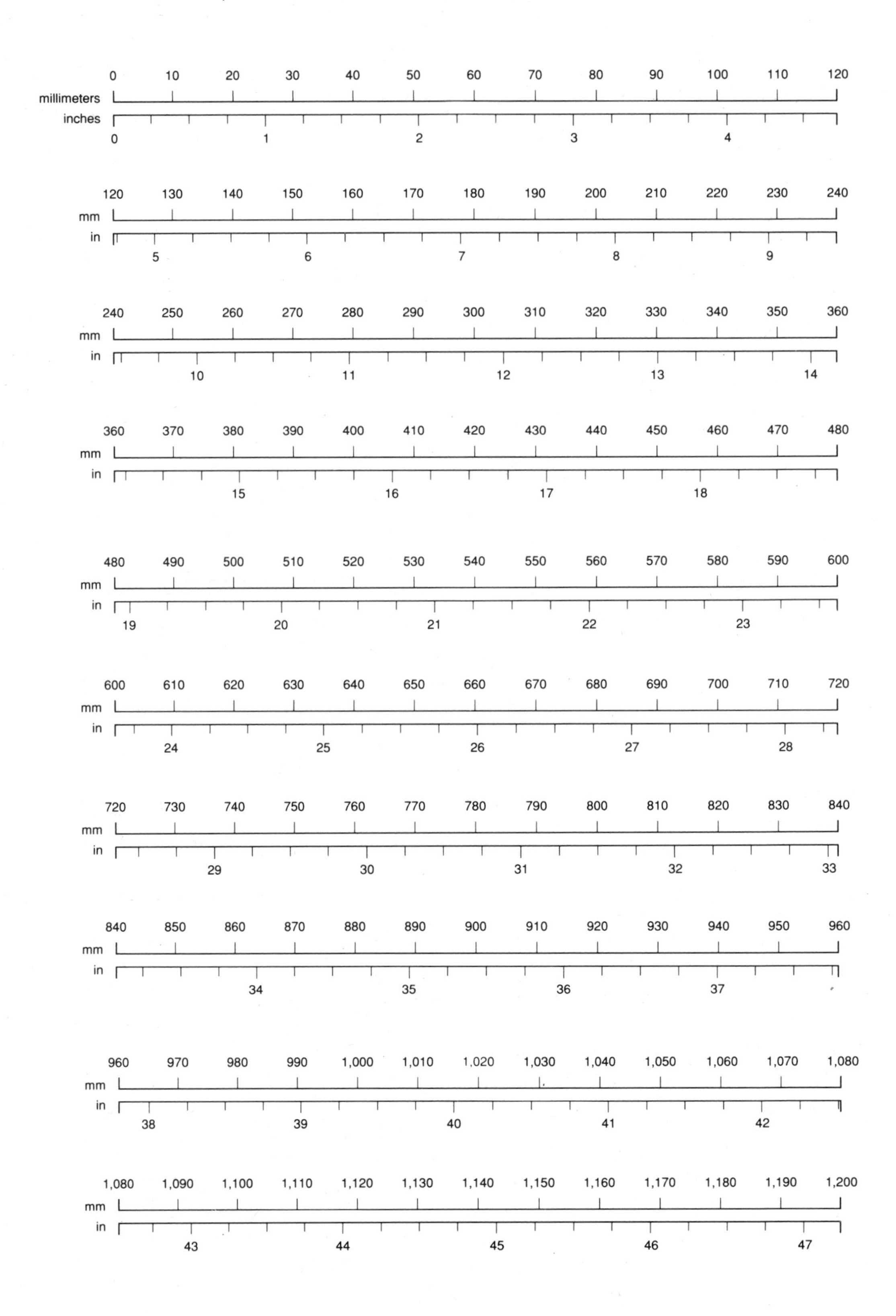

LENGTH

scales for comparison of metric and U.S. units of measurement

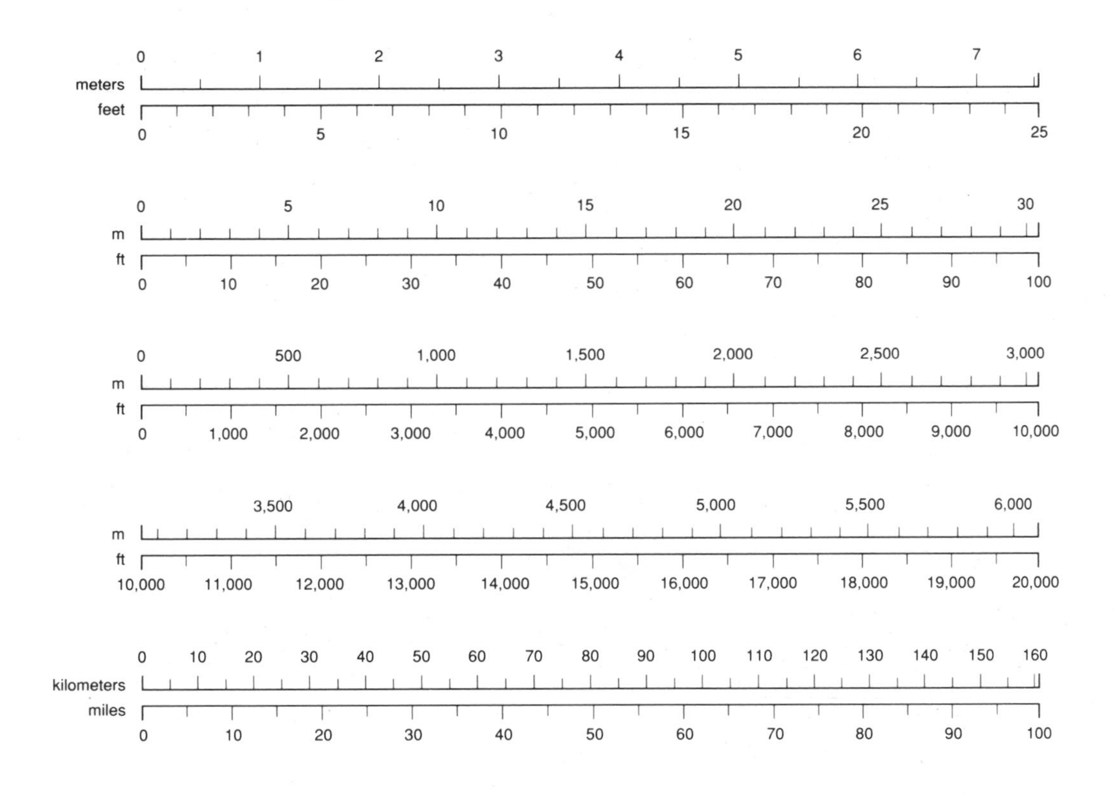

CONVERSION TABLES

	U.S. to Metric		Metric to U.S.	
	to convert	*multiply by*	*to convert*	*multiply by*
LENGTH	in. to mm.	25.4	mm. to in.	0.039
	in. to cm.	2.54	cm. to in.	0.394
	ft. to m.	0.305	m. to ft.	3.281
	yd. to m.	0.914	m. to yd.	1.094
	mi. to km.	1.609	km. to mi.	0.621
AREA	sq. in. to sq. cm.	6.452	sq. cm. to sq. in.	0.155
	sq. ft. to sq. m.	0.093	sq. m. to sq. ft.	10.764
	sq. yd. to sq. m.	0.836	sq. m. to sq. yd.	1.196
	sq. mi. to ha.	258.999	ha. to sq. mi.	0.004
VOLUME	cu. in. to cc.	16.387	cc. to cu. in.	0.061
	cu. ft. to cu. m.	0.028	cu. m. to cu. ft.	35.315
	cu. yd. to cu. m.	0.765	cu. m. to cu. yd.	1.308
CAPACITY (liquid)	fl. oz. to liter	0.03	liter to fl. oz.	33.815
	qt. to liter	0.946	liter to qt.	1.057
	gal. to liter	3.785	liter to gal.	0.264
MASS (weight)	oz. avdp. to g.	28.35	g. to oz. avdp.	0.035
	lb. avdp. to kg.	0.454	kg. to lb. avdp.	2.205
	ton to t.	0.907	t. to ton	1.102
	l. t. to t.	1.016	t. to l. t.	0.984

Abbreviations

avdp.	avoirdupois
cc.	cubic centimeter(s)
cm.	centimeter(s)
cu.	cubic
ft.	foot, feet
g.	gram(s)
gal.	gallon(s)
ha.	hectare(s)
in.	inch(es)
kg.	kilogram(s)
lb.	pound(s)
l. t.	long ton(s)
m.	meter(s)
mi.	mile(s)
mm.	millimeter(s)
oz.	ounce(s)
qt.	quart(s)
sq.	square
t.	metric ton(s)
yd.	yard(s)

WEIGHT

scales for comparison of metric and U.S. units of measurement

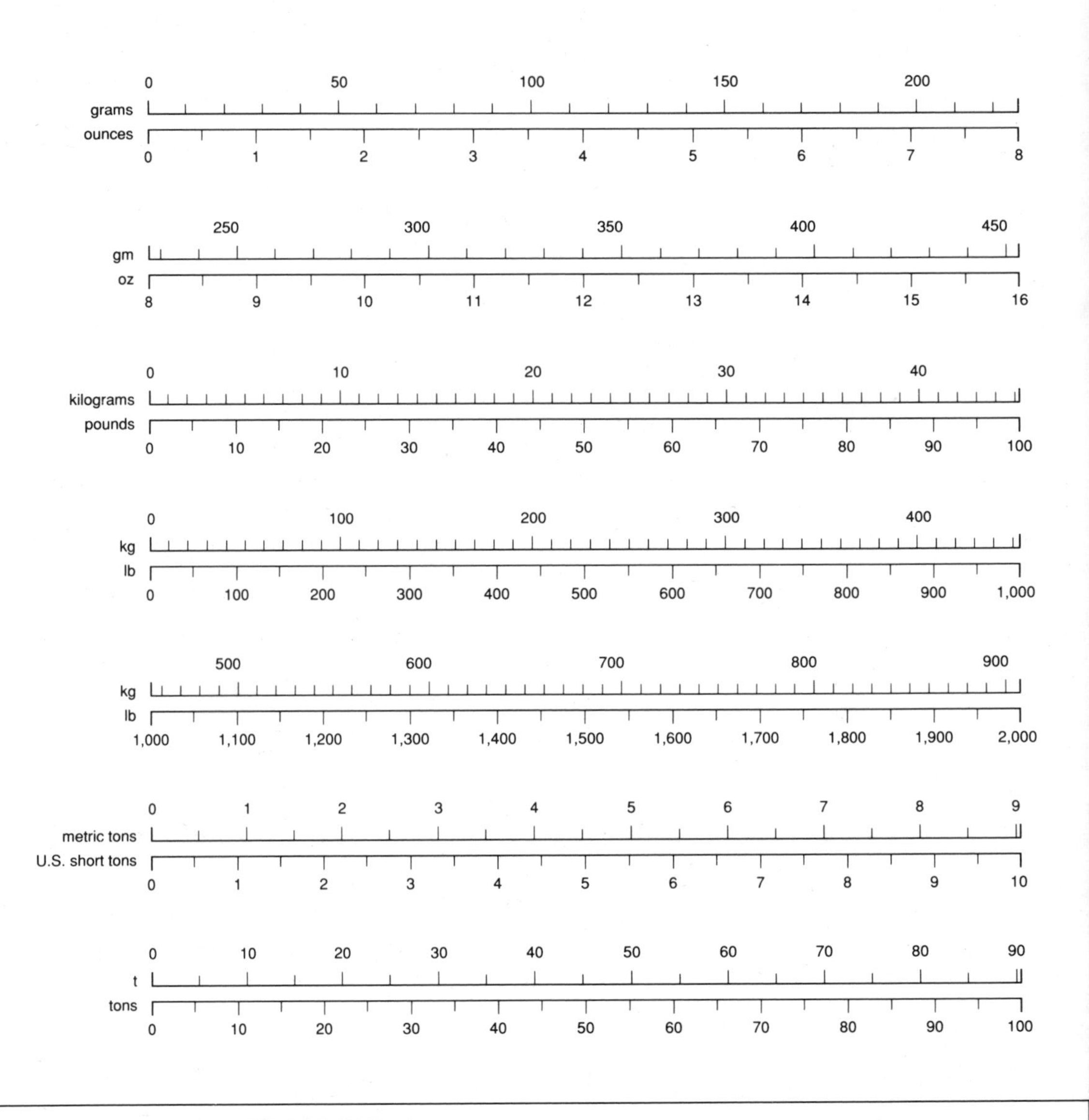

TEMPERATURE

scales for comparison of metric and U.S. units of measurement

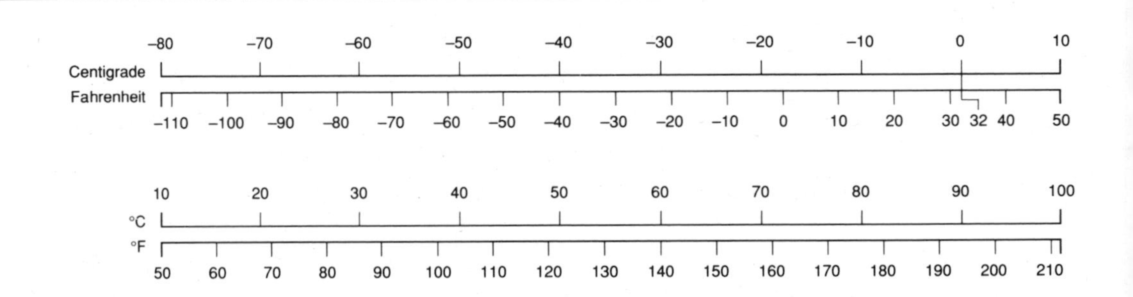

AREA

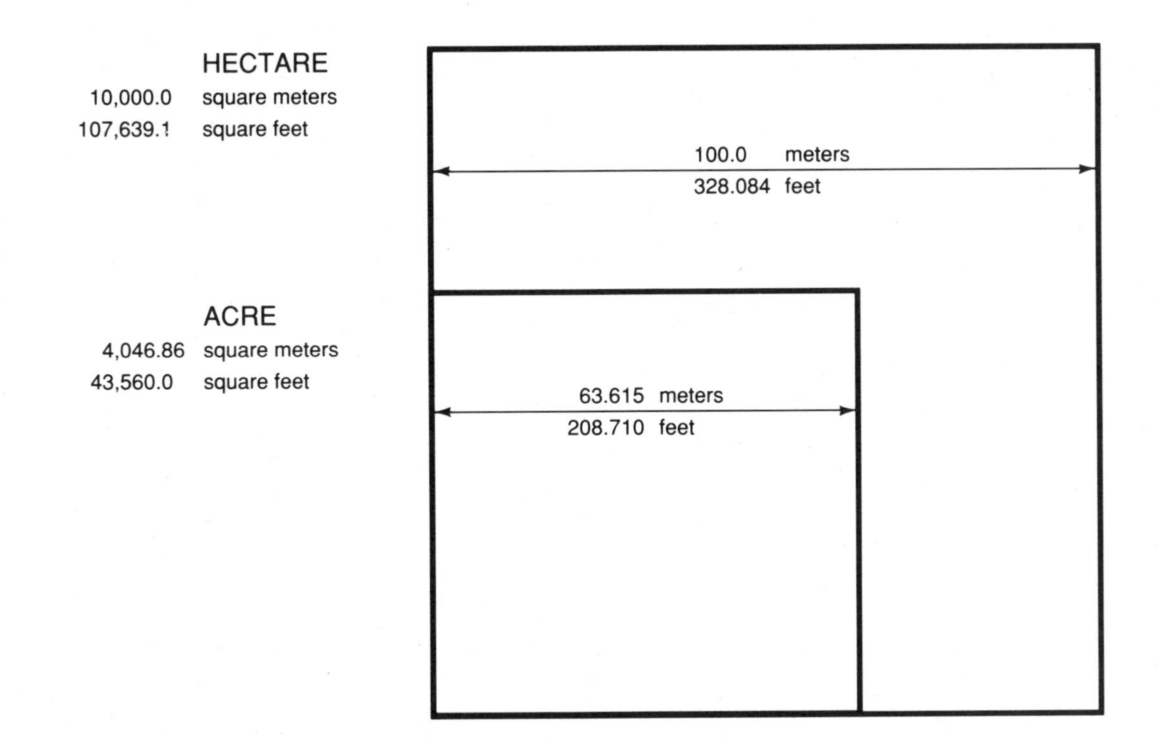

CONVERSION TABLES

	U.S. to Metric: *to convert*	*multiply by*	Metric to U.S.: *to convert*	*multiply by*
LENGTH	in. to mm.	25.4	mm. to in.	0.039
	in. to cm.	2.54	cm. to in.	0.394
	ft. to m.	0.305	m. to ft.	3.281
	yd. to m.	0.914	m. to yd.	1.094
	mi. to km.	1.609	km. to mi.	0.621
AREA	sq. in. to sq. cm.	6.452	sq. cm. to sq. in.	0.155
	sq. ft. to sq. m.	0.093	sq. m. to sq. ft.	10.764
	sq. yd. to sq. m.	0.836	sq. m. to sq. yd.	1.196
	sq. mi. to ha.	258.999	ha. to sq. mi.	0.004
VOLUME	cu. in. to cc.	16.387	cc. to cu. in.	0.061
	cu. ft. to cu. m.	0.028	cu. m. to cu. ft.	35.315
	cu. yd. to cu. m.	0.765	cu. m. to cu. yd.	1.308
CAPACITY (liquid)	fl. oz. to liter	0.03	liter to fl. oz.	33.815
	qt. to liter	0.946	liter to qt.	1.057
	gal. to liter	3.785	liter to gal.	0.264
MASS (weight)	oz. avdp. to g.	28.35	g. to oz. avdp.	0.035
	lb. avdp. to kg.	0.454	kg. to lb. avdp.	2.205
	ton to t.	0.907	t. to ton	1.102
	l. t. to t.	1.016	t. to l. t.	0.984

Abbreviations

avdp.	avoirdupois
cc.	cubic centimeter(s)
cm.	centimeter(s)
cu.	cubic
ft.	foot, feet
g.	gram(s)
gal.	gallon(s)
ha.	hectare(s)
in.	inch(es)
kg.	kilogram(s)
lb.	pound(s)
l. t.	long ton(s)
m.	meter(s)
mi.	mile(s)
mm.	millimeter(s)
oz.	ounce(s)
qt.	quart(s)
sq.	square
t.	metric ton(s)
yd.	yard(s)

Literature Cited

A

Acha, P. N., and A. M. Alba. 1988. Economic losses due to *Desmodus rotundus*. *In* Greenhall and Schmidt (1988), pp. 207–14.

Adams, M., P. R. Baverstock, C. H. S. Watts, and T. Reardon. 1987*a*. Electrophoretic resolution of species boundaries in Australian Microchiroptera. I. *Eptesicus* (Chiroptera: Vespertilionidae). Austral. J. Biol. Sci. 40:143–62.

———. 1987*b*. Electrophoretic resolution of species boundaries in Australian Microchiroptera. II. The *Pipistrellus* group (Chiroptera: Vespertilionidae). Austral. J. Biol. Sci. 40:163– 70.

Adams, M., T. R. Reardon, P. R. Baverstock, and C. H. S. Watts. 1988. Electrophoretic resolution of species boundaries in Australian Microchiroptera. IV. The Molossidae (Chiroptera). Austral. J. Biol. Sci. 41:315–26.

Aellen, V. 1973. Un *Rhinolophus* nouveau d'Afrique Centrale. Period. Biol. 75:101–5.

Aggundey, I. R., and D. A. Schlitter. 1984. Annotated checklist of the mammals of Kenya. I. Chiroptera. Ann. Carnegie Mus. 53:119–61.

Alberico, M. S. 1987. Notes on distribution of some bats from southwestern Colombia. Fieldiana Zool., n.s., 39:133–35.

Allen, G. M. 1939*a*. Bats. Harvard Univ. Press, Cambridge, 368 pp.

———. 1939*b*. A checklist of African mammals. Bull. Mus. Comp. Zool. 83:1–763.

———. 1942. Extinct and vanishing mammals of the Western Hemisphere with the marine species of all the oceans. Spec. Publ. Amer. Comm. Internatl. Wildl. Protection, no. 11, xv + 620 pp.

Allen, J. A. 1911. Mammals from Venezuela collected by Mr. M. A. Carriker, Jr., 1909–11. Bull. Amer. Mus. Nat. Hist. 30:239–73.

Anand Kumar, T. C. 1965. Reproduction in the rat-tailed bat *Rhinopoma kinneari*. J. Zool. 147:147–55.

Andersen, K. 1912. Catalogue of the Chiroptera in the collection of the British Museum. I. Megachiroptera. British Mus. (Nat. Hist.), London, ci + 854 pp.

Anderson, J. W., and W. A. Wimsatt. 1963. Placentation and fetal membranes of the Central American noctilionid bat, *Noctilio labialis minor*. Amer. J. Anat. 112:181–201.

Anderson, S. 1969. *Macrotus waterhousii*. Mammalian Species, no. 1, 4 pp.

Anderson, S., and J. K. Jones, Jr., eds. 1984. Orders and families of Recent mammals of the world. John Wiley & Sons, New York, xii + 686 pp.

Anderson, S., K. F. Koopman, and G. K. Creighton. 1982. Bats of Bolivia: an annotated checklist. Amer. Mus. Novit., no. 2750, 24 pp.

Anderson, S., and W. D. Webster. 1983. Notes on Bolivian mammals. 1. Additional records of bats. Amer. Mus. Novit., no. 2766, 3 pp.

Ansell, W. F. H. 1960. Mammals of Northern Rhodesia. Government Printer, Lusaka, xxxi + 155 + 24 pp.

———. 1974. Some mammals from Zambia and adjacent countries. Occas. Pap. Natl. Parks and Wildl. Serv. Zambia, Suppl., no. 1, 48 pp.

———. 1986. Some Chiroptera from south-central Africa. Mammalia 50:507–19.

Ansell, W. F. H., and R. J. Dowsett. 1988. Mammals of Malawi. An annotated check list and atlas. Trendrine Press, Cornwall, 170 pp.

Arata, A. A., and J. B. Vaughan. 1970. Analyses of the relative abundance and reproductive activity of bats in southwestern Colombia. Caldasia 10:517–28.

Arita, H. T., and D. E. Wilson. 1987. Long-nosed bats and agaves: the tequila connection. Bats 5(4):3–5.

Arroyo-Cabrales, J., and J. K. Jones, Jr. 1988*a*. *Balantiopteryx plicata*. Mammalian Species, no. 301, 4 pp.

———. 1988*b*. *Balantiopteryx io* and *Balantiopteryx infusca*. Mammalian Species, no. 313, 3 pp.

Atallah, S. I. 1977. Mammals of the eastern Mediterranean region; their ecology, systematics and zoogeographical relationships. Saugetierk. Mitt. 25:241–320.

Aulagnier, S., and R. Destre. 1985. Introduction a l'etude des chiropteres du Tafilalt (sud-est marocain). Mammalia 49:329–37.

Aulagnier, S., and P. Mein. 1985. Note sur la presence d'*Otonycteris hemprichi* Peters, 1859 au Maroc. Mammalia 49:582–84.

Ayala, S. C., and A. D'Alessandro. 1973. Insect feeding of some Colombian fruit-eating bats. J. Mamm. 54:266–67.

B

Baeten, B., V. Van Cakenberghe, and F. De Vree. 1984. An annotated inventory of a collection of bats from Rwanda. Rev. Zool. Afr. 98:183–96.

Bain, J. R. 1978. The breeding system of *Nycticeius humeralis* in Florida. Amer. Soc. Mamm., Abstr. Tech. Pap., 58th Ann. Mtg., p. 29.

Baker, H. G., and B. J. Harris. 1957. The pollination of *Parkia* by bats and its attendant evolutionary problems. Evolution 11: 449–60.

Baker, R. H. 1974. Records of mammals from Ecuador. Michigan State Univ. Mus. Publ., Biol. Ser., 5:129–46.

Baker, R. J. 1984. A sympatric cryptic species of mammal: a new species of *Rhogeessa* (Chiroptera: Vespertilionidae). Syst. Zool. 33:178–83.

Baker, R. J., P. V. August, and A. A. Steuter. 1978. *Erophylla sezekorni*. Mammalian Species, no. 115, 5 pp.

Baker, R. J., J. W. Bickham, and M. L. Arnold. 1985. Chromosomal evolution in *Rhogeessa* (Chiroptera: Vespertilionidae): possible speciation by centric fusions. Evolution 39:233– 43.

Baker, R. J., C. G. Dunn, and K. Nelson. 1988. Allozymic study of the relationships of *Phylloderma* and four species of *Phyllostomus*. Occas. Pap. Mus. Texas Tech Univ., no. 125, 14 pp.

Baker, R. J., H. H. Genoways, and J. C. Patton. 1978. Bats of Guadeloupe. Occas. Pap. Mus. Texas Tech Univ., no. 50, 16 pp.

Baker, R. J., H. H. Genoways, P. A. Seyfarth. 1981. Results of the Alcoa Foundation–Suriname Expeditions. VI. Additional chromosomal data for bats (Mammalia: Chiroptera) for Suriname. Ann. Carnegie Mus. 50:333–44.

Baker, R. J., R. L. Honeycutt, and R. A. Bass. 1988. Genetics. *In* Greenhall and Schmidt (1988), pp. 31–39.

Baker, R. J., and J. K. Jones, Jr. 1975. Additional records of bats from Nicaragua, with a revised checklist of Chiroptera. Occas. Pap. Mus. Texas Tech Univ., no. 32, 13 pp.

Baker, R. J., J. K. Jones, Jr., and D. C. Carter, eds. 1976. Biology of bats of the New World family Phyllostomatidae. Part I. Spec. Publ. Mus. Texas Tech Univ., no. 10, 218 pp.

———. 1977. Biology of bats of the New World family Phyllostomatidae. Part II. Spec. Publ. Mus. Texas Tech Univ., no. 13, 364 pp.

———. 1979. Biology of bats of the New World family Phyllostomatidae. Part III. Spec. Publ. Mus. Texas Tech Univ., no. 16, 441 pp.

Baker, R. J., J. C. Patton, H. H. Genoways, and J. W. Bickham. 1988. Genic studies of *Lasiurus* (Chiroptera: Vespertilionidae). Occas. Pap. Mus. Texas Tech Univ., no. 117, 15 pp.

Banfield, A. W. F. 1974. The mammals of Canada. Univ. Toronto Press, xxv + 438 pp.

Barbour, R. W., and W. H. Davis. 1969. Bats of America. Univ. Press of Kentucky, Lexington, 286 pp.

———. 1974. Mammals of Kentucky. Univ. Press of Kentucky, Lexington, xii + 322 pp.

Barghoorn, S. F. 1977. New material of *Vespertiliavus* Schlosser (Mammalia, Chiroptera) and suggested relationships of emballonurid bats based on cranial morphology. Amer. Mus. Novit., no. 2618, 29 pp.

Barquez, R. M., and R. A. Ojeda. 1979. Nueva subespecie de *Phylloderma* (Chiroptera Phyllostomidae). Neotrópica 25:83–89.

Barquez, R. M., and C. C. Olrog. 1980. Tres nuevas especies de *Vampyrops* para Bolivia. Neotrópica 26:53–56.

Bateman, G. C., and T. A. Vaughan. 1974. Nightly activities of mormoopid bats. J. Mamm. 55:45–65.

Baud, F. J. 1979. *Myotis aelleni*, nov. spec., chauve-souris nouvelle d'Argentine (Chiroptera: Vespertilionidae). Rev. Suisse Zool. 86:267–78.

———. 1982. Presence de *Rhinophylla alethina* (Mammalia, Chiroptera) en Equateur et repartition actuelle du genre en Amerique du Sud. Rev. Suisse Zool. 89: 815–21.

Baverstock, P. R., M. Adams, T. Reardon, and C. H. S. Watts. 1987. Electrophoretic resolution of species boundaries in Australian Microchiroptera. III. The Nycticeiini–*Scotorepens* and *Scoteanax* (Chiroptera: Vespertilionidae). Austral. J. Biol. Sci. 40: 417–33.

Beck, A. J., and Lim Boo Liat. 1973. Reproductive biology of *Eonycteris spelaea*, Dobson *(Megachiroptera)* in West Malaysia. Acta Tropica 30:251–60.

Bell, G. P. 1987. Evidence of a harem social system in *Hipposideros caffer* (Chiroptera: Hipposideridae) in Zimbabwe. J. Tropical Ecol. 3:87–90.

Bennett, S., L. J. Alexander, R. H. Crozier, and A. G. Mackinlay. 1988. Are megabats flying primates? Contrary evidence from a mitochondrial DNA sequence. Austral. J. Biol. Sci. 41:327–32.

Bergmans, W. 1973. New data on the rare African fruit bat *Scotonycteris ophiodon* Pohle, 1943. Z. Saugetierk. 38:285–89.

———. 1975*a*. A new species of *Dobsonia* Palmer, 1898 (Mammalia, Megachiroptera) from Waigeo, with notes on other members of the genus. Beaufortia 23:1–13.

———. 1975*b*. On the differences between sympatric *Epomops franqueti* (Tomes, 1860) and *Epomops buettikoferi* (Matschie, 1899), with additional notes on the latter species (Mammalia, Megachiroptera). Beaufortia 23:141–52.

———. 1976. A revision of the African genus *Myonycteris* Matschie, 1899 (Mammalia, Megachiroptera). Beaufortia 24:189–216.

———. 1977. Notes on new material of *Rousettus madagascariensis* Grandidier, 1929 (Mammalia, Megachiroptera). Mammalia 41:67–74.

———. 1978*a*. On *Dobsonia* Palmer 1898 from the Lesser Sunda Islands (Mammalia: Megachiroptera). Senckenberg. Biol. 59: 1–18.

———. 1979*a*. Taxonomy and zoogeography of the fruit bats of the People's Republic of Congo, with notes on their reproductive biology (Mammalia: Megachiroptera). Bijdragen Tot de Dierkunde 48:161–86.

———. 1979*b*. Taxonomy and zoogeography of *Dobsonia* Palmer, 1898, from the Louisiade Archipelago, the D'Entrecasteaux Group, Trobriand Island and Woodlark Island (Mammalia: Megachiroptera). Beaufortia 29:199–214.

———. 1979*c*. First records of *Epomops dobsonii* (Bocage, 1889) from Tanzania and Rwanda, with a note on its size range (Mammalia: Megachiroptera). Z. Saugetierk. 44: 240–41.

———. 1980. A new fruit bat of the genus *Myonycteris* Matschie, 1899, from eastern Kenya and Tanzania (Mammalia, Megachiroptera). Zool. Meded. 55:171–81.

———. 1982. Noteworthy extensions of known ranges of three African fruit bat species (Mammalia, Megachiroptera). Bull. Zool. Mus. Univ. Amsterdam 8:157–63.

———. 1988. Taxonomy and biogeography of African fruit bats (Mammalia, Megachiroptera). I. General introduction; material and methods; results: the genus *Epomophorus* Bennett, 1836. Beaufortia 38: 75–146.

Bergmans, W., L. Bellier, and J. Vissault. 1974. A taxonomical report on a collection of Megachiroptera (Mammalia) from the Ivory Coast. Rev. Zool. Afr. 88:18–48.

Bergmans, W., and J. E. Hill. 1980. On a new species of *Rousettus* Gray, 1821, from Sumatra and Borneo (Mammalia: Megachiroptera). Bull. British Mus. (Nat. Hist.) Zool. 38:95–104.

Bergmans, W., and H. Jachmann. 1983. Bat records from Malawi (Mammalia, Chiroptera). Bull. Zool. Mus. Univ. Amsterdam 9:117–22.

Bergmans, W., and F. G. Rozendaal. 1982. Notes on *Rhinolophus* Lacépède, 1799 from Sulawesi, Indonesia, with the description of a new species (Mammalia, Microchiroptera). Bijd. Dierkunde 52:169–74.

———. 1988. Notes on collections of fruit bats from Sulawesi and some off-lying islands (Mammalia, Megachiroptera). Zool. Verh., no. 248, 74 pp.

Bergmans, W., and S. Sarbini. 1985. Fruit bats of the genus *Dobsonia* Palmer, 1898 from the islands of Biak, Owii, Numfoor and Yapen, Irian Jaya (Mammalia, Megachiroptera). Beaufortia 34:181–89.

Bergmans, W., and P. J. H. Van Bree. 1972. The taxonomy of the African bat *Megaloglossus woermanni* Pagenstecher, 1885 (Megachiroptera: Macroglossinae). Biol. Gabonica 3–4:291–99.

———. 1986. On a collection of bats and rats from the Kangean Islands, Indonesia (Mammalia: Chiroptera and Rodentia). Z. Saugetierk. 51:329–44.

Bernard, R. T. F. 1982. Female reproductive cycle of *Nycteris thebaica* (Microchiroptera) from Natal, South Africa. Z. Saugetierk. 47:12–18.

Bernard, R. T. F., and J. A. J. Meester. 1982. Female reproduction and the female reproductive cycle of *Hipposideros caffer caffer* (Sundevall, 1846) in Natal, South Africa. Ann. Transvaal Mus. 33:131–44.

Best, T. L. 1988. Morphologic variation in the spotted bat *Euderma maculatum*. Amer. Midl. Nat. 119:244–52.

Bhat, H. R., M. A. Sreenivasan, and P. G. Jacob. 1980. Breeding cycle of *Eonycteris spelaea* (Dobson, 1871) (Chiroptera, Pteropodidae, Macroglossinae) in India. Mammalia 44:341–47.

Bickham, J. W. 1987. Chromosomal variation among seven species of lasiurine bats (Chiroptera: Vespertilionidae). J. Mamm. 68:837–42.

Birney, E. C., J. B. Bowles, R. M. Timm, and S. L. Williams. 1974. Mammalian distributional records in Yucatan and Quintana Roo, with comments on reproduction, structure, and status of peninsular populations. Occas. Pap. Bell Mus. Nat. Hist., no. 13, 25 pp.

Blood, B. R., and D. A. McFarlane. 1988. Notes on some bats from northern Thailand, with comments on the subgeneric status of *Myotis altarium*. Z. Saugetierk. 53:276–80.

Boeadi, B., and W. Bergmans. 1987. First record of *Dobsonia minor* (Dobson, 1879) from Sulawesi, Indonesia (Mammalia, Megachiroptera). Bull. Zool. Mus. Univ. Amsterdam 11:69–75.

Boeadi, B., and J. E. Hill. 1986. A new subspecies of *Aethalops alecto* (Thomas, 1923) (Chiroptera: Pteropodidae) from Java. Mammalia 50:263–66.

Bogan, M. A. 1972. Observations on parturition and development in the hoary bat, *Lasiurus cinereus*. J. Mamm. 53:611–14.

———. 1978. A new species of *Myotis* from the Islas Tres Marias, Nayarit, Mexico, with comments on variation in *Myotis nigricans*. J. Mamm. 59:519–30.

Bowles, J. B. 1972. Notes on reproduction in four species of bats from Yucatan, Mexico. Trans. Kansas Acad. Sci. 75:271–72.

Brack, V., Jr., and J. C. Carter. 1985. Use of an underground burrow by *Lasionycteris*. Bat Research News 26:28–29.

Bradbury, J. W. 1977. Lek mating behavior in the hammer-headed bat. Z. Tierpsychol. 45:225– 55.

Bradbury, J. W., and L. H. Emmons. 1974. Social organization of some Trinidad bats. I. Emballonuridae. Z. Tierpsychol. 36:137–83.

Bradbury, J. W., and S. L. Vehrencamp. 1977. Social organization and foraging in emballonurid bats. I. Field studies. Behav. Ecol. Sociobiol. 1:337–81.

Brooke, A. P. 1987. Tent construction and social organization in *Vampyressa nymphaea* (Chiroptera: Phyllostomidae) in Costa Rica. J. Tropical Ecol. 3:171–75.

Brosset, A. 1962. La reproduction des chiroptères de l'ouest et du centre de l'Inde. Mammalia 26:176–213.

———. 1976. Social organization in the African bat, *Myotis boccagei*. Z. Tierpsychol. 42:50– 56.

———. 1984. Chiroptères d'altitude du Mont Nimba (Guinee). Description d'une espèce nouvelle, *Hipposideros lamottei*. Mammalia 48:545–55.

Brosset, A., L. Barbe, J.-C. Beaucournu, C. Faugier, H. Salvayre, and Y. Tupinier. 1988. La raréfaction du rhinolophe euryale (*Rhinolophus euryale* Blasius) en France. Recherche d'une explication. Mammalia 52: 101–22.

Brosset, A., and H. Saint Girons. 1980. Cycles de reproduction des microchiropteres troglophiles du nord-est du Gabon. Mammalia 44:225–32.

Brown, P. 1976. Vocal communication in the pallid bat, *Antrozous pallidus*. Z. Tierpsychol. 41:34–54.

Brown, P., T. W. Brown, and A. D. Grinnell. 1983. Echolocation, development, and vocal communication in the lesser bulldog bat, *Noctilio albiventris*. Behav. Ecol. Sociobiol. 13:287–98.

Bruner, P. L., and H. D. Pratt. 1979. Notes on the status and natural history of Micronesian bats. Elepaio 40:1–4.

Buden, D. W. 1976. A review of the endemic West Indian genus *Erophylla*. Proc. Biol. Soc. Washington 89:1–16.

———. 1977. First records of bats of the genus *Brachyphylla* from the Caicos Islands with notes on geographic variation. J. Mamm. 58:221–25.

———. 1985. Additional records of bats from the Bahama Islands. Caribbean J. Sci. 21:19–25.

Burton, M. 1955. Bulldog bats. Ill. London News 226:28.

C

Cabrera, A. 1957, 1961. Catálogo de los mamiferos de América del Sur. Rev. Mus. Argentino Cien. Nat. "Bernardo Rivadavia," 4:1–732.

Caroll, J. B. 1984. The conservation and wild status of the Rodrigues fruit bat *Pteropus rodricensis*. Myotis 21–22:148–54.

Carter, C. H., H. H. Genoways, R. S. Loregnard, and R. J. Baker. 1981. Observations on bats from Trinidad, with a checklist of species occurring on the island. Occas. Pap. Mus. Texas Tech Univ., no. 72, 27 pp.

Carter, D. C., and J. K. Jones, Jr. 1978. Bats from the Mexican state of Hidalgo. Occas. Pap. Mus. Texas Tech Univ., no. 54, 12 pp.

Caubere, B., P. Gaucher, and J. F. Julien. 1984. Un record mondial de longévité *in natura* pour un chiroptère insectivore? Rev. Ecol. 39:351–53.

Ceballos, G., and R. A. Medellín L. 1988. *Diclidurus albus*. Mammalian Species, no. 316, 4 pp.

Chasen, F. N. 1940. A handlist of Malaysian mammals. Bull. Raffles Mus., Singapore, no. 15, xx + 209 pp.

Cheeseman, C. L., and R. B. Mitson, eds. 1982. Telemetric studies of vertebrates. Academic Press, London, 368 pp.

Cheke, A. S., and J. F. Dahl. 1981. The status of bats on western Indian Ocean islands, with special reference to *Pteropus*. Mammalia 45:205–38.

Chimimba, C. T., and D. J. Kitchener. 1987. Breeding in the Australian yellow-bellied sheath-tailed bat, *Saccolaimus flaviventris* (Peters, 1867) (Chiroptera: Emballonuridae). Rec. W. Austral. Mus. 13:241–48.

Choe, J. C., and R. M. Timm. 1985. Roosting site selection by *Artibeus watsoni* (Chiroptera: Phyllostomidae) on *Anthurium ravenii* (Araceae) in Costa Rica. J. Tropical Ecol. 1:241–47.

Churchill, S. K., L. S. Hall, and P. M. Helman. 1984. Observations on long-eared bats (Vespertilionidae: *Nyctophilus*) from northern Australia. Austral. Mamm. 7: 17–28.

Churchill, S. K., P. M. Helman, and L. S. Hall. 1987. Distribution, populations and status of the orange horseshoe bat, *Rhinonicteris aurantius* (Chiroptera: Hipposideridae). Austral. Mamm. 11:27–33.

Clawson, R. L. 1987. Indiana bats: down for the count. Endangered Species Tech. Bull. 12(9):9–11.

Coates-Estrada, R., and A. Estrada. 1985. Occurrence of the white bat, *Diclidurus virgo* (Chiroptera: Emballonuridae), in the region of "Los Tuxtlas," Veracruz. Southwestern Nat. 30:322–23.

Cockrum, E. L. 1969. Migration in the guano bat, *Tadarida brasiliensis*. Univ. Kansas Mus. Nat. Hist. Misc. Publ., no. 51, pp. 303–36.

———. 1973. Additional longevity records for American bats. J. Arizona Acad. Sci. 8:108–10.

Corbet, G. B. 1978. The mammals of the Palaearctic Region: a taxonomic review. British Mus. (Nat. Hist.), London, 314 pp.

Corbet, G. B., and J. E. Hill. 1986. A world list of mammalian species. British Mus. (Nat. Hist.), London, 254 pp.

Cox, P. M. 1983. Observations on the natural history of Samoan bats. Mammalia 47:519–23.

Cranbrook, Earl of. 1984. New and interesting records of mammals from Sarawak. Sarawak Mus. J. 33:137–44.

Crichton, E. G., and P. H. Krutzsch. 1987. Reproductive biology of the female little mastiff bat, *Mormopterus planiceps* (Chiroptera: Molossidae) in southeast Australia. Amer. J. Anat. 178:369–86.

Czaplewski, N. J. 1983. *Idionycteris phyllotis*. Mammalian Species, no. 208, 4 pp.

Czekala, N. M., and K. Benirschke. 1974. Observations on a twin pregnancy in the African long-tongued fruit bat *(Megaloglossus woermanni)*. Bonner Zool. Beitr. 25:220–30.

D

Dalquest, W. W. 1957. Observations on the sharpnosed bat, *Rhynchiscus naso* (Maximilian). Texas J. Sci. 9:219–26.

Daniel, M. J. 1975. First record of an Australian fruit bat (Megachiroptera: Pteropodidae) reaching New Zealand. New Zealand J. Zool. 2:227–31.

———. 1976. Feeding by the short-tailed bat *(Mystacina tuberculata)* on fruit and possibly nectar. New Zealand J. Zool. 3:391–98.

———. 1979. The New Zealand short-tailed bat, *Mystacina tuberculata:* a review of present knowledge. New Zealand J. Zool. 6:357–70.

Daniel, M. J., and G. R. Williams. 1983. Observations of a cave colony of the long-tailed bat *(Chalinolobus tuberculatus)* in North Island, New Zealand. Mammalia 47:71– 80.

———. 1984. A survey of the distribution, seasonal activity and roost sites of New Zealand bats. New Zealand J. Ecol. 7:9–25.

Davis, B. L., and R. J. Baker. 1974. Morphometrics, evolution, and cytotaxonomy of mainland bats of the genus *Macrotus* (Chiroptera: Phyllostomatidae). Syst. Zool. 23:26–39.

Davis, R. 1970. Carrying of young by flying female American bats. Amer. Midl. Nat. 83:186– 96.

Davis, W. B. 1966*a*. Review of South American bats of the genus *Eptesicus*. Southwestern Nat. 11:245–74.

———. 1970*a*. *Tomopeas ravus* Miller (Chiroptera). J. Mamm. 5 1:244–47.

———. 1970*b*. A review of the small fruit bats (genus *Artibeus*) of Middle America. Part II. Southwestern Nat. 14:389–402.

———. 1975. Individual and sexual variation in *Vampyressa bidens*. J. Mamm. 56: 262–65.

———. 1976*a*. Notes on the bats *Saccopteryx canescens* Thomas and *Micronycteris hirsuta* (Peters). J. Mamm. 57:604–7.

———. 1976*b*. Geographic variation in the lesser noctilio, *Noctilio albiventris* (Chiroptera). J. Mamm. 57:687–707.

———. 1980. New *Sturnira* (Chiroptera: Phyllostomidae) from Central and South America, with key to currently recognized species. Occas. Pap. Mus. Texas Tech Univ., no. 70, 5 pp.

———. 1984. Review of the large fruit-eating bats of the *Artibeus "lituratus"* complex (Chiroptera: Phyllostomidae) in Middle America. Occas. Pap. Mus. Texas Tech Univ., no. 93, 16 pp.

Davis, W. B., and D. C. Carter. 1978. A review of the round-eared bats of the *Tonatia silvicola* complex, with descriptions of three new taxa. Occas. Pap. Mus. Texas Tech Univ., no. 53, 12 pp.

Davis, W. B., and J. R. Dixon. 1976. Activity of bats in a small village clearing near Iquitos, Peru. J. Mamm. 57:747–49.

Davis, W. H., and J. W. Hardin. 1967. New records of mammals from Mesa Verde National Park, Colorado. J. Mamm. 48:322–23.

DeBlase, A. F. 1980. The bats of Iran: systematics, distribution, ecology. Fieldiana Zool., n.s., no. 4, xvii + 424 pp.

DeFrees, S. L., and D. E. Wilson. 1988. *Eidolon helvum*. Mammalian Species, no. 312, 5 pp.

De Jong, N., and W. Bergmans. 1981. A revision of the fruit bats of the genus *Dobsonia* Palmer, 1898 from Sulawesi and some nearby islands (Mammalia, Megachiroptera, Pteropodinae). Zool. Abhandl. (Dresden) 37:209–24.

De Vree, F. 1972. Description of a new form of *Pipistrellus* from Ivory Coast. Rev. Zool. Bot. Afr. 85:412–16.

———. 1973. New data on *Scotophilus gigas* Dobson, 1875 (Microchiroptera–Vespertilionidae). Z. Saugetierk. 38:189–96.

Dinerstein, E. 1985. First records of *Lasiurus castaneus* and *Antrozous dubiaquercus* from Costa Rica. J. Mamm. 411–12.

Dolan, P. G., and D. C. Carter. 1979. Distributional notes and records for Middle American Chiroptera. J. Mamm. 60:644–49.

Douglas, A. M. 1967. The natural history of the ghost bat, *Macroderma gigas* (Microchiroptera, Megadermatidae), in Western Australia. W. Austral. Nat. 10:125–37.

Dwyer, P. D. 1966. Observations on *Chalinolobus dwyeri* (Chiroptera: Vespertilionidae) in Australia. J. Mamm. 47:716–18.

———. 1968. The biology, origin, and adaption of *Miniopterus australis* (Chiroptera) in New South Wales. Austral. J. Zool. 16:49–68.

———. 1970. Latitude and breeding season in a polyestrous species of *Myotis*. J. Mamm. 51:405–10.

———. 1975*a*. Notes on *Dobsonia moluccensis* (Chiroptera) in the New Guinea highlands. Mammalia 39:113–18.

E

Easterla, D. A. 1973. Ecology of the 18 species of Chiroptera at Big Bend National Park, Texas. Part II. Northwest Missouri State Univ. Studies 34:54–165.

———. 1976. Notes on the second and third newborn of the spotted bat, *Euderma maculatum*, and comments on the species in Texas. Amer. Midl. Nat. 96:499–501.

Easterla, D. A., and L. C. Watkins. 1970. Breeding of *Lasionycteris noctivagans* and *Nycticeius humeralis* in southwestern Iowa. Amer. Midl. Nat. 84:254–55.

Eger, J. L. 1977. Systematics of the genus *Eumops* (Chiroptera: Molossidae). Roy. Ontario Mus. Life Sci. Contrib., no. 110, 69 pp.

Eger, J. L., and R. L. Peterson. 1979. Distribution and systematic relationship of *Tadarida bivittata* and *Tadarida ansorgei* (Chiroptera: Molossidae). Can. J. Zool. 57: 1887–95.

Ellerman, J. R., and T. C. S. Morrison-Scott. 1966. Checklist of Palaearctic and Indian mammals. British Mus. (Nat. Hist.), London, 810 pp.

El-Rayah, M. A. 1981. A new species of bat of the genus *Tadarida* (family Molossidae) from West Africa. Roy. Ontario Mus. Life Sci. Occas. Pap., no. 36, 10 pp.

Engstrom, M. D., T. E. Lee, and D. E. Wilson. 1987. *Bauerus dubiaquercus*. Mammalian Species, no. 282, 3 pp.

Engstrom, M. D., and D. E. Wilson. 1981. Systematics of *Antrozous dubiaquercus* (Chiroptera: Vespertilionidae), with comments on the status of *Bauerus* Van Gelder. Ann. Carnegie Mus. 50:371–83.

F

Fairon, J. 1980. Deux nouvelles espèces de chiroptères pour la faune du Massif de l'Air (Niger): *Otonycteris hemprichi* Peters, 1859 et *Pipistrellus nanus* (Peters, 1852). Bull. Inst. Roy. Soc. Nat. Belg. 52(17):1–7.

Fayenuwo, J. O., and L. B. Halstead. 1974. Breeding cycle of straw-colored fruit bat, *Eidolon helvum*, Ile-Ife, Nigeria. J. Mamm. 55:453–54.

Feiler, A. 1980. *Taphozous saccolaimus* Temminck, 1841 auf Sulawesi (Celebes) (Mammalia, Chiroptera, Emballonuridae). Zool. Abhandl. (Dresden) 36:225–28.

Felten, H. 1964. Flughunde der Gattung *Pteropus* von Neukaledonien und den Loyalty-Inseln (Mammalia, Chiroptera). Senckenberg. Biol. 45:671–83.

Felten, H., and D. Kock. 1972. Weitere Flughunde der Gattung *Pteropus* von den Neuen Hebriden, sowie den Banks- und Torres-Inseln, Pazifischer Ozean. Senckenberg. Biol. 53:179–88.

Fenton, M. B. 1975. Observations on the biology of some Rhodesian bats, including a key to the Chiroptera of Rhodesia. Roy. Ontario Mus. Life Sci. Contrib., no. 104, 27 pp.

Fenton, M. B., R. M. Brigham, A. M. Mills, and I. L. Rautenbach. 1985. The roosting and foraging areas of *Epomophorus wahlbergi* (Pteropodidae) and *Scotophilus viridis* (Vespertilionidae) in Kruger National Park, South Africa. J. Mamm. 66:461–68.

Fenton, M. B., C. L. Gaudet, and M. L. Leonard. 1983. Feeding behaviour of the bats *Nycteris grandis* and *Nycteris thebaica* (Nycteridae) in captivity. J. Zool. 200: 347–54.

Fenton, M. B., and T. H. Kunz. 1977. Movements and behavior. *In* Baker, Jones, and Carter (1977), pp. 351–64.

Fenton, M. B., and R. L. Peterson. 1972. Further notes on *Tadarida aloysiisabaudiae* and *Tadarida russata* (Chiroptera: Molossidae–Africa). Can. J. Zool. 50: 19–24.

Fenton, M. B., D. C. Tennant, and J. Wyszecki. 1987. Using echolocation calls to measure the distribution of bats: the case of *Euderma maculatum*. J. Mamm. 68:142–44.

Fenton, M. B., D. W. Thomas, and R. Sasseen. 1981. *Nycteris grandis* (Nycteridae): an African carnivorous bat. J. Zool. 194: 461–65.

Filewood, L. W. 1983. The possible occurrence in New Guinea of the ghost bat (*Macroderma gigas;* Chiroptera, Megadermatidae). Austral. Mamm. 6:35–36.

Findley, J. S. 1972. Phenetic relationships among bats of the genus *Myotis*. Syst. Zool. 21:31–52.

Findley, J. S., and D. E. Wilson. 1974. Observations on the neotropical disk-winged bat, *Thyroptera tricolor* Spix. J. Mamm. 55:562–71.

Finley, R. B., Jr., and J. Creasy. 1982. First specimen of the spotted bat *(Euderma maculatum)* from Colorado. Great Basin Nat. 42:360.

Fitch, J. H., and K. A. Shump, Jr. 1979. *Myotis keenii*. Mammalian Species, no. 121, 3 pp.

Flannery, T. F. 1987*b*. An historic record of the New Zealand greater short-tailed bat, *Mystacina robusta* (Microchiroptera: Mystacinidae) from the South Island, New Zealand. Austral. Mamm. 10:45–46.

Fleming, T. H. 1988. The short-tailed fruit bat. A study in plant-animal interactions. Univ. Chicago Press, xvi + 365 pp.

Fleming, T. H., E. T. Hooper, and D. E. Wilson. 1972. Three Central American bat communities: structure, reproductive cycles, and movement patterns. Ecology 53: 555–69.

Foster, M. S., and R. M. Timm. 1976. Tent-making by *Artibeus jamaicensis* (Chiroptera: Phyllostomatidae) with comments on plants used by bats for tents. Biotrópica 8:265–69.

Francis, C. M., and J. E. Hill. 1986. A review of the Bornean *Pipistrellus* (Mammalia: Chiroptera). Mammalia 50:43–55.

Freeman, P. W. 1981. A multivariate study of the family Molossidae (Mammalia, Chiroptera): morphology, ecology, and evolution. Fieldiana Zool., n.s., no. 7, vii + 173 pp.

Fujita, M. S., and T. H. Kunz. 1984. *Pipistrellus subflavus*. Mammalian Species, no. 228, 6 pp.

G

Gaisler, J. 1983. Nouvelles données sur les chiroptères du nord Algerien. Mammalia 47:359–69.

Gaisler, J., G. Madkour, and J. Pelikan. 1972. On the bats (Chiroptera) of Egypt. Acta Sci. Nat. Brno 6:1–40.

Gallagher, M. D., and D. L. Harrison. 1977. Report on the bats (Chiroptera) obtained by the Zaire River Expedition. Bonner Zool. Beitr. 28:19–32.

Gardner, A. L. 1976. The distributional status of some Peruvian mammals. Occas. Pap. Mus. Zool. Louisiana State Univ., no. 48, 18 pp.

———. 1977. Feeding habits. *In* Baker, Jones, and Carter (1977), pp. 293–350.

———. 1986. The taxonomic status of *Glossophaga morenoi* Martinez and Villa, 1938 (Mammalia: Chiroptera: Phyllostomidae). Proc. Biol. Soc. Washington 99: 489–92.

Gardner, A. L., R. K. LaVal, and D. E. Wilson. 1970. The distributional status of some Costa Rican bats. J. Mamm. 51:712–29.

Gardner, A. L., and J. P. O'Neill. 1969. The taxonomic status of *Sturnira bidens* (Chiroptera: Phyllostomatidae) with notes on its karyotype and life history. Occas. Pap. Mus. Zool. Louisiana State Univ., no. 38, 8 pp.

———. 1971. A new species of *Sturnira* (Chiroptera: Phyllostomatidae) from Peru. Occas. Pap. Mus. Zool. Louisiana State Univ., no. 42, 7 pp.

Gardner, A. L., and J. L. Patton. 1972. New species of *Philander* (Marsupialia: Didelphidae) and *Mimon* (Chiroptera: Phyllostomatidae) from Peru. Occas. Pap. Mus. Zool. Louisiana State Univ., no. 43, 12 pp.

Gauckler, A., and M. Kraus. 1970. Kennzeichen und Verbreitung von *Myotis brandti* (Eversman, 1845). Z. Saugetierk. 35:113–24.

Geluso, K. N., J. S. Altenbach, and D. E. Wilson. 1976. Bat mortality: pesticide poisoning and migratory stress. Science 194: 184–86.

———. 1981. Organochlorine residues in young Mexican free-tailed bats from several roosts. Amer. Midl. Nat. 105:249–57.

Genoways, H. H., and R. J. Baker. 1972. *Stenoderma rufum*. Mammalian Species, no. 18, 4 pp.

———. 1988. *Lasiurus blossevillii* (Chiroptera: Vespertilionidae) in Texas. Texas J. Sci. 40:111–13.

Genoways, H. H., and S. L. Williams. 1979*a*. Notes on bats (Mammalia: Chiroptera) from Bonaire and Curacao, Dutch West Indies. Ann. Carnegie Mus. 48:311–21.

———. 1979*b*. Records of bats (Mammalia: Chiroptera) from Suriname. Ann. Carnegie Mus. 48:323–35.

———. 1980. Results of the Alcoa Foundation–Suriname Expeditions. I. A new species of bat of the genus *Tonatia* (Mammalia: Phyllostomatidae). Ann. Carnegie Mus. 49:203–11.

———. 1984. Results of the Alcoa Foundation–Suriname Expeditions. IX. Bats of the genus *Tonatia* (Mammalia: Chiroptera) in Suriname. Ann. Carnegie Mus. 53:327–46.

———. 1986. Results of the Alcoa Foundation–Suriname Expeditions. XI. Bats of the genus *Micronycteris* (Mammalia: Chiroptera) in Suriname. Ann. Carnegie Mus. 55:303–24.

Gerell, R., and K. Lundberg. 1985. Social organization in the bat *Pipistrellus pipistrellus*. Behav. Ecol. Sociobiol. 16:177–84.

Ghose, R. K., and D. K. Ghosal. 1984. Record of the fulvous fruit bat, *Rousettus leschenaulti* (Desmarest, 1820) from Sikkim, with notes on its interesting feeding habit and status. J. Bombay Nat. Hist. Soc. 81: 178–79.

Godin, A. J. 1977. Wild mammals of New England. Johns Hopkins Univ. Press, Baltimore, xii + 304 pp.

Goehring, H. H. 1972. Twenty-year study of *Eptesicus fuscus* in Minnesota. J. Mamm 53:201–7.

Goodwin, G. G., and A. M. Greenhall. 1961. A review of the bats of Trinidad and Tobago. Bull. Amer. Mus. Nat. Hist. 122:187–302.

Goodwin, H. A., and J. M. Goodwin. 1973. List of mammals which have become extinct or are possibly extinct since 1600. Internatl. Union Conserv. Nat. Occas. Pap., no. 8, 20 pp.

Goodwin, M. K. 1979. Notes on caravan and play behavior in young captive *Sorex cinereus*. J. Mamm. 60:411–13.

Goodwin, R. E. 1970. The ecology of Jamaican bats. J. Mamm. 51:571–79.

———. 1979. The bats of Timor: systematics and ecology. Bull. Amer. Mus. Nat. Hist. 163:75–122.

Gopalakrishna, A. 1955. Observations on the breeding habits and ovarian cycle in the Indian sheath-tailed bat, *Taphozous longimanus* (Hardwicke). Proc. Natl. Inst. Sci. India 21B:29–41.

Gopalakrishna, A., and P. N. Choudhari. 1977. Breeding habits and associated phenomena in some Indian bats, Part 1–*Rousettus leschenaulti* (Desmarest)–Megachiroptera. J. Bombay Nat. Hist. Soc. 74:1–16.

Gopalakrishna, A., and K. B. Karim. 1980. Female genital anatomy and the morphogenesis of foetal membranes of Chiroptera and their bearing on the phylogenetic relationships of the group. Natl. Acad. Sci. India Golden Jubilee Commemoration Vol., pp. 379–428.

Gopalakrishna, A., R. S. Thakur, and A. Madhavan. 1975. Breeding biology of the southern dwarf pipistrelle, *Pipistrellus mimus mimus* (Wroughton) from Maharashtra, India. Dr. B. S. Chauhan Commemoration Vol., pp. 225–40.

Gosnell, M. 1977. Carlsbad's famous bats are dying off. Natl. Wildl. 15(4):28–33.

Gould, E. 1977. Foraging behavior of *Pteropus vampyrus* on the flowers of *Durio zibethinus*. Malayan Nat. J. 30:53–57.

———. 1978*a*. Rediscovery of *Hipposideros ridleyi* and seasonal reproduction in Malaysian bats. Biotrópica 10:30–32.

Graham, G. L. 1987. Seasonality of reproduction in Peruvian bats. Fieldiana Zool., n.s., 39:173–86.

Graham, G. L., and L. J. Barkley. 1984. Noteworthy records of bats from Peru. J. Mamm. 65:709–11.

Green, R. H., and J. L. Rainbird. 1984. The bat genus *Eptesicus* Gray in Tasmania. Tasmanian Nat. 76:1–5.

Greenbaum, I. F., R. J. Baker, and D. E. Wilson. 1975. Evolutionary implications of the karyotypes of the stenodermine genera *Ardops, Ariteus, Phyllops,* and *Ectophylla*. Bull. S. California Acad. Sci. 74:156–59.

Greenbaum, I. F., and J. K. Jones, Jr. 1978. Noteworthy records of bats from El Salvador, Honduras, and Nicaragua. Occas. Pap. Mus. Texas Tech Univ., no. 55, 7 pp.

Greenhall, A. M. 1968. Notes on the behavior of the false vampire bat. J. Mamm. 49:337–40.

———. 1972. The biting and feeding habits of the vampire bat, *Desmodus rotundus.* J. Zool. 168:451–61.

Greenhall, A. M., G. Joermann, U. Schmidt, and M. R. Seidel. 1983. *Desmodus rotundus.* Mammalian Species, no. 202, 6 pp.

Greenhall, A. M., and U. Schmidt, eds. 1988. Natural history of vampire bats. CRC Press, Boca Raton, Florida, 272 pp.

Greenhall, A. M., U. Schmidt, and G. Joermann. 1984. *Diphylla ecaudata.* Mammalian Species, no. 227, 3 pp.

Greer, J. K. 1965. Mammals of Malleco Province Chile. Michigan State Univ. Mus. Publ., Biol. Ser., 3:49–152.

Griffiths, T. A. 1982. Systematics of the New World nectar-feeding bats (Mammalia, Phyllostomidae), based on the morphology of the hyoid and lingual regions. Amer. Mus. Novit., no. 2742, 45 pp.

———. 1985. Molar cusp patterns in the bat genus *Brachyphylla:* some functional and systematic observations. J. Mamm. 66:544–49.

Grzimek, B., ed. 1975. Grzimek's animal life encyclopedia. Mammals, I–IV. Van Nostrand Reinhold, New York, vols. 10–13.

H

Haiduk, M. W., and R. J. Baker. 1982. Cladistical analysis of g-banded chromosomes of nectar feeding bats (Glossophaginae: Phyllostomidae). Syst. Zool. 31: 252–65.

Hall, E. R. 1981. The mammals of North America. John Wiley & Sons, New York, 2 vols.

Hall, E. R., and W. W. Dalquest. 1963. The mammals of Veracruz. Univ. Kansas Publ. Mus. Nat. Hist. 14:165–362.

Hall, L. S. 1987. Identification, distribution and taxonomy of Australian flying-foxes (Chiroptera: Pteropodidae). Austral. Mamm. 10:75–79.

Hamilton, R. B., and D. T. Stalling. 1972. *Lasiurus borealis* with five young. J. Mamm. 53:190.

Hamilton-Smith, E. 1979. Endangered and threatened Chiroptera of Australia and the Pacific region. *In* Tyler (1979), pp. 85–91.

Hanák, V., and A. Elgadi. 1984. On the bat fauna (Chiroptera) of Libya. Vest. Cesk. Spol. Zool. 48:165–87.

Hanák, V., and J. Gaisler. 1983. *Nyctalus leisleri* (Kuhl, 1818), une espèce nouvelle pour le continent africain. Mammalia 47: 585–87.

Hand, S. J. 1985. New Miocene megadermatids (Chiroptera: Megadermatidae) from Australia with comments on megadermatid phylogenetics. Austral. Mamm. 8:5–43.

Handley, C. O., Jr. 1976. Mammals of the Smithsonian Venezuelan Project. Brigham Young Univ. Sci. Bull., Biol. Ser., 20(5): 1–89.

———. 1980*a*. Inconsistencies in formation of family-group and subfamily-group names in Chiroptera. Proc. 5th Internatl. Bat Res. Conf., pp. 9–13.

———. 1984. New species of mammals from northern South America: a long-tongued bat, genus *Anoura* Gray. Proc. Biol. Soc. Washington 97:513–21.

———. 1987. New species of mammals from northern South America: fruit-eating bats, genus *Artibeus* Leach. Fieldiana Zool., n.s., 39:163–72.

Happold, D. C. D. 1987. The mammals of Nigeria. Clarendon Press, Oxford, xvii + 402 pp.

Happold, D. C. D., and M. Happold. 1978*a*. The fruit bats of western Nigeria. Nigerian Field 43:30–37.

———. 1978*b*. The fruit bats of western Nigeria. Part 2. Nigerian Field 43:72–77.

Harrison, D. L. 1964, 1968, 1972. The mammals of Arabia. Ernest Benn, London, 3 vols.

———. 1975*a*. *Macrophyllum macrophyllum.* Mammalian Species, no. 62, 3 pp.

———. 1975*b*. A new species of African free-tailed bat (Chiroptera: Molossidae) obtained by the Zaire River Expedition. Mammalia 39:313–18.

———. 1979. A new species of pipistrelle bat (*Pipistrellus:* Vespertilionidae) from Oman, Arabia. Mammalia 43:573–76.

Harrison, D. L., and P. J. J. Bates. 1984. Mammals of Saudi Arabia. On the occurrence of the European free-tailed bat, *Tadarida teniotis* Rafinesque, 1814 (Chiroptera: Molossidae) in Saudi Arabia with a zoogeographical review of the molossids of the Kingdom. Fauna Saudi Arabia 6: 551–56.

Harrison, D. L., and M. C. Jennings. 1980. Occurrence of the noctule, *Nyctalus noctula* Schreber, 1774 (Chiroptera: Vespertilionidae) in Oman, Arabia. Mammalia 44: 409–10.

Hayman, R. W. 1946. A new genus of fruit-bat and a new squirrel, from Celebes. Ann. Mag. Nat. Hist., ser. 11, 12:766–75.

Hayward, B. J., and E. L. Cockrum. 1971. The natural history of the western long-nosed bat *Leptonycteris sanborni.* Western New Mexico State Univ. Res. Sci. 1:75–123.

Heaney, L. R., and A. C. Alcala. 1986. Flat-headed bats (Mammalia, *Tylonycteris*) from the Philippine Islands. Silliman J. 33:117–23.

Heaney, L. R., and P. D. Heideman. 1987. Philippine fruit bats: endangered and extinct. Bats 5(1):3–5.

Heaney, L. R., P. D. Heideman, and K. M. Mudar. 1981. Ecological notes on mammals in the Lake Balinsasayao region, Negros Oriental, Philippines. Silliman J. 28:122–31.

Heaney, L. R., and R. L. Peterson. 1984. A new species of tube-nosed fruit bat *(Nyctimene)* from Negros Island, Philippines (Mammalia: Pteropodidae). Occas. Pap. Mus. Zool. Univ. Michigan, no. 708, 16 pp.

Heaney, L. R., and D. S. Rabor. 1982. Mammals of Dinagat and Siargao Islands, Philippines. Occas. Pap. Mus. Zool. Univ. Michigan, no. 699, 30 pp.

Heideman, P. D. 1988. The timing of reproduction in the fruit bat *Haplonycteris fischeri* (Pteropodidae): geographic variation and delayed development. J. Zool. 215:577–95.

Heithaus, E. R., and T. H. Fleming. 1978. Foraging movements of a frugivorous bat, *Carollia perspicillata* (Phyllostomatidae). Ecol. Monogr. 48:127–43.

Heller, K.-G., and M. Volleth. 1984. Taxonomic position of "*Pipistrellus societatis*" Hill, 1972 and the karyological characteristics of the genus *Eptesicus* (Chiroptera: Vespertilionidae). Z. Zool. Syst. Evol. 22:65–77.

Hensley, A. P., and K. T. Wilkins. 1988. *Leptonycteris nivalis*. Mammalian Species, no. 307, 4 pp.

Henson, O. W., Jr., and A. Novick. 1966. An additional record of the bat *Phyllonycteris aphylla*. J. Mamm. 47:351–52.

Herbert, H. 1985. Echoortungsverhalten des Flughundes *Rousettus aegyptiacus* (Megachiroptera). Z. Saugetierk. 50:141–52.

Hermanson, J. W., and T. J. O'Shea. 1983. *Antrozous pallidus*. Mammalian Species, no. 213, 8 pp.

Hernandez, C. S., C. B. C. Tapia, A. N. Garduno, E. C. Corona, and M. A. G. Hidalgo. 1985. Notes on distribution and reproduction of bats from coastal regions of Michoacan, Mexico. J. Mamm. 66:549–53.

Hernandez-Camacho, J., and A. Cadena. 1978. Notas para la revisión del genero *Lonchorhina* (Chiroptera, Phyllostomidae). Caldasia 12:199–251.

Herselman, J. C., and P. M. Norton. 1985. The distribution and status of bats (Mammalia: Chiroptera) in the Cape Province. Ann. Cape Prov. Mus. (Nat. Hist.) 16:73–126.

Hershkovitz, P. 1975*a*. The scientific name of the lesser *Noctilio* (Chiroptera), with notes on the chauve-souris de la Vallee d'Ylo (Peru). J. Mamm. 56:242–47.

Hill, J. E. 1956. The mammals of Rennell Island. *In* Wolff, T., The natural history of Rennell Island, British Solomon Islands, vol. 1, Univ. Copenhagen, pp. 73–84.

———. 1958. Some observations on the fauna of the Maldive Islands. II. Mammals. J. Bombay Nat. Hist. Soc. 55:3–10.

———. 1961*a*. Fruit bats from the Federation of Malaya. Proc. Zool. Soc. London 136:629–42.

———. 1961*b*. Indo-Australian bats of the genus *Tadarida*. Mammalia 25:29–56.

———. 1962. Notes on some insectivores and bats from upper Burma. Proc. Zool. Soc. London 139:119–37.

———. 1963*a*. A revision of the genus *Hipposideros*. Bull. British Mus. (Nat. Hist.) Zool. 11:1–129.

———. 1963*b*. Notes on some tube-nosed bats, genus *Murina*, from southeastern Asia, with descriptions of a new species and a new subspecies. Fedn. Mus. J., n.s., 8:48–59.

———. 1965. Asiatic bats of the genera *Kerivoula* and *Phoniscus* (Vespertilionidae), with a note on *Kerivoula aerosa* Tomes. Mammalia 29:524–56.

———. 1966. A review of the genus *Philetor* (Chiroptera: Vespertilionidae). Bull. British Mus. (Nat. Hist.) Zool. 14:371–87.

———. 1971*a*. A note on *Pteropus* (Chiroptera: Pteropodidae) from the Andaman Islands. J. Bombay Nat. Hist. Soc. 68:1–8.

———. 1971*b*. The status of *Vespertilio brachypterus* Temminck, 1840 (Chiroptera: Vespertilionidae). Zool. Meded. 45:139–46.

———. 1971*c*. Bats from the Solomon Islands. J. Nat. Hist. 5:573–81.

———. 1972*a*. A note on *Rhinolophus rex* Allen, 1923 and *Rhinomegalophus paradoxolophus* Bourret, 1951 (Chiroptera: Rhinolophidae). Mammalia 36:428–34.

———. 1972*b*. The Gunong Benom Expedition 1967. 4. New records of Malayan bats, with taxonomic notes and the description of a new *Pipistrellus*. Bull. British Mus. (Nat. Hist.) Zool. 23:21–42.

———. 1974*a*. A review of *Laephotis* Thomas, 1901 (Chiroptera: Vespertilionidae). Bull. British Mus. (Nat. Hist.) Zool. 27:73–82.

———. 1974*b*. New records of bats from southeastern Asia, with taxonomic notes. Bull. British Mus. (Nat. Hist.) Zool. 27: 127–38.

———. 1974*c*. A review of *Scotoecus* Thomas, 1901 (Chiroptera: Vespertilionidae). Bull. British Mus. (Nat. Hist.) Zool. 27:169–88.

———. 1974*d*. A new family, genus and species of bat (Mammalia: Chiroptera) from Thailand. Bull. British Mus. (Nat. Hist.) Zool. 27:301–36.

———. 1976*a*. A note on *Pipistrellus rusticus* (Tomes, 1861) (Chiroptera: Vespertilionidae). Rev. Zool. Afr. 90:626–33.

———. 1976*b*. Bats referred to *Hesperoptenus* Peters, 1869 (Chiroptera: Vespertilionidae) with the description of a new subgenus. Bull. British Mus. (Nat. Hist.) Zool. 30:1–28.

———. 1977*a*. A review of the Rhinopomatidae (Mammalia: Chiroptera). Bull. British Mus. (Nat. Hist.) Zool. 32:29–43.

———. 1977*b*. African bats allied to *Kerivoula lanosa* (A. Smith, 1847). Rev. Zool. Afr. 91:623–33.

———. 1979. The flying fox *Pteropus tonganus* in the Cook Islands and on Niue Island, Pacific Ocean. Acta Theriol. 24:115–17.

———. 1980*a*. A note on *Lonchophylla* (Chiroptera: Phyllostomatidae) from Ecuador and Peru, with the description of a new species. Bull. British Mus. (Nat. Hist.) Zool. 38:233–36.

———. 1980*b*. The status of *Vespertilio borbonicus* E. Geoffroy, 1803 (Chiroptera: Vespertilionidae). Zool. Meded. 55:287–95.

———. 1982*a*. Records of bats from Mount Nimba, Liberia. Mammalia 46:116–20.

———. 1982*b*. A review of the leaf-nosed bats *Rhinonycteris, Cloeotis* and *Triaenops* (Chiroptera: Hipposideridae). Bonner Zool. Beitr. 33:165–86.

———. 1983*a*. Further records of bats from the Central African Republic. Ann. Carnegie Mus. 52:55–58.

———. 1983*b*. Bats (Mammalia: Chiroptera) from Indo-Australia. Bull. British Mus. (Nat. Hist.) Zool. 45:103–208.

———. 1985*a*. Records of bats (Chiroptera) from New Guinea, with the description of a new *Hipposideros* (Hipposideridae). Mammalia 49:525–35.

———. 1985*b*. The status of *Lichonycteris degener* Miller, 1931 (Chiroptera: Phyllostomidae). Mammalia 49:579–82.

———. 1986. A note on *Rhinolophus pearsonii* Horsfield, 1851 and *Rhinolophus yunanensis* Dobson, 1872 (Chiroptera: Rhinolophidae). J. Bombay Nat. Hist. Soc. 83:12–18.

Hill, J. E., and W. N. Beckon. 1978. A new species of *Pteralopex* Thomas, 1888 (Chiroptera: Pteropodidae) from the Fiji Islands. Bull. British Mus. (Nat. Hist.) Zool. 34: 65–82.

Hill, J. E., and B. Boeadi. 1978. A new species of *Megaerops* from Java (Chiroptera: Pteropodidae). Mammalia 42:427–34.

Hill, J. E., and M. J. Daniel. 1985. Systematics of the New Zealand short-tailed bat *Mystacina* Gray, 1843 (Chiroptera: Mystacinidae). Bull. British Mus. (Nat. Hist.) Zool. 48:279–300.

Hill, J. E., and C. M. Francis. 1984. New bats (Mammalia: Chiroptera) and new records of bats from Borneo and Malaya. Bull. British Mus. (Nat. Hist.) Zool. 47:305–29.

Hill, J. E., and D. L. Harrison. 1987. The baculum in the Vespertilioninae (Chiroptera: Vespertilionidae) with a systematic review, a synopsis of *Pipistrellus* and *Eptesicus*, and the descriptions of a new genus and subgenus. Bull. British Mus. (Nat. Hist.) Zool. 52:225– 305.

Hill, J. E., D. L. Harrison, and T. S. Jones. 1988. New records of bats (Microchiroptera) from Nigeria. Mammalia 52:590–92.

Hill, J. E., and K. F. Koopman. 1981. The status of *Lamingtona lophorhina* McKean and Calaby, 1968 (Chiroptera: Vespertilionidae). Bull. British Mus. (Nat. Hist.) Zool. 41:275–78.

Hill, J. E., and P. Morris. 1971. Bats from Ethiopia collected by the Great Abbai Expedition, 1968. Bull. British Mus. (Nat. Hist.) Zool. 21:25–49.

Hill, J. E., and T. K. Pratt. 1981. A record of *Nyctophilus timorensis* (Geoffroy, 1806) (Chiroptera: Vespertilionidae) from New Guinea. Mammalia 45:264–66.

Hill, J. E., and D. A. Schlitter. 1982. A record of *Rhinolophus arcuatus* (Chiroptera: Rhinolophidae) from New Guinea, with the description of a new subspecies. Ann. Carnegie Mus. 51:455–64.

Hill, J. E., and S. E. Smith. 1981. *Craseonycteris thonglongyai*. Mammalian Species, no. 160, 4 pp.

Hill, J. E., and K. Thonglongya. 1972. Bats from Thailand and Cambodia. Bull. British Mus. (Nat. Hist.) Zool. 22:171–96.

Hill, J. E., and G. Topal. 1973. The affinities of *Pipistrellus ridleyi* Thomas, 1898 and *Glischropus rosseti* Oey, 1951 (Chiroptera: Vespertilionidae). Bull. British Mus. (Nat. Hist.) Zool. 24:447–51.

Hill, J. E., and S. Yenbutra. 1984. A new species of the *Hipposideros bicolor* group (Chiroptera: Hipposideridae) from Thailand. Bull. British Mus. (Nat. Hist.) Zool. 47:77–82.

Hill, J. E., and M. Yoshiyuki. 1980. A new species of *Rhinolophus* (Chiroptera, Rhinolophidae) from Iriomote Island, Ryukyu Islands, with notes on the Asiatic members of the *Rhinolophus pusillus* group. Bull. Natl. Sci. Mus. (Tokyo), ser. A, 6:179–89.

Hill, J. E., A. Zubaid, and G. W. H. Davison. 1985. *Hipposideros lekaguli*, a new leaf-nosed bat recorded in peninsular Malaya. Malayan Nat. J. 39:147–48.

———. 1986. The taxonomy of leaf-nosed bats of the *Hipposideros bicolor* group (Chiroptera: Hipposideridae) from southeastern Asia. Mammalia 50:535–40.

Hitchcock, H. B. 1965. Twenty-three years of bat banding in Ontario and Quebec. Can. Field- Nat. 79:4–14.

Hoffman, R. S., J. K. Jones, Jr., and J. A. Campbell. 1987. First record of *Myotis auricularis* from Guatemala. Southwestern Nat. 32:391.

Hoffmeister, D. F. 1986. Mammals of Arizona. Univ. Arizona Press, xix + 602 pp.

Hollander, R. R., and J. K. Jones, Jr. 1988. Northernmost record of the tropical brown bat, *Eptesicus furinalis*. Southwestern Nat. 33:100.

Homan, J. A., and J. K. Jones, Jr. 1975*a*. *Monophyllus redmani*, Mammalian Species, no. 57, 3 pp.

———. 1975*b*. *Monophyllus plethodon*, Mammalian Species, no. 58. 2 pp.

Honacki, J. H., K. E. Kinman, and J. W. Koeppl. 1982. Mammal species of the world: a taxonomic and geographic reference. Allen Press, Lawrence, Kansas, ix + 694 pp.

Honeycutt, R. L., R. J. Baker, and H. H. Genoways. 1980. Results of the Alcoa Foundation–Suriname Expeditions. III. Chromosomal data for bats (Mammalia: Chiroptera) from Suriname. Ann. Carnegie Mus. 49:237–50.

Honeycutt, R. L., and V. M. Sarich. 1987. Albumin evolution and subfamilial relationships among New World leaf-nosed bats (family Phyllostomidae). J. Mamm. 68: 508–17.

Hood, C. S., and J. K. Jones, Jr. 1984. *Noctilio leporinus*. Mammalian Species, no. 216, 7 pp.

Hood, C. S., and J. Pitocchelli. 1983. *Noctilio albiventris*. Mammalian Species, no. 197, 5 pp.

Horacek, I. 1975. Notes on the ecology of the bats of the genus *Plecotus* Geoffroy, 1818 (Mammalia: Chiroptera). Vest. Cesk. Spol. Zool. 39:195–210.

Horacek, I., and V. Hanák. 1984. Comments on the systematics and phylogeny of *Myotis nattereri* (Kuhl, 1818). Myotis 21–22:20–29.

———. 1986. Generic status of *Pipistrellus savii* and comments on classification of the genus *Pipistrellus* (Chiroptera, Vespertilionidae). Myotis 23–24:9–16.

Howe, H. F. 1974. Additional records of *Phyllonycteris aphylla* and *Ariteus flavescens* from Jamaica. J. Mamm. 55:662–63.

Hudson, W. S., and D. E. Wilson. 1986. *Macroderma gigas*. Mammalian Species, no. 260, 4 pp.

Humphrey, S. R. 1975. Nursery roosts and community diversity of Nearctic bats. J. Mamm. 56:321–46.

———. 1978. Status, winter habitat, and management of the endangered Indiana bat, *Myotis sodalis*. Florida Sci. 41:65–76.

Humphrey, S. R., and L. N. Brown. 1986. Report of a new bat (Chiroptera: *Artibeus jamaicensis*) in the United States is erroneous. Florida Sci. 49:262–63.

Humphrey, S. R., and T. H. Kunz. 1976. Ecology of a Pleistocene relict, the western big- eared bat (*Plecotus townsendii*), in the southern Great Plains. J. Mamm. 57:470–94.

Humphrey, S. R., A. R. Richter, and J. B. Cope. 1977. Summer habitat and ecology of the endangered Indiana bat, *Myotis sodalis*. J. Mamm. 58:334–46.

Husson, A. M. 1978. The mammals of Suriname. E. J. Brill, Leiden, xxxiv + 569 pp.

Hyndman, D., and J. I. Menzies. 1980. *Aproteles bulmerae* (Chiroptera: Pteropodidae) of New Guinea is not extinct. J. Mamm. 61:159–60.

I

Ibáñez, C. 1980. Descripcion de un nuevo genero de quiroptero neotropical de la familia Molossidae. Doñana Acta Vert. 7:104–11.

———. 1985. Notes on *Amorphochilus schnablii* Peters (Chiroptera, Furipteridae). Mammalia 49:584–87.

———. 1988. Notes on bats from Morocco. Mammalia 52:278–81.

Ibáñez, C., and R. Fernandez. 1985*a*. Murcielagos (Mammalia, Chiroptera) de las Canarias. Doñana Acta Vert. 12:307–15.

———. 1985*b*. Systematic status of the long-eared bat *Plecotus teneriffae* Barret-Hamilton, 1907 (Chiroptera; Vespertilionidae). Saugetierk. Mitt. 32:143–49.

Ibáñez, C., and J. Ochoa G. 1985. Distribucion y taxonomia de *Molossops temminckii* (Chiroptera, Molossidae) en Venezuela. Doñana Acta Vert. 12:141–50.

Ibáñez, C., and J. A. Valverde. 1985. Taxonomic status of *Eptesicus platyops* (Thomas, 1901) (Chiroptera, Vespertilionidae). Z. Saugetierk. 50:241–42.

Izor, R. J. 1979. Winter range of the silver-haired bat. J. Mamm. 60:641–43.

J

Jacobsen, N. H. G., and E. Du Plessis. 1976. Observations on the ecology and biology of the Cape fruit bat *Rousettus aegyptiacus leachi* in the eastern Transvaal. S. Afr. J. Sci. 72:270–73.

Jeanne, R. L. 1970. Note on a bat *(Phylloderma stenops)* preying upon the brood of a social wasp. J. Mamm. 51:624–25.

Jenkins, P. D., and J. E. Hill. 1981. The status of *Hipposideros galeritus* Cantor, 1846 and *Hipposideros cervinus* (Gould, 1854) (Chiroptera: Hipposideridae). Bull. British Mus. (Nat. Hist.) Zool. 41:279–94.

Jiménez M., P., and J. Pefaur. 1982. Aspectos sistemáticos y ecologicos de *Platalina genovensium* (Chiroptera: Mammalia). Actas Congr. Latinoamer. Zool. 8:707–18.

Jones, C. 1971*a*. The bats of Rio Muni, West Africa. J. Mamm. 52:121–40.

———. 1972. Comparative ecology of three pteropid bats in Rio Muni, West Africa. J. Zool. 167:353–70.

———. 1977. *Plecotus rafinesquii*. Mammalian Species, no. 69, 4 pp.

Jones, F. W. 1923–25. The mammals of South Australia. A. B. Jones, Government Printer, Adelaide, 458 pp.

Jones, G. S. 1983. Ecological and distributional notes on mammals from Vietnam, including the first record of *Nyctalus*. Mammalia 47:339–44.

Jones, J. K., Jr. 1966. Bats from Guatemala. Univ. Kansas Publ. Mus. Nat. Hist. 16: 439–72.

———. 1977. *Rhogeessa gracilis*. Mammalian Species, no. 76, 2 pp.

Jones, J. K., Jr., J. Arroyo-Cabrales, and R. D. Owen. 1988. Revised checklist of bats (Chiroptera) of Mexico and Central America. Occas. Pap. Mus. Texas Tech Univ., no. 120, 34 pp.

Jones, J. K., Jr., and R. J. Baker. 1979. Notes on a collection of bats from Montserrat, Lesser Antilles. Occas. Pap. Mus. Texas Tech Univ., no. 60, 6 pp.

Jones, J. K., Jr., and D. C. Carter. 1976. Annotated checklist, with keys to subfamilies and genera. *In* Baker, Jones, and Carter (1976), pp. 7–38.

———. 1979. Systematic and distributional notes. *In* Baker, Jones, and Carter (1979), pp. 7– 11.

Jones, J. K., Jr., D. C. Carter, H. H. Genoways, R. S. Hoffmann, D. W. Rice, and C. Jones. 1986. Revised checklist of North American mammals north of Mexico, 1986. Occas. Pap. Mus. Texas Tech Univ., no. 107, 22 pp.

Jones, J. K., Jr., J. R. Choate, and A. Cadena. 1972. Mammals from the Mexican State of Sinaloa. II. Chiroptera. Occas. Pap. Mus. Nat. Hist. Univ. Kansas, no. 6, 29 pp.

———. 1973. *Ardops nichollsi*. Mammalian Species, no. 24, 2 pp.

Jones, J. K., Jr., H. H. Genoways, and R. J. Baker. 1971. Morphological variation in *Stenoderma rufum*. J. Mamm. 52:244–47.

Jones, J. K., Jr., and J. A. Homan. 1974. *Hylonycteris underwoodi*. Mammalian Species, no. 32, 2 pp.

Jones, J. K., Jr., J. D. Smith, and H. H. Genoways. 1973. Annotated checklist of mammals of the Yucatan Peninsula, Mexico. I. Chiroptera. Occas. Pap. Mus. Texas Tech Univ., no. 13, 31 pp.

Jones, J. K., Jr., J. D. Smith, and R. W. Turner. 1971. Noteworthy records of bats from Nicaragua, with a checklist of the chiropteran fauna of the country. Occas. Pap. Mus. Nat. Hist. Univ. Kansas, no. 2, 35 pp.

Jones, J. K., Jr., P. Swanepoel, and D. C. Carter. 1977. Annotated checklist of the bats of Mexico and Central America. Occas. Pap. Mus. Texas Tech Univ., no. 47, 35 pp.

Jones, M. L. 1982. Longevity of captive mammals. Zool. Garten 52:113–28.

Juarez G., J., T. Jimenez A., and D. Navarro L. 1988. Additional records of *Bauerus dubiaquercus* (Chiroptera: Vespertilionidae) in Mexico. Southwestern Nat. 33:385.

K

Keen, R., and H. B. Hitchcock. 1980. Survival and longevity of the little brown bat *(Myotis lucifugus)* in southeastern Ontario. J. Mamm. 61:1–7.

Kerridge, D. C., and R. J. Baker. 1978. *Natalus micropus*. Mammalian Species, no. 114, 3 pp.

Kingdon, J. 1974*a*. East African mammals. An atlas of evolution in Africa. II(A). Insectivores and bats. Academic Press, London, xi + 341 + 1 pp.

Kitchener, D. J. 1973. Reproduction in the common sheath-tailed bat, *Taphozous georgianus* (Thomas) (Microchiroptera: Emballonuridae), in Western Australia. Austral. J. Zool. 21:375– 89.

———. 1975. Reproduction in female Gould's wattled bat, *Chalinolobus gouldii* (Gray) (Vespertilionidae), in Western Australia. Austral. J. Zool. 23:29–42.

———. 1976. Further observations on reproduction in the common sheath-tailed bat, *Taphozous georgianus* Thomas, 1915 in Western Australia, with notes on the gular pouch. Rec. W. Austral. Mus. 4:335–47.

———. 1980*a*. *Taphozous hilli* sp. nov. (Chiroptera: Emballonuridae), a new sheath-tailed bat from Western Australia and Northern Territory. Rec. W. Austral. Mus. 8:161–69.

Kitchener, D. J., and N. Caputi. 1985. Systematic revision of Australian *Scoteanax* and *Scotorepens* (Chiroptera: Vespertilionidae), with remarks on relationships to other Nycticeiini. Rec. W. Austral. Mus. 12:85–146.

Kitchener, D. J., N. Caputi, and B. Jones. 1986. Revision of Australo-Papuan *Pipistrellus* and of *Falsistrellus* (Microchiroptera: Vespertilionidae). Rec. W. Austral. Mus. 12:435–95.

Kitchener, D. J., and P. Coster. 1981. Reproduction in female *Chalinolobus morio* (Gray) (Vespertilionidae) in south-western Australia. Austral. J. Zool. 29:305–20.

Kitchener, D. J., and S. Foley. 1985. Notes on a collection of bats (Mammalia: Chiroptera) from Bali I., Indonesia. Rec. W. Austral. Mus. 12:223–32.

Kleiman, D. G., and P. A. Racey. 1969. Observations on noctule bats *(Nyctalus noctula)* breeding in captivity. Lynx 10:65–77.

Klingener, D., and G. K. Creighton. 1984. On small bats of the genus *Pteropus* from the Philippines. Proc. Biol. Soc. Washington 97:395–403.

Klingener, D., H. H. Genoways, and R. J. Baker. 1978. Bats from southern Haiti. Ann. Carnegie Mus. 47:81–97.

Koch-Weser, S. 1984. Fledermause aus Obervolta, W-Afrika (Mammalia: Chiroptera). Senckenberg. Biol. 64:255–311.

Kock, D. 1969*a*. *Dyacopterus spadiceus* (Thomas 1890) auf den Philippinen (Mammalia, Chiroptera). Senckenberg. Biol. 50: 1–7.

———. 1969*b*. Eine neue Gattung und Art cynopteriner Flughunde von Mindanao, Philippinen (Mammalia, Chiroptera). Senckenberg. Biol. 50:319–27.

———. 1969*c*. Eine bemerkenswerte neue Gattung und Art Flughunde von Luzon, Philippinen. Senckenberg. Biol. 50:329–38.

———. 1974*b*. Egyptian tomb bat *Taphozous perforatus* E. Geoffroy 1818: first record from Uganda. E. Afr. Nat. Hist. Soc. Bull., July, p. 130.

———. 1975. Ein originalexemplar von *Nyctinomus ventralis* Heuglin 1861 (Mammalia: Chiroptera: Molossidae). Stuttgarter Beitr. Naturkunde, ser. A, no. 272, 9 pp.

———. 1978*a*. A new fruit bat of the genus *Rousettus* Gray 1821, from the Comoro Islands, western Indian Ocean (Mammalia: Chiroptera). Proc. 4th Internatl. Bat Res. Conf., Nairobi, pp. 205–16.

———. 1981*a*. *Philetor brachypterus* auf Neu-Britannien und den Philippinen (Mammalia: Chiroptera: Vespertilionidae). Senckenberg. Biol. 61:313–19.

———. 1981*b*. Zur Chiropteren-Fauna von Burundi. Senckenberg. Biol. 61:329–36.

———. 1987. *Micropterus intermedius* Hayman 1963 und andere Fledermause vom unteren Zaire. Senckenberg. Biol. 67:219–24.

Kock, D., and H. Felten. 1980. Zwei Fledermause neu fur Pakistan. Senckenberg. Biol. 61:1– 9.

Koepcke, J. 1984. "Blattzelte" als Schafplatze der Fledermaus *Ectophylla macconnelli* (Thomas, 1901) (Phyllostomidae) im tropischen Regenwald von Peru. Saugetierk. Mitt. 31:123– 26.

Koepcke, J., and R. Kraft. 1984. Cranial and external characters of the larger fruit bats of the genus *Artibeus* from Amazonian Peru. Spixiana 7:75–84.

Koop, B. F., and R. J. Baker. 1983. Electrophoretic studies of relationships of six species of *Artibeus* (Chiroptera: Phyllostomidae). Occas. Pap. Mus. Texas Tech Univ., no. 83, 12 pp.

Koopman, K. F. 1971. Taxonomic notes on *Chalinolobus* and *Glauconycteris* (Chiroptera, Vespertilionidae). Amer. Mus. Novit., no. 2451, 10 pp.

———. 1972. *Eudiscopus denticulus*. Mammalian Species, no. 19, 2 pp.

———. 1973. Systematics of Indo-Australian *Pipistrellus*. Period. Biol. 75: 113–16.

———. 1975. Bats of the Sudan. Bull. Amer. Mus. Nat. Hist. 154:353–444.

———. 1978*a*. Zoogeography of Peruvian bats with special emphasis on the role of the Andes. Amer. Mus. Novit., no. 2651, 33 pp.

———. 1978*b*. The genus *Nycticeius* (Vespertilionidae), with special reference to tropical Australia. Proc. 4th Internatl. Bat Res. Conf., Nairobi, pp. 165–71.

———. 1979. Zoogeography of mammals from islands off the northeastern coast of New Guinea. Amer. Mus. Novit., no. 2690, 17 pp.

———. 1982*a*. Results of the Archbold Expeditions. No. 109. Bats from eastern Papua and the East Papuan Islands. Amer. Mus. Novit., no. 2747, 34 pp.

———. 1982*b*. Biogeography of the bats of South America. Pymatuning Lab. Ecol. Spec. Publ. 6:273–302.

———. 1983. A significant range extension for *Philetor* (Chiroptera, Vespertilionidae) with remarks on geographical variation. J. Mamm. 64:525–26.

———. 1984*a*. Taxonomic and distributional notes of tropical Australian bats. Amer. Mus. Novit., no. 2778, 48 pp.

———. 1984*b*. A synopsis of the families of bats. Part VII. Bat Research News 25: 25–27.

———. 1984*c*. Bats. *In* Anderson and Jones (1984), pp. 145–86.

———. 1986. Sudan bats revisited: an update of "Bats of the Sudan." Cimbebasia, ser. A, 8:9- -13.

———. 1988. Systematics and distribution. *In* Greenhall and Schmidt (1988), pp. 8–17.

Koopman, K. F., and F. Gudmundsson. 1966. Bats in Iceland. Amer. Mus. Novit., no. 2262, 6 pp.

Koopman, K. F., and J. K. Jones, Jr. 1970. Classification of bats. *In* Slaughter, B. H., and D. W. Walton, eds., About bats, S. Methodist Univ. Press, Dallas, pp. 22–28.

Koopman, K. F., R. E. Mumford, and J. F. Heisterberg. 1978. Bat records from Upper Volta, West Africa. Amer. Mus. Novit., no. 2643, 6 pp.

Kowalski, K. 1955. Our bats and their protection. Polish Acad. Sci. Nat. Protection Res. Cent. Publ., no. 11, 111 pp.

Krishna, A., and C. J. Dominic. 1981. Reproduction in the vespertilionid bat, *Scotophilus heathi* Horsefield. Arch. Biol. (Bruxelles) 92:247–58.

Krutzsch, P. H., and E. G. Crichton. 1985. Observations on the reproductive cycle of female *Molossus fortis* (Chiroptera: Molossidae) in Puerto Rico. J. Zool. 207:137–50.

Krzanowski, A. 1977. Contribution to the history of bats on Iceland. Acta Theriol. 22:272– 73.

Kunz, T. H. 1973*a*. Population studies of the cave bat *(Myotis velifer)*: reproduction, growth, and development. Occas. Pap. Mus. Nat. Hist. Univ. Kansas, no. 15, 43 pp.

———. 1973*b*. Resource utilizations: temporal and spatial components of bat activity in central Iowa. J. Mamm. 54:14–32.

———. 1982. *Lasionycteris noctivagans*. Mammalian Species, no. 172, 5 pp.

Kunz, T. H., P. V. August, and C. D. Burnett. 1983. Harem social organization in cave roosting *Artibeus jamaicensis* (Chiroptera: Phyllostomidae). Biotrópica 15:133–38.

L

Lang, H., and J. P. Chapin. 1917. Notes on the distribution and ecology of Central African Chiroptera. Bull. Amer. Mus. Nat. Hist. 37:479–563.

Largen, M. J., D. Kock, and D. W. Yalden. 1974. Catalogue of the mammals of Ethiopia. 1. Chiroptera. Italian J. Zool., Suppl., n.s., 5:221–98.

Laurie, E. M. 0., and J. E. Hill. 1954. List of land mammals of New Guinea, Celebes and adjacent islands, 1758–1952. British Mus. (Nat. Hist.), London, 175 pp.

LaVal, R. K. 1973*a*. A revision of the neotropical bats of the genus *Myotis*. Los Angeles Co. Nat. Hist. Mus. Sci. Bull., no. 15, 54 pp.

———. 1973*b*. Systematics of the genus *Rhogeessa* (Chiroptera: Vespertilionidae). Occas. Pap. Mus. Nat. Hist. Univ. Kansas, no. 19, 47 pp.

———. 1973*c*. Observations on the biology of *Tadarida brasiliensis cynocephala* in southeastern Louisiana. Amer. Midl. Nat. 89:112–20.

———. 1977. Notes on some Costa Rican bats. Brenesia 10/11:77–83.

LaVal, R. K., and H. S. Fitch. 1977. Structure, movements and reproduction in three Costa Rican bat communities. Occas. Pap. Mus. Nat. Hist. Univ. Kansas, no. 69, 28 pp.

LaVal, R. K., and M. L. LaVal. 1977. Reproduction and behavior of the African banana bat, *Pipistrellus nanus*. J. Mamm. 58:403–10.

Lawrence, B. 1939. Collections from the Philippine Islands. Mammals. Bull. Mus. Comp. Zool. 86:28–73.

Lazell, J. D., Jr., and K. F. Koopman. 1985. Notes on bats of Florida's lower keys. Florida Sci. 48:37–42.

Legendre, S. 1984. Etude odontologique des representants actuels du groupe *Tadarida* (Chiroptera, Molossidae). Implications phylogeniques, systematiques et zoogeographiques. Rev. Suisse Zool. 91:399–442.

Lekagul, B., and J. A. McNeely. 1977. Mammals of Thailand. Sahakarnbhat, Bangkok, li + 758 pp.

Lemke, T. O. 1984. Foraging ecology of the long-nosed bat, *Glossophaga soricina*, with respect to resource availability. Ecology 65:538–48.

———. 1986. Distribution and status of the sheath-tailed bat *(Emballonura semicaudata)* in the Mariana Islands. J. Mamm. 67:743–46.

Lemke, T. O., A. Cadena, R. H. Pine, and J. Hernandez-Camacho. 1982. Notes on opossums, bats, and rodents new to the fauna of Colombia. Mammalia 46:225–34.

Lemke, T. O., and J. R. Tamsitt. 1979. *Anoura cultrata* (Chiroptera: Phyllostomatidae) from Colombia. Mammalia 43: 579–81.

Leonard, M. L., and M. B. Fenton. 1983. Habitat use by spotted bats (*Euderma maculatum*, Chiroptera: Vespertilionidae): roosting and foraging behavior. Can. J. Zool. 61:1487–91.

Lidicker, W. Z., Jr., and A. C. Ziegler. 1968. Report on a collection of mammals from eastern New Guinea including species keys for fourteen genera. Univ. California Publ. Zool. 87:i–v + 1–64.

Lim Boo Liat, Chai Koh Shin, and I. Muul. 1972. Notes on the food habit of bats from the fourth division, Sarawak, with special reference to a new record of Bornean bat. Sarawak Mus. J. 20:351–57.

Linares, O. J. 1973. Présence de l'oreillard d'Amerique du Sud dans les Andes Vénézuéliennes (Chiroptères, Vespertilionidae). Mammalia 37:433–38.

Lopez-Forment, W. 1980. Longevity of wild *Desmodus rotundus* in Mexico. Proc. 5th Internatl. Bat Res. Conf., pp. 143–44.

Lowery, G. H., Jr. 1974. The mammals of Louisiana and its adjacent waters. Louisiana State Univ. Press, xxiii + 565 pp.

Lunney, D., and J. Barker. 1986. The occurrence of *Phoniscus papuensis* (Dobson) (Chiroptera: Vespertilionidae) on the south coast of New South Wales. Austral. Mamm. 9:57–58.

Lunney, D., J. Barker, and D. Priddel. 1985. Movements and day roosts of the chocolate wattled bat *Chalinolobus morio* (Gray) (Microchiroptera: Vespertilionidae) in a logged forest. Austral. Mamm. 8:313–17.

M

McBee, K., D. A. Schlitter, and R. L. Robbins. 1987. Systematics of the African bats of the genus *Eptesicus* (Mammalia: Vespertilionidae). 2. Karyotypes of African species and their generic relationships. Ann. Carnegie Mus. 56:213–22.

McCarthy, T. J. 1982*b*. Bat records from the Caribbean lowlands of El Peten, Guatemala. J. Mamm. 63:683–85.

———. 1987. Distributional records of bats from the Caribbean lowlands of Belize and adjacent Guatemala and Mexico. Fieldiana Zool., n.s., 39:137–62.

McCarthy, T. J., and N. A. Bitar. 1983. New bat records (*Enchisthenes* and *Myotis*) from the Guatemalan central highlands. J. Mamm. 64:526–27.

McCarthy, T. J., and M. Blake. 1987. Noteworthy bat records from the Maya Mountains Forest Reserve, Belize. Mammalia 51:161–64.

McCarthy, T. J., A. Cadena G., and T. O. Lemke. 1983. Comments on the first *Tonatia carrikeri* (Chiroptera: Phyllostomatidae) from Colombia. Lozania (Acta Zool. Colombiana), no. 40, 6 pp.

McCarthy, T. J., and C. O. Handley, Jr. 1987. Records of *Tonatia carrikeri* (Chiroptera: Phyllostomidae) from the Brazilian Amazon and *Tonatia schulzi* in Guyana. Bat Research News 28:20–23.

McCarthy, T. J., P. Robertson, and J. Mitchell. 1988. The occurrence of *Tonatia schulzi* (Chiroptera: Phyllostomidae) in French Guiana with comments on the female genitalia. Mammalia 52:583–84.

McCracken, G. F. 1984. Communal nursing in Mexican free-tailed bat maternity colonies. Science 223:1090–91.

McCracken, G. F., and J. W. Bradbury. 1981. Social organization and kinship in the polygynous bat *Phyllostomus hastatus*. Behav. Ecol. Sociobiol. 8:11–34.

McCracken, G. F., and M. K. Gustin. 1987. Batmom's daily nightmare. Nat. Hist. 96(10): 66– 73.

McFarlane, D. A. 1986. Cave bats in Jamaica. Oryx 20:27–30.

McFarlane, D. A., and B. R. Blood. 1986. Taxonomic notes on a collection of Rhinolophidae (Chiroptera) from northern Thailand, with a description of a new subspecies of *Rhinolophus robinsoni*. Z. Saugetierk. 51:218–23.

McKean, J. L. 1972. Notes on some collections of bats (order Chiroptera) from Papua New Guinea and Bougainville Island. CSIRO Div. Wildl. Res. Tech. Pap., no. 26, 5 pp.

———. 1975. The bats of Lord Howe Island with the description of a new nyctophiline bat. Austral. Mamm. 1:329–32.

McKean, J. L., and J. H. Calaby. 1968. A new genus and two new species of bats from New Guinea. Mammalia 32:372–78.

McKean, J. L., and G. R. Friend. 1979. *Taphozous kapalgensis*, a new species of sheath-tailed bat from the Northern Territory, Australia. Victorian Nat. 96:239–41.

McKean, J. L., and L. S. Hall. 1964. Notes on microchiropteran bats. Victorian Nat. 81:36– 37.

McKean, J. L., and W. J. Price. 1967. Notes on some Chiroptera from Queensland, Australia. Mammalia 31:101–19.

McKean, J. L., G. C. Richards, and W. J. Price. 1978. A taxonomic appraisal of *Ep-*

tesicus (Chiroptera: Mammalia) in Australia. Austral. J. Zool. 26:529–37.

McLellan, L. J. 1986. Notes on bats of Sudan. Amer. Mus. Novit., no. 2839, 12 pp.

McMahon, E. E., C. C. Oakley, and S. P. Cross. 1981. First record of the spotted bat *(Euderma maculatum)* from Oregon. Great Basin Nat. 41:270.

McWilliam, A. N. 1987*a*. The reproductive and social biology of *Coleura afra* in a seasonal environment. *In* Fenton, M. B., P. Racey, and J. M. V. Rayner, eds., Recent advances in the study of bats, Cambridge Univ. Press, London, pp. 324–50.

———. 1987*b*. Territorial and pair behavior of the African false vampire bat, *Cardioderma cor* (Chiroptera: Megadermatidae), in coastal Kenya. J. Zool. 213:243–52.

———. 1988. Social organization of the bat *Tadarida (Chaerephon) pumila* (Chiroptera: Molossidae) in Ghana, West Africa. Ethology 77:115–24.

Maddock, T. H., and A. McLeod. 1974. Polyoestry in the little brown bat, *Eptesicus pumilus*, in central Australia. S. Austral. Nat. 48:50, 63.

Madhavan, A. 1971. Breeding habits in the Indian vespertilionid bat, *Pipistrellus ceylonicus chrysothrix* (Wroughton). Mammalia 25:283–306.

Madhavan, A., D. R. Patil, and A. Gopalakrishna. 1978. Breeding habits and associated phenomena in some Indian bats. Part IV–*Hipposideros fulvus fulvus* (Gray)–Hipposideridae. J. Bombay Nat. Hist. Soc. 75:96–103.

Maeda, K. 1980. Review on the classification of little tube-nosed bats, *Murina aurata* group. Mammalia 44:531–51.

———. 1982. Studies on the classification of *Miniopterus* in Eurasia, Australia and Melanesia. Honyurui Kagaku (Mammalian Science), Suppl., no. 1, 176 pp.

———. 1983. Classificatory study of the Japanese large noctule, *Nyctalus lasiopterus aviator* Thomas, 1911. Zool. Mag. 92:21–36.

Manning, R. W., C. Jones, R. R. Hollander, and J. K. Jones, Jr. 1987. An unusual number of fetuses in the pallid bat. Prairie Nat. 19(4):261.

Marinkelle, C. J., and A. Cadena. 1972. Notes on bats new to the fauna of Colombia. Mammalia 36:50–58.

Marques, S. A. 1986. Activity cycle, feeding and reproduction of *Molossus ater* (Chiroptera: Molossidae) in Brazil. Bol. Mus. Paraense Emilio Goeldi, Ser. Zool., 2:159–79.

Marques, S. A., and D. C. Oren. 1987. First Brazilian record for *Tonatia schulzi* and *Sturnira bidens* (Chiroptera: Phyllostomidae). Bol. Mus. Par. Emilio Goeldi, Ser. Zool., 3:159–60.

Marshall, A. G., and A. N. McWilliam. 1982. Ecological observations on epomorphorine fruit-bats (Megachiroptera) in West African savanna woodland. J. Zool. 198:53–67.

Martin, C. O., and D. J. Schmidly. 1982. Taxonomic review of the pallid bat, *Antrozous pallidus* (Le Conte). Spec. Publ. Mus. Texas Tech Univ., no. 18, 48 pp.

Martin, R. A. 1972. Synopsis of late Pliocene and Pleistocene bats of North America and the Antilles. Amer. Midl. Nat. 87:326–35.

Medellín L., R. A. 1983. *Tonatia bidens* and *Mimon crenulatum* in Chiapas, Mexico. J. Mamm. 64:150.

———. 1989. *Chrotopterus auritus*. Mammalian Species, no. 343, 5 pp.

Medellín L., R. A., D. Navarro L., W. B. Davis, and V. J. Romero. 1983. Notes on the biology of *Micronycteris brachyotis* (Dobson) (Chiroptera), in southern Veracruz, Mexico. Brenesia 21:7–11.

Medway, Lord. 1972. Reproductive cycles of the flat-headed bats *Tylonycteris pachypus* and *T. robustula* (Chiroptera: Vespertilioninae) in a humid equatorial environment. Zool. J. Linnean Soc. 51:33–61.

———. 1973. The taxonomic status of *Tylonycteris malayana* Chasen 1940 (Chiroptera). J. Nat. Hist. 7:125–31.

———. 1977. Mammals of Borneo. Monogr. Malaysian Branch Roy. Asiatic Soc., no. 7, xii + 172 pp.

———. 1978. The wild mammals of Malaya (peninsular Malaysia) and Singapore. Oxford Univ. Press, Kuala Lumpur, xxii + 128 pp.

Medway, Lord, and A. G. Marshall. 1972. Roosting associations of flat-headed bats, *Tylonycteris* species (Chiroptera: Vespertilionidae) in Malaysia. J. Zool. 168:463–82.

Meester, J., and H. W. Setzer. 1977. The mammals of Africa: an identification manual. Smithson. Inst. Press, Washington, D.C.

Meester, J. A. J., I. L. Rautenbach, N. J. Dippenaar, and C. M. Baker. 1986. Classification of southern African mammals. Transvaal Mus. Monogr., no. 5, x + 359 pp.

Mein, P., and Y. Tupinier. 1977. Formule dentaire et position systématique du Minoptere (Mammalia, Chiroptera). Mammalia 41:207–11.

Meirte, D. 1983. New data on *Casinycteris argynnis* Thomas 1910 (Megachiroptera, Pteropodidae). Ann. Mus. Roy. Afr. Cent., Sec. Zool., 237:9–17.

Menu, H. 1984. Revision du statut de *Pipistrellus subflavus* (F. Cuvier, 1832). Proposition d'un taxon generique nouveau: *Perimyotis* nov. gen. Mammalia 48:409–16.

Menzies, J. I. 1971. The lobe-lipped bat (*Chalinolobus nigrogriseus* Gould) in New Guinea. Rec. Papua and New Guinea Mus. 1:6–8.

———. 1973. A study of leaf-nosed bats (*Hipposideros caffer* and *Rhinolophus landeri*) in a cave in northern Nigeria. J. Mamm. 54:930–45.

———. 1977. Fossil and subfossil fruit bats from the mountains of New Guinea. Austral. J. Zool. 25:329–36.

Miller, G. S., Jr. 1907. The families and genera of bats. Bull. U.S. Natl. Mus. 57: i–xvii + 1–282.

Mills, R. S., G. W. Barrett, and M. P. Farrell. 1975. Population dynamics of the big brown bat *(Eptesicus fuscus)* in southwestern Ohio. J. Mamm. 56:591–604.

Mitchell, R. M. 1980. New records of bats (Chiroptera) from Nepal. Mammalia 44: 339–42.

Mohr, C. E. 1972. The status of threatened species of cave-dwelling bats. Bull. Natl. Speleol. Soc. 34:33–47.

Molinari, J., and P. J. Soriano. 1987. *Sturnira bidens*. Mammalian Species, no. 276, 4 pp.

Molnar, R. E., L. S. Hall, and J. H. Mahoney. 1984. New fossil localities for *Macroderma* Miller, 1906 (Chiroptera: Megadermatidae) in New South Wales and its past and present distribution in Australia. Austral. Mamm. 7:63–73.

Moore, N. W. 1975. The diurnal flight of the Azorean bat *(Nyctalus azoreum)* and the avifauna of the Azores. J. Zool. 177:483–86.

Morrison, D. W. 1978. Influence of habitat on the foraging distance of the fruit bat, *Artibeus jamaicensis.* J. Mamm. 59:622–24.

———. 1979. Apparent male defense of tree hollows in the fruit bat, *Artibeus jamaicensis.* J. Mamm. 60:11–15.

———. 1980. Foraging and day-roosting dynamics of canopy fruit bats in Panama. J. Mamm. 61:20–29.

Mudar, K. M., and M. S. Allen. 1986. A list of bats from northeastern Luzon, Philippines. Mammalia 50:219–25.

Mumford, R. E. 1973. Natural history of the red bat *(Lasiurus borealis)* in Indiana. Period. Biol. 75:155–58.

Mumford, R. E., and D. M. Knudson. 1978. Ecology of bats at Vicosa, Brazil. Proc. 4th Internatl. Bat Res. Conf., Nairobi, pp. 287–95.

Musser, G. G., K. F. Koopman, and D. Califia. 1982. The Sulawesian *Pteropus arquatus* and *P. argentatus* are *Acerodon celebensis;* the Philippine *P. leucotis* is an *Acerodon.* J. Mamm. 63:319–28.

Mutere, F. A. 1967. The breeding biology of equatorial vertebrates: reproduction in the fruit bat, *Eidolon helvum,* at latitude 0°20′ N. J. Zool. 153:153–61.

———. 1970. The breeding biology of equatorial vertebrates: reproduction in the insectivorous bat, *Hipposideros caffer,* living at 0°27′ N. Bijdragen Tot de Dierkunde 40:56–58.

———. 1973*a*. Reproduction in two species of equatorial free-tailed bats (Molossidae). E. Afr. Wildl. J. 11:271–80.

———. 1973*b*. A comparative study of reproduction in two populations of the insectivorous bats, *Otomops martiensseni,* at latitudes 1°5′ S and 2°30′ S. J. Zool. 171: 79–92.

Myers, P. 1977*a*. Patterns of reproduction of four species of vespertilionid bats in Paraguay. Univ. California Publ. Zool. 107: 1–41.

Myers, P., and R. M. Wetzel. 1983. Systematics and zoogeography of the bats of the Chaco Boreal. Misc. Publ. Mus. Zool. Univ. Michigan, no. 165, iv + 59 pp.

Myers, P., R. White, and J. Stallings. 1983. Additional records of bats from Paraguay. J. Mamm. 64:143–45.

N

Nader, I. A. 1975. On the bats (Chiroptera) of the Kingdom of Saudi Arabia. J. Zool. 176:331–40.

———. 1982. New distributional records of bats from the Kingdom of Saudi Arabia (Mammalia: Chiroptera). J. Zool. 198: 69–82.

Nader, I. A., and D. Kock. 1980. First record of *Tadarida nigeriae* (Thomas 1913) from the Arabian Peninsula. Senckenberg. Biol. 60:131–35.

———. 1983*a*. A new slit-faced bat from central Saudi Arabia (Mammalia: Chiroptera: Nycteridae). Senckenberg. Biol. 63: 9–15.

———. 1983*b*. *Rhinopoma microphyllum asirensis* n. subsp. from southwestern Saudi Arabia (Mammalia: Chiroptera: Rhinopomatidae). Senckenberg. Biol. 63:147–52.

Nagorsen, D., and J. R. Tamsitt. 1981. Systematics of *Anoura cultrata, A. brevirostrum,* and *A. werckleae.* J. Mamm. 62: 82–100.

Nellis, D. W., and C. P. Ehle. 1977. Observations on the behavior of *Brachyphylla cavernarum* (Chiroptera) in Virgin Islands. Mammalia 41:403–9.

Nelson, J. E., and E. Hamilton-Smith. 1982. Some observations on *Notopteris macdonaldi* (Chiroptera: Pteropodidae). Austral. Mamm. 5:247–52.

Neuhauser, H. N., and A. F. DeBlase. 1974. Notes on bats (Chiroptera: Vespertilionidae) new to the faunal lists of Afghanistan and Iran. Fieldiana Zool. 62:85–96.

Neuweiler, G. 1969. Verhaltensbeobachtungen an einer indischen Flughundkolonie (*Pteropus g. giganteus* Brünn). Z. Tierpsychol. 26:166–99.

Nicoll, M. E., and P. A. Racey. 1981. The Seychelles fruit bat, *Pteropus seychellensis seychellensis.* Afr. J. Ecol. 19:361–64.

Nicoll, M. E., and J. M. Suttie. 1982. The sheath-tailed bat, *Coleura seychellensis* (Chiroptera: Emballonuridae) in the Seychelles Islands. J. Zool. 197:421–26.

Norberg, U. M., and M. B. Fenton. 1988. Carnivorous bats? Biol. J. Linnean Soc. 33: 383–94.

Novacek, M. J. 1985. Evidence for echolocation in the oldest known bats. Nature 315: 140–41.

Novick, A., and B. A. Dale. 1971. Foraging behavior in fishing bats and their insectivorous relatives. J. Mamm. 52:817–18.

O

Ochoa G., J. 1984. Presencia de *Nyctinomops aurispinosa* en Venezuela (Chiroptera: Molossidae). Acta Cien. Venezolana 35:147–50.

Ochoa G., J., and C. Ibáñez. 1982. Nuevo murcielago del genero *Lonchorhina* (Chiroptera: Phyllostomidae). Mem. Soc. Ciencias Nat. La Salle 42:145–59.

———. 1985. Distributional status of some bats from Venezuela. Mammalia 49:65–73.

O'Farrell, M. J., and E. H. Studier. 1973. Reproduction, growth, and development in *Myotis thysanodes* and *M. lucifugus* (Chiroptera: Vespertilionidae). Ecology 54: 18–30.

Ognev, S. I. 1962–64. Mammals of eastern Europe and northern Asia. Israel Progr. Sci. Transl., Jerusalem, 8 vols.

Ojasti, J., and O. J. Linares. 1971. Adiciones a la fauna de murciélagos de Venezuela con notas sobre las especies del género *Diclidurus* (Chiroptera). Acta Biol. Venezuelica 7:421– 41.

Ojasti, J., and C. J. Naranjo. 1974. First record of *Tonatia nicaraguae* in Venezuela. J. Mamm. 55:248–49.

Okia, N. O. 1974*a*. Breeding in Franquet's bat, *Epomops franqueti* (Tomes), in Uganda. J. Mamm. 55:462–65.

———. 1974*b*. The breeding pattern of the eastern epauletted bat, *Epomophorus anurus* Heuglin, in Uganda. J. Reprod. Fert. 37:27–31.

Osgood, W. H. 1932. Mammals of the Kelley-Roosevelts and Delacour Asiatic Expeditions. Field Mus. Nat. Hist. Publ., Zool. Ser., 18:193–339.

O'Shea, T. J., and T. A. Vaughan. 1977. Nocturnal and seasonal activities of the pallid bat, *Antrozous pallidus.* J. Mamm. 58: 269–84.

Ottenwalder, J. A., and H. H. Genoways. 1982. Systematic review of the Antillean bats of the *Natalus micropus*–complex (Chiroptera: Natalidae). Ann. Carnegie Mus. 51:17–38.

Owen, R. D. 1987. Phylogenetic analyses of the bat subfamily Stenodermatinae (Mammalia: Chiroptera). Spec. Publ. Mus. Texas Tech Univ., no. 26, 65 pp.

Owen, R. D., and W. D. Webster. 1983. Morphological variation in the Ipanema bat, *Pygoderma bilabiatum*, with description of a new subspecies. J. Mamm. 64:146–49.

P

Pagels, J. F., and C. Jones. 1974. Growth and development of the free-tailed bat, *Tadarida brasiliensis cynocephala* (Le Conte). Southwestern Nat. 19:267–76.

Palmeirim, J. M. 1982. On the presence of *Nyctalus lasiopterus* in North Africa (Mammalia: Chiroptera). Mammalia 46:401–3.

Paradiso, J. L. 1971. A new subspecies of *Cynopterus sphinx* (Chiroptera: Pteropodidae) from Serasan (South Natuna) Island, Indonesia. Proc. Biol. Soc. Washington 84:293–300.

Paradiso, J. L., and A. M. Greenhall. 1967. Longevity records for American bats. Amer. Midl. Nat. 78:251–52.

Parsons, H. J., D. A. Smith, and R. F. Whittam. 1986. Maternity colonies of silver-haired bats, *Lasionycteris noctivagans*, in Ontario and Saskatchewan. J. Mamm. 67:598–600.

Patten, D. R., and L. T. Findley. 1970. Observations and records of *Myotis (Pizonyx) vivesi* Menegaux (Chiroptera: Vespertilionidae). Los Angeles Co. Mus. Nat. Hist. Contrib. Sci., no. 183, 9 pp.

Payne, J., C. M. Francis, and K. Phillipps. 1985. A field guide to the mammals of Borneo. Sabah Society with World Wildlife Fund Malaysia, 332 pp.

Payne, N. 1986. The trade in Pacific fruit bats. Traffic Bull. 8(2):25–27.

Peracchi, A. L. 1968. Sobre os habitos de *"Histiotus velatus"* (Geoffroy, 1824) (Chiroptera, Vespertilionidae). Rev. Brasil. Biol. 28:469–73.

Perez, G. S. A. 1972. Observations of Guam bats. Micronesia 8:141–49.

———. 1973. Notes on the ecology and life history of Pteropodidae on Guam. Period. Biol. 75:163–68.

Peterson, R. L. 1965*a*. A review of the flat-headed bats of the family Molossidae from South America and Africa. Roy. Ontario Mus. Life Sci. Contrib., no. 64, 32 pp.

———. 1965*b*. A review of the bats of the genus *Ametrida*, family Phyllostomidae. Roy. Ontario Mus. Life Sci. Contrib., no. 65, 13 pp.

———. 1969. Notes on the Malaysian fruit bats of the genus *Dyacopterus*. Roy. Ontario Mus. Life Sci. Occas. Pap., no. 13, 4 pp.

———. 1971*a*. The African molossid bat *Tadarida russata*. Can. J. Zool. 49:297–301.

———. 1971*b*. Notes on the African long-eared bats of the genus *Laephotis* (family Vespertilionidae). Can. J. Zool. 49:885–88.

———. 1971*c*. The systematic status of the African molossid bats *Tadarida bemmeleni* and *Tadarida cistura*. Can. J. Zool. 49: 1347–54.

———. 1972. Systematic status of the African molossid bats *Tadarida congica*, *T. niangarae* and *T. trevori*. Roy. Ontario Mus. Life Sci. Contrib., no. 85, 32 pp.

———. 1973. The first known female of the African long-eared bat *Laephotis wintoni* (Vespertilionidae: Chiroptera). Can. J. Zool. 51:601–3.

———. 1974*a*. Variation in the African bat, *Tadarida lobata*, with notes on habitat and habits (Chiroptera: Molossidae). Roy. Ontario Mus. Life Sci. Occas. Pap., no. 24, 8 pp.

———. 1981. Systematic variation in the *tristis* group of the bent-winged bats of the genus *Miniopterus* (Chiroptera: Vespertilionidae). Can. J. Zool. 59:828–43.

———. 1982. A new species of *Glauconycteris* from the east coast of Kenya (Chiroptera: Vespertilionidae). Can. J. Zool. 60: 2521–25.

———. 1985. A systematic review of the molossid bats allied with the genus *Mormopterus* (Chiroptera: Molossidae). Acta Zool. Fennica 170:205–8.

Peterson, R. L., and M. B. Fenton. 1970. Variation in the bats of the genus *Harpyionycteris*, with the description of a new race. Roy. Ontario Mus. Life Sci. Occas. Pap., no. 17, 15 pp.

Peterson, R. L., and D. L. Harrison. 1970. The second and third known specimens of the African molossid bat, *Tadarida lobata*. Roy. Ontario Mus. Life Sci. Occas. Pap., no. 16, 6 pp.

Peterson, R. L., and D. A. Smith. 1973. A new species of *Glauconycteris* (Vespertilionidae, Chiroptera). Roy. Ontario Mus. Life Sci. Occas. Pap., no. 22, 9 pp.

Pettigrew, J. D. 1986. Flying primates? Megabats have the advanced pathway from eye to midbrain. Science 231:1304–6.

———. 1987. Are flying foxes (Chiroptera: Pteropodidae) really primates? Austral. Mamm. 10:119–24.

Phillips, C. J. 1966. A new species of bat of the genus *Melonycteris* from the Solomon Islands. J. Mamm. 47:23–27.

———. 1967. A collection of bats from Laos. J. Mamm. 48:633–36.

———. 1968. Systematics of megachiropteran bats in the Solomon Islands. Univ. Kansas Mus. Nat. Hist. Publ. 16:777–837.

Phillips, C. J., and E. C. Birney. 1968. Taxonomic status of the vespertilionid genus *Anamygdon* (Mammalia; Chiroptera). Proc. Biol. Soc. Washington 81:491–98.

Phillips, C. J., and J. K. Jones, Jr. 1971. A new subspecies of the long-nosed bat, *Hylonycteris underwoodi*, from Mexico. J. Mamm. 52:77–80.

Phillips, W. R., and S. J. Inwards. 1985. The annual activity and breeding cycles of Gould's long-eared bat, *Nyctophilus gouldi* (Microchiroptera: Vespertilionidae). Austral. J. Zool. 33:111–26.

Piechocki, R. 1966. Uber die Nachweise der Langohr-Fledermause, *Plecotus auritus L.* und *Plecotus austriacus* Fischer in mitteldeutschen Raum. Hercynia 3:407–15.

Pienaar, U. D. V. 1972. A new bat record for the Kruger National Park. Koedoe 15: 91–93.

Pierson, E. D., V. M. Sarich, J. M. Lowenstein, M. J. Daniel, and W. E. Rainey. 1986. A molecular link between the bats of New Zealand and South America. Nature 323: 60–63.

Pine, R. H. 1972*b*. The bats of the genus *Carollia*. Texas Agric. Exp. Sta. Tech. Monogr., no. 8, 125 pp.

Pine, R. H., D. C. Carter, and R. K. LaVal. 1971. Status of *Bauerus* Van Gelder and its relationships to other nyctophiline bats. J. Mamm. 52:663–69.

Pine, R. H., S. D. Miller, and M. L. Schamberger. 1979. Contributions to the mammalogy of Chile. Mammalia 43:339–76.

Pine, R. H., and A. Ruschi. 1976. Concerning certain bats described and recorded from Espirito Santo, Brazil. An. Inst. Biol. Univ. Nal. Autón. Mexico, Ser. Zool., 47:183–96.

Poché, R. M., and G. L. Bailie. 1974. Notes on the spotted bat *(Euderma maculatum)* from southwest Utah. Great Basin Nat. 34:254–56.

Polaco, O. J. 1987. First record of *Noctilio albiventris* (Chiroptera: Noctilionidae) in Mexico. Southwestern Nat. 32:508–9.

Porter, F. L. 1978. Roosting patterns and social behavior in captive *Carollia perspicillata*. J. Mamm. 59:627–30.

———. 1979*a*. Social behavior in the leaf-nosed bat, *Carollia perspicillata*. I. Social organization. Z. Tierpsychol. 49:406–17.

———. 1979*b*. Social behavior in the leaf-nosed bat, *Carollia perspicillata*. II. Social communication. Z. Tierpsychol. 50:1–8.

Porter, F. L., and G. F. McCracken. 1983. Social behavior and allozyme variation in a captive colony of *Carollia perspicillata*. J. Mamm. 64:295–98.

Prociv, P. 1983. Seasonal behaviour of *Pteropus scapulatus* (Chiroptera: Pteropodidae). Austral. Mamm. 6:45–46.

Q

Qumsiyeh, M. B. 1980. New records of bats from Jordan. Saugetierk. Mitt. 40:36–39.

———. 1983. Occurrence and zoogeographical implications of *Myotis blythi* (Tomes, 1857) in Libya. Mammalia 47:429–30.

———. 1985. The bats of Egypt. Spec. Publ. Mus. Texas Tech Univ., no. 23, 102 pp.

Qumsiyeh, M. B., and J. K. Jones, Jr. 1986. *Rhinopoma hardwickii* and *Rhinopoma muscatellum*. Mammalian Species, no. 263, 5 pp.

Qumsiyeh, M. B., and D. A. Schlitter. 1982. The bat fauna of Jabal Al Akhdar, northeast Libya. Ann. Carnegie Mus. 51: 377–89.

R

Rabor, D. S. 1952. Two new mammals from Negros Island, Philippines. Nat. Hist. Misc. (Chicago Acad. Sci.), no. 96, 7 pp.

Racey, P. A. 1973. The time of onset of hibernation in pipistrelle bats, *Pipistrellus pipistrellus*. J. Zool. 171:465–67.

Ramirez-Pulido, J., and W. Lopez-Forment. 1979. Additional records of some Mexican bats. Southwestern Nat. 24:541–44.

Rasweiler, J. J., IV. 1973. Care and managenent of the long-tongued bat, *Glossophaga soricina* (Chiroptera: Phyllostomatidae), in the laboratory, with observations on estivation induced by food deprivation. J. Mamm. 54:391–404.

Rautenbach, I. L., and I. W. Espie. 1982. First records of occurrence for two species of bats in the Kruger National Park. Koedoe 25:111–12.

Rautenbach, I. L., M. B. Fenton, and L. E. O. Braack. 1985. First records of five species of insectivorous bats from the Kruger National Park. Koedoe 28:73–80.

Rautenbach, I. L., D. A. Schlitter, and L. E. O. Braack. 1984. New distributional records of bats for the Republic of South Africa, with special reference to the Kruger National Park. Koedoe 27:131–35.

Ray, C. E., O. J. Linares, and G. S. Morgan. 1988. Paleontology. *In* Greenhall and Schmidt (1988), pp. 19–30.

Reduker, D. W., T. L. Yates, and I. F. Greenbaum. 1983. Evolutionary affinities among southwestern long-eared *Myotis* (Chiroptera: Vespertilionidae). J. Mamm. 64: 666-77.

Richardson, E. G. 1977. The biology and evolution of the reproductive cycle of *Miniopterus schreibersii* and *M. australis* (Chiroptera: Vespertilionidae). J. Zool. 183: 353-75.

Rick, A. M. 1968. Notes on bats from Tikal, Guatemala. J. Mamm. 49:516-20.

Ride, W. D. L. 1970. A guide to the native mammals of Australia. Oxford Univ. Press, Melbourne, xiv + 249 pp.

Robbins, C. B. 1978. Taxonomic identifcation and history of *Scotophilus nigrita* (Schreber) (Chiroptera: Vespertilionidae). J. Mamm. 59:212–13.

———. 1980. Small mammals of Togo and Benin. I. Chiroptera. Mammalia 44:83–88.

———. 1984. A new high forest species in the African bat genus *Scotophilus* (Vespertilionidae). Ann. Mus. Roy. Afr. Cent., Sec. Zool., 237:19–24.

Robbins, C. B., F. De Vree, and V. Van Cakenberghe. 1985. A systematic revision of the African bat genus *Scotophilus* (Vespertilionidae). Zool. Wetenschappen-Ann. 246:53–84.

Robbins, L. W., and V. M. Sarich. 1988. Evolutionary relationships in the family Emballonuridae (Chiroptera). J. Mamm. 69:1–13.

Roberts, T. J. 1977. The mammals of Pakistan. Ernest Benn, London, xxvi + 361 pp.

Robson, S. K. 1984. *Myotis adversus* (Chiroptera: Vespertilionidae): Australia's fish-eating bat. Austral. Mamm. 7:51–52.

Rookmaaker, L. C., and W. Bergmans. 1981. Taxonomy and geography of *Rousettus amplexicaudatus* (Geoffroy, 1810) with comparative notes on sympatric congeners (Mammalia, Megachiroptera). Beaufortia 31:1–29.

Rosatte, R. C. 1988*a*. Bat rabies in Canada: history, epidemiology and prevention. Can. Vet. J. 28:754–56.

———. 1988*b*. Rabies in Canada: history, epidemiology and control. Can. Vet. J. 29: 362–65.

Rosevear, D. R. 1965. The bats of West Africa. British Mus. (Nat. Hist.), London, xvii + 418 pp.

Rozendaal, F. G. 1984. Notes on macroglossine bats from Sulawesi and the Moluccas, Indonesia, with the description of a new species of *Syconycteris* Matschie, 1899 from Halmahera (Mammalia: Megachiroptera). Zool. Meded. 58:187–212.

Ryan, M. J., and M. D. Tuttle. 1983. The ability of the frog-eating bat to discriminate among novel and potentially poisonous frog species using acoustic cues. Anim. Behav. 31:827–33.

Ryan, M. J., M. D. Tuttle, and R. M. R. Barclay. 1983. Behavioral responses of the frog- eating bat, *Trachops cirrhosus*, to sonic frequencies. J. Comp. Physiol., ser. A, 150:413– 18.

Ryan, R. M. 1966. A new and some imperfectly known Australian *Chalinolobus* and the taxonomic status of African *Glauconycteris*. J. Mamm. 47:86–91.

Ryberg, O. 1947. Studies on bats and bat parasites. Univ. Lund and Zool. Lab. Agric., Dairy, and Hort. Inst. Alnarp, Stockholm, 330 pp.

Rydell, J. 1986. Feeding territoriality in female northern bats, *Eptesicus nilssoni*. Ethology 72:329–37.

Rzebik-Kowalska, B., B. W. Woloszyn, and A. Nadachowski. 1978. A new bat, *Myotis nattereri* (Kuhl, 1818) (Vespertilionidae), in the fauna of Iraq. Acta Theriol. 23:541–50.

S

Sailler, H., and U. Schmidt. 1978. Die sozialen Laute der Gemeinen vampirfledermaus *Desmodus rotundus* bei Konfrontation am Futterplatz unter experimentellen Bedingungen. Z. Saugetierk. 43:249–61.

Sanborn, C. C. 1931. Bats from Polynesia, Melanesia, and Micronesia. Field Mus. Nat. Hist. Publ., Zool. Ser., 18:7–29.

———. 1950. New Philippine fruit bats. Proc. Biol. Soc. Washington 63:189–90.

Sanborn, C. C., and A. J. Nicholson. 1950. Bats from New Caledonia, the Solomon Islands, and New Hebrides. Fieldiana Zool. 31:313–38.

Sandhu, S. 1984. Breeding biology of the Indian fruit bat, *Cynopterus sphinx* (Vahl) in central India. J. Bombay Nat. Hist. Soc. 81:600–611.

Sazima, I. 1976. Observations on the feeding habits of phyllostomatid bats (*Carollia, Anoura,* and *Vampyrops*) in southeastern Brazil. J. Mamm. 57:381–82.

———. 1978. Vertebrates as food items of the woolly false vampire, *Chrotopterus auritus*. J. Mamm. 59:617–18.

Sazima, I., and W. Uieda. 1977. O morcego *Promops nasutus* no sudeste Brasileiro (Chiroptera, Molossidae). Ciencia e Cultura 29:312–14.

Sazima, I., L. D. Vizotto, and V. A. Taddei. 1978. Uma nova especie de *Lonchophylla* da serra do Cipo, Minas Gerais, Brasil (Mammalia, Chiroptera, Phyllostomatidae). Rev. Brasil. Biol. 38:81–89.

Schliemann, H., and B. Maas. 1978. *Myzopoda aurita*. Mammalian Species, no. 116, 2 pp.

Schlitter, D. A., and I. R. Aggundey. 1986. Systematics of African bats of the genus *Eptesicus* (Mammalia: Vespertilionidae). 1. Taxonomic status of the large serotines of eastern and southern Africa. Cimbebasia, ser. A, 8:167–74.

Schlitter, D. A., I. R. Aggundey, M. B. Qumsiyeh, K. Nelson, and R. L. Honeycutt. 1986. Taxonomic and distributional notes on bats from Kenya. Ann. Carnegie Mus. 55:297–302.

Schlitter, D. A., and S. B. McLaren. 1981. An additional record of *Myonycteris relicta* Bergmans, 1980, from Tanzania (Mammalia: Chiroptera). Ann. Carnegie Mus. 50:385–89.

Schlitter, D. A., L. W. Robbins, and S. A. Buchanan. 1982. Bats of the Central African Republic (Mammalia: Chiroptera). Ann. Carnegie Mus. 51:133–55.

Schlitter, D. A., S. L. Williams, and J. E. Hill. 1983. Taxonomic review of Temminck's trident bat, *Aselliscus tricuspidatus* (Temminck, 1834) (Mammalia: Hipposideridae). Ann. Carnegie Mus. 52:337–58.

Schmidly, D. J. 1977. The mammals of trans-Pecos Texas. Texas A & M Univ. Press, College Station, xii + 225 pp.

Schmidt, U. 1988. Reproduction. *In* Greenhall and Schmidt (1988), pp. 99–109.

Schmidt, U., C. Schmidt, W. Lopez-Forment, and R. F. Crespo. 1978. Rückfunde beringter Vampirfledermause *Desmodus rotundus* in Mexiko. Z. Saugetierk. 43:65–70.

Schowalter, D. B., J. R. Gunson, and L. D. Harder. 1979. Life history characteristics of little brown bats (*Myotis lucifugus*) in Alberta. Can. Field-Nat. 93:243–51.

Schowalter, D. B., L. D. Harder, and B. H. Treichel. 1978. Age composition of some vespertilionid bats as determined by dental annuli. Can. J. Zool. 56:355–58.

Schwartz, C. W., and E. R. Schwartz. 1959. The wild mammals of Missouri. Univ. Missouri Press, vi + 341 pp.

Setzer, H. W. 1971. New bats of the genus *Laephotis* from Africa (Mammalia: Chiroptera). Proc. Biol. Soc. Washington 84:259–64.

Seymour, C., and R. W. Dickerman. 1982. Observations on the long-legged bat, *Macrophyllum macrophyllum*, in Guatemala. J. Mamm. 63:530–32.

Shaimardanov, R. T. 1982. *Otonycteris hemprichi* and *Barbastella leucomelas* (Chiroptera) in Kazakhstan. Zool. Zhur. 61: 1765.

Short, H. L. 1961. Age at sexual maturity of Mexican free-tailed bats. J. Mamm. 42: 533–36.

Shryer, J., and D. L. Flath. 1980. First record of the pallid bat (*Antrozous pallidus*) from Montana. Great Basin Nat. 40:115.

Silva Taboada, G., and R. H. Pine. 1969. Morphological and behavioral evidence for the relationship between the bat genus *Brachyphylla* and the Phyllonycterinae. Biotrópica 1:10–19.

Sinha, Y. P., and S. Chakraborty. 1971. Taxonomic status of the vespertilionid bat, *Nycticeius emarginatus* Dobson. Proc. Zool. Soc. Calcutta 24:53–59.

Sly, G. R. 1975. Second record of the bronzed tube-nosed bat (*Murina aenea*) in peninsular Malaysia. Malayan Nat. J. 28:217.

Smit, C. J., and A. Van Wijngaarden. 1981. Threatened mammals in Europe. Akademische Verlagsgesellschaft, Wiesbaden, 259 pp.

Smith, J. D. 1972. Systematics of the chiropteran family Mormoopidae. Univ. Kansas Mus. Nat. Hist. Misc. Publ., no. 56, 32 pp.

———. 1977. On the nomenclatorial status of *Chilonycteris gymnonotus* Natterer, 1843. J. Mamm. 58:245–46.

Smith, J. D., and H. H. Genoways. 1974. Bats of Margarita Island, Venezuela, with zoogeographic comments. Bull. S. California Acad. Sci. 73:64–79.

Smith, J. D., and J. E. Hill. 1981. A new species and subspecies of bat of the *Hipposideros bicolor* group from Papua New Guinea, and the systematic status of *Hipposideros calcaratus* and *Hipposideros cupidus* (Mammalia: Chiroptera: Hipposideridae). Los Angeles Co. Mus. Nat. Hist. Contrib. Sci., no. 331, 19 pp.

Smith, J. D., and C. S. Hood. 1981. Preliminary notes on bats from the Bismarck Archipelago (Mammalia: Chiroptera). Sci. New Guinea 8:81–121.

———. 1983. A new species of tube-nosed fruit bat (*Nyctimene*) from the Bismarck Archipelago, Papua New Guinea. Occas. Pap. Mus. Texas Tech Univ., no. 81, 14 pp.

———. 1984. Genealogy of the New World nectar-feeding bats reexamined: a reply to Griffiths. Syst. Zool. 33:435–60.

Smithers, R. H. N. 1971. The mammals of Botswana. Trustees Natl. Museums and Monuments Rhodesia Mus. Mem., no. 4, 340 pp.

———. 1983. The mammals of the southern African subregion. Univ. Pretoria, xxii + 736 pp.

———. 1986. South African red data book–terrestrial mammals. S. Afr. Natl. Sci. Programmes Rept., no. 125, ix + 216 pp.

Soota, T. D., and Y. Chaturvedi. 1980. New locality record of *Pipistrellus camortae* Miller from Car Nicobar and its systematic status. Rec. Zool. Surv. India 77:83–87.

Soriano, P. J., and J. Molinari. 1987. *Sturnira aratathomasi*. Mammalian Species, no. 284, 4 pp.

Sreenivasan, M. A., H. R. Bhat, and G. Geevarghese. 1973. Breeding cycle of *Rhi-*

nolophus rouxi Temminck, 1835 (Chiroptera: Rhinolophidae), in India. J. Mamm. 54:1013–17.

———. 1974. Observations on the reproductive cycle of *Cynopterus sphinx sphinx* Vahl, 1797 (Chiroptera: Pteropodidae). J. Mamm. 55:200–202.

Starrett, A. 1972. *Cyttarops alecto*. Mammalian Species, no. 13, 2 pp.

Starrett, A., and R. S. Casebeer. 1968. Records of bats from Costa Rica. Los Angeles Co. Mus. Nat. Hist. Contrib. Sci., no. 148, 21 pp.

Start, A. N. 1972*a*. Notes on *Dyacopterus spadiceus* from Sarawak. Sarawak Mus. J. 20:367–69.

———. 1972*b*. Some bats of Bako National Park, Sarawak. Sarawak Mus. J. 20:371–76.

———. 1975. Another specimen of *Dyacopterus spadiceus* from Sarawak. Sarawak Mus. J. 23:267.

Stebbings, R. E. 1982. Radio tracking greater horseshoe bats with preliminary observations on flight patterns. *In* Cheeseman and Mitson (1982), pp. 161–73.

Stebbings, R. E., and F. Griffith. 1986. Distribution and status of bats in Europe. Inst. Terr. Ecol., Nat. Environ. Res. Council, 142 pp.

Strahan, R., ed. 1983. The Australian Museum complete book of Australian mammals. Angus & Robertson Publ., London, xxi + 530 pp.

Strelkov, P. P. 1986. The Gobi bat (*Eptesicus gobiensis* Bobrinskoy, 1926), a new species of Chiroptera of Palaearctic fauna. Zool. Zhur. 65:1103–8.

Subbaraj, R., and M. K. Chandrashekaran. 1977. 'Rigid' internal timing in the circadian rhythm of flight activity in a tropical bat. Oecologia 29:341–48.

Sung, C. V. 1976. New data on morphology and biology of some rare small mammals from North Vietnam. Zool. Zhur. 55:1880–85.

Suthers, R. A., and J. M. Fattu. 1973. Fishing behaviour and acoustic orientation by the bat *(Noctilio labialis)*. Anim. Behav. 21:61–66.

Swanepoel, P., and H. H. Genoways. 1978. Revision of the Antillean bats of the genus *Brachyphylla* (Mammalia: Phyllostomatidae). Bull. Carnegie Mus. Nat. Hist., no. 12, 53 pp.

———. 1983*a*. *Brachyphylla cavernarum*. Mammalian Species, no. 205, 6 pp.

———. 1983*b*. *Brachyphylla nana*. Mammalian Species, no. 206, 3 pp.

T

Taddei, V. A. 1976. The reproduction of some Phyllostomatidae (Chiroptera) from the northwestern region of the state of Sao Paulo. Bol. Zool. Univ. Sao Paulo 1:313–30.

———. 1979. Phyllostomidae (Chiroptera) do norte-ocidental do estado de Sao Paulo. III–Stenodermatinae. Ciencia e Cultura 31: 900–914.

Taddei, V. A., and V. Garutti. 1981. The southernmost record of the free-tailed bat, *Tadarida aurispinosa*. J. Mamm. 62:851–52.

Taddei, V. A., L. D. Vizotto, and S. M. Martins. 1976. Notas taxónomicas e biológicas sobre *Molossops brachymeles cerastes* (Thomas, 1901) (Chiroptera--Molossidae). Naturalia 2:61–69.

Taddei, V. A., L. D. Vizotto, and I. Sazima. 1983. Uma nova especie de *Lonchophylla* do Brasil e chave para identificacao das especies do genero (Chiroptera, Phyllostomidae). Ciencia e Cultura 35:625–30.

Tamsitt, J. R., A. Cadena, and E. Villarraga. 1986. Records of bats (*Sturnira magna* and *Sturnira aratathomasi*) from Colombia. J. Mamm. 67:754–57.

Tamsitt, J. R., and C. Hauser. 1985. *Sturnira magna*. Mammalian Species, no. 240, 4 pp.

Tamsitt, J. R., and D. Nagorsen. 1982. *Anoura cultrata*. Mammalian Species, no. 179, 5 pp.

Tate, G. H. H. 1934. Bats from the Pacific Islands, including a new fruit bat from Guam. Amer. Mus. Novit., no. 713, 3 pp.

———. 1942. Results of the Archbold Expeditions. No. 47. Review of the vespertilionine bats, with special attention to genera and species of the Archbold Collections. Bull. Amer. Mus. Nat. Hist. 80:221–97.

———. 1951*a*. *Harpyionycteris*, a genus of rare fruit bats. Amer. Mus. Novit., no. 1522, 9 pp.

Tate, G. H. H., and R. Archbold. 1939. A revision of the genus *Emballonura* (Chiroptera). Amer. Mus. Novit., no. 1035, 14 pp.

Taylor, E. H. 1934. Philippine land mammals. Philippine Bur. Sci. Monogr., no. 30, 548 pp.

Thomas, D. W. 1983. The annual migrations of three species of West African fruit bats (Chiroptera: Pteropodidae). Can. J. Zool. 61:2266–72.

Thomas, D. W., and A. G. Marshall. 1984. Reproduction and growth in three species of West African fruit bats. J. Zool. 202:265–81.

Thomas, M. E., and D. N. McMurray. 1974. Observations on *Sturnira aratathomasi* from Colombia. J. Mamm. 55: 834–36.

Thomas, O. 1910*b*. A new genus of fruit-bats and two new shrews from Africa. Ann. Mag. Nat. Hist., ser. 8, 6:111–14.

———. 1915. On bats of the genera *Nyctalus, Tylonycteris,* and *Pipistrellus*. Ann. Mag. Nat. Hist., ser. 8, 15:225–32.

Thompson, B. G. 1982. Records of *Eptesicus vulturnus* (Thomas) (Vespertilionidae: Chiroptera) from the Alice Springs area, Northern Territory. Austral. Mamm. 5:69–70.

Thonglongya, K. 1973. First record of *Rhinolophus paradoxolophus* (Bourret, 1951) from Thailand, with the description of a new species of the *Rhinolophus philippinensis* group (Chiroptera, Rhinolophidae). Mammalia 37:587–97.

Thornback, J., and M. Jenkins. 1982. The IUCN mammal red data book. Part 1: Threatened mammalian taxa of the Americas and the Australasian zoogeographic region (excluding Cetacea). Internatl. Union Conserv. Nat., Gland, Switzerland, xl + 516 pp.

Tidemann, C. R. 1986. Morphological variation in Australian and island populations of Gould's wattled bat, *Chalinolobus gouldii* (Gray) (Chiroptera: Vespertilionidae). Austral. J. Zool. 34:503–14.

———. 1987*a*. Notes on the flying-fox, *Pteropus melanotus* (Chiroptera: Pteropodidae), on Christmas Island, Indian Ocean. Austral. Mamm. 10:89–91.

———. 1987*b*. Flying-foxes (Chiroptera: Pteropodidae) and bananas: some interactions. Austral. Mamm. 10:133–35.

Tidemann, C. R., D. M. Priddel, J. E. Nelson, and J. D. Pettigrew. 1985. Foraging behaviour of the Australian ghost bat, *Macroderma gigas* (Microchiroptera: Megadermatidae). Austral. J. Zool. 33:705–13.

Timm, R. M. 1982. *Ectophylla alba*. Mammalian Species, no. 166, 4 pp.

———. 1984. Tent construction by *Vampyressa* in Costa Rica. J. Mamm. 65:166–67.

———. 1985. *Artibeus phaeotis*. Mammalian Species, no. 235, 6 pp.

———. 1987. Tent construction by bats of the genera *Artibeus* and *Uroderma*. Fieldiana Zool., n.s., 39:187–212.

Timm, R. M., and J. Mortimer. 1976. Selection of roost sites by Honduran white bats, *Ectophylla alba* (Chiroptera: Phyllostomatidae). Ecology 57:385–89.

Toop, J. 1985. Habitat requirements, survival strategies and ecology of the ghost bat *Macroderma gigas* Dobson (Microchiroptera, Megadermatidae) in central coastal Queensland. Macroderma 1:37–41.

Topal, G. 1970*a*. The first record of *Ia io* Thomas, 1902 in Vietnam and India, and some remarks on the taxonomic position of *Parascotomanes beaulieui* Bourret, 1942, *Ia longimana* Pen, 1962, and the genus *Ia* Thomas, 1902 (Chiroptera: Vespertilionidae). Opusc. Zool. Budapest 10:341–47.

———. 1970*b*. On the systematic status of *Pipistrellus annectans* Dobson, 1871 and *Myotis primula* Thomas, 1920 (Mammalia). Ann. Hist.-Nat. Mus. Natl. Hung., Zool., 62:373–79.

———. 1974. Field observations on Oriental bats. Vert. Hung. 15:83–94.

Trajano, E. 1982. New records of bats from southeastern Brazil. J. Mamm. 63:529.

Troughton, E. Le G. 1931. Three new bats of the genera *Pteropus*, *Nyctimene*, and *Chaerephon* from Melanesia. Proc. Linnaean Soc. New South Wales 56:204–9.

Tupinier, Y. 1977. Description d'une chauve-souris nouvelle: *Myotis nathalinae* nov. sp. (Chiroptera--Vespertilionidae). Mammalia 41:327–40.

Turner, D. C. 1975. The vampire bat: a field study in behavior and ecology, Johns Hopkins Univ. Press, Baltimore, viii + 145 pp.

Tuttle, M. D. 1970. Distribution and zoogeography of Peruvian bats, with comments on natural history. Univ. Kansas Sci. Bull. 49:45–86.

———. 1976*a*. Population ecology of the gray bat *(Myotis grisescens)*: philopatry, timing and patterns of movement, weight loss during migration, and seasonal adaptive strategies. Occas. Pap. Mus. Nat. Hist. Univ. Kansas, no. 54, 38 pp.

———. 1976*b*. Population ecology of the gray bat *(Myotis grisescens)*: factors influencing growth and survival of newly volant young. Ecology 57:587–95.

———. 1979. Status, causes of decline, and management of endangered gray bats. J. Wildl. Mgmt. 43:1–17.

———. 1987. Endangered gray bat benefits from protection. Endangered Species Tech. Bull. 12(3):4–5.

Tuttle, M. D., and S. J. Kern. 1981. Bats and public health. Milwaukee Pub. Mus. Publ. Biol. Geol., no. 48, 11 pp.

Tyler, M. J., ed. 1979. The status of endangered Australian wildlife. Proc. Cent. Symp. Roy. Zool. Soc. S. Australia, Adelaide, ix + 210 pp.

U

Uieda, W., I. Sazima, and A. Storti Filho. 1980. Aspectos da biologia do morcego *Furipterus horrens* (Mammalia, Chiroptera, Furipteridae). Rev. Brasil. Biol. 40:59–66.

V

Valdez, R., and R. K. LaVal. 1971. Records of bats from Honduras and Nicaragua. J. Mamm. 52:247–50.

Van Cakenberghe, V., and F. De Vree. 1985. Systematics of African *Nycteris* (Mammalia: Chiroptera). Proc. Internatl. Symp. Afr. Vert., Bonn, pp. 53–90.

Van der Merwe, M. 1975. Preliminary study on the annual movements of the Natal clinging bat. S. Afr. J. Sci. 71:237–41.

———. 1978. Postnatal development and mother-infant relationships in the Natal clinging bat *Miniopterus schreibersi natalensis* (A. Smith 1834). Proc. 4th Internatl. Bat Res. Conf., Nairobi, pp. 309–22.

Van der Merwe, M., and I. L. Rautenbach. 1986. Multiple births in Schlieffen's bat, *Nycticeius schlieffenii* (Peters, 1859) (Chiroptera: Vespertilionidae) from the southern African subregion. S. Afr. J. Zool. 21: 48–50.

Van der Merwe, M., I. L. Rautenbach, and W. J. Van der Colf. 1986. Reproduction in females of the little free-tailed bat, *Tadarida (Chaerephon) pumila*, in the eastern Transvaal, South Africa. J. Reprod. Fert. 77:355–64.

Van Deusen, H. M. 1968. Carnivorous habits of *Hypsignathus monstrosus*. J. Mamm. 49:335–36.

———. 1969. Results of the 1958–1959 Gilliard New Britain Expedition. 5. A new species of *Pteropus* (Mammalia, Pteropodidae) from New Britain, Bismarck Archipelago. Amer. Mus. Novit., no. 2371, 16 pp.

Van Deusen, H. M., and K. F. Koopman. 1971. Results of the Archbold Expeditions. No. 95. The genus *Chalinolobus* (Chiroptera, Vespertilionidae). Taxonomic review of *Chalinolobus picatus*, *C. nigrogriseus*, and *C. rogersi*. Amer. Mus. Novit., no. 2468, 30 pp.

Van Zyll de Jong, C. G. 1984. Taxonomic relationships of nearctic small-footed bats of the *Myotis leibii* group (Chiroptera: Vespertilionidae). Can. J. Zool. 62:2519–26.

Varty, N., and J. E. Hill. 1988. Notes on a collection of bats from the riverine forests of the Jubba Valley, southern Somalia, including two species new to Somalia. Mammalia 52:533– 92.

Vaughan, T. A. 1976. Nocturnal behavior of the African false vampire bat *(Cardioderma cor)*. J. Mamm. 57:227–48.

Vaughan, T. A., and G. C. Bateman. 1970. Functional morphology of the forelimb of mormoopid bats. J. Mamm. 51:217–35.

Vaughan, T. A., and T. J. O'Shea. 1976. Roosting ecology of the pallid bat, *Antrozous pallidus*. J. Mamm. 57:19–41.

Vaughan, T. A., and R. P. Vaughan. 1986. Seasonality and the behavior of the African yellow- winged bat. J. Mamm. 67:91–102.

———. 1987. Parental behavior in the African yellow-winged bat. J. Mamm. 68:217–23.

Vehrencamp, S. L., F. G. Stiles, and J. W. Bradbury. 1977. Observations on the foraging behavior and avian prey of the neotropical carnivorous bat, *Vampyrum spectrum*. J. Mamm. 58:469–78.

Verschuren, J. 1985. Note sur les mammiferes des Seychelles: un facteur de mortalite de la rousette endemique, *Pteropus seychellensis*. Mammalia 49:424–26.

Vestjens, W. J. M., and L. S. Hall. 1977. Stomach contents of forty-two species of bats from the Australian region. Austral. Wildl. Res. 4:25–35.

Vizotto, L. D., and V. A. Taddei. 1976. Notas sobre *Molossops temminckii temminckii* e *Molossops planirostris* (Chiroptera–Molossidae). Naturalia 2:47–59.

W

Walley, H. D., and W. L. Jarvis. 1971. Longevity record for *Pipistrellus subflavus*. Trans. Illinois Acad. Sci. 64:305.

Wallin, L. 1969. The Japanese bat fauna. Zool. Bidrag Uppsala 37:223–440.

Walton, R., and B. J. Trowbridge. 1983. The use of radio-tracking in studying the foraging behaviour of the Indian flying fox *(Pteropus giganteus)*. J. Zool. 201:575–79.

Wang Yingxiang. 1982. New subspecies of the pipistrels (Chiroptera, Mammalia) from Yunnan, China. Zool. Res., Suppl., 3:343–48.

Watkins, L. C. 1972. *Nycticeius humeralis*. Mammalian Species, no. 23, 4 pp.

———. 1977. *Euderma maculatum*. Mammalian Species, no. 77, 4 pp.

Watkins, L. C., J. K. Jones, Jr., and H. H. Genoways. 1972. Bats of Jalisco, Mexico. Spec. Publ. Mus. Texas Tech Univ., no. 1, 44 pp.

Webster, W. D., and C. O. Handley. 1986. Systematics of Miller's long-tongued bat, *Glossophaga longirostris*, with description of two new subspecies. Occas. Pap. Mus. Texas Tech Univ., no. 100, 22 pp.

Webster, W. D., and J. K. Jones, Jr. 1980*a*. Noteworthy records of bats from Bolivia. Occas. Pap. Mus. Texas Tech Univ., no. 68, 6 pp.

———. 1980*b*. Taxonomic and nomenclatorial notes on bats of the genus *Glossophaga* in North America, with description of a new species. Occas. Pap. Mus. Texas Tech Univ., no. 71, 12 pp.

———. 1982. A new subspecies of *Glossophaga commissarisi* (Chiroptera: Phyllostomidae) from western Mexico. Occas. Pap. Mus. Texas Tech Univ., no. 76, 6 pp.

———. 1983. First record of *Glossophaga commissarisi* (Chiroptera: Phyllostomidae) from South America. J. Mamm. 64:150.

———. 1984*a*. Notes on a collection of bats from Amazonian Ecuador. Mammalia 48: 247–52.

———. 1984*b*. *Glossophaga leachii*. Mammalian Species, no. 226, 3 pp.

———. 1984*c*. A new subspecies of *Glossophaga mexicana* (Chiroptera: Phyllostomidae) from southern Mexico. Occas. Pap. Mus. Texas Tech Univ., no. 91, 5 pp.

———. 1985. *Glossophaga mexicana*. Mammalian Species, no. 245, 2 pp.

———. 1987. A new subspecies of *Glossophaga commissarisi* (Chiroptera: Phyllostomidae) from South America. Occas. Pap. Mus. Texas Tech Univ., no. 109, 6 pp.

Webster, W. D., J. K. Jones, Jr., and R. J. Baker. 1980. *Lasiurus intermedius*. Mammalian Species, no. 132, 3 pp.

Webster, W. D., and W. B. McGillivray. 1984. Additional records of bats from French Guiana. Mammalia 48:463–65.

Webster, W. D., and R. D. Owen. 1984. *Pygoderma bilabiatum*. Mammalian Species, no. 220, 3 pp.

Webster, W. D., L. W. Robbins, R. L. Robbins, and R. J. Baker. 1982. Comments on the status of *Musonycteris harrisoni* (Chiroptera: Phyllostomidae). Occas. Pap. Mus. Texas Tech Univ., no. 78, 5 pp.

Wenstrup, J. J., and R. A. Suthers. 1984. Echolocation of moving targets by the fish-catching bat, *Noctilio leporinus*. J. Comp. Physiol., ser. A, 155:75–89.

Wheeler, M. E., and C. F. Aguon. 1978. The current status and distribution of the Marianas fruit bat on Guam. Guam Aquatic and Wildl. Res. Div. Tech. Rept., no. 1, 29 pp.

Whitaker, J. O., Jr., and H. Black. 1976. Food habits of cave bats from Zambia, Africa. J. Mamm. 57:199–204.

Wickler, W., and U. Seibt. 1976. Field studies of the African fruit bat *Epomophorus wahlbergi* (Sundevall), with special reference to male calling. Z. Tierpsychol. 40: 345–76.

Wiles, G. J., and N. H. Payne. 1986. The trade in fruit bats *Pteropus* spp. on Guam and other Pacific islands. Biol. Conserv. 38:143–61.

Wilkins, K. T. 1987. *Lasiurus seminolus*. Mammalian Species, no. 280, 5 pp.

Wilkinson, G. S. 1985*a*. The social organization of the common vampire bat. I. Pattern and cause of association. Behav. Ecol. Sociobiol. 17:111–21.

———. 1985*b*. The social organization of the common vampire bat. II. Mating system, genetic structure, and relatedness. Behav. Ecol. Sociobiol. 17:123–34.

———. 1988. Social organization and behavior. *In* Greenhall and Schmidt (1988), pp. 85–97.

Williams, C. F. 1986. Social organization of the bat, *Carollia perspicillata* (Chiroptera: Phyllostomidae). Ethology 71:265–82.

Williams, D. F. 1978*b*. Taxonomic and karyologic comments on small brown bats, genus *Eptesicus*, from South America. Ann. Carnegie Mus. 47:361–83.

———. 1986. Mammalian species of special concern in California. California Dept. Fish and Game, 112 pp.

Williams, D. F., J. D. Druecker, and H. L. Black. 1970. The karyotype of *Euderma maculatum* and comments on the evolution of the plecotine bats. J. Mamm. 51:602–6.

Williams, S. L., and H. H. Genoways. 1980*a*. Results of the Alcoa Foundation–Suriname Expeditions. II. Additional records of bats (Mammalia: Chiroptera) from Suriname. Ann. Carnegie Mus. 49:213–36.

———. 1980*b*. Results of the Alcoa Foundation–Suriname Expeditions. IV. A new species of bat of the genus *Molossops* (Mammalia: Molossidae). Ann. Carnegie Mus. 49:487–98.

Williams, S. L., H. H. Genoways, and J. A. Groen. 1983. Results of the Alcoa Foundation–Suriname Expeditions. VII. Records of mammals from central and southern Suriname. Ann. Carnegie Mus. 52:329–36.

Williams, T. C., L. C. Ireland, and J. M. Williams. 1973. High altitude flights of the free-tailed bat, *Tadarida brasiliensis*, observed with radar. J. Mamm. 54:807–21.

Willig, M. R. 1983. Composition, microgeographic variation, and sexual dimorphism in caatingas and cerrado bat communities from northeast Brazil. Bull. Carnegie Mus. Nat. Hist., no. 23, 131 pp.

———. 1985*a*. Ecology, reproductive biology, and systematics of *Neoplatymops mattogrossensis* (Chiroptera: Molossidae). J. Mamm. 66:618–28.

———. 1985*b*. Reproductive patterns of bats from caatingas and cerrado biomes in northeast Brazil. J. Mamm. 66:668–81.

Willig, M. R., and R. R. Hollander. 1987. *Vampyrops lineatus*. Mammalian Species, no. 275, 4 pp.

Willig, M. R., and J. K. Jones, Jr. 1985. *Neoplatymops mattogrossensis*. Mammalian Species, no. 244, 3 pp.

Wilson, D. E. 1971. Ecology of *Myotis nigricans* (Mammalia: Chiroptera) on Barro Colorado Island, Panama Canal Zone. J. Zool. 163:1–13.

———. 1976. The subspecies of *Thyroptera discifera* (Lichtenstein and Peters). Proc. Biol. Soc. Washington 89:305–12.

———. 1979. Reproductive patterns. *In* Baker, Jones, and Carter (1979), pp. 317–78.

Wilson, D. E., and E. L. Tyson. 1970. Longevity records for *Artibeus jamaicensis* and *Myotis nigricans*. J. Mamm. 51:203.

Wilson, P. 1985. Maeda's *Miniopterus* taxonomy. Macroderma 1:29–36.

Wodzicki, K., and H. Felten. 1975. The peka, or fruit bat *(Pteropus tonganus tonganus)* (Mammalia, Chiroptera), of Niue Island, South Pacific. Pacific Sci. 29:131–38.

———. 1981. Fruit bats of the genus *Pteropus* from the islands Rarotonga and Mangaia, Cook Islands, Pacific Ocean. Senckenberg. Biol. 61:143–51.

Wolton, R. J., P. A. Arak, H. C. J. Godfray, and R. P. Wilson. 1982. Ecological and behavioural studies of the Megachiroptera at Mount Nimba, Liberia, with notes on Microchiroptera. Mammalia 46:419–48.

Woodsworth, G. C., G. P. Bell, and M. B. Fenton. 1981. Observations of the echolocation, feeding behaviour, and habitat use of *Euderma maculatum* (Chiroptera: Vespertilionidae) in southcentral British Columbia. Can. J. Zool. 59:1099–1102.

Y

Yates, T. L., D. J. Schmidly, and K. L. Culbertson. 1976. Silver-haired bat in Mexico. J. Mamm. 57:205.

Yenbutra, S., and H. Felten. 1983. A new species of the fruit bat genus *Megaerops* from SE-Asia (Mammalia: Chiroptera: Pteropodidae). Senckenberg. Biol. 64:1–11.

Yoshiyuki, M. 1979. A new species of the genus *Ptenochirus* (Chiroptera, Pteropodidae) from the Philippine Islands. Bull. Natl. Sci. Mus. (Tokyo), ser. A, 5:75–81.

———. 1983. A new species of *Murina* from Japan (Chiroptera, Vespertilionidae). Bull. Natl. Sci. Mus. (Tokyo), ser. A, 9: 141–50.

———. 1984. A new species of *Myotis* (Chiroptera, Vespertilionidae) from Hokkaido, Japan. Bull. Natl. Sci. Mus. (Tokyo), ser. A, 10:153–58.

Z

Ziegler, A. C. 1982*a*. The Australo-Papuan genus *Syconycteris* (Chiroptera: Pteropodidae) with the description of a new Papua New Guinea species. Occas. Pap. Bishop Mus. 25(5):1–22.

Zubaid, A. 1988. The second record of a trident horseshoe bat, *Aselliscus stoliczkanus* (Hipposiderinae) from peninsular Malaysia. Malayan Nat. J. 42:29–30.

Index

The scientific names of orders, families, and genera, which have titled accounts in the text, are in **boldface,** *as are the page numbers on which such accounts begin. Other scientific names, and vernacular names, appear in* roman.

Library of Congress Cataloging-in-Publication Data

Nowak, Ronald M.
Walker's bats of the world / Ronald M. Nowak ; introduction by Thomas H. Kunz and Elizabeth D. Pierson.
p. cm.
"Portions of this book have been adapted from Walker's mammals of the world, 5th edition, by Ronald M. Nowak"—Verso t.p.
Includes bibliographical references (p.) and index.
ISBN 0-8018-4986-1 (pbk. : alk. paper)
1. Bats. 2. Bats—Classification. I. Walker, Ernest P. (Ernest Pillsbury), 1891– Walker's mammals of the world. II. Title. III. Title: Bats of the world.
QL737.C5N69 1994
599.4—dc20

94-30262
CIP